AF386601

Life Cycle Management from Global to Local

Marzia Traverso · Rose Nangah Mankaa ·
Davide Bonaffini · Antonio Covais
Editors

Life Cycle Management from Global to Local

 Springer

Editors
Marzia Traverso
Institute of Sustainability in Civil
Engineering
RWTH Aachen University
Aachen, Germany

Davide Bonaffini
Hitachi Rail
Pistoia, Italy

Rose Nangah Mankaa
Institute of Sustainability in Civil
Engineering
RWTH Aachen University
Aachen, Germany

Antonio Covais
Circular s.r.l
Palermo, Italy

ISBN 978-3-032-17986-9   ISBN 978-3-032-17987-6 (eBook)
https://doi.org/10.1007/978-3-032-17987-6

This Springer imprint is published by the registered company Springer Nature Switzerland AG
The registered company address is: Gewerbestrasse 11, 6330 Cham, Switzerland

If disposing of this product, please recycle the paper.

Acknowledgments

We are thankful to all who contributed to the successful publication of this book. The authors responded with incredible enthusiasm, making available quality research to the scientific community at large. In addition, their contributions provide valuable insights to what is going on in different parts of the world, covering LCM topics in four continents.

We do acknowledge the huge support of the scientific committee: Thomas Albert, Paola Federica Albizzati, Gabriella Arcese, Umberto Arena, Jana Gerta Backes, Martin Baitz, Christian Bauer, Enrico Benetto, Markus Berger, Catherine Benoit-Norris, Gian Andrea Blengini, Carlo Brondi, Nicolás Héctor Carreño Gómez, Guy Castelan, Maurizio Cellura, Juan Felipe Cerdas Marin, Kostas Chatziioanou, Takunda Yeukai Chitaka, Tamara Coello García, Archana Datta, Gianluca De Nardi, Pascal Di Croce, Yahong Dong, Tomas Ekvall, Alexandra Evans, Matthias Finkbeiner, Matthias Fischer, Alexander Forell, Ann Francis, Laura Garcia Herrero, Claudia Giacovelli, Mark Goedkoop, Katharina Gompf, Yamini Gopalapillai, Kathrin Greiff, Zivile Gudeliunaite, Jessica Hanafi, Franziska Hesser, Maike Juliane Horlacher, Diego Iribarren, Nils Jäger, Davis Jose, Stephan Krinke, Christian Krüger, Luiz Kulay, Agustin Laveglia, Lucyna Lekawska-Andrinopoulou, Davide Lo Presti, Federico Lo Presti, Anna Luthin, Diego C Magalhaes, Giulia Maggiora, Alessandro Manzardo, Alessandro Marson, Antonino Marvuglia, Thomas Matiz, Scott McAlister, Prasad Modak, Stephane Morel, Emilia Moreno Ruiz, Stéphanie Muller, Monia Niero, Gregory A. Norris, Martina Orefice, Lavanya Pawar, Luigia Petti, Stephan Pfister, Martina Prox, Meinolf Rameil, Andreas Redmann, Lindsey Roche, Sophia Roy, Tomas Vilhelm Rydberg, Mateo Saavedra del Oso, Peter Rudolf Hans Saling, Thomas Schaubroeck, Stefan Schmitz, Purvish Shah, Luka Smajila, Philip Strothmann, Julian Suer, Ladji Tikana, Olubukola Olumuyiwa Tokede, Ketan Vaidya, Sonia Valdivia, Antonio Valente, Tobias Viere, Samuel Vionnet, Bo Weidema, Jonas Wigger, Anna Wikström, Alessandra Zamagni, Luca Zampori, Silvia Zingale.

LCM 2025 and hence the content of this open-access publication would not have been possible without the financial support of our Platinum (PRé, Makersite), Gold (ecoinvent, Sphera), Silver (Weloop, Trashy Bags Africa, Siemens, International EPD System (EPD), iPoint, Edison, Life Cycle Indonesia (LCI), Arche Consulting,

Minviro, Mérieux NutriSciences—Blonk Ecomatters) and Bronze (International Copper Association, Coccadoro, Riolo, l'altroparlante, G. Milazzo) sponsors.

The vibrant team at the Institute of Sustainability in Civil Engineering, RWTH Aachen University, ensured a seamless and timely review process. We would like to express our gratitude to Alejandra Torres Silva and Davis Jose for their overwhelming support in editing and excellent liaison with authors.

Our special thanks also go to Martina Creutzfeldt and the MCC Team accompanying us in this exciting journey.

Last but not least, we would like to commend the patience and unflinching support of our respective families and friends.

<table>
<tr><td>Aachen, Germany</td><td>Prof. Dr. Marzia Traverso</td></tr>
<tr><td>Palermo, Italy</td><td>Dr.-Ing. Rose Nangah Mankaa</td></tr>
<tr><td></td><td>Ing. Davide Bonaffini</td></tr>
<tr><td></td><td>Dr.-Ing. Antonio Covais</td></tr>
</table>

Introduction

The present book compiles a selection of the best studies presented at the 12th edition of the Life Cycle Management (LCM) Conference series (LCM 2025). Out of 980 abstracts submitted, 39 articles were selected. The LCM conference series is a global forum dedicated to environmental, economic, and social sustainability. Held biennially, each edition is organised by leading research institutions and industry experts. Its main focus is to present practical solutions for integrating life cycle approaches into strategic and operational decision-making across science, industry, non-governmental organisations (NGOs), and public institutions. A central objective is to bring together professionals and the scientific community. For this reason, at least 50% of participants are expected to come from industry, small- and medium-sized enterprises, and the government sector. The theme of the 2025 conference, "Global to Local", highlighted the challenge of applying sustainability across different activities, sectors, and regions. The aim was to foster interest in a synaptic ecosystem around sustainable solutions. LCM 2025 welcomed more than 1015 participants from 51 countries who, for the first time in the history of the series, had the opportunity to attend sessions dedicated to healthcare, artificial intelligence, and regional LCA. At LCM, sustainability in all three dimensions is examined through a life cycle perspective. In particular, the conference highlights advancements in research and practical implementation related to methodologies such as Life Cycle Assessment (LCA), Social Life Cycle Assessment (S-LCA), Life Cycle Costing (LCC), and Life Cycle Sustainability Assessment (LCSA). LCA is one of the most widely used tools for evaluating the environmental impacts of products and services throughout their life cycle, from raw material extraction to final disposal. S-LCA extends this approach by assessing social and socio-economic impacts, considering aspects such as labour rights, human health, and community well-being. LCC addresses the economic dimension by calculating the total costs of a product or service, including investment, operation, maintenance, and end-of-life. LCSA integrates all three perspectives—environmental, social, and economic—into a holistic assessment of sustainability that supports decision-making in business, policy, and research.

LCM 2025 captured research guided by recent and future regulatory developments. At the international level, the latest ISO standards show how the scope of life cycle approach is spreading. The release of ISO 14075:2024 on social life cycle assessment (S-LCA) provided a standardised framework for assessing social and socio-economic impacts throughout the life cycle of products. In parallel, and integrating the third pillar of sustainability, the ISO/TS 14076:2025 standard on environmental techno-economic assessment (eTEA) contributes to more informed decision-making in the business sector.

At the EU level, the LCM framework is supported by several recent initiatives. The Ecodesign for Sustainable Products Regulation (ESPR) enhances the criteria for durability, repairability, energy and resource efficiency, and recyclability, and introduces new measures such as green public procurement and digital product passports. In parallel, the Omnibus packages together with the EU Taxonomy Regulation aim to simplify sustainability reporting frameworks and strengthen the role of corporate disclosure in guiding decision-making. The "Safe and Sustainable by Design" (SSbD) framework further expands the scope of LCM by providing criteria for innovation in chemicals and materials, promoting the use of life cycle tools already at the design stage. Finally, the proposal on green claims addresses the problems of "greenwashing", emphasising transparency and comparability in environmental communication.

The joint development of these and other regulatory initiatives contributes to making LCM an increasingly robust methodology for implementation. Within this context, this book brings together the current state of the art on methods, tools, and best practices in enabling a sustainable transition. The selection of contributions prioritises regional and global works with a strong industrial perspective, as well as methodological advances that address emerging challenges in both political and scientific contexts.

The chapters are organised into eight parts, covering the diversity of topics addressed at the conference. The following parts provide an overview of the themes and structure of each part, outlining the key insights that shape them.

Methodological and Conceptual Advancements in LCA and LCSA

This part brings together chapters aimed at strengthening the methodological basis and expanding the capacity of LCA and LCSA as tools for responding to the challenges of the transition to more sustainable production models.

Particular emphasis is placed on the dynamic and forward-looking character of LCA when applied alongside complementary frameworks. For instance, integrating LCA with Natural Capital Accounting (NCA) or material criticality assessment shows how it can go beyond its traditional scope and open new perspectives. In an era where environmental sustainability is a priority, each of these approaches integrated with LCA reinforces its potential as a tool to guide scientific research, sustainable business practices, and public policy.

The potential of LCSA to support decision-making in the industrial sector, especially in the early stages of research and development, is also examined. The combination of environmental, economic, and social assessments facilitates the identification of systemic risks (risk assessment), the validation of sustainability claims (green claims), and the anticipation of regulatory requirements.

Digitalisation, Artificial Intelligence Enhancing Data Availability, Traceability Robustness in LCM

This part explores the incorporation of advanced digital technologies to strengthen LCM and expand its capabilities in the face of sustainability challenges. Among the approaches presented, the growing use of machine learning as a tool to optimise various phases of analysis stands out, from filling data gaps and extracting information in literature to predicting indicators, constructing ontologies for knowledge representation, detecting outliers, and visualising data. Alongside these traditional methods, applications based on more advanced models are emerging, such as graphical neural networks, which expand the potential for analysis and automation in LCM through artificial intelligence.

The contributions also highlight the role of the Digital Product Passport (DPP) as a tool for collecting and disseminating comprehensive information on sustainability and circularity throughout the product life cycle, promoting more responsible consumption decisions and encouraging producers to improve the durability, design, and efficiency of their products.

Furthermore, attention is given to the adaptation of verification processes in Life Cycle Assessment (LCA), the Product Environmental Footprint (PEF), and Environmental Product Declarations (EPDs). To meet the growing demand for evaluating complete systems, these processes are evolving from a traditional focus on individual products towards scalable approaches that integrate digitalisation and automation. Aligned with regulations such as the Corporate Sustainability Reporting Directive (CSRD) and international standards (e.g. ISO 14071:2014 on critical review processes and reviewer competencies, and ISO/IEC 17029:2019 on requirements for validation and verification bodies), they are becoming essential tools for managing data with greater transparency.

Carbon and Environmental Footprinting in Industry and Business

Decarbonisation and environmental management in industry require useful tools for decision-making. This part brings together contributions that show how sectors such as the chemical and automotive industries, construction, and regional management are adopting advanced methodologies to quantify their impacts and guide sustainability strategies.

Efforts to standardise product carbon footprints (PCFs) and facilitate the exchange of verified data are highlighted. Similarly, the importance of converting methodological innovation in LCA into applicable solutions is emphasised. Innovation and scientific advancement are essential to ensure that LCA remains up to date and capable of addressing emerging environmental challenges. However, the true value of these advances can only be realised when they are effectively integrated into the practical decision-making processes of industry, business, financial institutions, and policymakers.

Life Cycle Evaluations of Energy Systems and Technologies

The global transition to renewable energy sources and sustainable technologies in energy systems brings new challenges. The choices made today will have consequences for the entire energy system, affecting environmental impact, economic viability, access, and energy security at the social level. This part provides an in-depth look at the role that LCA and LCM can play in the context of Multi-Criteria Decision-Making (MCDM) for energy technologies and systems.

The cases presented include the development of methods to assess the environmental performance of thermal energy storage (TES) systems in different configurations, the production of hydrogen and solid carbon from biomethane with negative emission potentials, and the assessment of liquid hydrogen tanks in aircraft made from carbon fibre reinforced polymers. It also compares first- and second-life lithium iron phosphate (LFP) batteries in grid-scale storage and analyses the replacement of natural gas with syngas obtained from automotive scrap waste in steel reheating furnaces.

Embedding Circularity in LCM

One of the main objectives in value chains seeking sustainability is to achieve circularity. Products should deliver value even after fulfilling the purpose for which they were created. Some industries have already focused their efforts and workforce on creating circular systems. This part therefore reviews research advances in some sectors and highlights the importance of regulation and regulatory frameworks. It also explores communication and awareness-raising about the circular economy with robust life cycle assessments. Advances in LCM for the bioeconomy are analysed through innovative methods and practices, offering an alternative to fossil fuel-based industries, such as the synthesis of olefins from waste. In addition, the LCA of circular and low-carbon plastic value chains is carried out. To accelerate the path towards greater circularity, resource reduction can be achieved through efficiency measures, reuse concepts, and recycling of post-consumer and pre-consumer plastic waste. The case studies included reflect progress in this area, which assess the environmental impacts of a new physical recycling technology for polyvinyl chloride (PVC), recovering high-quality recycled PVC comparable to virgin PVC.

Products on the market should be endorsed by a thorough sustainability assessment to guarantee their suitability. Similarly, regulatory frameworks should be in place to verify this. This part includes the case of e-mobility. The corresponding chapter presents and analyses two policy measures to improve the circularity of electric traction motors by applying a life cycle approach. In addition to the regulatory aspect, communicating the results to specific stakeholders is also important to create transparency in the process, but also to promote and raise awareness. One of the contributions discusses the consequences of modelling decisions in communicating the different LCA results by analysing the circularity of winter sports equipment.

LCM Supporting Product and Organizational Reporting

Sustainability constitutes the transition to value chains where transparency and accountability are increasingly part of the process. In this sense, LCM emerges as a fundamental tool for environmental and organisational reporting. Here the central focus of reducing the carbon footprint of companies, institutions, and even governments is not only carried out but also verified, as each decision impacts supply chains, the products offered, but also society.

The application of life cycle approaches improves the quality and credibility of corporate reports, providing a solid basis for climate commitments such as reducing emissions in health, integrating emerging products into regulatory policies, communicating circular business models, and accounting for indirect emissions from the value chain (Scope 3).

A case from the pharmaceutical industry is presented, showing how product LCA can support sustainability objectives while also enhancing scope 3 emissions reporting. Another case study explores how supplier-specific Product Carbon Footprints (PCFs) can be efficiently aligned and integrated into Environmental Product Declarations (EPDs), helping to reduce redundancies, improve data quality, and increase reporting clarity. In addition, one research work demonstrates the use of augmented reality to visualise life cycle results, making the methodology more transparent and accessible.

Social Sustainability Performance in Industry

The social dimension is one of the three fundamental pillars of sustainability. It includes aspects of the value chain that stand out for their direct impact on people and communities. This part addresses methodological advances and improvements in data availability for S-LCA. The chapter also seeks to explore and address, within S-LCA, topics that are still insufficiently addressed in LCA, such as consequential and dynamic approaches, prospective assessments, and the incorporation of social aspects into long-term planning. Ultimately, the goal is to contribute to the advancement of the S-LCA methodology and improve its effectiveness as a support for informed decision-making at both the industrial and political levels.

Global sustainability challenges, such as labour rights, economic security, and resource management, require local solutions that strengthen social sustainability. In this regard, this part brings together research that advances the integration of environmental and social assessment methods across different value chains, including food, water supply, the metallurgical industry, and the plastics sector.

Policy, Regulatory Frameworks and Standards

The LCM has been established as a comprehensive analytical framework for sustainable transformation, fostering the development and implementation of environmental regulations. While it originated in scientific and technical domains, it has also become a key instrument for strengthening policies, regulatory frameworks, and standards. In particular, its application supports the integration and interpretation of instruments such as EPDs and the PEF across different sectors.

This part brings together experiences from multiple regions and industries, illustrating how LCA informs regulatory guidelines, standardisation frameworks, green public procurement criteria, and policies aimed at reducing environmental impacts. At the same time, it underscores the challenges that remain, including inconsistent data, methodological harmonisation, limited transparency, and gaps in knowledge and skills.

Across the chapters, sustainability in construction practices is explored in depth, with a focus on the use of environmentally responsible materials such as biogenic products and asphalt mixtures. Contributions also address raw materials in particular aluminium, and explores LCA communities promoting the development and reinforcement of regulatory frameworks in countries such as New Zealand, Australia, and South Africa.

The contributions compiled here clearly demonstrate the vitality of the field and the progress made in advancing methodologies, applying life cycle thinking across diverse industries, and linking research with practical decision-making. Yet, they also reveal persisting challenges and emerging opportunities that call for continued academic and professional attention.

Future research on life cycle management should prioritise developing forward-looking and systemic applications. Many current approaches remain retrospective, yet addressing today's sustainability challenges requires methods that can anticipate future risks, capture interdependencies, and support long-term transitions. Additionally, digitalisation and artificial intelligence offer enormous potentials in improving data availability, transparency, and traceability, but their effective implementation requires clear frameworks that address governance, interoperability, and ethical concerns. Without these safeguards, their transformative capacity will remain limited.

At the same time, advancing the social dimension of sustainability remains a critical point. Despite methodological progress in S-LCA, the integration into global supply chains is still lacking. Similarly, while life cycle methods are increasingly embedded in policy and regulatory frameworks, much of this progress remains compliance-driven rather than transformative. To achieve systemic impact, LCM must support forward-looking policy agenda and just transitions, while also overcoming practical barriers that hinder adoption by small- and medium-sized enterprises. Expanding accessibility, strengthening capacity, and ensuring inclusivity will be essential for LCM to fulfil its role as a cornerstone of sustainable transformation.

It is also important to acknowledge the limitations of the present volume. Geographical coverage remains uneven, with case studies disproportionately concentrated in Europe and other industrialised regions, while experiences from the Global South are less represented. Moreover, the social and economic perspectives are less prominently featured.

In conclusion, the achievements documented in this book reflect the significant progress made by the LCM community in developing tools, methods, and practices that advance sustainability performance across sectors. Yet the path forward demands continued innovation, interdisciplinarity, and global inclusivity. By extending its methodological frontiers, deepening integration across the three dimensions of sustainability, and enhancing its role in both governance and industry, life cycle management can continue to evolve as a cornerstone of sustainable transformation. The outlook for research and practice must therefore remain ambitious, ensuring that LCM is equipped to guide science, business, and policy towards a more resilient and sustainable future.

Contents

Contributors

Naeem Adibi WeLOOP, Lambersart, France

Silvia Agrafojo Zabala Innovation Consulting, Mutilva Alta, Navarra, Spain

Beatriz Amante ENMA (Environmental Engineering), ESEIAAT (The School of Industrial, Aerospace and Audiovisual Engineering of Terrassa), Universitat Politècnica de Catalunya (UPC), Terrassa, Spain

Angelos Amditis Institute of Communication and Computer Systems, Athens, Greece

Daniel Anyanya University College London, London, UK

Katy Armstrong Safety, Environmental and Regulatory Science (SERS), Unilever, Sharnbrook, UK

Simon Aumônier Aumônier Consulting Limited, Oxford, UK

Jens Bachmann Institute of Lightweight Systems (SY), DLR German Aerospace Center, Braunschweig, Germany

Jana Gerta Backes Junior Professor for Safety, Security and Sustainability Evaluations in Foresight Research, Aachen, Germany

Martin Baitz Sphera Solutions GmbH, Stuttgart, Germany

Semra Bakkaloglu Department of Chemical Engineering, Imperial College London, London, UK

Tim Becker Sphera Solutions GmbH, Stuttgart, Germany

Sabina Bednářová Environmental Research and Innovation (ERIN) Department, Environmental Sustainability Assessment and Circularity (SUSTAIN) Unit, Luxembourg Institute of Science and Technology (LIST), Esch-Sur-Alzette, Luxembourg

Enrico Benetto Environmental Research and Innovation (ERIN) Department, Environmental Sustainability Assessment and Circularity (SUSTAIN) Unit, Luxembourg Institute of Science and Technology (LIST), Esch-Sur-Alzette, Luxembourg

Gabriela Benveniste Catalonia Institute for Energy Research, Sant Adrià de Besòs, Spain

Silvia Bobba European Commission—Joint Research Centre (JRC), Ispra, Italy

Isabella Bulfaro Catalonia Institute for Energy Research, Sant Adrià de Besòs, Spain;
ENMA (Environmental Engineering), ESEIAAT (The School of Industrial, Aerospace and Audiovisual Engineering of Terrassa), Universitat Politècnica de Catalunya (UPC), Terrassa, Spain

Christian Bülow Institute of Lightweight Systems (SY), DLR German Aerospace Center, Braunschweig, Germany

Stefane Caldeira Postgraduate Program in Product and Process Technology, Federal Center for Technological Education of Minas Gerais State, Belo Horizonte, Brazil;
Department Decarbonization and Circularity, ArcelorMittal Long Carbon Steel, Belo Horizonte, Brazil

Aoife Calnan ERM—Environmental Resources Management, London, UK

Benjamin Canaguier Schneider Electric, Rueil-Malmaison, France

Daniele Candelaresi Seidor Italy S.R.L, Milan, Italy

Nicolás Héctor Carreño Gómez VINCI Construction Shared Services GmbH, Bottrop, Germany

Sara Casale Head Sport GmbH, Schwechat, Austria

Wesley Cavalcante Department Decarbonization and Circularity, ArcelorMittal Long Carbon Steel, Belo Horizonte, Brazil

Aurora Cavaliere University College London, London, UK

Monica Celotto Electrolux Italia SpA, Porcia, PN, Italy

Kolobe Chaba CSIR, Stellenbosch, South Africa

Kostas Chatziioannou Institute of Communication and Computer Systems, Athens, Greece

Charnett Chau University College London, London, UK

Justin Ningwei Chiu Department of Energy Technology, School of Industrial Engineering and Management, KTH Royal Institute of Technology, Stockholm, Sweden

Michael Collins ERM—Environmental Resources Management, London, UK

Cecilia Makishi Colodel Sphera Solutions GmbH, Stuttgart, Germany

Manuela D'Eusanio Department of Economic Studies (DEC), University "G. d'Annunzio", Pescara, Italy;
Department for the Promotion of Human Sciences and Quality of Life, San Raffaele Roma University, Rome, Italy

Rannvá Danielsen NORSUS—Norwegian Institute of Sustainability Research, Fredrikstad, Norway

Vsevolod Dengin Electrolux AB, Stockholm, Sweden

Marion Devienne SCORE LCA, Villeurbanne, France

Carla Di Biccari Università degli Studi del Salento, Lecce, Italy

Sabrina Diniz Institute of Lightweight Systems (SY), DLR German Aerospace Center, Braunschweig, Germany

Nahikari Diosdado Zabala Innovation Consulting, Mutilva Alta, Navarra, Spain

Haoyang Dong Department of Energy Technology, School of Industrial Engineering and Management, KTH Royal Institute of Technology, Stockholm, Sweden

Javier Martin Echazarreta Instituto Nacional de Tecnología Industrial (INTI), Buenos Aires, Argentina

Widiene Essouid University of Bordeaux, CNRS, INP, ISM, Talence, France; University of Bordeaux, IAE School of Management - Bordeaux, IRGO, Bordeaux, France

Alexandra Evans Remote Sensing Applications, Flemish Institute for Technological Research (VITO), Mol, Belgium

Umberto Eynard European Commission—Joint Research Centre (JRC), Ispra, Italy

Mikel Fadul Catalonia Institute for Energy Research, Sant Adrià de Besòs, Spain

Ewald Fauster Processing of Composites and Design for Recycling, Technical University Leoben, Leoben, Austria

Victor José Ferreira Catalonia Institute for Energy Research, Sant Adrià de Besòs, Spain

Lorraine Ferris Henry Royce Institute, University of Manchester, Manchester, UK

Silvia Fiorini London, UK

Christian Fontana Ecoinnovazione srl, Bologna, Italy

Rosa Cuellar Franca Department of Chemical Engineering, University of Manchester, Manchester, UK

Sebastian Freund Institute of Lightweight Systems (SY), DLR German Aerospace Center, Braunschweig, Germany

Gioia Garavini Ecoinnovazione, Bologna, Italy

Stefan Gärtner Meo Carbon Solutions GmbH, Cologne, Germany

Stella Georgali Institute of Communication and Computer Systems, Athens, Greece

Andreas Gess Institute for Acoustics and Building Physics, University of Stuttgart, Stuttgart, Germany

Thomas Gibon Environmental Research and Innovation (ERIN) Department, Environmental Sustainability Assessment and Circularity (SUSTAIN) Unit, Luxembourg Institute of Science and Technology (LIST), Esch-Sur-Alzette, Luxembourg

Anastasia Globa Sydney School of Architecture, Design and Planning, The University of Sydney, Sydney, Australia

Taahira Goga CSIR, Stellenbosch, South Africa

Marvin Gornik Sustainability Department, EurA AG, Erfurt, Germany

Shashank Goyal Sustainability Department, EurA AG, Erfurt, Germany

Herbert Gruber Head Sport GmbH, Schwechat, Austria

Emilie Guilvert WeLOOP, Lambersart, France

Saman Nimali Gunasekara Department of Energy Technology, School of Industrial Engineering and Management, KTH Royal Institute of Technology, Stockholm, Sweden

Caireen Hargreaves AstraZeneca, Macclesfield, UK

Zaid Hashash Institute of Sustainability in Civil Engineering, RWTH Aachen University, Aachen, Germany

Pamela Haverkamp Institute of Sustainability in Civil Engineering, RWTH Aachen University, Aachen, Germany

Maryna Henrysson Department of Energy Technology, School of Industrial Engineering and Management, KTH Royal Institute of Technology, Stockholm, Sweden

David Hernando Sustainable Pavements and Asphalt Research (SUPAR), Faculty of Applied Engineering, University of Antwerp, Antwerp, Belgium

Sun Hea Hong Institute for Acoustics and Building Physics, University of Stuttgart, Stuttgart, Germany

Rafael Horn Fraunhofer-Institute for Building Physics IBP, Stuttgart, Germany; Institute for Acoustics and Building Physics, University of Stuttgart, Stuttgart, Germany

Marina Isasa TECNALIA, Basque Research and Technology Alliance (BRTA), Elexalde Derio, Spain

Florian Ansgar Jaeger Siemens AG, Berlin, Germany

Nils Jäger Innovation and Quality Center—Sustainable Steel Production, Thyssenkrupp Steel Europe AG, Duisburg, Germany

Winter Jesse Institute of Social Neuroscience, Ivanhoe, Australia

Knut Jøssang Pipelife Norge AS, Surnadal, Norway

Shaffie Juma Sustainable Pavements and Asphalt Research (SUPAR), Faculty of Applied Engineering, University of Antwerp, Antwerp, Belgium

Alexandra Kalnev TRACTO-Technik GmbH & Co. KG, Lennestadt, Germany

Ulrike Kirschnick Processing of Composites and Design for Recycling, Technical University Leoben, Leoben, Austria

Markus Kleineberg Institute of Lightweight Systems (SY), DLR German Aerospace Center, Braunschweig, Germany

Sandra Köhler Resource Lab, Institute of Materials Resource Management, University of Augsburg, Augsburg, Germany

Anish Koyamparambath Groupe Analyse du Cycle de Vie Et Chimie Durable (CyVi), L'Institut des Sciences Moléculaires (ISM), Université de Bordeaux, Bordeaux, Talence, France

Gijs Krekel Institute of Sustainability in Civil Engineering, RWTH Aachen University, Aachen, Germany; Innovation and Quality Center—Sustainable Steel Production, Thyssenkrupp Steel Europe AG, Duisburg, Germany

Karina Kroos Institute of Lightweight Systems (SY), DLR German Aerospace Center, Braunschweig, Germany

Frédéric Lai Bureau de Recherches Géologiques et Minières, BRGM, Orleans, France

Vincenzo Lariccia Electrolux Italia SpA, Porcia, PN, Italy

Gustavo Larrea-Gallegos Luxembourg Institute of Science and Technology (LIST), Esch-Sur-Alzette, Luxembourg

Lucyna Lekawska-Andrinopoulou Institute of Communication and Computer Systems, Athens, Greece

Angélique Léonard Chemical Engineering Research Unit, PEPs-Product, Environment, and Processes Group, University of Liège, Liege, Belgium

Paola Lettieri University College London, London, UK

Bernadette Sidonie Libom Department of Civil, Environmental, and Architectural Engineering, DICEA, Padova University, Padova, Italy

Philippe Loubet University of Bordeaux, CNRS, INP, ISM, Talence, France

Luciana Magalhães Department Decarbonization and Circularity, ArcelorMittal Long Carbon Steel, Belo Horizonte, Brazil

Stefan Majer DBFZ GmbH, Leipzig, Germany

Mattia Mangia Università degli Studi del Salento, Lecce, Italy

Alessandro Manzardo Department of Civil, Environmental, and Architectural Engineering, DICEA, Padova University, Padova, Italy;
CESQA (Quality and Environmental Research Centre), Department of Civil, Environmental and Architectural Engineering, University of Padova, Padova, Italy

Natalia Lopez Marin Electrolux Italia SpA, Porcia, PN, Italy

Alessandro Marson CESQA (Quality and Environmental Research Centre), Department of Civil, Environmental and Architectural Engineering, University of Padova, Padova, Italy;
Bluegreen Ecoinnovations Srl Società Benefit, Trieste, Italy

Antonino Marvuglia Luxembourg Institute of Science and Technology (LIST), Esch-Sur-Alzette, Luxembourg

Massimiliano Materazzi University College London, London, UK

Fabrice Mathieux European Commission—Joint Research Centre (JRC), Ispra, Italy

Thibaut Maury European Commission—Joint Research Centre (JRC), Ispra, Italy

Lionel Thellier Mercier Sphera Solutions GmbH, Stuttgart, Germany

Jean Metzmacher Chemical Engineering Research Unit, PEPs-Product, Environment, and Processes Group, University of Liège, Liege, Belgium

Anna Birgitte Milford NIBIO—Norwegian Institute of Bioeconomy Research, Division of Food Production and Society, Bergen, Norway

Ben Moins Energy and Materials in Infrastructure and Buildings (EMIB), Faculty of Applied Engineering, University of Antwerp, Antwerp, Belgium

Matthias Müller Institute for Acoustics and Building Physics, University of Stuttgart, Stuttgart, Germany

Fadzai Mundembe Groupe Analyse du Cycle de Vie Et Chimie Durable (CyVi), L'Institut des Sciences Moléculaires (ISM), Université de Bordeaux, Bordeaux, Talence, France

Anton Nahman CSIR, Stellenbosch, South Africa

Barbara Nebel thinkstep-anz, Porirua, New Zealand

Malina Nikolic Meo Carbon Solutions GmbH, Cologne, Germany

Benedetta Nucci European Aluminium, Brussels, Belgium

Patrick Ober ISCC System GmbH, Cologne, Germany

Steffen Opitz Institute of Lightweight Systems (SY), DLR German Aerospace Center, Braunschweig, Germany

Martina Orefice European Commission—Joint Research Centre (JRC), Ispra, Italy

Mary Osorio Sustainability Department, EurA AG, Erfurt, Germany

Philippe Osset SCORE LCA, Villeurbanne, France

Denise Ott Sustainability Department, EurA AG, Erfurt, Germany

Simon Panik Institute for Acoustics and Building Physics, University of Stuttgart, Stuttgart, Germany

Andrea Paulillo University College London, London, UK

Luigia Petti Department of Economic Studies (DEC), University "G. d'Annunzio", Pescara, Italy

Claudia Peña PINDA LCT SpA, Santiago de Chile, Chile

Magnus Piotrowski Sphera Solutions GmbH, Stuttgart, Germany

Roland Pomberger Waste Processing Technology and Waste Management, Technical University Leoben, Leoben, Austria

Stefania Presta Department of Civil, Environmental, and Architectural Engineering, DICEA, Padova University, Padova, Italy

Antonia Rahn Institute of Maintenance, Repair and Overhaul (MO), DLR German Aerospace Center, Hamburg, Germany

Francesca Reale Ecoinnovazione srl, Bologna, Italy

Donald Reid ERM—Environmental Resources Management, London, UK

Patrícia Rezende Postgraduate Program in Product and Process Technology, Federal Center for Technological Education of Minas Gerais State, Belo Horizonte, Brazil

Leonardo Ribeiro ArcelorMittal Maizieres Research SA, Maizières-Lès-Metz, France

Julio Rivera ArcelorMittal Maizieres Research SA, Maizières-Lès-Metz, France

Rachael Rothman Grantham Centre for Sustainable Futures, School of Chemical, Materials and Biological Engineering, Sheffield, UK

Matthias Rudolf Sphera Solutions GmbH, Stuttgart, Germany

Valentina Russo CSIR, Stellenbosch, South Africa

Anna Sánchez Catalonia Institute for Energy Research, Sant Adrià de Besòs, Spain

Manel Sansa Schneider Electric, Rueil-Malmaison, France

Stefan Schmitz TRACTO-Technik GmbH & Co. KG, Lennestadt, Germany

Dieuwertje Schrijvers WeLOOP, Lambersart, France

Maximilian Schüler Technische Hochschule Luebeck, Luebeck, Germany

Matt Seiler Schneider Electric, Rueil-Malmaison, France

Nilay Shah Department of Chemical Engineering, Imperial College London, London, UK

Federica Silveri Department of Economic Studies (DEC), University "G. d'Annunzio", Pescara, Italy

Luka Smajila Department of Energy Technology, School of Industrial Engineering and Management, KTH Royal Institute of Technology, Stockholm, Sweden

Chloë Smithers AstraZeneca, Macclesfield, UK

Guido Sonnemann University of Bordeaux, CNRS, INP, ISM, Talence, France; Groupe Analyse du Cycle de Vie Et Chimie Durable (CyVi), L'Institut des Sciences Moléculaires (ISM), Université de Bordeaux, Bordeaux, Talence, France

William Stafford CSIR, Stellenbosch, South Africa

Sharon Stauffert Institute for Acoustics and Building Physics, University of Stuttgart, Stuttgart, Germany

Julian Suer Innovation and Quality Center—Sustainable Steel Production, Thyssenkrupp Steel Europe AG, Duisburg, Germany

Nacef Tazi European Commission—Joint Research Centre (JRC), Ispra, Italy

Thodoris Theodoropoulos Institute of Communication and Computer Systems, Athens, Greece

Andrea Thorenz Resource Lab, Institute of Materials Resource Management, University of Augsburg, Augsburg, Germany

Olubukola Olumuyiwa Tokede School of Architecture and Built Environment, Geelong Waterfront Campus, Deakin University, Geelong, Australia

Marzia Traverso Institute of Sustainability in Civil Engineering, RWTH Aachen University, Aachen, Germany

Stéphane Trébucq University of Bordeaux, IAE School of Management - Bordeaux, IRGO, Bordeaux, France

Georgios Tsimiklis Institute of Communication and Computer Systems, Athens, Greece

Axel Tuma Chair for Production and Supply Chain Management, University of Augsburg, Augsburg, Germany

Clara Valente NORSUS—Norwegian Institute of Sustainability Research, Fredrikstad, Norway

Wim Van den bergh Sustainable Pavements and Asphalt Research (SUPAR), Faculty of Applied Engineering, University of Antwerp, Antwerp, Belgium

Monu George Varghese Institute of Sustainability in Civil Engineering, RWTH Aachen University, Aachen, Germany

Stuart Walker Grantham Centre for Sustainable Futures, School of Chemical, Materials and Biological Engineering, Sheffield, UK

Xinyuan Wang University of Illinois Urbana-Champaign, Champaign, USA

Julia Weißert Institute for Acoustics and Building Physics, University of Stuttgart, Stuttgart, Germany

Hugh Whetherly ERM—Environmental Resources Management, London, UK

Dan Whitaker Institute for the Development of Environmental Economic Accounting Group, Melbourne, Australia

Andy Whiting AstraZeneca, Macclesfield, UK

Simen Wilsher-Lohre NIBIO—Norwegian Institute of Bioeconomy Research, Division of Food Production and Society, Bergen, Norway

Johannes Wunderlich Siemens AG, Berlin, Germany

Wenxin Yang University of California Berkeley, Berkeley, USA

Alessandra Zamagni Ecoinnovazione, Bologna, Italy

Sanaa Zinbi European Aluminium, Brussels, Belgium

Johannes Zobel Resource Lab, Institute of Materials Resource Management, University of Augsburg, Augsburg, Germany

Stefano Zuin Electrolux Italia SpA, Porcia, PN, Italy

Methodological and Conceptual Advancements in LCA and LCSA

Streamlining Life Cycle Assessment and Natural Capital Accounting: Identifying Synergies and Challenges for Enhanced Sustainability Evaluation

Rafael Horn, Alexandra Evans, Andreas Gess, Dan Whitaker, Marina Isasa, Sharon Stauffert, and Sun Hea Hong

Abstract Assessing the impact of products and organizations on nature is essential, particularly under the European Sustainability Reporting Standards (ESRS). At the product level, environmental impacts are effectively captured through life cycle assessment (LCA) which focuses on the completeness and comparability of life cycle impacts. At the corporate level, various approaches exist, with Natural Capital Accounting (NCA) quantifying natural capital in terms of stocks (e.g., ecosystems) and flows (e.g., ecosystem services). While LCA and NCA frameworks are accepted, they are often applied separately, leading to inconsistencies and redundant work patterns. Harmonized application requires identifying interlinkages and barriers for streamlined decision-making. We investigate properties and features as well as assumptions of LCA and NCA to identify synergies through their joint application, covering key terminology, system understanding, and boundary conditions. Building on identified barriers and opportunities we suggest a harmonized assessment framework for complementary application. We provide recommendations for resolving inconsistencies while reflecting the differences in goal and overall assumptions.

R. Horn (✉)
Fraunhofer-Institute for Building Physics IBP, Stuttgart, Germany
e-mail: rafael.horn@ibp.fraunhofer.de

R. Horn · A. Gess · S. Stauffert · S. H. Hong
Institute for Acoustics and Building Physics, University of Stuttgart, Stuttgart, Germany

A. Evans
Remote Sensing Applications, Flemish Institute for Technological Research (VITO), Mol, Belgium

D. Whitaker
Institute for the Development of Environmental Economic Accounting Group, Melbourne, Australia

M. Isasa
TECNALIA, Basque Research and Technology Alliance (BRTA), Elexalde Derio, Spain

© The Author(s) 2026

M. Traverso et al. (eds.), *Life Cycle Management from Global to Local*,
https://doi.org/10.1007/978-3-032-17987-6_1

1 Introduction

Assessing the impact of products and organizations on nature has become a key requirement, especially in the context of European Sustainability Reporting Standards (ESRS) and sustainability disclosure obligations (D'Amato et al. 2024). At a product level, environmental impacts are well captured through life cycle assessment (LCA), which is agnostic to sectors and environmental areas of protection, focusing on completeness and transparency of life cycle impacts (Office of the European Union 2024). At a corporate level different approaches are available focusing on specific applications and environmental realms. Natural capital accounting (NCA) is an approach tailored to quantify natural capital in terms of stocks (e.g., ecosystems, water bodies) and flows (e.g., ecosystem services, flows of water) using the System of Environmental Economic Accounting-Ecosystem Accounting (SEEA-EA) (United Nations 2021). While both LCA and NCA frameworks are harmonized and accepted, they are oftentimes applied separately causing both inconsistencies and unnecessary double work (Cordella et al. 2022). While the frameworks also refer to different levels of assessment and application areas, their application offers synergies. To facilitate this, a harmonized application requires the specification of interlinkages, synergies, and potential barriers to allow streamlined integration in decision-making and accounting (D'Amato et al. 2024). LCA focuses on completeness and transparency of environmental impacts across the life cycle, often requiring generalized models and averaged data. NCA emphasizes natural capital stock and condition, tied more closely to specific corporate activities and locations, often making different baseline assumptions than LCA. Each framework operates with different preconditions and boundary conditions, which can complicate integration. In the following, a systematic investigation on both frameworks is provided to derive and suggest a complementary approach paving the way toward future harmonized application.

1.1 Life Cycle Assessment

LCA is a standardized method used to evaluate the environmental aspects and potential impacts associated with a product, process, or service throughout its life cycle (European Commission 2021). It is divided into several iterative key phases—goal and scope definition, life cycle inventory (LCI), life cycle impact assessment (LCIA) and interpretation—each of which contributes to a comprehensive understanding of a system's environmental impacts (ISO 2006).

The goal and scope definition phase includes defining the goal of the study, but also calculation principles such as functional unit and definition of the system boundary: The system boundary is defined based on a set of criteria that determine which process modules are part of the analyzed system (ISO 2006). Figure 1 shows this for the technical system and its life cycle phases.

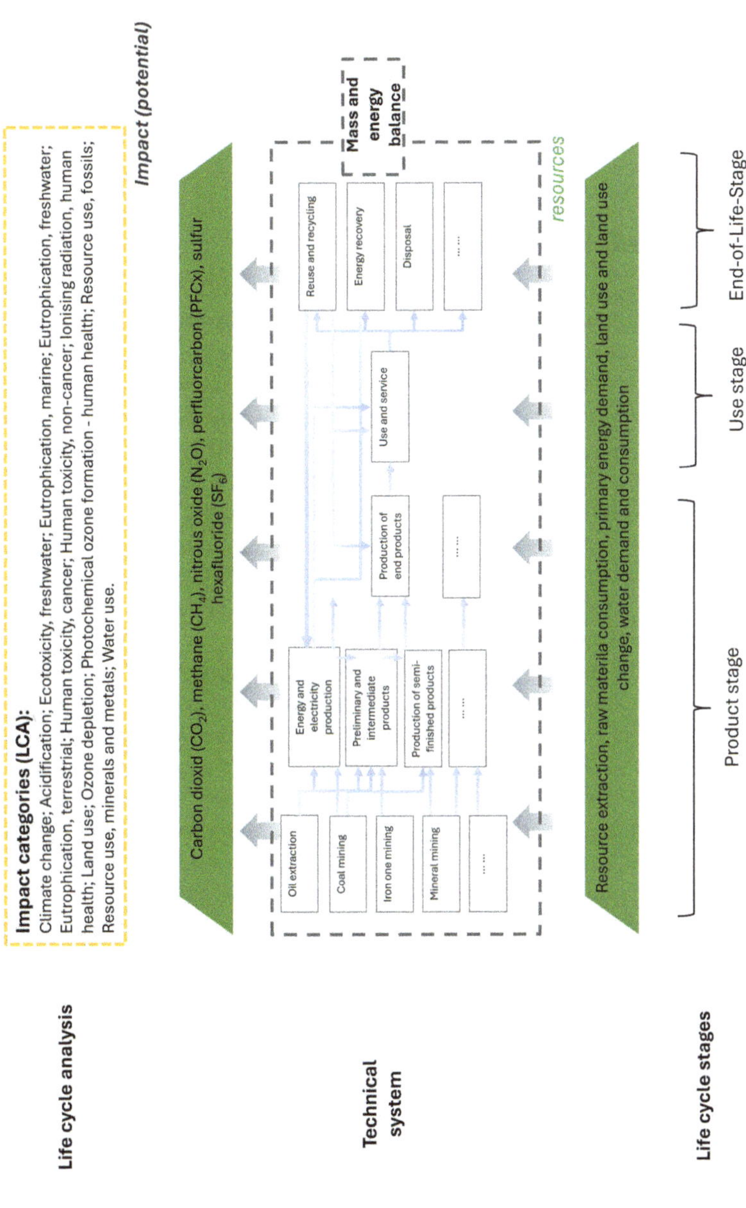

Fig. 1 Summary of the life cycle assessment concept (Held and Albrecht 2015)

In the life cycle inventory phase, inputs and outputs of the product system are collected and linked. These involve mass flows and energy flows such as raw materials, energy, water, land, emissions to air/land/water, and waste generation. The data—obtained as primary or secondary data—is used to model the product system, a representation of the product's life cycle, incorporating all relevant processes and interactions.

The interrelationships within the technical system make it possible to determine which elementary flows occur for the respective product and form the basis for determining the potential environmental impacts in life cycle impact assessment. Here, all inventory flows are classified and characterized to receive aggregated environmental results on each selected impact category per functional unit.

1.2 Natural Capital Accounting

NCA is the process of compiling consistent and comparable data on natural capital. That is natural capital stocks, relevant abiotic flows, and flows of ecosystem services generated by natural capital stocks. Natural capital accounts show the contributions of the environment to the economy and the impact of economic units on the environment. Natural capital accounting is grounded in spatial and temporal information on specific natural capital "assets" (e.g., a specific natural capital stock in a specific location) and can be applied at multiple spatial scales by both the public and private sectors. Figure 2 illustrates an example of a procedure for compiling environmental economic data in NCA. The specific approach is used in this case as it shows a certain analogy to the four phases on an LCA presented above. The exact sequence of NCA is flexible and always application-specific.

The UN-SEEA and its two major conceptual and methodological frameworks published by the UN Statistical Division provide a unifying framework for natural capital accounting. The frameworks were designed to be consistent with the System of National Accounts (SNA), a measurement framework for economic activity. The UN-SEEA can also be customized to suit the varying policy needs of stakeholders and integrates environmental and economic information in both physical and monetary terms. These frameworks include the SEEA-Central Framework (SEEA-CF) and the SEEA-Ecosystem Accounting (SEEA-EA) (United Nations 2021).

The SEEA-CF defines and assesses the "interactions between the economy and environment," including stocks, and changes in stocks, of environmental assets. It complements the methodological and conceptual framework of the SEEA-CF and expands on the definition of a natural capital asset to include ecosystems. It considers the extent (the area shares of ecosystem types within an area of study), condition (the state characteristics these ecosystems are in, consisting from different variables, e.g., species numbers, tree cover or soil pH), and resulting services/benefits of ecosystems as assets in a spatially based statistical framework. Assets are classified by different ecosystem types, such as forests or wetlands. The SEEA-EA also applies

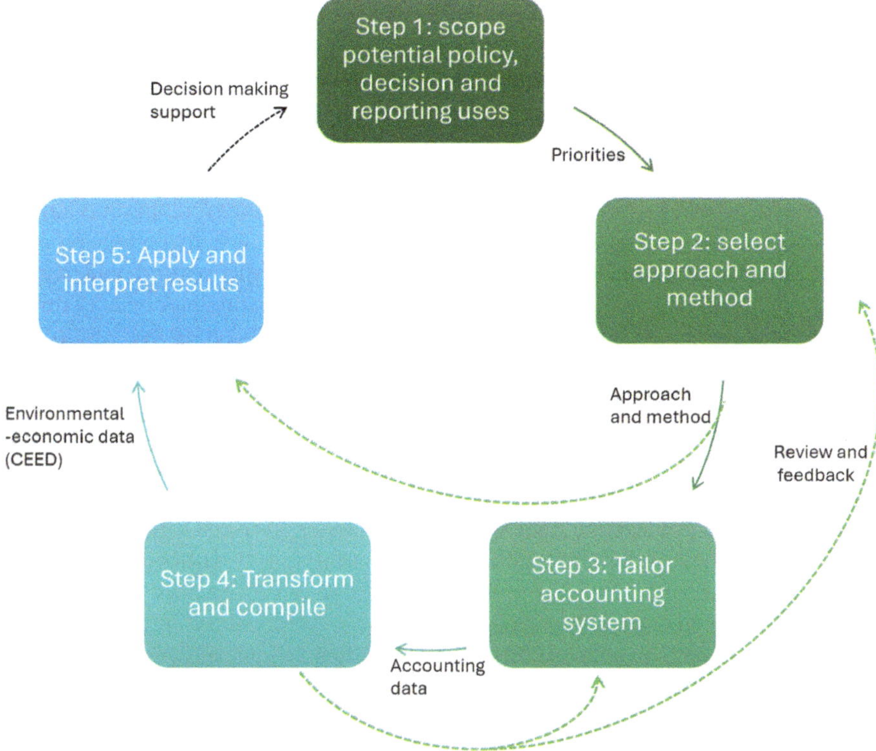

Fig. 2 Natural capital accounting process building on SEEA as developed by IDEEA group (own elaboration based on IDEEA 2021)

the accounting principles of the System of National Accounts (SNA) (United Nations 2021).

1.3 Existing Approaches in Scientific Literature

While both concepts are frequently applied, their joint application is rarely stated or investigated scientifically, as systematically investigated in the Horizon Project A-Track (Isasa Saralde et al. 2025).

Practical examples of this integration often focus on companies aiming to reduce and/or offset their environmental impact (Rugani et al. 2023). In such cases, LCA is used to assess the harmful environmental impacts, while NCA helps identify potential ecosystem gains (and not just losses) following the steps of the mitigation hierarchy.

The added value of this mainly is NCA can provide LCA with spatial information on ecosystem extent and condition, which can be used to better understand related dependencies and impacts triggered by business along the value chains. On the other hand, LCA can equip NCA with the value chain perspective necessary to better understand full impacts along the life cycle. Additionally, the use of a systematic and harmonized monitoring of biodiversity and ecosystem services along the entire value could be an instrument not only for accounting and target setting, but also for reporting.

While this combined application shows clear potential, several challenges remain according to current literature that currently limit its widespread adoption (Cordella et al. 2022; Isasa Saralde et al. 2025; Rugani et al. 2023). While considerable progress can be stated, LCA still lacks an agreed methodological framework to quantify and assess impacts on ecosystem services and biodiversity. Notably, biodiversity impact assessment remains underdeveloped in LCA due to limited data availability and the lack of standardized ready-to-use methods (Damiani et al. 2023). Furthermore, existing approaches often fail to comprehensively cover all drivers of biodiversity loss and neglect key aspects of biodiversity, such as genetic or ecosystem diversity. While there exist several initiatives working on these issues, they are still far from reaching consensus and widespread adoption (Damiani et al. 2023).

From a spatial perspective, another challenge arises from the different scales of information that are relevant for products, organizations (e.g., traditionally focused on product and organizational level), and public accounting frameworks (e.g., typically focused on national scale). This disparity in scale systems makes it challenging to establish a seamless link between these two assessments, hindering the integration of micro-level organizational data with macro-level national accounting frameworks.

Despite the complexities of the current landscape, the scientific community recognizes that integrating LCA with NCA holds significant potential for enhancing the biodiversity and ecosystem services footprint assessment (Cordella et al. 2022). This linkage can make the assessment more comprehensive and robust in capturing the intricate relationships between human activities and environmental impacts across different scales and along value chains.

2 Comparative Analysis of Terminology, Assumptions, and Applications

While there are many similarities in the applications of NCA and LCA, they generally build on different terminologies and differ in terms of underlying assumptions and principles.

2.1 Key Terminology

In the following, key terms used within LCA are compared with similar concepts and definitions within NCA (Table 1). Note that the set of terms included is not

Table 1 Key terms and their meaning in LCA and NCA

Concept	LCA (European Commission 2021; ISO 2006)	NCA (Ecosystem Accounting) (United Nations 2021)
Life cycle	Consecutive and interlinked stages of a product system, from raw material acquisition or generation from natural resources to final disposal	NCA does not explicitly address product life cycles
Impact	Environmental burdens across the life cycle (e.g., GHG emissions, eutrophication, biodiversity risks of 1 kg of avocado production), quantified through classification and characterization of elementary flows	Quantified in terms of changes in natural capital stocks (extent and/or condition) which can then lead to changes in flows of ecosystem services
System boundaries	Product centric: quantifies all environmental impacts from all processes required to produce, transport, use, and dispose a product (which can be any good or service)	Location-centric: quantifies extent and condition of ecosystem assets and the flows of ecosystem services provided by those assets
Inventory model	Life Cycle Inventory (LCI): quantified inputs/ outputs per process (materials, energy, emissions) of which the elementary flows are used to calculate the life cycle impacts	Registers of natural capital assets can be provided detailing their extent, condition, and other characteristics
Safeguard objects	Environmental areas of protection, (e.g., human health, ecosystems) specified through midpoint impact categories (e.g., global warming, eutrophication, land use—soil quality)	Natural capital assets and related ecosystem services can be considered on a case-by-case basis. Linkages to human health can be drawn
Assessment result	Environmental impact profile across impact categories and life cycle stages	State, trends, and value of natural capital
Valuation	Often midpoint or endpoint impact indicators (not monetized), with optional weighting and normalization	Quantification of ecosystem services in biophysical and/or monetary terms

intended to be exhaustive and were chosen as a selection of terms that are of critical importance within LCA to which comparisons with NCA can be drawn.

2.2 Assumptions and Boundary Conditions

In addition to these explicit terminology-based differences, several key differences in terms of underlying conception, system understanding including implicit preconditions, explicit boundary conditions, and modeling requirements can be identified (Table 2). Again, these are collected with a focus on LCA practice and do not cover all relevant terms but reflect the author's selection of key assumptions and properties of the two concepts.

Table 2 Assumptions and boundary conditions

Aspect	LCA	NCA
System perspective	Assumes a technosphere-centered view: impacts are driven by industrial activities and their resource use as well as emissions to air, water, and soil	Assumes a nested systems view: the economy operates within a broader society that operates within the natural environment
Key motivation	Primarily designed for comparative product assessment and impact minimization	Provides a framework for the organization and management of data on natural capital that can enable an understanding of stocks and flows and be linked to economic units
Responsibility scope	Impact responsibility is distributed across value chain actors and usually not depicted. Instead impacts are distributed along life cycle phases	Responsibility scope is not a concept, accounting area may be defined based on the interests of the compiler of accounts
Context consideration	Assumes impacts are context-independent once characterized (e.g., CO_2 has the same global impact)	Takes local context into account—e.g., condition indicators on water in a dry region will be chosen to reflect the local environment
Life cycle perspective	LCA usually requires full product life cycle consideration (comparable to scope 3 in GHG reporting)	In NCA and asset, focus is undertaken (compared to scope 1 in GHG reporting)
Temporal scope	Land using aspects are modeled in a simplified approach, assuming only one land use change (transformation) and stable conditions during occupation	Data is collected at defined accounting periods, variables are collected on a regular basis (e.g., annually)
Geographical scope	Global value chains are evaluated using background data models, granularity for specific activities is usually limited to country average values	Data is always spatially embedded, and usually spatially explicit
Absolute or relative approach	Impacts are linearized to the relative relevance of a products life cycle on a specific impact category. Absolute quality is usually not depicted	Data acquisition can cover both absolute and relative numbers
Reference system selection	For land use impacts, the reference situation is mostly specified as potential natural vegetation	Reference condition is a benchmark to which ecosystem condition is measured over time
Allocation	Allocation is used to handle multi-output processes when inevitable. Economic allocation causes mass inconsistency in product models	Flows are associated with both suppliers and users; it may be possible for a single flow to be allocated to various users

3   Discussion: A Perspective Toward Harmonized Application

The complementary purpose, application contexts, and assumptions of NCA and LCA show a potential for joint application, especially in the context of intertwined investigations of both corporate- and product-level assessment and of ecosystem service as well as biodiversity relevant assets, products, and business models. This requires a clear framework to define the purpose of each method, as well as distinct consideration of differences in perspective and assumptions.

In the A-Track project, a general Biodiversity and Ecosystem Service (BES) Footprinting Framework is developed aiming for complementary application of NCA and LCA in the context of product and corporate footprinting. This framework aims to allow for flexible application in key areas on both corporate and product assessments of ecosystems and biodiversity, and therefore builds on existing approaches and standards such as ISO 14026, carbon footprinting, and water footprinting (ISO 2017). However, these approaches do not cover challenges in the abovementioned sections, especially on system boundaries and assumptions. The harmonization of LCA and NCA is seen as fundamental to the development of a BES footprint framework, and as outlined above this requires the consideration and recognition of differences in shaping harmonized operational solutions.

The identified differences in key terminology are seen as technical challenges that are to be resolved when developing a harmonized method. This mainly requires transparency as well as mutual recognition and mapping to allow a clear understanding for practitioner, in which of the methods they are operating in.

For the key assumptions and boundary conditions, the harmonization task shows more fundamental challenges. In general, the considerations of some key assumptions from NCA on regionality and temporal scope would strengthen the current LCA framework especially with land use focus. As this would come with additional requirements for practitioners, it requires a flexible scope and level of comprehensiveness in applications to reflect the necessary granularity on temporal and spatial scale. On the other hand, life cycle thinking might bring additional decision support and advanced overall understanding of the impact of systems in addition to their direct implications in the area under study.

Figure 3 shows a simplified synthesis on complementary application of SEEA, providing a systematic structure for data acquisition on ecosystem extent and condition. This allows to consistently build NCA and LCA models while still reflecting and accepting specificities of both concepts. Information on ecosystem extent is used to inform inventory modeling of land use flows for occupation and transformation. In addition, the calculation of specific characterization factors in impact assessment is supported by information on ecosystem condition, such as the provisioning of soil organic carbon content to calculate specific land use impacts on SOC that then feed into the calculation of environmental impacts on soil quality.

The suggested A-Track BES footprint framework is building on a multi-tiered approach, building on a comprehensive assessment with full-scale LCA as well as

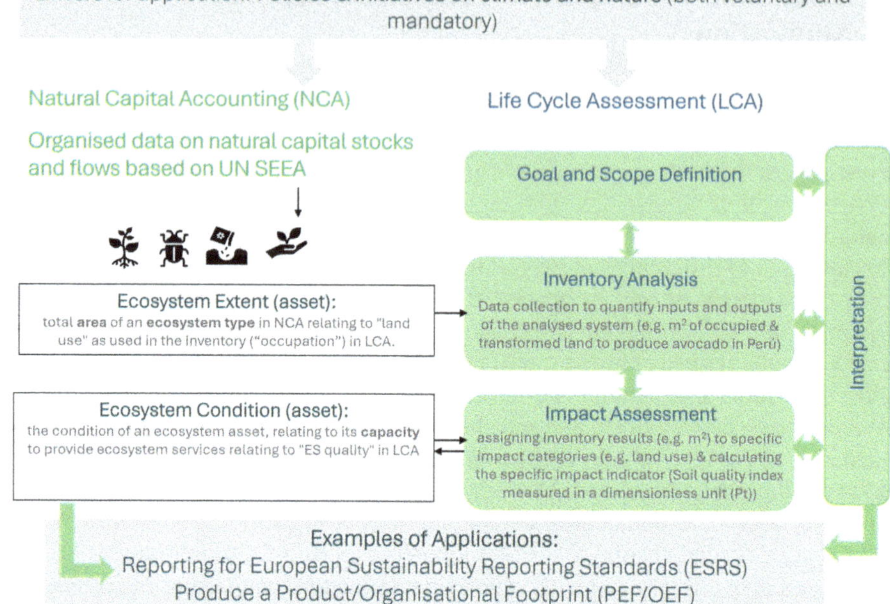

Fig. 3 A-track BES footprint framework. An LCA-based biodiversity footprint framework grounded in ecosystem accounting principles for land use impact assessment. The figure shows how concepts from NCA to describe stocks of ecosystem assets (extent and condition), based on the System of Environmental Economic Accounting (SEEA) Ecosystem Accounting (EA) (SEEA-EA), can inform the inventory (extent) and the impact assessment (condition) phases of an LCA, leading to the calculation of a BES Footprint grounded on NCA data (Isasa Saralde et al. 2025)

comprehensive NCA data consideration on ecosystem extent and condition. For screening purposes, a simplified approach is suggested using generic data for the life cycle modeling and reduced spatial information required to reflect on the ecosystem extent and condition. Furthermore, a partial assessment is suggested to allow for asset-focused investigations only covering parts of the life cycle (such as farm-level assessment in agriculture). These tiers are subject of ongoing development in the operationalization phase following the framework publication (Isasa Saralde et al. 2025).

Acknowledgements This work was created within the Horizon 2020 project A-Track (A-Track) that has received funding from the European Union's Horizon Europe research and innovation programme under grant agreement no. 101082268, as well as from UK Research and Innovation (UKRI) and Switzerland's State Secretariat for Education, Research and Innovation (SERI).

References

Cordella M, Gonzalez-Redin J, Lodeiro RU, Garcia DA (2022) Assessing impacts to biodiversity and ecosystems: understanding and exploiting synergies between life cycle assessment and natural capital accounting. Procedia CIRP 134–139. https://doi.org/10.1016/j.procir.2022.02.023

D'Amato D, La Notte A, Damiani M, Sala S (2024) Biodiversity and ecosystem services in business sustainability: toward systematic, value chain-wide monitoring that aligns with public accounting. J Ind Ecol. https://doi.org/10.1111/jiec.13521

Damiani M, Sinkko T, Caldeira C, Tosches D, Robuchon M, Sala S (2023) Critical review of methods and models for biodiversity impact assessment and their applicability in the LCA context. Environ Impact Assess Rev 101. https://doi.org/10.1016/j.eiar.2023.107134

European Commission C (2021) 9332 final. Commission recommendation of 16.12.2021 on the use of the environmental footprint methods to measure and communicate the life cycle environmental performance of products and organisations

Held M, Albrecht S (2015) Ganzheitliche Bilanzierung - mit dem Lebenszyklusansatz zu ressourceneffizienten Produkten. wt Werkstattechnik online. Jahrgang 105(7/8). https://doi.org/10.37544/1436-4980-2015-07-08-88

IDEEA Group (2021) Natural capital accounting design and implementation protocol V 1.1, Melborne

Isasa Saralde M et al (2025) A-Track D3.1 a biodiversity and ecosystem services footprint: the A-track approach, EU Horizon Europe project A-Track (GA No 101082268). https://a-track.info/a-track/resource/biodiversity-and-ecosystem-services-footprint-track-approach

ISO (2006) ISO 14040:2006 Environmental management—Life cycle assessment—Principles and framework, International Organization for Standardization, Geneva

ISO (2017) ISO 14026:2017 Environmental labels and declarations. Principles, requirements and guidelines for communication of footprint information. International Organization for Standardization, Geneva

Office of the European Union (2024) Regulation (EU) 2024/1781 of the European parliament and of the council of 13 June 2024 establishing a framework for the setting of ecodesign requirements for sustainable products, amending Directive (EU) 2020/1828 and Regulation (EU) 2023/1542 and repealing Directive 2009/125/EC (Text with EEA relevance), Luxembourg. http://data.europa.eu/eli/reg/2024/1781/oj

Rugani B, Osset P, Blanc O, Benetto E (2023) Environmental footprint neutrality using methods and tools for natural capital accounting in life cycle assessment. Land. https://doi.org/10.3390/land12061171

United Nations (2021) System of environmental-economic accounting-ecosystem accounting (SEEA EA). White cover publication, pre-edited text subject to official editing. https://seea.un.org/ecosystem-accounting

Sustainability Assessment in Industrial R&D: Challenges and Contributions to Decision-Making

Denise Ott, Marvin Gornik, Shashank Goyal, Mary Osorio, Nahikari Diosdado, and Silvia Agrafojo

Abstract Sustainability assessments are becoming essential tools for decision-making and R&D as industry faces increasing pressure to address sustainability impacts. Supported by EU policy, Life Cycle Sustainability Assessment (LCSA) techniques are becoming more necessary in funded R&D projects, particularly for impact optimisation in the early stages. But in the early stages of development, supply chain opacity, data gaps, and regulatory complexity pose significant obstacles. This study investigates the ways in which sustainability evaluations can support risk assessment, verification of green claims, and regulatory anticipation. Using case studies from AID4GREENEST (steel), ZDZW (manufacturing), and ECO2LIB (battery), common obstacles and useful insights are highlighted. To facilitate sustainable technological advancement, bringing innovation and regulation into alignment through reflective evaluations is needed.

1 Introduction

In recent years, in the European Union (EU), sustainability is now required by law rather than being a voluntary objective. The goal of laws such as the Sustainable Finance Disclosure Regulation (SFDR) (EC 2019), EU Taxonomy Regulation (EC 2020a), and Corporate Sustainability Reporting Directive (CSRD) (EC 2025a) is to focus innovation and capital on socially and environmentally responsible endeavours. In addition to reporting cycles, these frameworks increasingly require due diligence, transparency, and quantifiable impact throughout the entire life cycle of products and innovations (Mazijn and Revéret 2015).

This change brings both opportunity and responsibility for R&D activities. Although there is still a gap between R&D practice and regulatory compliance,

D. Ott (✉) · M. Gornik · S. Goyal · M. Osorio
Sustainability Department, EurA AG, Erfurt, Germany
e-mail: denise.ott@eura-ag.de

N. Diosdado · S. Agrafojo
Zabala Innovation Consulting, Mutilva Alta, Navarra, Spain

M. Traverso et al. (eds.), *Life Cycle Management from Global to Local*,
https://doi.org/10.1007/978-3-032-17987-6_2

LCSA methods are useful for evaluating environmental, economic, and social trade-offs during design (Englhardt et al. 2025). R&D teams frequently encounter difficulties implementing sustainability criteria because of their complexity, data gaps, and early-stage uncertainty, even though these criteria are now codified in EU funding programmes such as Horizon Europe (González-Varona et al. 2023). This disparity can be seen in the lack of feedback loops to guide more flexible policymaking, the restricted methodological capacity, and the misalignment of compliance frameworks with the iterative nature of innovation (De Grandis et al. 2023).

Yet, there is great potential to bridge this gap. Embedding sustainability assessment early in the innovation process can act as a strategic enabler for both regulatory alignment and competitive advantage. It can support the identification of systemic risks (e.g. resource dependencies, social hotspots), the validation of green claims, and the anticipation of regulatory shifts (Jain et al. 2021). This paper investigates the practical use of LCSA in industrial R&D, the challenges encountered, and the ensuing policy insights using case studies from EU projects ECO2LIB, ZDZW, and AID4GREENEST. It seeks to promote a more reflective relationship between innovation and regulation, where sustainability evaluations promote change rather than obstruct it. An overview of projects is given in Table 1.

Table 1 Brief insight into the EU projects discussed herein

Project acronym/G.A. no./ run-time	Full name	Aim
ECO2LIB/ H2020 875514/ 2020–2024	Ecologically and Economically viable Production and Recycling of Lithium-Ion Batteries	Development of affordable battery materials for energy storage that consider every stage of the battery life cycle, from cell production and material selection to use and recycling
ZDZW/HEU 101057404/ 2022–2025	Non-Destructive Inspection Services for Digitally Enhanced Zero Defect and Zero Waste Manufacturing	Six pilots, including injection moulding and welding, in the automotive, energy, e-health, and food sectors showcase how innovation in digital NDI technologies can increase productivity and sustainability in important European industries
AID4GREENEST/HEU 101091912/2023–2026	AI-Powered Characterisation and Modelling for Green Steel Technology	Development of an AI-based characterisation and modelling tools for the steel industry that address performance assessment, steel, process, and product design

2 Life Cycle Assessment (Environmental Aspect)

Life Cycle Assessment (LCA) is a systematic method standardised by ISO 14040 and 14044 for evaluating environmental impacts associated with a product, process, or system across its entire life cycle (ISO 14040 2006; ISO 14044 2006). It enables early comparison of different design options and helps identify environmental trade-offs, already early in the innovation process. LCA has become an indispensable part of industrial and publicly funded R&D. In early-stage development, LCA allows researchers and engineers to explore the environmental performance of emerging technologies and to optimise material selection, manufacturing processes, and system configurations. Particularly during the configuration and pre-commercial phases of product development, LCA helps to embed environmental performance as a design parameter rather than a post hoc evaluation criterion (Cucurachi et al. 2022). In Horizon Europe and related funding programmes, conducting an LCA is increasingly mandatory to demonstrate sustainability potential and risk management capability. LCA aligns closely with several key EU sustainability frameworks:

- The European Green Deal sets the overarching ambition for climate neutrality and resource efficiency, reinforcing LCA's role in assessing progress towards these goals (Sanyé-Mengual and Sala 2022). Sustainable Products Initiative (SPI) and the upcoming Eco-design for Sustainable Products Regulation (ESPR) institutionalise LCA-based criteria (e.g. carbon footprint, durability) for market access (Sanyé-Mengual and Sala 2022).
- Environmental Footprint Methods (PEF/OEF) promoted by the European Commission further standardise LCA metrics for comparability and regulatory reporting (EC 2024).
- CSRD obligations cover the ecological impact of a product or services along the entire value chain. Here, LCA is a reliable tool for reporting on environmental impacts in a credible manner (Schneider et al. 2024).
- Taxonomy Regulation, as a classification system for assessing economic activities as ecologically sustainable names LCA or LCA-related metrics, depending on the activity, as one of the requirements for a taxonomy alignment (Canfora et al. 2022).
- The Empowering Consumers Directive (EmpCo) (EU 2024) and the drafted Green Claims Directive (EC 2023) explicitly state that environmental claims for marketing purposes must be substantiated by instruments such as LCA.

A real-world example of implementing methodological and regulatory requirements in the development of sustainable lithium-ion batteries is provided by the ECO2LIB project. It concentrated on recycling, cell design, and water-based electrode processing. The LCA complied with the JRC's CFB-IND regulations (Andreasi Bassi et al. 2025), the EU Battery Regulation (EU 2023), and the PEFCR guidelines (Recharge 2020). Working in a pre-commercial R&D setting, ECO2LIB demonstrated common early-stage technological challenges while testing and adapting important CFB-IND components, such as system boundaries, functional units, and EF 3.1 alignment.

One major obstacle is the regulatory requirement for primary data from primary manufacturing processes, which is hard to satisfy in the early stages of development because of unrepresentative pilot-scale data and unfinalised process routes. By identifying important parameters, such as energy consumption during electrode processing and cathode precursor supplier information, ECO2LIB's LCA activities assist future carbon footprint declarations. R&D was directed towards materials and procedures that satisfy crucial energy thresholds for environmental benefits by comparing experimental cells to a reference. The combined LCA and LCC results are displayed in Fig. 1 (see also Chap. 3).

AID4GREENEST is targeting to develop a series of new rapid characterisation methods and modelling tools based on artificial intelligence for the steel sector. In this context, conducting an LCA of industrial products such as turbine shafts poses certain challenges, as the components require high-strength and high-precision materials and are manufactured with specific alloys due to their highly demanding mechanical functions.

One tonne of steel is frequently utilised as the functional unit in the literature, with cradle-to-gate system boundaries (Li et al. 2021; 2022; Lostado-Lorza et al. 2023; Yang et al. 2023). Large-scale testing is expensive and restricts standardisation because of alloy specificity. This project creates machine learning-based tools for alloy design and processing optimisation to minimise trial-and-error and

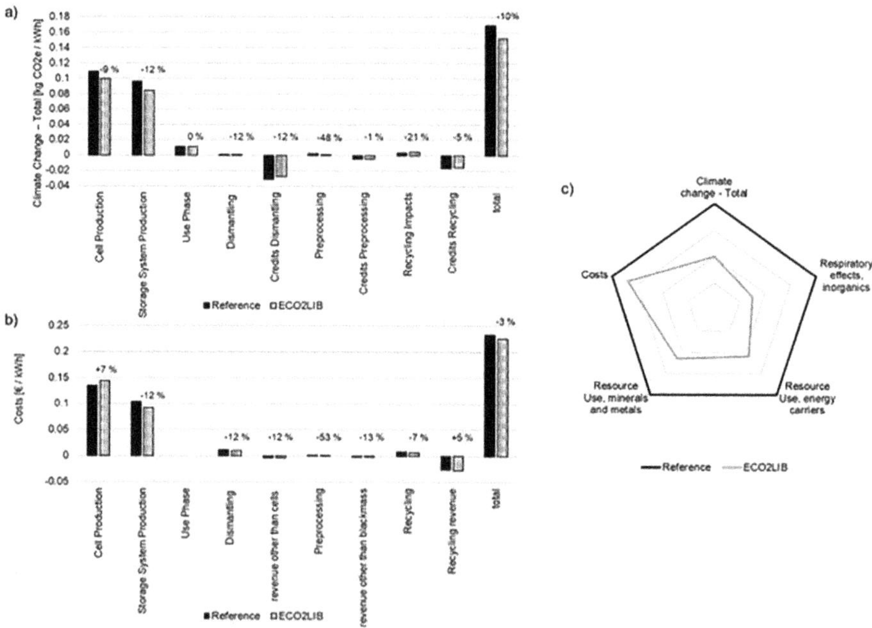

Fig. 1 Cradle-to-grave LCA and LCC results per kWh energy provided over the lifetime of the energy storage system **a** Climate change—total **b** Costs **c** Spider chart for all assessed life cycle indicators.

support sustainability goals. To increase R&D efficiency and satisfy growing regulatory requirements, an online platform encourages knowledge sharing and consistent LCA. Under the EU's Carbon Border Adjustment Mechanism (CBAM), incorporated carbon in imported steel must be reported and priced (EC 2025b). This forces manufacturers to use low-emission technologies and open evaluations like life cycle assessment (LCA). As a key sector in the green transition, steel must prioritise recyclability, low-carbon inputs, and energy-efficient methods such as hydrogen-based processes (Kazmi et al. 2023).

Non-destructive inspection (NDI) is used in the EU-funded ZDZW project to cut down on manufacturing waste and flaws. Although life cycle assessment (LCA) facilitates data-driven environmental insights and directs sustainable redesign, there are obstacles to its use in industrial R&D, including a lack of data, proprietary procedures, and methodological uncertainties. R&D's iterative nature necessitates quick, adaptable evaluations, which are frequently limited by available resources. Nevertheless, through standardisation and transparency, LCA improves communication, supports sustainability claims, and strengthens decision-making.

3 Life Cycle Cost Analysis (Economic Aspect)

The economic aspect of sustainability makes sure that new innovations are not only good for the environment and society but also make money. Knowing how new technologies affect the economy helps you weigh short-term costs against long-term benefits, which helps resources be used more efficiently and become more resilient. Cost–Benefit Analysis (CBA) looks at the overall value (McClenaghan et al. 2023), Economic Impact Analysis (EIA) looks at the effects on a local or regional level (Joseph et al. 2020), and Techno-Economic Analysis (TEA) combines technical and financial data to look at early-stage technologies (Aminzadeh et al. 2023).

The entire cost of an asset or project, including purchase, use, maintenance, and disposal, is evaluated using life cycle costing (LCC). In contrast to conventional analyses, LCC supports long-term, well-informed decisions by taking a total cost of ownership perspective (Bachmann et al. 2024). It affects R&D by directing decisions about investment, materials, and processes, particularly in circular economic contexts where longer-term savings outweigh higher upfront costs. By considering both economic and environmental effects, LCC improves integrated sustainability assessments and is frequently used in conjunction with LCA (Miah Koh and Stone 2017). The strategic importance of cost analysis is increasingly acknowledged in EU policy:

- Under the Green Public Procurement (GPP) guidelines, LCC is increasingly seen as central to sustainable sourcing, although its potential value for sustainable procurement is not yet fully understood (De Giacomo et al. 2019; EC 2014).

- Circular Economy Action Plan (CEAP) promotes lifecycle thinking, including cost considerations, to foster business models based on reuse, repair, and remanufacturing (EU 2020b).
- EU Taxonomy indirectly incentivizes the use of LCC by favouring investments with long-term cost efficiency and economic resilience.
- Innovation Fund and other R&D funding schemes encourage cost analyses and cost efficiency calculations to improve resource efficiency and reduce dependence on volatile inputs.

The European Commission (EC) has developed a series of sector-specific Life Cycle costing (LCC) calculation tools that aim to facilitate the use of LCC among public procurers in line with Article 68 of Directive 2014/24/EU and Article 83(2) of Directive 2014/25/EU (EC 2014).

By including an economic perspective on battery development, manufacturing, and end-of-life, the LCC in ECO2LIB enhanced the LCA. One important policy source is the EU Strategic Energy Technology Plan (EC 2016), which establishes cost targets of 150 €/kWh for stationary systems and 75 €/kWh for EV batteries by 2030. These targets are crucial for maintaining global competitiveness and are backed by international comparisons (Tarvydas et al. 2018). To evaluate these goals, ECO2LIB used a reference battery for scenario analysis. A modified EverBatt 2023 tool (Argonne National Laboratory 2023) was used to identify cost hotspots and scaling effects across the entire life cycle (see Fig. 1b). With cell production as the main cost driver—due to components like silicon-doped anodes—industry input helped refine assumptions, define thresholds (e.g. minimum recycling efficiency), and support decisions on cost and recyclability.

The ZDZW project carried out a systematic review to determine a comprehensive approach addressing environmental, economic, and social impacts to assess the sustainability of non-destructive inspection (NDI) technologies. Significant gaps still exist even though research on NDI and sustainability assessments has increased dramatically (Du Toit 2024). A lot of NDI evaluations separate environmental factors without considering how they interact with social and economic factors. Each sustainability pillar was reviewed separately using the PRISMA method (Page et al. 2021). For environmental considerations, LCA was the most often utilised tool; for economic considerations, LCC and occasionally Cost–Benefit Analysis (CBA) were employed (McClenaghan et al. 2023). Environmental Life Cycle Costing (eLCC), which combines LCA results with LCC in monetary terms, is the preferred method for more integrated assessments (Swarr et al. 2011). This illustrates how long-term savings could balance higher upfront expenses. This goal is depicted in Fig. 2.

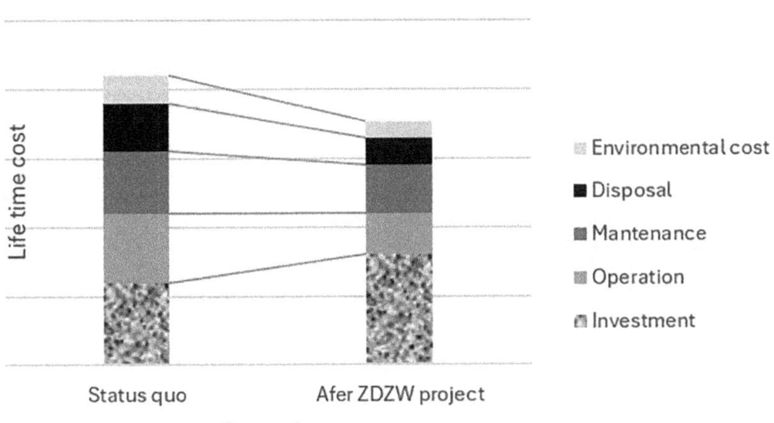

Fig. 2 eLCC before and after implementation of an NDI technology

4 Social Life Cycle Assessment (Social Aspect)

Using the UNEP/SETAC methodology, and ISO 14075:2024, as a guide, Social Life Cycle Assessment (SLCA) assesses the social and socioeconomic effects of a product over its whole life cycle, with a particular emphasis on labour rights, community well-being, and health and safety. SLCA is being used more in R&D to help manage social risks in early-stage innovation, particularly in high-risk or global supply chains. It promotes conformity to stakeholder expectations, reduces reputational hazards, and emphasises social co-benefits like equitable access and job creation. SLCA's relevance is growing rapidly considering emerging EU legislation:

- The Corporate Sustainability Due Diligence Directive (CSDDD) will require companies to assess and address adverse human rights and environmental impacts in their operations and supply chains (EC 2025c).
- The CSRD requires consideration and assessment of social risks and value chain dependencies, for which SLCA can serve as a robust data source.
- Just Transition Mechanism (JTM) and EU Pillar of Social Rights promote inclusive growth and fair treatment in green and digital transitions.
- The EU Strategy on Decent Work Worldwide puts additional pressure on companies to validate social claims with robust assessments.
- The EmpCo prohibits misleading statements about the social aspects of a product or service without credible verification.

The ZDZW project's PRISMA analysis showed that when it comes to sustainability evaluations of NDI technologies, the social dimension is the least developed.

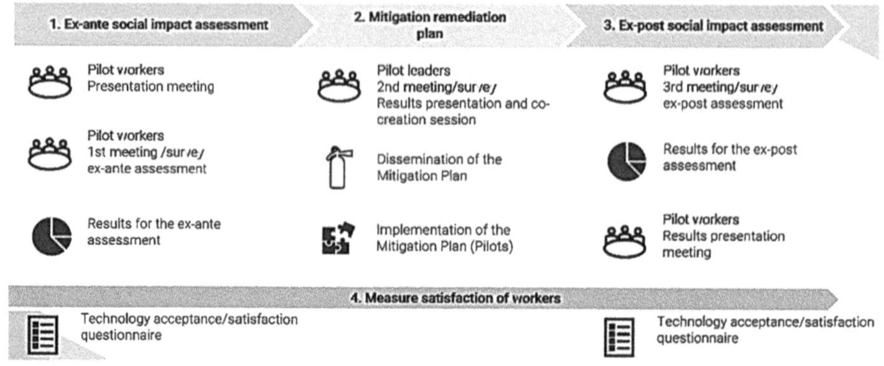

Fig. 3 Design for ZDZW project SLCA implementation plan

Social impact evaluation is supported by existing tools like Social Return on Investment, Participatory Assessment, and SLCA, which is renowned for its stakeholder focus (Ruben et al. 2018; Mancini and Sala 2018). However, because the effects are qualitative, it is still challenging to quantify social sustainability (Manik et al. 2013). Research indicates that because of stakeholder, regulatory, and data challenges, worker-related factors are frequently disregarded (Rimbau-Gilabert 2019). Based on surveys conducted before and after implementation, a targeted framework was created to address this issue using 31 indicators spanning seven dimensions (e.g. task changes, job security, and well-being) (Springer et al. 2024; Tan and Hsu 2018). The framework allows for comparison and the identification of best practices and is adaptable across pilots. The procedure, depicted in Fig. 3, is simple to incorporate into business operations.

5 Conclusion and Recommendations

As tools like LCA, LCC, and SLCA transform from reporting mechanisms to strategic design tools, incorporating sustainability assessments into industrial R&D presents both a challenge and an opportunity. Early, iterative use of these tools can guide innovation in a technical, economic, and social way, as demonstrated by case studies from AID4GREENEST, ZDZW, and ECO2LIB. Even though early regulatory alignment is frequently limited, these evaluations aid in guiding development beyond simple compliance. Data gaps, supplier transparency, and a lack of methodological harmonisation are still major obstacles. Stronger coordination between R&D and policymakers is required to address this, allowing for feedback loops and adaptive guidance. If evaluations are prompt, pertinent, and scaled to the innovation journey, integrating sustainability into R&D is ultimately a mentality change that can improve credibility, lower risk, and open new opportunities.

References

Aminzadeh A., Dimitrova M., Meiabadi M.S., Sattarpanah Karganroudi S., Taheri H., Ibrahim H., Wen Y., Non-contact inspection methods for wind turbine blade maintenance, J Nondestruct Eval, 42(2), 2023, https://doi.org/10.1007/s10921-023-00967-5.

Andreasi Bassi S., Ardente F., Candelaresi D., et al., Rules for the calculation of the carbon footprint of industrial batteries without external storage (CFB-IND), Publications Office of the European Union, 2025, https://data.europa.eu/doi/https://doi.org/10.2760/6346639.

Argonne National Laboratory, EverBatt, Argonne's closed-loop battery life-cycle model (2023). Accessed 10.06.2025. https://www.anl.gov/amd/everbatt

Bachmann et al (2024) Life cycle costing as part of a life cycle sustainability assessment of products: methodology and case studies. Int J Life Cycle Assess 29:1863–1879. https://doi.org/10.1007/s11367-024-02347-1

Canfora P, Arranz Padilla M, Polidori O, Pickard Garcia N, Ostojic S, Dri M (2022) Development of the EU sustainable finance taxonomy–a framework for defining substantial contribution for environmental objectives 3–6, EUR 30999 EN, Publications Office of the European Union, Luxembourg. https://doi.org/10.2760/256390. ISBN 978-92-76-47898-0

Cucurachi S, Steubing B, Siebler F, Navarre N, Caldeira C, Sala S (2022) Prospective LCA methodology for novel and emerging technologies for bio-based products–the PLANET BIO project, Publications Office of the European Union. Luxembourg. https://doi.org/10.2760/167543

De Giacomo MR, Testa F, Iraldo F, Formentini M (2019) Does green public procurement lead to life cycle costing (LCC) adoption? J Purch Supply Manag 25(3):100500. https://doi.org/10.1016/j.pursup.2018.05.001

De Grandis G, Brass I, Farid SS (2023) Is regulatory innovation fit for purpose? A Case Study of Adaptive Regulation for Advanced Biotherapeutics, Regul Gov 17(3):810–832. https://doi.org/10.1111/rego.12496

Du Toit E (2024) Thirty years of sustainability reporting: insights, gaps and an agenda for future research through a systematic literature review. Sustainability 16(23):10750. https://doi.org/10.3390/su162310750

EC (2014) Life-cycle costing, Energy, Climate Change, Environment. Accessed 10 Jun 2025. https://green-forum.ec.europa.eu/green-public-procurement/life-cycle-costing_en

EC (2016) SET-Plan ACTION n°7–Declaration of intent. Become competitive in the global battery sector to drive e-mobility forward. Accessed 10 Jun 2025. https://setis.ec.europa.eu/document/download/8cb5c098-1f3a-4ab3-96d8-e9c06d978137_en?filename=action7_declaration_of_intent_0.pdf

EC (2019) Sustainability-related disclosure in the financial services sector, Directorate-General for Financial Stability, Financial Services and Capital Markets Union. Accessed 12 Jun 2025. https://finance.ec.europa.eu/sustainable-finance/disclosures/sustainability-related-disclosure-financial-services-sector_en

EC (2020) EU taxonomy for sustainable activities, Directorate-General for Financial Stability, Financial Services and Capital Markets Union. Accessed 13 Jun 2025. https://finance.ec.europa.eu/sustainable-finance/tools-and-standards/eu-taxonomy-sustainable-activities_en

EC (2023) Directive of the European Parliament and of the Council on substantiation and communication of explicit environmental claims (Green Claims Directive

EC (2024) Environmental footprint methods–Benefits for companies–Overview of the environmental footprint methods, Publications Office of the European Union. https://doi.org/10.2779/8944075.

EC (2025a) Corporate sustainability reporting, Directorate-General for Financial Stability, Financial Services and Capital Markets Union. Accessed 13 Jun 2025. https://finance.ec.europa.eu/capital-markets-union-and-financial-markets/company-reporting-and-auditing/company-reporting/corporate-sustainability-reporting_en

EC (2025b) Carbon border adjustment mechanism, Directorate-General for Taxation and Customs Union. Accessed 13 Jun 2025. https://taxation-customs.ec.europa.eu/carbon-border-adjustment-mechanism_en

EC (2025c) Corporate sustainability due diligence, Directorate-general for communication. Accessed 10 Jun 2025. https://commission.europa.eu/business-economy-euro/ng-business-eu/sustainability-due-diligence-responsible-business/corporate-sustainability-due-diligence_en

Englhardt Z, Hähnlein F, Mei Y, Lin T, Sun CM, Zhang Z, Schulz A, Patel S, Iyer V (2025) Incorporating sustainability in electronics design: obstacles and opportunities

EU (2020) Circular economy action plan. Accessed 06 Jun 2025. https://www.eu2020.de/resource/blob/2429146/156d2d98b66b2ff28b6990161eed91e9/12-17-kreislaufwirtschaftsaktionsplan-bericht-en-data.pdf

EU (2024) Directive (EU) 2024/825 of the European Parliament and of the Council, Off J Eur Union, 1–16

EU (2023) Regulation (EU) 2023/1542. Accessed 10 Jun 2025. https://eur-lex.europa.eu/legal-content/EN/TXT/?uri=CELEX%3A32023R1542

González-Varona JM, Martín-Cruz N, Acebes F, Pajares J (2023) How public funding affects complexity in R&D projects: an analysis of team project perceptions. J Bus Res 158:113672. https://doi.org/10.1016/j.jbusres.2023.113672

ISO (2006) ISO 14040:2006/AMD 1:2020–Environmental management—Life cycle assessment—Principles and framework, https://www.iso.org/standard/37456.html

ISO (2006) ISO 14044:2006/AMD 2:2020–Environmental management—Life cycle assessment—Requirements and guidelines, https://www.iso.org/standard/38498.html

Jain P, Chou MC, Fan F, Santoso MP (2021) Embedding sustainability in the consumer goods innovation cycle and enabling tools to measure progress and capabilities. Sustainability 13(12):6662. https://doi.org/10.3390/su13126662

Joseph C, Gunton T, Knowler D, Broadbent S (2020) The role of cost-benefit analysis and economic impact analysis in environmental assessment: the case for reform. Impact Assess Proj Apprais 38(6):491–501. https://doi.org/10.1080/14615517.2020.1767954

Kazmi B, Taqvi SAA, Juchelková D (2023) State-of-the-art review on the steel decarbonization technologies based on process system engineering perspective. Fuel 347:128459. https://doi.org/10.1016/j.fuel.2023.128459

Li F, Chu M, Tang J, Liu Z, Wang J, Li S (2021) Life-cycle assessment of the coal gasification-shaft furnace-electric furnace steel production process. J Clean Prod. https://doi.org/10.1016/j.jclepro.2020.125075

Li F, Chu M, Tang J, Liu Z, Zhao Z, Liu P, Yan R (2022) Quantifying the energy saving potential and environmental benefit of hydrogen-based steelmaking process: status and future prospect. Appl Therm Eng. https://doi.org/10.1016/j.applthermaleng.2022.118489

Lostado-Lorza R, Corral-Bobadilla M, Íñiguez-Macedo S, Somovilla-Gómez F (2023) Characterization, LCA and FEA for an efficient ecodesign of novel stainless steel woven wire mesh reinforced recycled aluminum alloy matrix composite. J Clean Prod. https://doi.org/10.1016/j.jclepro.2023.137380

Mancini L, Sala S (2018) Social impact assessment in the mining sector: review and comparison of indicators frameworks. Resour Policy 57:98–111. https://doi.org/10.1016/j.resourpol.2018.02.002

Manik Y, Leahy J, Halog A (2013) Social life cycle assessment of palm oil biodiesel: a case study in Jambi Province of Indonesia. Int J Life Cycle Assess 18(7):1386–1392. https://doi.org/10.1007/s11367-013-0581-5

Mazijn B, Revéret JP (2015) Life cycle sustainability assessment: a tool for exercising due diligence in life cycle management, pp 51–63. https://doi.org/10.1007/978-94-017-7221-1_5

McClenaghan A, Gopsill J, Ballantyne R, Hicks B (2023) Cost benefit analysis for digital twin model selection at the time of investment. Procedia CIRP 120:1197–1202. https://doi.org/10.1016/j.procir.2023.09.148

Miah JH, Koh SCL, Stone D (2017) A hybridised framework combining integrated methods for environmental life cycle assessment and life cycle costing. J Clean Prod 168:846–866. https://doi.org/10.1016/j.jclepro.2017.08.187

Page MJ, McKenzie JE, Bossuyt PM, Boutron I, Hoffmann TC, Mulrow CD et al (2021) The PRISMA 2020 statement: an updated guideline for reporting systematic reviews. PLoS Med 18(3):e1003583. https://doi.org/10.1371/journal.pmed.1003583

Recharge, PEFCR (2020) Product environmental footprint category rules for high specific energy rechargeable batteries for mobile applications. Accessed 10 Jun 2025. https://wayback.archive-it.org/org-1495/20221006222329mp_/https:/ec.europa.eu/environment/eussd/smgp/pdf/PEFCR_Batteries_Feb%202020-2.pdf

Rimbau-Gilabert E (2019) Digitalización y bienestar de los trabajadores. IUSLabor. Revista d'anàlisi de Dret del Treball. https://doi.org/10.31009/IUSLABOR.2019.I02.01

Ruben R, Ben, Menon P, Sreedharan R (2018) Development of a social life cycle assessment framework for manufacturing organizations. In: proceeding 2018 international conference on production operations management society (POMS 2018) (2018). https://doi.org/10.1109/POMS.2018.8629496.

Sanyé-Mengual E, Sala S (2022) Life cycle assessment support to environmental ambitions of EU policies and the sustainable development goals. Integr Environ Assess Manag 18(5):1221–1232. https://doi.org/10.1002/ieam.4586

Schneider D, Woerle M, Kagermeier J, Zaeh MF, Reinhart G (2024) Sustainability risk assessment in manufacturing: a life cycle assessment-based failure mode and effects analysis approach. Sustain Prod Consum 47:617–631. https://doi.org/10.1016/j.spc.2024.04.030

Springer SK, Wulf C, Zapp P (2024) Potential social impacts regarding working conditions of fuel cell electric vehicles. Int J Hydrogen Energy 52:618–632. https://doi.org/10.1016/j.ijhydene.2023.04.034

Swarr TE, Hunkeler D, Klöpffer W, Pesonen H-L, Ciroth A, Brent AC, Pagan R (2011) Environmental life-cycle costing: a code of practice. Int J Life Cycle Assess 16(5):389–391. https://doi.org/10.1007/s11367-011-0287-5

Tan PJB, Hsu M-H (2018) Management of educational needs of employees in the electronics industry using English e-learning website programs. Manag Inf Syst. https://doi.org/10.5772/intechopen.79391

Tarvydas D, Tsiropoulos I, Lebedeva N (2018) Li-ion batteries for mobility and stationary storage applications: scenarios for costs and market growth. Publications Office of the European Union. https://doi.org/10.2760/87175

Yang L, Hu H, Yang S, Wang S, Chen F, Guo Y (2023) Life cycle carbon footprint of electric arc furnace steelmaking processes under different smelting modes in China. Sustain Mater Technol. https://doi.org/10.1016/j.susmat.2022.e00564

Integration of Life Cycle and Criticality Assessments: Case-Based Investigation of Lithium-Ion Battery Technologies

Naeem Adibi, Emilie Guilvert, Anish Koyamparambath, Frédéric Lai, Fadzai Mundembe, Marion Devienne, Philippe Osset, Guido Sonnemann, and Dieuwertje Schrijvers

Abstract Critical Raw Materials (CRM) are gaining increasing attention in both European policy and corporate strategies due to their essential role in emerging technologies and associated supply risks. Life Cycle Assessment (LCA) has been employed in various criticality studies to evaluate the environmental impacts of CRMs, while also offering tools to assess the sustainability of resource use, including material dissipation and short-term geopolitical supply vulnerabilities. Based on a comprehensive review of methodologies for CRM assessment and resource evaluation in LCA and a case study on lithium-ion battery production, recommendations are done in this SCORE LCA project to integrate LCA and criticality frameworks to support sustainable resource management and inform policy development.

1 Introduction

Due to their vital role in today's technological advancements, economic growth, and the green and digital transitions, critical raw materials (CRMs) are gaining increasing attention. RMs are necessary for essential applications, ranging from rare earths in electronics to lithium and cobalt in batteries. To emphasise the significance of CRMs, policy frameworks like the European Union's Critical Raw Materials Act (CRMA)

N. Adibi (✉) · E. Guilvert · D. Schrijvers
WeLOOP, Lambersart, France
e-mail: n.adibi@weloop.org

A. Koyamparambath · F. Mundembe · G. Sonnemann
Groupe Analyse du Cycle de Vie Et Chimie Durable (CyVi), L'Institut des Sciences Moléculaires (ISM), Université de Bordeaux, Bordeaux, Talence, France

F. Lai
Bureau de Recherches Géologiques et Minières, BRGM, Orleans, France

M. Devienne · P. Osset
SCORE LCA, Villeurbanne, France

© The Author(s) 2026

M. Traverso et al. (eds.), *Life Cycle Management from Global to Local*,
https://doi.org/10.1007/978-3-032-17987-6_3

highlight the necessity of securing supply chains and reducing reliance on vulnerable sources.

It remains challenging and resource-intensive to assess the 'criticality' of materials, which is determined by supply risk and economic or technical importance (Schrijvers et al. 2020), particularly for small- and medium-sized enterprises. On the other hand, life cycle assessment (LCA) has become a widely adopted and standardised approach for assessing the environmental impacts of materials and products. Impacts assessed in LCA include those related to the use of natural resources, such as their depletion, scarcity, and dissipation, which are mostly related to impacts on their availability. With LCA, businesses may gain insights into the environmental impacts of using raw materials, but often lack information on their accessibility. Several studies have addressed this issue, proposing various methods, and some have also evaluated the compatibility of these methods with LCA (Berger et al. 2020; Luthin et al. 2023; Hackenhaar et al. 2024). However, the debate still lies in whether and how LCA and criticality assessment (CA) can be integrated, especially since none of the proposed methods have been implemented in an LCA software yet. Hence, a significant gap exists in decision-making due to the lack of an integrated tool or a holistic approach that is widely accessible. Exploring means to improve accessibility and effectively integrating CA with LCA is essential to close this gap.

This study presents the results of the SCORE LCA project 'LCA and critical raw materials' (SCORE LCA 2024), which investigated the concept of criticality, explored its relationship with LCA, and illustrated the potential for synergy between the two using a practical case study. Consequently, recommendations are made to support resource management strategies that are more resilient and cognizant, and to aid both economic and environmental objectives.

2 State of the Art

Criticality assessments are 'outside-in' in nature, assessing the potential effects of external factors such as political unrest on the system under study. This is in contrast to LCA's 'inside-out' approach, which models how a system's actions impact resource or environmental domains. This distinction is vital: while criticality assessments evaluate and support resilience alternatives such as material substitution, supply diversification, or an increase in recycling potential, LCA quantifies potential environmental impacts.

2.1 Environmental Impacts in Criticality Assessments

Environmental impacts are increasingly recognised as essential factors in assessing the criticality of raw materials. Schrijvers et al. (2021, 2020) outline four distinct perspectives on how environmental impacts relate to material criticality, which are

(1) Environmental Regulation as Risk Factor: Environmental impacts increase the likelihood of regulatory actions (e.g. restrictions or bans), which can disrupt the supply of a material.
(2) Corporate Reputation and Social License to Operate: Use of materials with high environmental or social impacts can damage a company's reputation or social license to operate.
(3) Direct Environmental Burden of Use: The actual environmental impact is associated with using the material itself (e.g. emissions, resource depletion).
(4) Environmental Consequences of Supply Disruption: A disruption in the supply of a material leads to environmental harm, either through substitution with more damaging alternatives or halting environmentally beneficial applications.

In addition, there are underlying causal mechanisms behind the four perspectives mentioned on environmental impacts in assessing the criticality of raw materials, which influence whether they are classified as 'outside-in' or 'inside-out' and determine the appropriate LCA approach, as depicted in Fig. 1.

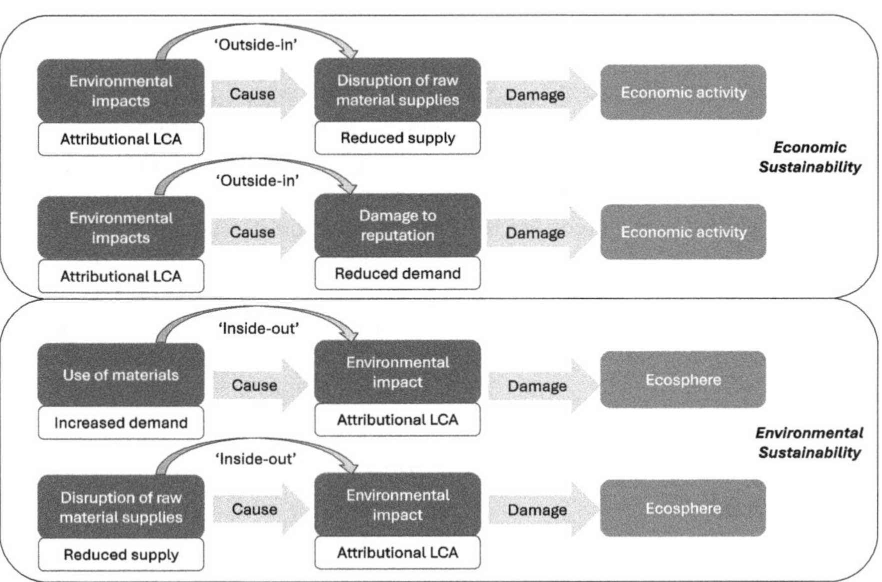

Fig. 1 Illustration of the underlying causal mechanism of four perspectives regarding environmental impacts in the assessment of the criticality of raw materials, their classification into 'outside-in' or 'inside-out', and the corresponding LCA approach (adapted from (Schrijvers et al. 2021))

2.2   Resources in LCA

A variety of resource indicators have been developed to assess the environmental and economic impacts of resource use in LCA. Examining the traditional and dissipation-based methods reveals their individual focus areas and relevance in evaluating the fate of CRMs from extraction to their entry into the economy. The Abiotic Depletion Potential (ADP) (van Oers et al. 2020a) represents a traditional method that focuses on extraction rates and geological scarcity. In contrast, dissipation-based indicators, such as Average Dissipation Rate (ADR) (Charpentier Poncelet et al. 2021) and Environmental Dissipation Potential (EDP) (van Oers et al. 2020b), assess the permanent loss and irreversibility of resource emissions. Others, including the Lost Potential Service Time (LPST) (Charpentier Poncelet et al. 2021), integrate economic dimensions to evaluate the future availability and value of resources permanently lost. Overall, the dissipation-based indicators offer a more nuanced and future-oriented perspective on resource availability, addressing both environmental and economic consequences.

However, there is a general consensus that 'abiotic resources' are not directly linked to environmental concerns (Sonderegger et al. 2017). Methods developed to quantify damage in the 'resource use' area of protection primarily focus on reducing the value creation potential in alternative product systems. However, the AoP 'resources' is not merged with the economic AoP 'prosperity' in the new framework for life cycle sustainability assessment (LCSA), which was developed in the ORIENTING project (Hackenhaar et al. 2024). Moreover, one could argue that the value obtained from resources leads to the satisfaction of human needs, leading to increased human well-being, which is another distinct AoP in the framework of (Hackenhaar et al. 2024). Consequently, there is still no consensual framework for assessing resources alongside environmental LCA, and a parallel assessment of environmental impacts and resource use impacts is therefore the current best practice.

2.3   Integrating Criticality Assessment and Life Cycle Assessment

There are two main approaches to integrating CA into LCA: one at the results level and the other at the methodological level. The advantages, limitations, and inter-relations of these approaches have been widely discussed in the LCA community, particularly in the context of combining and integrating life cycle and environmental risk assessment (Guineé et al. 2017). Result-level integration maintains methodological clarity and is easier to apply, but outputs may not be directly comparable due to differing scopes, metrics, and assumptions. Methodological-level integration allows for more comprehensive assessments, enables joint interpretation under a single functional unit or spatial boundary, but requires more detailed data and there is a lack of standardised tools. Both approaches are complementary, and their combination can improve decision support in early design or regulatory contexts.

In the case of integrating criticality into LCA, this implies that either the results of criticality assessments for raw materials can complement the outcomes of a product system's LCSA, thereby providing a broader perspective on the sustainability implications of resource use, or elements of supply risk can be incorporated directly into the LCSA framework at the life cycle impact assessment by applying characterisation factors derived from a comparative risk assessment model, similar in approach to USEtox (Rosenbaum et al. 2011).

3 Case-Based Investigation of Lithium-Ion Battery Technologies

This section summarises the case study conducted within the SCORE LCA project (SCORE LCA 2024).

3.1 Goal and Scope Definition

Lithium-ion batteries (LIBs) play a central role in the transition towards a low-carbon society, powering electric vehicles (EVs) and facilitating the integration of renewable energy. They dominate the rechargeable battery market due to their high energy and power density and low self-discharge rate, and LIBs contain materials considered critical and/or strategic by the European Commission. By 2030, the most relevant chemistries are projected to be NMC (nickel–manganese–cobalt) and LFP (lithium iron phosphate) cathodes, both with graphite anodes (IEA 2025). This makes NMC811 and LFP batteries highly relevant for comparative LCA and criticality assessments. Furthermore, complete inventories are available in ecoinvent 3.10. To enable a fair comparison between the two battery types, their energy storage capacity must be equivalent. The average battery capacity for electric vehicles is approximately 65 kWh, which is used as the functional unit for the assessments.

3.2 Life Cycle Impact Assessment Methods

The analysis was performed using the Environmental Footprint method version 3.1 (EF 3.1), developed by the European Commission. A cradle-to-gate approach is applied with SimaPro v10.0.1.2. Inventory data were sourced from ecoinvent v3.10. Within EF 3.1, the mineral and metal resource use category is based on ADP, which relies on extraction rates and reference reserves (typically the ultimate reserve) (van

Oers et al. 2020a). To address additional aspects of resource use, complementary indicators are added: ADR reflects physical scarcity by considering depletion relative to known reserves, making it more responsive to short- and medium-term supply constraints and Potential Value Loss from Resource use (PVLR) introduces a socio-economic dimension by estimating the potential long-term economic loss due to resource extraction, taking into account both scarcity and market value.

3.3 Criticality Assessment Methods

The IRTC Tool (IRTC 2023) is a decision support tool that was developed as part of the IRTC Business project funded by EIT RawMaterials to address the lack of explicitly applied cause–effect mechanisms in earlier criticality assessments. It incorporates cause–effect chains that reflect business-relevant supply risks, identified through stakeholder workshops involving methodology developers, companies, and governmental institutions. The tool addresses three primary concerns faced by companies in relation to CRMs: physical accessibility issues, high price volatility, and reputational risks, which are addressed by 24 indicators. GeoPolRisk method (Gemechu et al. 2015) evaluates the geopolitical risk related to the supply of abiotic resources (energy and non-energy) for LCA. It assesses the supply risk of raw materials for an economic actor over a given year, providing both a dimensionless score and a CF for LCA resource elementary flows, expressed in kg Cu-eq/kg (Koyamparambath et al. 2024). The calculation comprises three components: the concentration of mining production, the share of imports weighted by political instability (using the Worldwide Governance Indicator, which measures political stability and the absence of violence), complemented by domestic production, and the traded price of the raw material.

ESSENZ plus (Yavor et al. 2021) is a comprehensive method that measures the multiple dimensions of resource efficiency of products, including supply risk, and integrates it into LCA. It covers 11 categories (including concentration of reserves and production, political instability, and trade barriers) related to socio-economic availability, each represented by specific indicators (Herfindahl–Hirschman index, Worldwide Governance Indicator, and Enabling Trade Index), set against target values. The default target values are based on expert opinion and stakeholder surveys, normalised using global production data.

3.4 Case Study Results

The cell and electronic components are the primary sources of environmental impacts for both LFP and NMC811 batteries, with copper being the most consistent contributor. In LFP batteries, the cathode has a limited impact, whereas in NMC811 batteries, the use of nickel and cobalt sulphates increases the environmental burden by introducing sulphur, in addition to copper. Gold and silver have minor contributions in both technologies, primarily through electronic components.

In the LCA, NMC811 batteries show slightly higher environmental impacts than LFP batteries, particularly for 'resource use, minerals and metals' as indicated by the ADP method. This is further confirmed by the ADR and PVLR methods, with NMC811 scoring higher on dissipation, particularly for ADR, which indicates irreversible losses of materials. ADP highlights key contributors such as copper, gold, silver, sulphur, and tellurium (the latter due to inventory artefacts), but only copper stands out across both ADP and dissipation analyses. In contrast, dissipation-based assessments highlight additional high-scoring materials, such as gallium, barium, and iron (the latter due to inventory artefacts), reinforcing the benefit of integrating these metrics.

The CAs further identify distinct hotspots (Table 1). For example, both GeoPolRisk from the French perspective and ESSENZ + identify iron as a key contributor in NMC811 (cradle-to-gate), but not in LFP, while copper is flagged only by GeoPolRisk, which can be explained by the price component included in the method.

GeoPolRisk, based on the full LCI and interpreted at the first aggregation level, allows inclusion of primary and energetic raw materials that are consumed both directly in the product system and in the background activities. It identifies nickel, cobalt, lithium, and copper as critical for NMC batteries. IRTC yields a different list of critical elements due to the inclusion of other indicators (e.g. export restrictions) and a different point of assessment (i.e. global supply versus import to France in the GeoPolRisk method). Furthermore, the IRTC method does not include price as a risk

Table 1 Comparison of three criticality assessment methods

	IRTC	GeoPolRisk	ESSENZ plus
Number of metals	More than 50	56	48
Indicators	24	1	21
Scope (assessed risks)	Risks of accessibility, price fluctuations, and reputation damage	Geopolitical supply risk	Socio-economic risks and social acceptability
Data used	Complete LCI	Complete LCI, but the results are presented at the first level of aggregation	Complete LCI
Most critical elements identified for NMC811	Cobalt, tantalum, antimony, graphite, rhodium, magnesium, manganese, silicon, aluminium	Nickel, cobalt, lithium, iron, copper, fluorspar	Iron, sulphur, nickel, magnesium, gallium, lithium, tantalum
Most critical elements identified for LFP	Tantalum, antimony, graphite, rhodium, aluminium	Lithium, copper, fluorspar	Aluminium, iron, selenium, gallium, lithium, tantalum

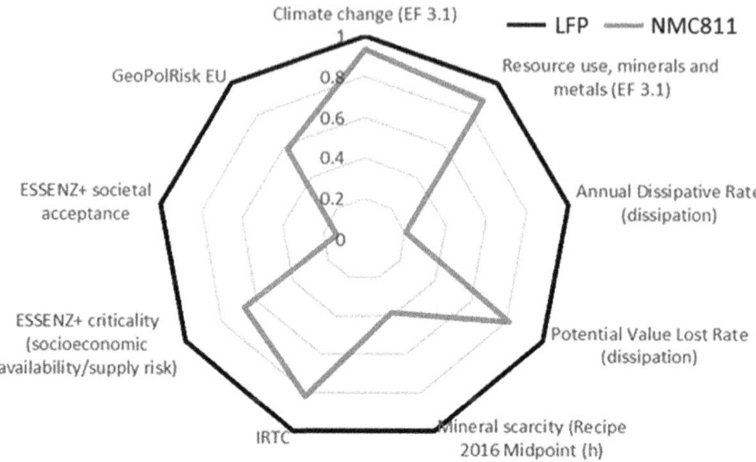

Fig. 2 Comparison of NMC 811 and LFP batteries based on LCA and criticality (SCORE LCA 2024)

factor, while this is predominant in the GeoPolRisk method. ESSENZ +, integrating criticality and social risks, highlights tantalum, cobalt, and lithium, with allocation effects influencing outcomes. This divergence highlights the complementary, yet distinct insights gained by combining LCA and criticality assessments.

To compare the overall criticality and environmental results, overall scores by indicator were calculated for each battery. These overall scores were then normalised and weighted per battery (Fig. 2), with each highest value assigned a score of 100%. Figure 2 shows that although the differences in values between the two batteries vary across all methods, the NMC811 battery consistently exhibits higher impacts across all methods.

4 Limitations

A major limitation in combining LCA with criticality assessments lies in the inappropriate merging of environmental impact metrics with supply risk indicators, which can lead to methodological inconsistencies (mismatched system boundaries, functional units, or temporal scopes) and misinterpretations (comparing inherently different types of indicators).

The assessments of criticality and environmental impacts in this study are based on the LCIs, in order to establish a link between LCA and criticality at the level of the application methods. The specific features of the inventories thus have an influence on the results and their interpretation. Allocation plays a key role in LCI and LCA, often based on economic, mass, or energy criteria. However, these practices can introduce elementary flows into the inventory that are neither part of the Bill of

Materials nor the intermediate stages of the process, nor the supporting services such as electricity generation, which was observed, for example, for tellurium. This is one of the limitations of using LCI databases and must be considered when interpreting the results of the impacts of resource use in LCA.

5 Criticality and LCA: Complementary or Contradictory?

While LCA and criticality assessments do not answer the same questions, these methods are complementary, and their cross-referencing is key to making sustainable and strategic decisions, as illustrated in Fig. 3.

The differences in scope, objectives, and underlying data make it difficult to directly merge the two approaches. However, LCA and criticality assessment can effectively complement each other when used in tandem. For instance, juxtaposing results can help identify materials that are both environmentally burdensome and at high risk of supply disruption—crucial information for sustainable decision-making. While LCA primarily calculates potential environmental impacts, criticality adds the layer of supply risks, both guiding mitigation efforts. In practice, the two methods are often used separately in different contexts—such as environmental reporting versus resource security planning—but efforts are also underway to integrate criticality metrics into LCA models. These integration efforts, though promising, face hurdles due to incompatible datasets and differing assumptions, underscoring the need for methodological clarity and cautious interpretation.

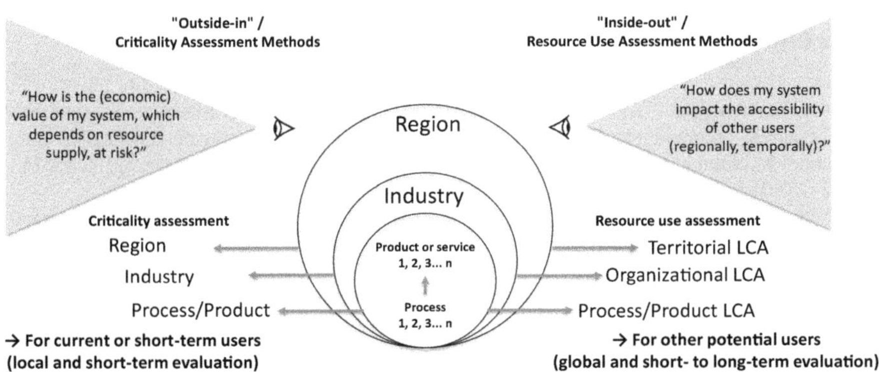

Fig. 3 The 'outside-in' approach to criticality and the 'inside-out' approach to assessing the impact of resource consumption in LCA (SCORE LCA 2024)

6   Conclusion and Outlook

This case study highlights the value of combining LCA and CA to support informed resource management decisions. While LCA highlights raw materials with high environmental impacts, such as nickel and cobalt, CA identifies supply risks and ethical concerns, revealing complementary insights. However, the analysis also exposes limitations in LCI, notably, the reliance on generic, global data that lacks regional or technological specificity. This reduces accuracy, especially for by-products with complex supply chains.

Conceptually, LCA and CA offer contrasting perspectives: LCA takes an 'inside-out' view of system impacts, whereas CA applies an 'outside-in' lens to external risks. Integration can occur either by juxtaposing results (for example, using IRTC or ESSENZ + alongside LCA) or by embedding characterisation factors (such as done by GeoPolRisk) directly into LCIA, though differences in data granularity and underlying assumptions challenge this latter approach. The merits of integrating CA with LCA are emphasised in this study.

Combining LCA and CA enables more strategic decisions. For instance, relocating lithium production to Europe reduces both geopolitical risk (CA) and emissions (LCA). In a short-term perspective, CA supports the identification of high-risk materials for procurement or policy focus. Therefore, the implementation of CA methods in LCA software is recommended. Integrating circularity strategies and communicating comparative LCA results can further improve environmental performance and resilience.

References

Berger M, Sonderegger T, Alvarenga R et al (2020) Mineral resources in life cycle impact assessment: part II–recommendations on application-dependent use of existing methods and on future method development needs. Int J Life Cycle Assess 25:798–813. https://doi.org/10.1007/s11 367-020-01737-5

Charpentier Poncelet A, Helbig C, Loubet P et al (2021) Life cycle impact assessment methods for estimating the impacts of dissipative flows of metals. J Ind Ecol 13136. https://doi.org/10.1111/ jiec.13136

Gemechu ED, Helbig C, Sonnemann G et al (2015) Import-based indicator for the geopolitical supply risk of raw materials in life cycle sustainability assessments. J Ind Ecol 20:154–165. https://doi.org/10.1111/jiec.12279

Guineé JB, Heijungs R, Vijver MG, Peijnenburg WJGM (2017) Setting the stage for debating the roles of risk assessment and life-cycle assessment of engineered nanomaterials. Nat Nanotechnol 12:727–733. https://doi.org/10.1038/NNANO.2017.135

Hackenhaar IC, Moraga G, Thomassen G et al (2024) A comprehensive framework covering life cycle sustainability assessment, resource circularity and criticality. Sustain Prod Consum 45:509–524. https://doi.org/10.1016/j.spc.2024.01.018

International energy agency (IEA) (2025) Global EV Outlook 2025. https://www.iea.org/reports/ global-ev-outlook-2025.

International round table on materials criticality (IRTC) (2023) IRTC Decision Tool. Accessed 19 Jan 2023. https://www.irtc-decision-tool.weloop.org/

Koyamparambath A, Loubet P, Young SB, Sonnemann G (2024) Spatially and temporally differentiated characterization factors for supply risk of abiotic resources in life cycle assessment. Resour Conserv Recycl 209:107801. https://doi.org/10.1016/j.resconrec.2024.107801

Luthin A, Traverso M, Crawford RH (2023) Circular life cycle sustainability assessment: an integrated framework. J Ind Ecol. https://doi.org/10.1111/JIEC.13446

Rosenbaum RK, Huijbregts MAJ, Henderson AD et al (2011) USEtox human exposure and toxicity factors for comparative assessment of toxic emissions in life cycle analysis: sensitivity to key chemical properties. Int J Life Cycle Assess 16:710–727. https://doi.org/10.1007/s11367-011-0316-4

Schrijvers D, Hool A, Blengini GA et al (2020) A review of methods and data to determine raw material criticality. Resour Conserv Recycl 155:104617. https://doi.org/10.1016/j.resconrec.2019.104617

Schrijvers D, Loubet P, Sonnemann G (2021) LCA for criticality assessments. In: Pradel M (ed.), Développement Durable, Presses des Mines

SCORE LCA Project No-2024–01–ACV et matériaux critiques (2024)

Sonderegger T, Dewulf J, Fantke P et al (2017) Towards harmonizing natural resources as an area of protection in life cycle impact assessment. Int J Life Cycle Assess. https://doi.org/10.1007/s11367-017-1297-8

van Oers L, Guinée JB, Heijungs R (2020a) Abiotic resource depletion potentials (ADPs) for elements revisited—updating ultimate reserve estimates and introducing time series for production data. Int J Life Cycle Assess 25:294–308. https://doi.org/10.1007/s11367-019-01683-x

van Oers L, Guinée JB, Heijungs R et al (2020b) Top-down characterization of resource use in LCA: from problem definition of resource use to operational characterization factors for dissipation of elements to the environment. Int J Life Cycle Assess 25:2255–2273. https://doi.org/10.1007/s11367-020-01819-4

Yavor KM, Bach V, Finkbeiner M (2021) Adapting the ESSENZ method to assess company-specific criticality aspects. Resources 10. https://doi.org/10.3390/resources10060056.

Variability of Electricity Mixes: Does Using a Time-Aware Database Make a Difference? How Timing of Electric Vehicle Charging Influences LCA Results

Sabina Bednářová, Thomas Gibon, and Enrico Benetto

Abstract Electricity mixes in European countries vary significantly by hour, day and season. This is mainly due to an increasing share of renewable energy in the electricity mix, meaning that relying on annual average electricity data may misrepresent life cycle impact assessment results for applications where electricity supply is a main contributor and can be discontinuous. To address this, we developed a tool to use live tracked electricity data and prepare them into ready-to-use *brightway* database. We demonstrate the importance of time-aware electricity databases through a small case study of electric vehicle charging at different times of the day, different seasons, and different years. Our results highlight the variability of electricity mixes and showcase that using annual average electricity mixes, LCIA results can either be over- or underestimated, underscoring the need for more temporally granular approaches.

1 Introduction

In current practice, the quality of a life cycle assessment (LCA) study strongly depends on the quality of data supplied by commercial databases, such as *ecoinvent* (Wernet Bauer et al. 2016) or *sphera* (Sphera 2022)—two state-of-the-art widely used repositories of life cycle inventories. These databases are usually used as *background* databases and offer a wide range of processes to be used in LCA studies. Electricity mixes constitute a subset thereof, consisting of key processes used in almost all applications; as the global economy is increasingly electrifying, these representations of electricity supply influence LCA results notably (Sacchi Bauer et al. 2022, Gibon 2024). Electricity mixes are modelled regionally from annual average electricity statistics. In LCA databases, these annual averages are representative of year [n-3], e.g. the *ecoinvent* release in November 2024 uses 2021 annual average data for most countries, and 2022 for USA, China and Switzerland (Wernet Bauer et al.

S. Bednářová (✉) · T. Gibon · E. Benetto
Environmental Research and Innovation (ERIN) Department, Environmental Sustainability Assessment and Circularity (SUSTAIN) Unit, Luxembourg Institute of Science and Technology (LIST), Esch-Sur-Alzette, Luxembourg
e-mail: sabina.bednarova@list.lu

M. Traverso et al. (eds.), *Life Cycle Management from Global to Local*,
https://doi.org/10.1007/978-3-032-17987-6_4

2016). For most applications, such as productions or general use, an annual average electricity mix is suitable. However, in some cases, there is a potential lack of accuracy when representing electricity-producing activities using annual average mixes, which is even greater with a higher share of renewable energy. Implications of these misrepresentations range from inaccurate life cycle impact assessment (LCIA) results, especially in electricity-intensive systems, to the incompatibility of current LCA databases with upcoming regulations, e.g. the EU Directive on renewable fuels of non-biological origin (European Commission 2023), or national rules regarding electricity-intensive infrastructure (Cloud Infrastructure Ireland, 2025).

Following decarbonisation targets, many countries have started to heavily rely on renewable energy. While significantly reducing their global warming potential, the variability of electricity mixes is greatly increased with the rise of renewable energy (Sander et al. 2022). Electricity mixes with a high share of solar and wind power typically vary from hour to hour, day to night, month to month and season to season. Variability, also called *ontic uncertainty* (Walker et al. 2003), is a part of uncertainty that in theory cannot be reduced; however, this stems from the belief that variability is random. In this case, electricity mixes are not *truly random*, and with live electricity tracking (Hirth Mühlenpfordt et al. 2018, Electricity Maps 2025, Fraunhofer Institute for Solar Energy Systems ISE 2025), we can gather precise data, up to even 15-min intervals, about the exact composition of the electricity consumption and production at a certain time in the past. The high variability of electricity mixes is directly related to a high uncertainty of LCIA results, which may decrease the robustness and reliability of LCA studies. Such high resolution not only improves the accuracy of results but also enables the optimisation of studied systems. For example, in scenario analysis, identifying the optimal time to charge an electric vehicle requires high-resolution temporal data. For such applications, annual average electricity mix cannot provide the necessary level of detail to identify time-specific optimisation. Instead, using high-resolution temporal data can reduce the uncertainty in LCIA and optimise studied systems.

The need for time-aware electricity database has been already discussed in the literature (Naumann Famiglietti et al. 2024). Several tools have already been developed in this direction, to name a few: *bentso* (Mutel and Azzi 2022), *Futura* (Joyce and Björklund 2022), *EcoDynElec* (Lédée Padey et al. 2023), *Peakachu* (Diepers 2024), however most lack direct applicability in LCI/LCIA, such as integration into already existing projects before executing final impact assessment, *ecoinvent*-like electricity production categories, or their compatibility with impact categories other than global warming potential. In this paper, we will introduce a tool to use live tracked electricity data to create time-aware electricity databases, and demonstrate the importance of such databases on a small study of electric vehicle charging.

2 Methods

Time-aware life cycle inventories (LCIs) are built using Python package *shrecc* (simple hourly resolution electricity consumption calculation), which in turn uses live electricity tracking data from the Energy Charts API (Fraunhofer Institute for Solar Energy Systems ISE 2025), is directly compatible with *brightway* (2 or 2.5, *shrecc* is *brightway*-agnostic), and can write an hourly electricity database into a *brightway* project. The raw data originates from numerous sources, among which the ENTSO-E Transparency Platform, on an hourly or daily basis (ENTSO-E 2025). This data can be downloaded in yearly batches for all countries available from the API and further processed. Datasets for production and load for each country as well as imports and exports are required. The package is accessible at https://pypi.org/pro ject/shrecc/.

To process the data, the exchange network graph is solved for all countries on the European grid, and every single hour at once. This is done by calculating the Leontief inverse of the adjacency matrix of trade (import and export) data. This operation relies on the same assumptions as flow-tracing algorithms (Tranberg Corradi et al. 2019), whereby the electricity mix of a given trade flow is set identical to the consumption mix of the exporting country or bidding zone. Solving the exchange network graph for the whole consumption matrix, which includes higher- and lower-voltage sources, means that *shrecc* does not follow the split into high-, medium- and low-voltage electricity as it is seen in *ecoinvent*, but rather "flattens" all flows at the same level. This approach can also track imports to country A from the original country X, even if no straightforward (neighbouring) connection between the two countries exists, thereby assuming re-exports.

After calculating the Leontief inverse, the data is stored in a *pandas* dataframe, which follows the ENTSO-E classification, and must be matched against the more detailed *ecoinvent* classification. For example, ENTSO-E only has one category for solar technologies, coined "solar". However, in ecoinvent, there is more than one source for solar power (at least three types of solar power technologies, mostly photovoltaics but not only). Therefore, we use ecoinvent to determine the ratio of multiple solar power sources per category of ENTSO-E. In some countries, new types of electricity production technologies have been implemented recently, and therefore ecoinvent does not carry the specific activities for those technologies. In that case, the technology split of one ENTSO-E category follows the overall European mix.

User-specified region and time intervals are supported. Time "intervals" can be a date of a period, single or recurrent, e.g. all working days from 9 to 5 during June, July and August. Then, *shrecc* creates an average database based on the selected time interval (in volume), and scales everything down to 1 kWh. During the scale-down, consumption is compared to load; in case it is lower, the difference is assigned as the market for *market group for electricity, low voltage (RER).*[1] Consumption mixes

---

[1] In *ecoinvent* terminology, a *market* is a linear combination of activities producing a same reference product (across sectors or regions), a *market group* is a market of markets. The acronym *RER* refers to the European region.

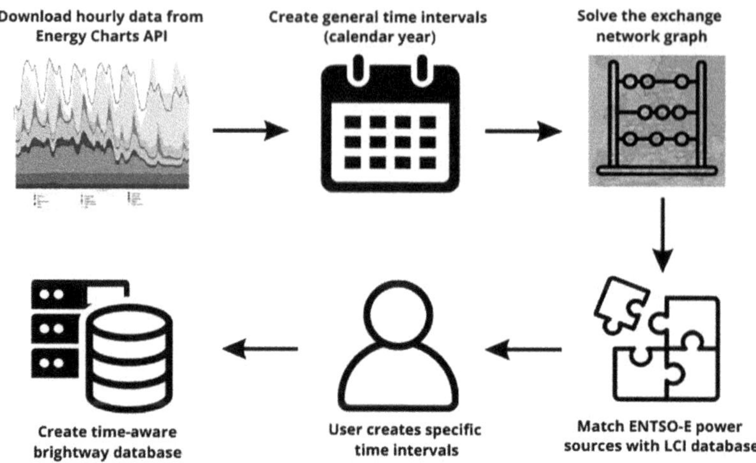

Fig. 1 Conceptual workflow of the shrecc package

may contain a large number of inputs; in this case, a cut-off may be applied to pool inputs smaller than a given threshold into one single input. The default cut-off value is 0.001 kWh/kWh but can be adjusted. Once grouped, the set of small inputs is then assigned to *"market group for electricity, high voltage, kWh, RER"*. The result is a structure similar to *ecoinvent*, except for the low/medium/high-voltage split. Users can create as many databases as they need.

Apart from selecting the time intervals, users can also select whether they want to include the distribution and transmission networks (land, sea and aerial cable) and market for sulphur hexafluoride. The different cables included follow ecoinvent's transmission network. For distribution network and market for sulphur hexafluoride, an average across all available countries is used. By default, both distribution and transmission network and market for sulphur hexafluoride are included. The whole workflow is summarised in Fig. 1.

3 Results and Discussion

To validate the model, we compare 2021 *shrecc* data with the newest *ecoinvent* 3.11 (see Fig. 2). Countries with missing production data are excluded from the selection (a later version of *shrecc* will correct this issue). Most western European countries are well tracked and can be compared to *ecoinvent* annual average electricity mixes. Switzerland is excluded from this comparison, as the *ecoinvent* 3.11 high-voltage dataset uses a market-based electricity model, against which the comparison with

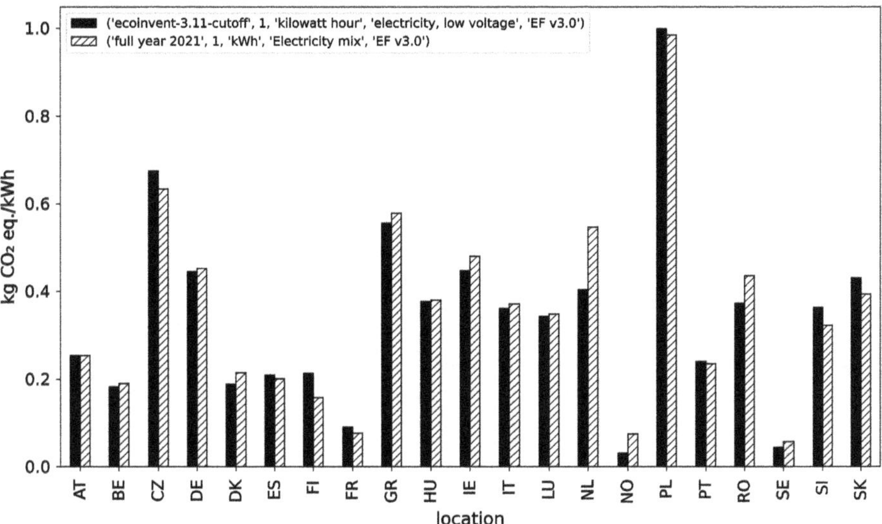

Fig. 2 Comparison between the greenhouse gas intensities of *ecoinvent* 3.11 (black) and shrecc (striped) national electricity mixes, annual average in 2021

our tool would not be appropriate.[2] In a few cases, such as for the Netherlands and Norway, anomalies were observed due to either inconsistencies in the original electricity data or modelling artefacts related to the cut-off system. While these issues are being addressed in ongoing work, users should be aware that country-specific data limitations and modelling choices can significantly influence results.

To illustrate the variability of LCIA results over different timeframes, a simple electric vehicle (EV) charging model was developed, assuming a slow charging rate of 3.7 kW over 8 h (29.6 kWh in total), during distinct times, namely: day, night, summertime and wintertime. In this example, we are using 2024 *shrecc* data and the newest (as of spring 2025) *ecoinvent* 3.11. We selected four countries with different characteristics: Austria, Czechia, France and Greece. In 2024, the Austrian electricity mix was low carbon, mostly using renewable energy in the grid (in 2024, the renewable energy share was almost 90%), whereas Czechia's consumption mix was relatively fossil fuel-heavy, with a low renewable energy share. France relied on a very low-carbon mix due to nuclear energy with another one-third of the mix consisting of renewable energy, while Greece depended on a relatively high fossil fuel-heavy electricity mix, but with more renewable energy (in 2024, the mix consists of 50% fossil fuel and 50% renewable energy) than Czechia. We chose these countries to test the following hypotheses:

---

[2] As a general remark, there is currently little evidence that renewable energy certificates (RECs) are "additional", e.g. it cannot be proven that they accelerate renewable infrastructure deployment (Langer, Brander et al. 2024). To preserve the biophysical nature of LCA modelling and avoid the risk of double counting, we therefore exclude any market-based accounting in *shrecc*.

1. Using a time-aware electricity database shows a clear difference when compared to annual average, with respect to LCIA results in specific applications.
2. Electricity mixes with a high share of renewable energy show a great LCIA variability, which is lost within the annual average mix.
3. Even low-carbon electricity mixes that rely mostly on (baseload) nuclear energy show some variability.
4. Using time-aware electricity database compared to an annual average mix means that, in some cases, LCIA results are *both* overestimated and underestimated.

Figure 3 shows LCIA results for impact category; *climate change, global warming potential (GWP100)*, we can see a great variability in all results. The first thing to notice is that using *ecoinvent* data, we are overestimating the LCIA results for all cases of charging the EV, both winter and summer and day and night. Secondly, we notice that in case of the French electricity mix, the variability is much lower, but still prominent—especially the seasonal one. Seasonal variety is usually governed by more solar power in the mix during summer, and hydropower to a lower extent. Especially because we are comparing 2024 (*shrecc*) data with 2021 (*ecoinvent*) data, many countries have increased their share of solar power. For example, in case of Greece, the share of solar power has doubled, from 10 to 20% (Fig. 5). In the case of Czechia, while the share of solar power also doubled, it is in much lower shares, from 2.8% to 5.8%. However, Czechia's overall share of renewable energy increased from 13.0% to 17.5%, just as the nuclear power had a slight increase as well, from 37.2% to 41.4% and the share of fossil fuel energy decreased from 49.8% to 41.1% (Fig. 5). In the case of Austria, variability is sourced purely from the fact that in 2024, 90% of the electricity mix is made of renewable energy. In 2021, the annual share of renewable energy in Austria was 80% of consumption. It is also important to note that in this overall comparison of sources of electricity production, we are comparing data from Energy Charts (Fraunhofer Institute for Solar Energy Systems ISE 2025), as we cannot share the data from *ecoinvent*.

Let us also have a look at a different impact category, *land use, soil quality index* (see Fig. 4). This is a category that, unlike with climate change, might be underestimating the LCIA impacts when using annual average data for time-specific applications, due to most countries raising share of renewable energy, which is usually land-use-heavy. In the case of Greece, we can see that the *shrecc* 2024 impacts are consistently lower than *ecoinvent* 2021 impacts. And while Greece did invest in renewable energy and did grow their overall share of renewable versus fossil fuel energy production, certain shifts between the shares of solar, wind and hydropower occurred. The share of hydropower decreased, while the share of solar and wind increased. Hydropower (reservoirs) is one of the most land-use-heavy renewable energy production technologies. This shift in shares provides a certain trade-off in terms of land use.

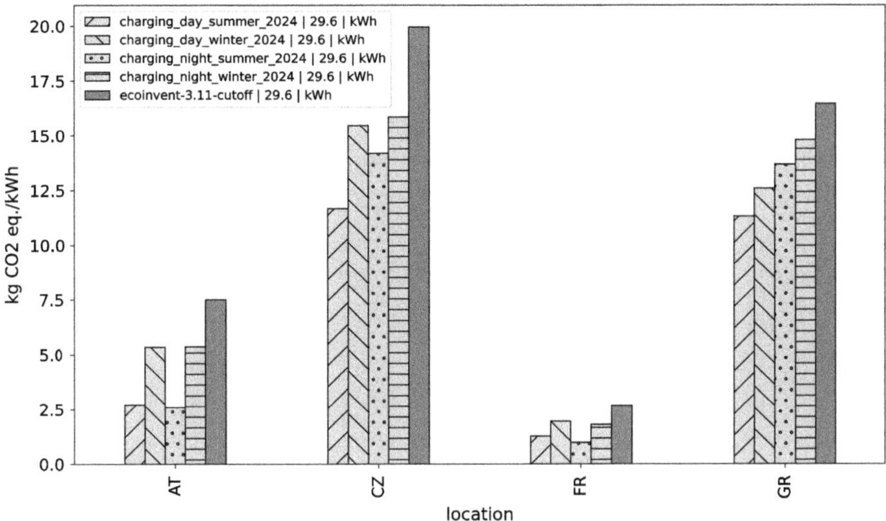

Fig. 3 Global warming potential impacts from charging an EV at 3.7 kW for 8 h, electricity consumption only, per country and timeframe

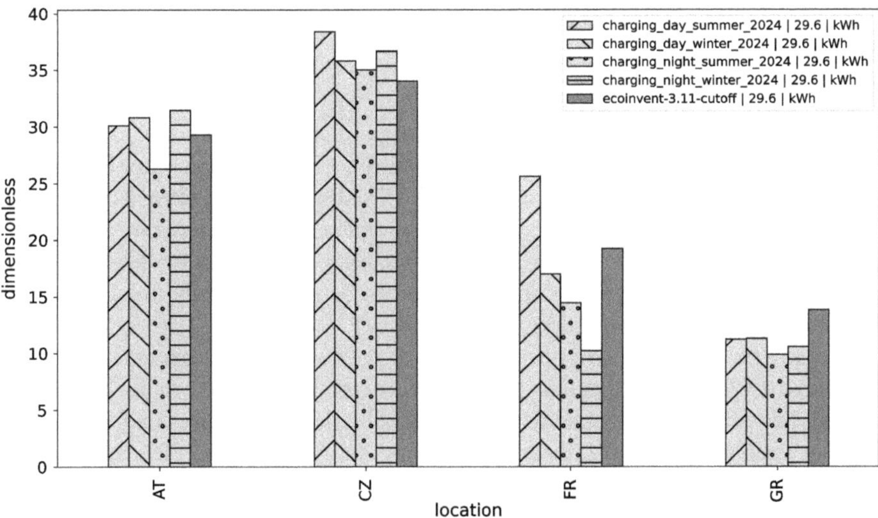

Fig. 4 Land-use impacts (soil quality index) from charging an EV at 3.7 kW for 8 h, electricity consumption only, per country and timeframe

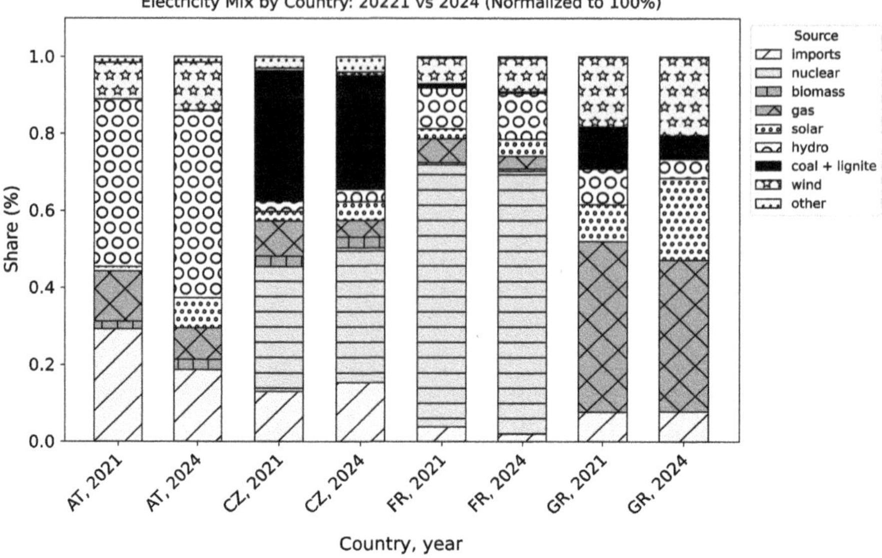

Fig. 5 Consumption mixes per country and year (2021 and 2024), results from *shrecc* calculation

4   Conclusions

In conclusion, this study highlights the variability of electricity mixes and showcases the importance of capturing it within a time-aware electricity database. Using electricity tracking data from Energy Charts API (Fraunhofer Institute for Solar Energy Systems ISE 2025), we built a tool to capture and seamlessly integrate time-aware electricity databases in *brightway*. This tool is easy to use, can be directly implemented into existing *brightway* projects, and can be used with any impact category available in *brightway* methods.

The variability between day and night, as well as across seasons, can greatly influence LCIA results, especially in cases where electricity use is a major contributor and can be temporally regulated—for instance, in electric vehicle charging. Such a tool can support LCA practitioners in reducing the variability of electricity-related LCIs (and LCIAs), as well as enabling the optimisation of system operation times. This ensures that electricity is consumed during periods with lower environmental impacts, mainly in time-sensitive applications.

The current implementation has limited compatibility with *brightway* and activity browser, with the option to export inventories from *brightway* in Simapro format. Additionally, electricity is modelled at a single voltage level, which may affect compatibility with certain applications, though it also raises questions about the current voltage differentiations modelling choices in LCA. Package *shrecc* uses only retrospective data, but future developments could incorporate predictive approaches, such as combining machine learning with weather forecasts to estimate electricity mixes. Data coverage also remains a challenge, especially for countries with incomplete or missing data on Energy Charts, such as those in Eastern Europe or cases like the Netherlands, where solar energy is underrepresented. While future versions of *shrecc* aim to address these issues, further challenges may arise depending on broader adoption and feedback.

Acknowledgements The authors are grateful to Tomás Navarrete Gutiérrez for the support in software development and packaging. This research was funded by Luxembourg National Research Fund through contract C21/SR116217831/IMMEC (FNR CORE 2021). Thomas Gibon thanks Romain Besseau and Thomas Auriel for early discussions.

References

Cloud infrastructure Ireland (2025) CRU large energy user connection policy

Diepers T (2024) Sentier Peakachu package

Electricity Maps (2025) Live 24/7 CO_2 emissions of electricity consumption, Electricity map

ENTSO-E (2025) ENTSO-E Transparency platform

European Commission (2023) Commission delegated regulation (EU) 2023/1184 of 10 February 2023 supplementing Directive (EU) 2018/2001…, C/2023/1087, European Union

Fraunhofer Institute for Solar Energy Systems ISE (2025) Energy-Charts.info

Gibon T (2024) Learning from the life cycle assessment of power-to-hydrogen systems. In: White RJ, Figueiredo MC (eds), Chemical Technologies in the Energy Transition, Royal Society of Chemistry, pp 33

Hirth L, Mühlenpfordt J, Bulkeley M (2018) The ENTSO-E transparency platform–a review of Europe's most ambitious electricity data platform. Appl Energy 225:1054–1067

Joyce PJ, Björklund A (2022) Futura: a new tool for transparent and shareable scenario analysis in prospective life cycle assessment. J Ind Ecol 26(1):134–144

Langer L, Brander M, Lloyd SM, Keles D, Matthews HD, Bjørn A (2024) Does the purchase of voluntary renewable energy certificates lead to emission reductions? A Rev Stud Quantifying Impact, J Clean Prod 478:143791

Lédée F, Padey P, Goulouti K, Lasvaux S, Beloin-Saint-Pierre D (2023) EcoDynElec: open python package to create historical profiles of environmental impacts from regional electricity mixes. SoftwareX 23:101485

Mutel C, Azzi E (2022) bentso: a library to process ENTSO-E electricity data for use in industrial ecology and life cycle assessment

Naumann G, Famiglietti J, Schropp E, Motta M, Gaderer M (2024) Dynamic life cycle assessment of European electricity generation based on a retrospective approach. Energy Convers Manage 311:118520

Sacchi R, Bauer C, Cox B, Mutel C (2022) When, where and how can the electrification of passenger cars reduce greenhouse gas emissions? Renew Sustain Energy Rev 162:112475

Schindler D, Sander L, Jung C (2022) Importance of renewable resource variability for electricity mix transformation: a case study from Germany based on electricity market data. J Clean Prod 379:134728

Sphera (2022) Sphera® GaBi–GaBi Databases & Modelling Principles 2022

Tranberg B, Corradi O, Lajoie B, Gibon T, Staffell I, Andresen GB (2019) Real-time carbon accounting method for the European electricity markets. Energ Strat Rev 26:100367

United nations framework convention on climate change (UNFCCC) (2015) The Paris Agreement, UNFCCC Secretariat

United Nations (2015) Transforming our World: The 2030 Agenda for Sustainable Development, United Nations

Walker WE, Harremoes P, Rotmans J, van der Sluijs JP, van Asselt MBA, Janssen P, Krayer von Krauss M (2003) Defining uncertainty: a conceptual basis for uncertainty management in model-based decision support. Integrated Assessment 4(1):5–17

Wernet G, Bauer C, Steubing B, Reinhard J, Moreno-Ruiz E, Weidema B (2016) The ecoinvent database version 3 (part I): overview and methodology. Int J Life Cycle Assess 21(9):1218–1230

Digitalisation, Artificial Intelligence Enhancing Data Availability, Traceability Robustness in LCM

Integrating Large Language Models into the In-Silico Environmental Assessment of New Chemicals and Materials for SSbD

Gustavo Larrea-Gallegos and Antonino Marvuglia

Abstract The introduction of the European Commission's 'Safe and Sustainable-by-Design' (SSbD) framework represents a catalyst in the wide adoption of SSbD practices and it has spurred an increasing interest in displaying practical implementations among policymakers, academia, and industry players, such as the chemical sector. Conducting such a type of assessment at early stages of the design process is far from trivial due to the lack of data or the need of lengthy modelling pipelines. This study proposes a novel modelling workflow that combines the power of Large Language Models and machine learning methods to bootstrap the data gap filling in the early stage of development of new chemicals by automating the generation of Life Cycle Inventories. The workflow is meant to produce ready-to-use proxy data that can be later used by any LCA computing engine to be enhanced or analysed in detail.

1 Introduction

The introduction of the European Commission's 'Safe and Sustainable-by-Design' (SSbD) framework represents a catalyst in the wide adoption of SSbD practices and it is a milestone in the transition towards more sustainable societal systems (Caldera et al. 2022). This has spurred an increasing interest in displaying practical implementations of SSbD among policymakers, academia, and industry players, such as the chemical sector. As far as the sustainability part is concerned, conducting such a type of assessment at early stages of the design process is far from trivial, since the lack of data and the need of lengthy modelling pipelines challenge the adoption of this framework in a consistent way. Thus, effective workflows and tools to fill this gap are needed. This study proposes a novel modelling workflow that combines the power of Large Language Models (LLMs) and conventional Machine Learning (ML) methods to bootstrap the data gap filling in the early stage of development of new chemicals

G. Larrea-Gallegos (✉) · A. Marvuglia
Luxembourg Institute of Science and Technology (LIST), Esch-Sur-Alzette, Luxembourg
e-mail: gustavo.larrea@list.lu

M. Traverso et al. (eds.), *Life Cycle Management from Global to Local*,
https://doi.org/10.1007/978-3-032-17987-6_5

by automating the generation of their Life Cycle Inventories (LCIs). The workflow consists of different modules that process the semantic information and molecular data provided by the user, relying on LLM and ML, respectively. Unlike existing LLM-based life cycle assessment (LCA) implementations (CarbonGraph 2025; Deng et al. 2023), our approach is focused on creating inventories for innovative chemicals, and it is designed to provide fast but flexible LCI models. For this, we opted for an AI multi-agent system capable of deciding which models or services to request based on the user query. The workflow is meant to produce ready-to-use structured data object (i.e. JSON format) that can be later used by any other LCA computation engine to be analysed or enhanced, such as the Brightway python framework. Finally, this multi-agent system was implemented as a tool called ChemLCI.

2 Current AI-Based Approaches for Forward-Looking LCA of New Chemicals and Materials

Explorative LCA and predictive (forward-looking) methodologies represent new trends and methods for conducting LCA studies. Recently, Larrea-Gallegos et al. (2025) presented the predictive LCA models for chemicals found in the literature as path 1 and path 2 models, being the former *chemical-to-impact*, and the latter *chemical-to-LCI*. In this study, we aim to develop *chemical-to-LCI* workflows since our aim is to produce LCI models that can be later plugged-in into a full LCA model so it can be used as the product development progress.

Chemical-to-LCI models are not abundant in the literature, and they are far from being a mainstream forward-looking LCA approach. Only two studies have focused on exploring this path by treating the life cycle data as a graph of interconnected processes and predicting the missing links and values when only little information is provided (Hou et al. 2018; Zhao et al. 2021). In these studies, the main challenge is to predict links and values relying only on a description of the chemical candidate. In fact, Zhao et al. (2021) results indicated that reasonable predictions can be obtained when no less than 10 non-zero values (i.e. LCI dataset flows) are known. Additionally, this approach is only valid when dealing with structured data formats (e.g. graph data as matrices), such as the ones provided by Ecoinvent. This means that any unstructured data entry like expert knowledge, diagrams, equations, or semantic information would need to be converted first into LCI data in a valid format (e.g. Ecoinvent-compliant ecospold). Current developments in AI (namely, LLMs) may shed light on the possible directions that *chemical-to-LCI* implementations could take. While not in the same way as described above, the literature (CarbonGraph 2025; Deng et al. 2023) has shown attempts to exploit the power of LLMs for LCA purposes. For instance, EcoScan (CarbonGraph 2025) is a commercial tool capable of automating the construction of LCI models from bills of materials (BOMs) that are built from semantic descriptions using chatbots. In another example, Balaji et al. (2025) used LLMs in combination with Retrieval Augmented Generation

(RAG) techniques to match the user's prompt to the most similar Ecoinvent process. Similarly, AutoPCF allows to calculate the carbon footprint of products using the embedded knowledge of foundation models and RAG techniques (Deng et al. 2023).

While the use of LLMs as reliable LCI modelling tools for novel products is still challenging, current state-of-the-art examples show that these tools can be used in the rapid drafting of LCI datasets (Tu et al. 2024). With the accelerated advancement in AI models and AI-based technologies, we can foretell the improvement and expansion of these kind of tools. In this sense, if we make an analogy with tools with similar characteristics, such as the Rapid Application Development (RAD) tools in software development, it is fair to envisage a new wave of similar tools in the LCA field. Adopting a more LCA-specific context, we can then coin these sort of tools as Rapid Inventory Modelling (RIM) tools, since we believe that their potential relies on the fast prototyping rather than the precise prediction. In this sense, this study presents ChemLCI, an automated LCI generation pipeline for chemicals at early stages of design.

3   Materials and Methods

3.1   General ChemLCI Architecture

ChemLCI is a RIM tool that serves as a bridge between the identification of a novel chemical candidate and the beginning of the iterative LCI modelling process. ChemLCI is a multi-agent AI system at its core and the user is meant to interact with it by sending prompts providing a targeted chemical accompanied by additional information of a potential production process (e.g. synthesis route). The user's prompt needs to be a semantic explanation of the query, meaning that the more information is provided, the more the obtained solutions can be expected to be reliable. In cases where no semantic description is provided, ChemLCI will attempt to predict possible production routes using a set of tools implemented in the system. ChemLCI is then designed to produce a valid JSON object that describes an LCI inventory draft of the user's prompt. This inventory is generated following a data model that can be interpreted by any LCA computation engine that uses Ecoinvent as background database, such as the python framework Brightway. Under the hood, ChemLCI has access to different tools that can be called by an LLM agent called 'conductor'. The 'conductor' parses the target chemical reference and the production context from the user prompt and it is capable of using tools or asking for user inputs whenever it is considered as necessary. The 'conductor's' objectives are introduced as a system prompt, and the tools are python functions that make requests to other services using APIs.

The first service is a RAG system that performs semantic searches using a PostgreSQL Ecoinvent 3.11 vector database. The second service is a retrosynthesis route finder engine relying on the AiZynthFInder open-source package. This service

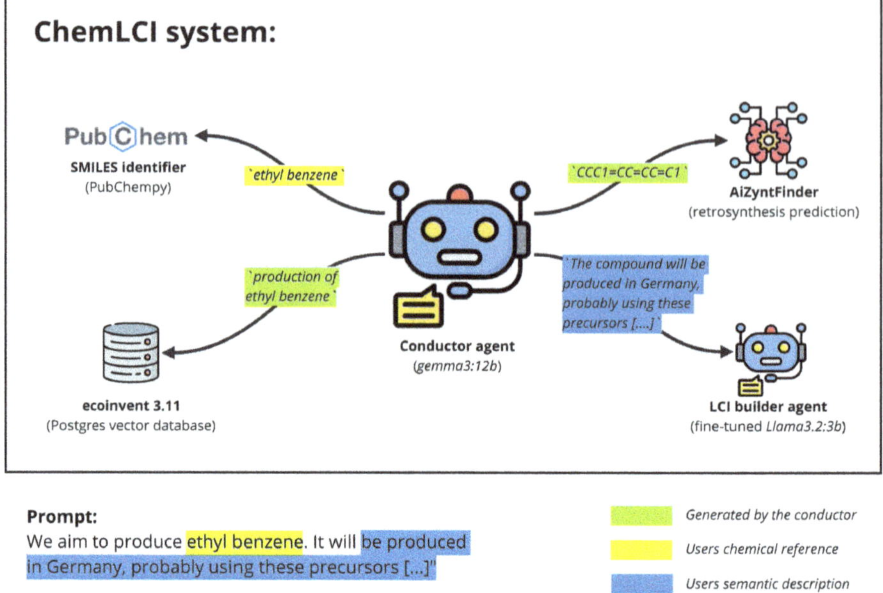

Fig. 1 ChemLCI agent system. The parsing and execution flow is orchestrated by a conductor agent that interacts with other services by making API requests. The conductor can parse the initial chemical reference (i.e. parts highlighted in yellow) and the production context (i.e. parts highlighted in blue) from the user's prompt and interpret and transform this information when it is convenient (i.e. parts highlighted in green)

receives a SMILES (simplified molecular input line entry system) string and returns a recommended synthesis route using a combination of Monte Carlo tree search and Deep Learning models. The third service hosts an LLM that was fine-tuned to produce JSON objects from semantic description of aimed LCI dataset. This fine-tuned model embeds Ecoinvent data into the LLM knowledge. Finally, a SMILES identification service is also available, so that the 'conductor' can consult it when no SMILES representation is provided (see Fig. 1).

3.2 Data Collection and Treatment

ChemLCI's pipeline relies on a combination of predictive models, LLMs and a knowledge base that provide the general LCA context to all the tasks. For this, we used the Ecoinvent v3.11 database as the main source of general LCA knowledge. We are interested in the semantic content and the meaning that the datasets represent so we could use them as part of fine-tuning steps and RAG systems. For this, we used the 'bw2data' python package to parse and convert them into python data types. Given that datasets are rich in metadata and various information, we built different data models that were used as schemas to build the RAG database, generate the fine-tuning training data, and for the communication among the different APIs.

The RAG database was created by defining a data model that was used to host the necessary information that could summarise an Ecoinvent dataset. This data model was used to write, vectorise, and query data in a vector database. For vectorisation purposes, the semantic summary was limited to 8000 tokens due to the restrictions of the selected embedding model. We opted for this approach in order to be able to rely on small and local models, which was convenient when addressing data privacy constraints and to maintain full control.

While LLMs can output data in almost any possible format, the selection of a proper data format or template can influence the robustness of the LLM tasks. In this sense, we defined a data model (i.e. 'SemanticNode') as the mandatory LCI dataset representation for all the LLM-related tasks. In other words, we simplified the original Ecoinvent data format stored to a suited and more convenient data model. The simplification implied the discarding of metadata, parameters, and biosphere flows given our interest in drafting foreground LCI models. The benefits of this effort are threefold. Firstly, by introducing and demanding precise data elements, we reduce the risk of noise ingestion and potential hallucinations. Secondly, given the necessity of local models and hardware limitations, we were constrained to contexts lengths of maximum 16,000 tokens (i.e. inputs and outputs combined). For this, the 'SemanticNode' provided enough information to describe an LCI dataset in a reduced length of tokens. Finally, the 'SemanticNode' allowed us to communicate LLM's outputs with any LCA computational engine capable of interpreting its schema and converting it to an Ecoinvent compatible format (e.g. an LCA computation engine built with Brightway).

Since the end of the pipe of ChemLCI is a JSON object that follows the 'SemanticNode' schema, the LCI templating LLM was expected to generate these objects from prompts with or without meaningful information in a zero-shot fashion. For this, we fine-tuned an LLM to take a semantic description of a production activity as input and produce a 'SemanticNode' compliant JSON file as output.

3.3 *Retrosynthesis API Service*

The retrosynthesis API is a service that executes predictions using the AiZynthFinder python package and is meant to be used when no information about the production processes is provided. AiZynthFinder performs a retrosynthesis process in which the target molecule is decomposed into purchasable precursors. The package uses a Monte Carlo tree search algorithm in combination with a neural-network model to predict synthesis routes (Genhden et al. 2020). The API receives a request containing the SMILES of the target chemical. This is then processed by the AiZynthFinder engine, and a list of adequate synthesis routes are then returned as a JSON object.

3.4 Ecoinvent RAG Database Service

Users can provide any sort of queries regarding the target chemical or the envisioned production process. For instance, chemicals could be described using different types of identifiers such as commercial names, CAS numbers, or SMILES descriptors. For some chemicals, the Ecoinvent database may already contain information about the specific or similar production processes. In this case, the 'conductor' requires a tool to consult if this information is, somehow, already available in the Ecoinvent database. For this purpose, the RAG database service allows searching fast and semantically over the thousands of Ecoinvent datasets. The search is conducted using a similarity approach (i.e. cosine distance) in which a vectorial representation of the query is compared with the vectorial representations of the Ecoinvent datasets. This semantic search technique has gained wide popularity in scientific literature and industrial applications because of its efficiency in providing additional and specific information to LLM systems (Gao et al. 2023). Vectorial representations are generated by using embedding models, which are LLMs trained to capture the semantic relationship among words and encode them into a numeric embedding space. We vectorised the dataset summaries using the nomic-embed-text-v1.5 LLM embedding model (Nussbaum et al. 2024). This model can take queries of maximum 8191 tokens and produces a vector of 768 elements. Finally, the RAG database service consists of a PostgreSQL database with a pgvector extension.

3.5 Ecoinvent LCI Template Builder

RIM tools found in the literature have attempted to automate the end-to-end prediction of carbon footprints by relying on the knowledge embedded in foundation models (Deng et al. 2023), or by fine-tuning LLMs with public articles and reports (Li et al. 2025). Other tools like Ecoscan (CarbonGraph 2025) generate LCI models using BOMs as reference, although it is not disclosed if the BOMs are generated by fine-tuned models, zero-shot predictions, or RAG systems. Differently from these approaches, our goal is not to provide end-to-end predictions, but to generate LCI dataset templates using the Ecoinvent database knowledge, predictive models' results, and the user's input. To this aim, we fine-tuned a local Llama 3.2 model to learn to generate JSON objects compliant with the 'SemanticNode' schema. The resulting model acts then as an LCI template builder agent that can generate JSON objects ready to be parsed by LCA computing engines. These templates contain a structured representation of the input prompt, but they also are complemented with activities coming from the Ecoinvent knowledge base. We fine-tuned the model using a technique called Quantized Low-Rank Adaptation (QLoRA) (Dettmers et al. 2023). This fine-tuning technique permits to improve a model's understanding of a specific task by modifying a small fraction of the parameters of a quantized version of the original LLM. By using this approach, we were able to fine-tune the model faster and with less computational resources, compared to conventional techniques.

The fine-tuning is done by creating a training dataset consisting of examples of the expected responses given a user's input in a sort of 'question' and 'answer'. The 'answers' are Ecoinvent datasets in JSON format compliant with 'SemanticNode' schema, while the 'questions' are synthetic user's queries that were meant to mimic the input of a human user. These synthetic queries were generated by using the open-sourced Google gemma3:27b model as a local summarisation model. A system prompt that contained the initial instructions that the LLM should consider before receiving the training data was added as a prefix to the 'question'-'answer' pair. We then prepared the training set by assigning roles to each prompt block: 'system' to the system prompt, 'user' to the semantic summary, and 'assistant' to the Ecoinvent dataset in 'SemanticNode' format. Finally, we fine-tuned a 4-bit quantized version of Llama 3.2:3b using the unsloth python framework and an Nvidia A100 GPU with 40 Gb of VRAM.

3.6   ChemLCI Conductor Agent

The different services previously described are accessible through REST APIs. To access them in an integrated and automated manner, we used an LLM agent labelled as the 'conductor' agent. The role of this agent is to understand the user query and execute a series of tasks by calling the different available services and respecting a predefined plan and objectives. The 'conductor' agent relied on a gemma3:12b model loaded on a local server, and we used plain python to program the agent execution loop. The available tools are provided to the agent's context as a JSON object with tools' specifications, and the action logic and objectives are defined as a comprehensive system prompt.

The 'conductor' is required to take different decisions along the action steps depending on the user's query or tools' responses. In this sense, agent's actions respond to the LLM's capacity to understand or 'obey' the instructions provided in the system prompt. We can attempt to represent the expected action plan in a diagram, as shown in Fig. 2. The plan starts with deciding if the user provided a valid SMILES in the query and calling the 'get_SMILES' tool if that was not the case. Then, the agent will look in the RAG database if the chemical is already available. The agent will show the available datasets and let the user decide if choosing one of them to finish the interaction, or to proceed with the pipeline. When no dataset is chosen, the agent will call the 'predict_retrosynthesis_tool' to obtain a retrosynthesis plan, this will be parsed and semantically summarised, to then be returned to the user. The user can use this information as input of the LCI template builder tool to obtain the aimed LCI dataset.

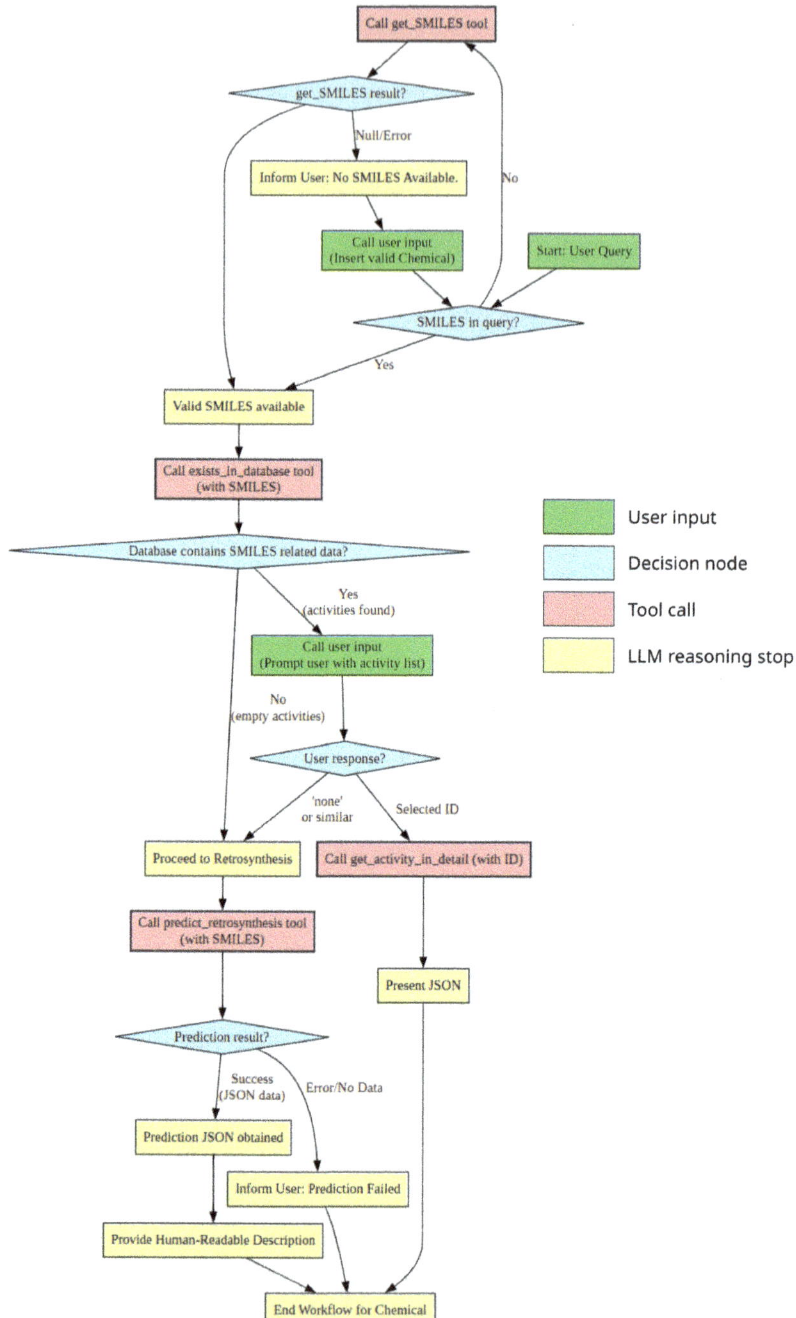

Fig. 2 Expected action diagram for the 'conductor' agent. Instructions are provided as a semantic description in the system prompt. Green blocks indicate the start and finish of the user–agent interaction. Light blue and yellow blocks represent decision and reasoning steps, part of the agent's inference. Red block represents tool calls to the different API services

4 Results

We verified the functioning of the LCI template builder by prompting different types of products as described in the Ecoinvent database. When analysing the model's response, we observed that in most of the cases the model could produce valid JSON objects containing elements like the original dataset. However, given the generative nature of the model, the responses were not always consistent even when reducing the LLM's temperature parameter. This was noted in the inclusion or exclusion of certain product inputs, the slight inconsistencies in product names, and the variations on the amounts of the exchanges. For example, for a query asking for 'production of ethyl benzene', the response indicated ~ 0.79 kg of benzene as input, while it is 0.74 kg in the real dataset. The response also included very small quantities of 'aluminium cast alloy', and 'aluminium, wrought alloy', which are not included in the original dataset, suggesting that the model hallucinates with some responses. In contrast to this example, when including a semantic summary of the output of the retrosynthesis path as context, the model produced more reliable LCI templates as data of the context are adequately included in the response. Namely, the precursor and energy that are in the query context and are also part of the response. Moreover, inputs such as 'chemical factory, organics', which are common in chemical Ecoinvent processes, are not part of the context but they were adequately included in the response.

We identified that model's inability to produce numeric estimations still represents an important limitation of our approach. Given that LLMs are not capable of performing complex math operations, tasks like yield estimation still require to be conducted using other methods. This, however, could be addressed by allowing the agent to access more precise modelling tools (e.g. process simulation). Even though we present ChemLCI as a RIM tool, the need of enhancing it towards more precise pipelines, such as adding precise means of validation, is still latent. However, this is not a trivial task, since evaluating the response of generative models with creative tasks is still a subject of research (Wang et al. 2022). Moreover, even if robust validation metrics existed, guaranteeing the reliability of the system in SSbD contexts will require understanding the reliability expectations of the potential user.

5 Conclusion

Our experimentation and development have shown the potential that agentic LLMs have as core technology in LCI modelling exercises. More specifically, when it concerns early stages of development, the use of RIM tools can bootstrap the preliminary evaluation of impacts in a more comprehensive manner than *chemical-to-impact* methods. Providing the user with means to introduce non-structured knowledge, such as semantic descriptions, can facilitate the exploration of different production plans or setups and support the SSbD adoption in chemical design. The technologies developed as part of the ChemLCI system are useful not only in the context of early stages of design, but also in other types of LCA tasks. For instance, the RAG database service could be easily incorporated as part of any LCA computation engine as a smart search engine. Similarly, the LCI template builder could also be used to generate templates for other types of products, since it uses the whole universe of Ecoinvent as training data.

Acknowledgements This study was conducted in the context of the PINK project that received funding from the European Union's Horizon Europe Research and Innovation programme under grant agreement no. 101137809. We would like to also thank Dr. Jonas Hoffmann for the insightful exchange of ideas that helped us to improve our prototype.

References

Balaji B, Ebrahimi F, Domingo NG, Vunnava VSG, Faridee AZ, Ramalingam S, et al (2025) Emission factor recommendation for life cycle assessments with generative AI. Environ Sci and Technol

Caldeira C, Farcal LR, Garmendia Aguirre I, Mancini L, Tosches D et al (2022) Safe and sustainable by design chemicals and materials–framework for the definition of criteria and evaluation procedure for chemicals and materials. Publications Office of the European Union

CarbonGraph (2024) Ecoscan

Deng Z, Liu J, Luo B, Yuan C, Yang Q, Xiao L et al (2023) AutoPCF: Efficient product carbon footprint accounting with large language models. arXiv

Dettmers T, Pagnoni A, Holtzman A, Zettlemoyer L (2023) Qlora: Efficient finetuning of quantized LLMs. arXiv

Gao Y, Xiong Y, Gao X, Jia K, Pan J et al (2023) Retrieval-augmented generation for large language models: a survey

Genheden S, Thakkar A, Chadimová V, Reymond JL, Engkvist O, Bjerrum E (2020) AiZynthFinder: a fast, robust and flexible open-source software for retrosynthetic planning. J Cheminformatics. 12:1–9

Hou P, Cai J, Qu S, Xu M (2018) Estimating missing unit process data in life cycle assessment using a similarity-based approach. Environ Sci Technol 52(9):5259–5267

Larrea Gallegos GM, Hofer S, Hofstätter N, Punz B, Watzek N et al. (2025) Integrate & balance aspects for safe and sustainable innovation: Needs analysis on SSbD categories and product development stage requirements to cover the entire life cycle. Comput Struct Biotechnol J

Li Z, Tang P, Wang X, Liu X, Mou P (2025) PCF-RWKV: large language model for product carbon footprint estimation. Sustainability 17(3):1321

Nussbaum Z, Morris JX, Duderstadt B, Mulyar A, Ai A, Ai N (2024) Nomic embed: training a reproducible long context text Embedder

Tu Q, Guo J, Li N, Qi J, Xu M (2024) Mitigating grand challenges in life cycle inventory modeling through the applications of large language models. Environ Sci Technol 58(44):19595–19603

Wang X, Wei J, Schuurmans D, Le Q, Chi E, Narang S, et al (2022) Self-consistency improves chain of thought reasoning in language models. arXiv

Zhao B, Shuai C, Hou P, Qu S, Xu M (2021) Estimation of unit process data for life cycle assessment using a decision tree-based approach. Environ Sci Technol 55(12):8439–8446

Dynamic Digital Product Passports: Advancing Circularity and Sustainable Innovation in the Steel Industry Supply Chain

Stella Georgali, Lucyna Lekawska-Andrinopoulou, Kostas Chatziioannou, Thodoris Theodoropoulos, Georgios Tsimiklis, and Angelos Amditis

Abstract The steel industry's value chain is being transformed through innovative digitalization aimed at achieving sustainability, decarbonization, and operational excellence (Clean Steel Partnership 2021). Central to this transformation is the development of dynamic Digital Product Passports (DPP), which offer comprehensive, continuously updated repositories of product data across a product's lifecycle. This research presents preparatory work for developing a dynamic DPP for the steel industry, examining enabling technologies, regulatory context, and implementation challenges. The dynamic DPPs will aggregate real-time data on material properties, production processes, and sustainability metrics, enabling seamless information exchange among steelmakers, end-users, and other stakeholders through an industrial dataspace. Live traceability methods embedded within DPPs will enhance transparency and resource optimization within the steel supply chain, aligning with the Ecodesign for Sustainable Products Regulation (ESPR) (European Commission 2024).

1 Introduction

The steel industry, a central pillar of the European economy, is under increasing pressure to decarbonize, digitalize, and transition to a circular economy. These demands are driven by regulatory initiatives such as the EU Green Deal (European Commission 2019) and the Circular Economy Action Plan (CEAP) (European Commission 2020), which aim to promote resource efficiency, lifecycle transparency, and reduced environmental impact.

In this context, the ESPR calls for the systematic documentation of product-related information across the value chain (European Commission 2024). Notably, steel has been identified as one of seven priority intermediate products in the first ESPR

S. Georgali (✉) · L. Lekawska-Andrinopoulou · K. Chatziioannou · T. Theodoropoulos ·
G. Tsimiklis · A. Amditis
Institute of Communication and Computer Systems, Athens, Greece
e-mail: stella.georgali@iccs.gr

© The Author(s) 2026

M. Traverso et al. (eds.), *Life Cycle Management from Global to Local*,
https://doi.org/10.1007/978-3-032-17987-6_6

Working Plan, underscoring the sector's strategic role in meeting EU sustainability targets (European Commission, JRC 2024).

In this context, DPPs are emerging as vital tools—especially dynamic DPPs that update over time—to enable lifecycle transparency, traceability, and sustainable resource management (Jensen et al. 2023). These capabilities are particularly crucial in complex value chains like steel, where data must flow seamlessly among producers, users, and other stakeholders. DPPs consolidate structured data on product origin, material composition, production processes, energy usage, and end-of-life options.

However, the adoption of DPPs in the steel industry remains limited, constrained by a combination of technical- and governance-related barriers. From a technical standpoint, the integration of real-time production data at industrial-scale poses significant challenges, including sensor deployment, data standardization, and system interoperability. Equally important are concerns related to data governance and confidentiality.

This paper presents preparatory work under the SMART Chain Horizon Europe project (https://www.smartchain-project.eu/) toward a dynamic Digital Material and Digital Product Passports (DMP/DPP) framework tailored to the steel industry. The envisioned system integrates in-line sensors and real-time data capture on material properties, production processes, and sustainability metrics, enabling end-to-end traceability and automated, secure information exchange within an industrial dataspace.

2 Method

SMARTChain adopts a bottom-up, requirements-driven approach that combines desk-based regulatory and technical analysis with an ongoing technical dialogue among participating steelmakers and downstream users. In this pre-deployment phase, the work concentrates on three objectives:

- Determining the data requirements for steel passports.
- Specifying a minimal yet interoperable information model.
- Designing a secure, modular architecture capable of supporting dynamic DMP/DPPs across heterogeneous plant-IT environments.

2.1 Regulatory Foundations and Industrial Alignment

The methodological starting point was a desk review of the ESPR, the CEAP, and Electronic Product Code Information Services (EPCIS) 2.0 (Tolcha et al. 2025).

These documents establish the regulatory scope, traceability depth, and interoperability targets that any DPP for steel must satisfy. Additionally, DPPs can potentially enable steel or manufacturing companies to meet new EU Corporate Sustainability Reporting Directive (CSRD) (European Commission 2022) requirements while enhancing operational insights.

To ensure that the resulting specifications reflect industrial realities, the project maintains an ongoing technical dialogue with participating steelmakers and downstream users. Regular project meetings and joint specification sessions are used to validate assumptions, identify data-sharing constraints, and agree on integration pathways that can accommodate the existing plant-IT landscapes.

This iterative process has crystallized three design priorities that now shape both the information model and the supporting architecture:

1. Cradle-to-cradle traceability. Material and sustainability attributes must be captured with sufficient granularity to meet forthcoming ESPR reporting obligations.
2. Data-sovereignty safeguards. The solution must guarantee that commercially sensitive information remains under the originator's control, even when shared across organizational boundaries.
3. Lightweight integration. New components should interface seamlessly with existing Enterprise Resource Planning (ERP) systems, minimizing deployment effort and downtime.

These priorities provide the foundation for the ontology presented in Sect. 2.2 and the secure, modular architecture described in Sects. 2.3–2.5.

2.2 Information/Data Model and Flow

A lean ontology expressed in JSON-LD (Sporny et al. 2014) captures the data fields required by ESPR while remaining extensible for future regulation. The schema distinguishes five logical domains:

- Material/product information.
- Material properties, based on references and standards requirements.
- Process and sustainability metrics.
- Circularity indicators.
- Compliance metadata: certificates and regulatory labels.

Each record is anchored to a persistent GlobalID that couples the steelmaker's heat number with a decentralized identifier, ensuring uniqueness and cross-pilot compatibility.

The defined data model aligns with best practices identified in recent literature (Jensen et al. 2023), structuring product lifecycle information into clearly defined and interoperable domains. Additionally, it accommodates product-level sustainability indicators—such as carbon footprint and recycled content—that are relevant to broader Environmental, Social, and Governance (ESG) frameworks and reporting needs, as well as ESPR.

2.3 End-to-End Data Flow Architecture

The DMP is a structured digital dataset that captures and shares essential information about steel material properties, processing history across its entire lifecycle. For steel producers and downstream manufacturers, the DMP serves as a standardized tool for recording and accessing key data—such as composition, manufacturing steps, certification status, and recyclability—linked directly to a batch or product ID. This allows all actors in the value chain, including recyclers and regulators, to access consistent and verified information, supporting sustainable product design, responsible sourcing, and compliance with circular economy requirements.

To enable seamless and secure data exchange, the DMP system leverages a blockchain-based common data layer. Each step in the steel production process—like smelting, alloying, heat treatment, and shaping—is logged as a timestamped event on the blockchain, forming a traceable and immutable data trail. A data retrieval interface allows authorized users to query this shared ledger using material IDs or lifecycle event types, retrieving only the data relevant to their role or permissions (see Fig. 1). This architecture ensures trust, transparency, and interoperability across the steel value chain, helping lifecycle managers and decision-makers work with a unified and up-to-date view of material data from cradle to reuse.

Real-time sensor streams and batch ERP uploads are stored in local unified databases at both the steel-producer (up-stream) and steel-user (down-stream) sites. At the steel-maker side, each event is normalized into a DMP (Panza et al. 2022); its hash and timestamp are anchored on a permissioned Hyperledger Fabric ledger (Miron and Hulea 2024). An International Data Space (IDS)-compliant gateway mediates policy-controlled exchange of full payloads, allowing the enriched DMP to be wrapped into a DPP that accompanies the product through service, repair, and end-of-life stages.

2.4 Traceability Architecture and Data Management

The traceability model relies on the EPCIS 2.0 event standard, a Global Standards One (GS1) standard designed to enable sharing of information about the movement and status of products and assets throughout the supply chain, in a standardized and interoperable way. It enables consistent, event-based tracking of materials and products across both steel suppliers and manufacturers. Key event types such as ObjectEvent and TransformationEvent capture lifecycle actions—from heat creation and processing to final assembly—allowing seamless data linkage throughout the supply chain (see Table 1).

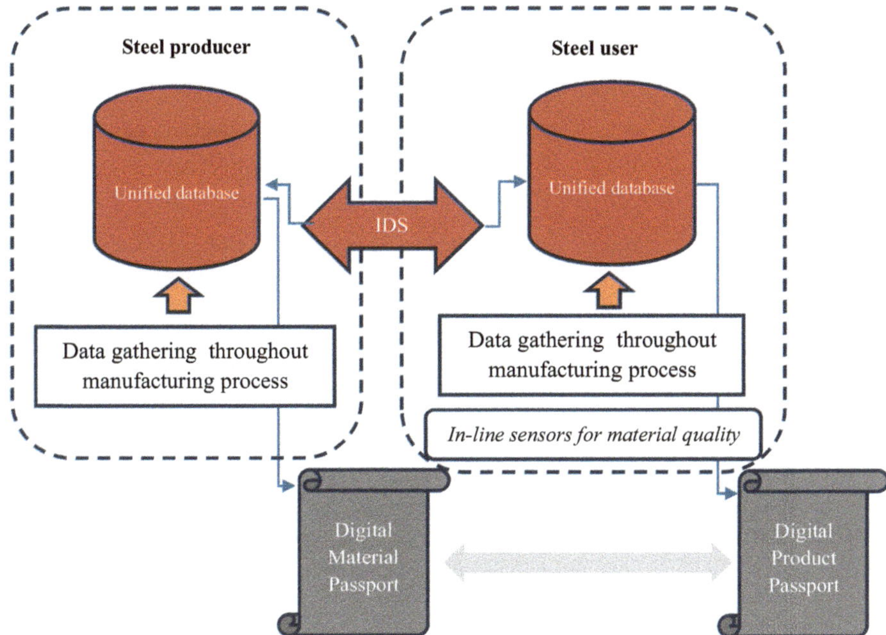

Fig. 1 Data flow for dynamic digital material and product passports in SMARTChain

Each update to a DMP corresponds to one or more EPCIS events and is recorded on a permissioned Hyperledger Fabric blockchain as a cryptographic hash paired with a timestamp. This on-chain fingerprint guarantees tamper-evidence and provenance across companies while maintaining the confidentiality of detailed data. Full data remain off-chain within company systems and are shared securely via IDS (Pettenpohl et al. 2022) connectors that control access based on user permissions. This architecture balances transparency with data sovereignty, allowing participants to control who can view sensitive information.

When the steel product leaves its last processing step, the DMP is finalized and wrapped into a DPP, which remains open for updates such as servicing, repair, or end-of-life events. This structure supports cradle-to-cradle traceability and enables compliance with evolving regulatory requirements.

2.5 *DMP/DPP Software Architecture Overview*

In the context of DMP/DPP and digital lifecycle systems, a standardized software system architecture has been designed comprising multiple layers that describe the main mechanisms set in place to communicate, process, and store data.

Table 1 Indicative EPCIs event types employed (varying between use cases)

Location	Event type	Action	Description	Purpose
Steel maker	ObjectEvent	ADD	Observation of heat output from steel furnace (e.g., Heat H123456)	Confirms creation and readiness of a specific heat
	TransformationEvent	TRANSFORM	Conversion of molten steel into strip or coil	Records the heat as it becomes a rolled product for dispatch
	ObjectEvent	OBSERVE	Observation at outbound shipping station	Marks readiness for shipment to steel user
Steel user	ObjectEvent	OBSERVE	Receipt of steel material at steel user's facility	Marks arrival of traceable material (linked to heat number)
	TransformationEvent	TRANSFORM	Transformation of steel material (e.g., steel strip) into steel product (e.g., razor parts) and assembly	Establishes material-to-product trace (e.g., steel → blade → razor)
	ObjectEvent	OBSERVE	Final packaging and outbound readiness	Registers finished product entering outbound process

The software architecture of the SMARTChain DPP system, designed to enable secure, interoperable, and real-time tracking of steel materials and products across complex supply chains is presented in Fig. 2.

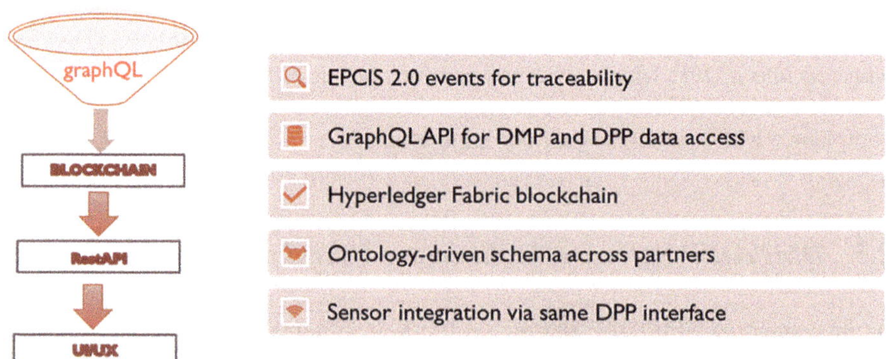

Fig. 2 SMARTChain DPP system software architecture

At the core, a GraphQL (Quiña-Mera et al. 2023) interface ingests data from multiple sources—including in-line sensors and plant-IT systems—ensuring structured, schema-compliant input. This data is then recorded on a permissioned Hyperledger Fabric blockchain, which stores cryptographic hashes of each passport update to guarantee immutability and cross-organizational provenance (Monrat et al. 2019).

A REST API layer exposes secure, standardized endpoints for external applications and services to query or update passport data. Finally, a user-friendly interface for engineers and auditors to visualize dynamic material and product passports, supporting transparency and operational decision-making will be created.

This stack incorporates several critical technological enablers: standardized EPCIS 2.0 event formats for traceability, an ontology-driven data schema to harmonize semantics across partners, and seamless sensor integration through the unified DPP interface. Together, these components form a modular and extensible platform that balances data security, usability, and regulatory compliance.

3 Results

The dynamic DMP/DPP architecture developed in this research integrates advanced structured data schemas (EPCIS 2.0 and JSON-LD), blockchain-based verification mechanisms (Hyperledger Fabric), and sensing technologies, aiming to facilitate secure, real-time, and interoperable data exchange across steel supply chains. Early-stage discussions with steel industry partners have confirmed the relevance and potential applicability of this architecture, highlighting the following outcomes:

Data Interoperability: Standardized formats and structured metadata tagging ensure seamless data portability among diverse IT systems and stakeholders. The implemented EPCIS 2.0 events and consistent JSON-LD schemas provide straightforward integration, even within complex, multi-party environments.

Enhanced Traceability: Industry dialogues emphasized the importance and perceived feasibility of tracking steel materials and products through critical stages, from initial forming and downstream processing to end-of-life handling, providing full lifecycle visibility.

Real-Time Attribute Monitoring: The architecture is designed to support continuous real-time recording and monitoring of essential steel attributes relevant to business-to-business transparency. Industry partners identified key metallurgical, process, and sustainability parameters at the steel heat level as particularly valuable for operational and compliance purposes.

Stakeholder Empowerment: Early dialogues with stakeholders indicate significant value in dynamic DMP/DPP solutions, particularly regarding transparency, simplified compliance reporting (such as ESG and ESPR requirements), and enhanced market differentiation through credible sustainability claims. While ESG indicators are typically defined at the corporate level, stakeholders see value in linking relevant company-level metrics—such as GHG intensity or responsible sourcing practices—to product-level data.

Overall, initial industry feedback underscores strong interest in adopting dynamic passport technologies, highlighting their potential for regulatory compliance and market competitiveness. Subsequent project phases will involve detailed technical validation and refinement of data attributes in alignment with stakeholder expectations and evolving regulatory frameworks.

4 Discussion

The steel industry stands to gain significantly from the adoption of DPPs. Unlike static product information systems, dynamic DMP/DPP solutions provide real-time visibility into material characteristics, environmental metrics, and product lifecycle data. This dynamic visibility can offer substantial competitive advantages for steel producers and users, especially in the areas of quality assurance, environmental reporting, and meeting regulatory compliance obligations. Notably, dynamic DPPs also have strong potential to support steel companies in addressing growing ESG reporting requirements.

Despite these promising benefits, several critical challenges must be addressed to realize widespread adoption. Challenges identified include the following:

(1) Data Ownership and Access Control: Ensuring that companies retain sovereignty and control over sensitive and commercially confidential information while sharing critical sustainability and lifecycle data.
(2) Standardization and Interoperability Barriers: Harmonizing data standards across companies and industries is necessary to allow seamless data exchange and interoperability among diverse systems.
(3) Integration with Legacy Systems: Effective implementation requires minimal disruption to established IT and production infrastructures. Thus, lightweight and adaptable integration approaches are needed.

These findings highlight the need for continued collaboration between technology providers, manufacturers, and regulators to ensure alignment on data formats, governance, and interoperability protocols and regulatory frameworks. Future research should explore how dynamic passport architectures can facilitate the connection between company-level ESG data and product-level information., further supporting steel companies in meeting evolving market expectations and sustainability commitments.

5   Conclusion

This study introduces a comprehensive and practical framework for the adoption and implementation of dynamic DMP/DPPs specifically within the steel industry. By integrating advanced sensing technologies, standardized data models, and blockchain-based verification (Hyperledger Fabric), dynamic DMP/DPPs significantly enhance transparency, improve traceability, and effectively promote circularity across the entire steel value chain. As the SMARTChain project progresses into its ninth month, continued proactive stakeholder engagement, robust technical validation, and iterative prototype development remain essential. This phase will culminate in initial prototypes, followed by thorough evaluation and full-scale implementation. These technological advancements not only facilitate secure, role-based data exchanges but also foster the potential for strategic business alliances and greater alignment with emerging regulatory frameworks, ultimately supporting the industry's transition toward sustainable and circular business models.

Acknowledgements This research has been supported by the European Union's Horizon Europe research and innovation program under Grant Agreement No. 101178919 (SMARTChain). The authors gratefully acknowledge the collaboration and technical input provided by the consortium partners, as well as the valuable insights and feedback received during ongoing project discussions and meetings. The contents of this paper reflect only the authors' views, and the European Commission is not responsible for any use that may be made of the information contained herein.

References

Clean Steel Partnership (CSP) (2021) Strategic research and innovation agenda (SRIA), CSP. https://www.estep.eu/assets/CSP/CSP_SRIA_Oct2021_clean.pdf (Accessed 09.06.2025)

European Commission, Directive (EU) 2022/2464… as regards corporate sustainability reporting, 2022. https://eur-lex.europa.eu/eli/dir/2022/2464/oj/eng (Accessed 09.06.2025)

European Commission, The European Green Deal (COM(2019) 640 final). https://eur-lex.europa.eu/resource.html?uri=cellar:b828d165-1c22-11ea-8c1f-01aa75ed71a1.0002.02/DOC_1&format=PDF (Accessed 09.06.2025)

European Commission, A new Circular Economy Action Plan: For a cleaner and more competitive Europe (COM(2020) 98 final), Brussels, 11 March 2020. https://eur-lex.europa.eu/legal-content/EN/TXT/?qid=1583933814386&uri=COM:2020:98:FIN, (Accessed 09.06.2025)

European Commission (2024) JRC, Ecodesign for sustainable products regulation: study on new product priorities. https://publications.jrc.ec.europa.eu/repository/handle/JRC138903 (Accessed 09.06.2025)

European Commission, Ecodesign for sustainable products regulation. https://eur-lex.europa.eu/legal-content/EN/TXT/PDF/?uri=OJ:L_202401781 (Accessed 09.06.2025)

Jensen SF, Kristensen JH, Adamsen S, Christensen A (2023) Digital product passports for a circular economy: data needs for product life cycle decision-making. Sustain Prod Consum 37:242–255

Miron R, Hulea M (2024) Digital product passport implementation based on hyperledger fabric technology. In: International conference innovation in engineering, Springer Nature Switzerland, Cham, 2024, 51–62

Monrat AA, Schelén O, Andersson K (2019) A survey of blockchain from the perspectives of applications, challenges, and opportunities. IEEE Access 7:117134–117151

Panza L, Bruno G, Lombardi F (2022) A collaborative architecture to support circular economy through digital material passports and internet of materials. IFAC-PapersOnLine 55(10):1491–1496

Pettenpohl H, Spiekermann M, Both JR (2022) International data spaces in a nutshell

Quiña-Mera A, Fernandez P, García JM, Ruiz-Cortés A (2023) GraphQL: a systematic mapping study. ACM Comput Surv 55(10):1–35

SMARTChain Project, SMARTChain Horizon Europe Project. https://www.smartchain-project.eu/ (Accessed 09.06.2025)

Sporny M, Longley D, Kellogg G, Lanthaler M, Lindström N (2014) JSON-LD 1.0, W3C recommendation 16:41

Tolcha YK, Park G, Kim D (2025) Building digital product passports: EPCIS 2.0 meets knowledge graphs. In: IEEE international conference on RFID (RFID), 2025, 1–6

A Manufacturer Experience: EPD as "PDF Anno 2018" to EPD as "DPP Anno 2025"

Knut Jøssang

Abstract This paper presents the journey of Pipelife Norway from publishing Environmental Product Declarations (EPDs) in static PDF format to enabling machine-readable Digital Product Passports (DPPs) in alignment with the 2025 digital construction landscape. It explores how EPDs, traditionally used for environmental transparency, are evolving into structured digital assets that integrate with Building Information Modeling (BIM) and support Life Cycle Assessment (LCA) and procurement. Through early adoption of data standards like ISO 23386, ISO 23387, and ISO 22057, Pipelife has structured its product and environmental data to enhance interoperability and automation across construction workflows. The case highlights technical and governance challenges—such as data quality, semantic consistency, and the "BIM gap"—that hinder industry-wide adoption. It also addresses the limitations of current Product Category Rules (PCRs) and the need for scalable EPDs for composite products. Ultimately, the study demonstrates that structured environmental data is critical to enabling environmental, transparent, and circular construction practices, and suggests that upcoming DPP requirements may serve as a unifying framework for digital transformation in the sector.

1 Introduction

Pipelife Norway (Pipelife n.d.) is a leading manufacturer of plastic pipe systems for infrastructure, buildings, and industry. As part of the Wienerberger group (Wienerberger n.d.), it combines local expertise with global experience to deliver innovative, environmentally focused solutions. With emphasis on quality, sustainability, and long-term performance, Pipelife supports Norway's shift toward resilient, future-proof infrastructure.

The construction and civil engineering sector is a major driver of global environmental challenges, responsible for significant resource use, emissions, and waste. Its impact spans land use, biodiversity, and water and air quality. As demand for new

K. Jøssang (✉)
Pipelife Norge AS, Surnadal, Norway
e-mail: Knut.Jossang@pipelife.com

© The Author(s) 2026

M. Traverso et al. (eds.), *Life Cycle Management from Global to Local*,
https://doi.org/10.1007/978-3-032-17987-6_7

infrastructure grows, so does the urgency to adopt more sustainable practices (World Economic Forum 2024).

The construction industry is digitally transforming how product information is created, managed, and exchanged. Traditionally shared via static PDFs, technical data is now shifting toward structured, machine-readable formats aligned with BIM and suited for digital workflows and automation.

For environmental data to support decision-making in construction, it must be available in the same structured, machine-readable formats as other product data. Although EPDs are key to assessing environmental impact, they are often isolated PDFs unsuited for digital tools. Embedding EPDs in standardized, BIM-aligned formats enables seamless integration into product selection, specification, and lifecycle analysis.

2 Background and Context

An EPD is a standardized document that communicates the environmental impact of a product throughout its life cycle, based on an LCA. It provides transparent, third-party verified data on factors such as greenhouse gas emissions, resource use, and waste generation. EPDs follow internationally recognized standards, including ISO 14025 (ISO 14025:2010) (Type III environmental declarations), ISO 21930 (ISO 21930:2017) (sustainability in construction works), and EN 15804 (EN 15804:2012) (defines core rules for construction products in Europe) (see Table 1). By providing a common language and structure, EPDs enable fair comparison and informed choices in environmental design and procurement processes.

2.1 Current Use of EPDs in Norway

In Norway, EPDs are widely used as documentation of environmental performance in construction projects, especially in response to increasing demands for climate- and resource-efficient solutions. Today, these declarations are primarily accessed

Table 1 Standards related to EPD in construction sector

Standard	Focus	Scope
ISO 14040/14044	LCA methodology	Global, all sectors
ISO 14025	Type III environmental declarations (EPD)	Global, all sectors
ISO 21930	EPDs for construction products	Global, construction sector
EN 15804	General rules for product category rules (PCR) for construction products	European standard, construction sector

and distributed as PDF documents, typically downloaded from the EPD program operator's database. While this ensures accessibility, the format limits integration with digital design tools and workflows. EPDs are often required in public procurement, by contractors, architects, and building owners aiming to meet environmental certification schemes [such as BREEAM-NOR, (Byggalliansen n.d.)] or national regulations. Institutions like Statsbygg (Statsbygg n.d.), Nye Veier (Nye Veier n.d.), and various municipalities increasingly mandate EPDs as part of tender requirements and climate accounting practices.

2.2 Digital Product Passports: Background and Purpose

DPPs are emerging as a key policy instrument under the European Green Deal and the EU's Circular Economy Action Plan. The concept is being formalized through the Ecodesign for Sustainable Products Regulation (European Commision n.d.), which will require certain product groups to provide standardized, digital information on composition, origin, environmental performance, and end-of-life handling. The primary goal of DPPs is to enhance transparency, traceability, and circularity across value chains and represent a significant step toward better lifecycle management and informed decision-making by enabling access to verified and structured product data. While the framework is still under development, pilot projects are already underway in several sectors, and implementation for selected product categories is expected to begin in the coming years.

2.3 Enabling Machine-Readable Product Information in the Construction Sector

In the context of DPPs for construction products, several international standards provide sufficient foundation for creating structured, machine-readable information. ISO 12006–3 (ISO 12006–3:2007) defines a framework for building object-oriented data dictionaries, enabling consistent classification and property referencing. ISO 23386 (ISO 23386:2020) specifies how to define and manage properties in data dictionaries to ensure semantic clarity and interoperability. ISO 23387 (ISO 23387:2020) builds on this by establishing Data Templates (DTs) that describe how product-specific data should be structured and exchanged digitally. These standards are crucial for aligning product data with BIM processes and ensuring compatibility with future DPP requirements. They enable manufacturers to digitize and share product information in a standardized way, facilitating integration into design tools, procurement systems, and regulatory compliance frameworks.

2.4 Linking EPDs and BIM: The Role of ISO 22057

To bridge the gap between EPDs and digital construction processes, ISO 22057 (ISO 22057:2022) provides a standardized approach for integrating EPD data into BIM environments, enabling the environmental characteristics of products to be used directly in digital planning, modeling, and analysis. By connecting EPDs with the data templates and dictionaries defined, ISO 22057 ensures that environmental data is both technically interoperable and practically usable across digital construction workflows (see Table 2).

3 Methodology

This study follows a qualitative, experience-based case study approach, grounded in the author's role as a professional at Pipelife Norway. The objective is to explore how digital environmental data and relevant data standards can be implemented in a real-world industrial setting. The paper is based on insights gained through several years of professional work with EPDs, data management, and digital product data modeling within the company.

The research does not involve external data collection such as surveys or interviews. Instead, the analysis is based on practical experience, internal documentation, structured product information, and direct involvement in pilot projects aimed at testing and applying emerging standards in a manufacturing context. The author has actively worked with the implementation of ISO 23386, ISO 23387, and ISO 22057, by evaluating and applying these standards to structure product and environmental data within the company's systems and workflows.

Key sources include project documentation, digital tools and platforms used in the data structuring process, feedback from industry partners and clients, and published EPDs. Publicly available material such as regulatory frameworks, standard texts, and sector-specific guidance documents have also been consulted to contextualize the case and assess alignment with ongoing industry developments.

Table 2 Standards linking EPDs and BIM in the construction sector

Standard	Focus	Scope
ISO 12006–3	Framework for building information—ontology and data dictionaries	Global, BIM, and data structure
ISO 23386	Attributes and properties in data templates—rules for creating and maintaining them	Global supports consistent data modeling
ISO 23387	Structure and content of data templates for construction objects (products, systems)	Global enables machine-readable BIM data
ISO 22057	Linking EDPs to BIM—digital environmental data integration	Global bridges LCA/EPD and BIM

The analytical approach is thematic and interpretive. Observations and documentation were reviewed with focus on recurring issues related to data consistency, system integration, and semantic alignment. Patterns and lessons learned were organized into key themes that form the basis for the results. The study emphasizes reflective practice and is not intended to be generalizable, but rather to offer transferable insights relevant to other manufacturers and stakeholders navigating similar digitalization processes.

4   Case: Pipelife Norway

4.1   *Pipelife Norway's History with EPDs and LCA*

Pipelife Norway started publishing verified EPDs in 2018 for four product groups from one facility. Since then, it has expanded to cover its entire portfolio across three sites: Surnadal, Stathelle, and Ringebu. To ensure high quality, traceable LCA data, the company systematically collects data on electricity and water use, raw materials, internal diesel consumption, production volumes linked to detailed bills of materials (including packaging), and transport distances and modes for materials and packaging.

Pipelife uses a verified EPD generator developed with LCA.no (LCA.no n.d.) for efficient EPD creation and third-party verification. It also produces project-specific EPDs reflecting actual material use and site-specific transport. These practices establish a transparent, robust environmental documentation process supporting improved climate performance throughout the value chain.

4.2   *Implementation of BIM-Compatible Digital Product Data at Pipelife Norway*

In 2018, Pipelife Norway entered a strategic collaboration with Cobuilder (Cobuilder n.d.) to begin structuring product information in alignment with emerging data standards for the construction sector. This work started before the publication of ISO 23386 (ISO 23386:2020) and ISO 23387 (ISO 23387:2020) in 2020, which define how product properties are managed in data dictionaries and how data templates for construction products should be structured. Pipelife actively participated in pilot projects during the standards' development, enabling early alignment with the principles of structured, machine-readable product data.

As part of this effort, Pipelife began using the data dictionary platform Define (Cobuilder n.d.), developed by Cobuilder, to standardize product properties. Product information was enriched and maintained in a dedicated database integrated with Define, ensuring consistency across the product portfolio.

At this early stage, environmental data from EPDs had to be manually extracted from PDF documents, and relevant properties were created and linked to individual products. Following the publication of ISO 22057 (ISO 22057:2022) in 2022—which defines how EPD data can be structured for integration into digital workflows—Pipelife updated its approach to align with the standard, improving interoperability and data usability. Alongside this effort, Pipelife Norway has contributed to the development of a national practice through active participation in the product domain group in buildingSMART Norway (buildingSMART Norway n.d.). Through this involvement, the company has both shared practical experience and acquired insight into the implementation of open standards for digital construction, supporting broader industry alignment. Since then, Pipelife Norway has scaled this structured data approach across all product groups, creating a comprehensive, digital product catalogue that supports both internal processes and external data exchange. This foundation is crucial to enable digital workflows in construction. It also positions Pipelife to meet future requirements such as DPPs.

5   Discussion: Challenges and Solutions

5.1   Benefits, Limitations, and Technical Challenges

The primary purpose of EPDs is to enable objective comparisons of environmental performance, allowing stakeholders in construction projects to make informed and environmental product choices. When EPDs are available in structured, machine-readable formats, they can be integrated directly into digital workflows and used during the design phase—not just as documentation after the fact. This enables better decision-making based on environmental performance and supports automated climate calculations, material optimization, and compliance.

While the PDF format ensures accessibility and third-party verification, it significantly limits the usability of EPDs in digital workflows. PDF files are static documents that are not suitable for search, machine processing, or integration with design and modeling tools. These files often exist in isolation from digital systems, meaning users may not be notified when an EPD becomes outdated or is withdrawn. Furthermore, PDF-based EPDs are rarely linked directly to digital product catalogues or building information models, which makes automated use and integration difficult. A single PDF document often covers an entire product group or a wide range of product variants, requiring manual adjustments to scale and adapt the data to the specific product used in a project. Another critical limitation is that the manufacturer, who is typically responsible for the entire EPD, usually has detailed knowledge only of the production stage. Other life cycle stages—such as transport, use, maintenance, and end-of-life—often rely on assumptions or generic default values that may not reflect the actual conditions in a given project.

Despite the availability of technical solutions, there are substantial practical and structural challenges that hinder the digitalization of environmental data and product information in general in the construction sector. The industry remains highly traditional, with many outdated IT systems and a continued reliance on manual data exchange methods such as emails, spreadsheets, and static documents. A key barrier is the so-called "BIM gap," which divides the value chain in two. On one side are clients, contractors, and designers—actors increasingly working in BIM environments. On the other side are manufacturers, wholesalers, and logistics providers—who are often disconnected from BIM workflows. This disconnect prevents product data, including environmental data, from flowing seamlessly into digital models and decision-making tools. In addition, there are significant issues related to semantics, data quality, and governance. Many attempts to implement ISO 22057 (ISO 22057:2022) take shortcuts, due to a limited understanding of the foundational standards ISO 23386 (ISO 23386:2020) and ISO 23387 (ISO 23387:2020). These define how product properties should be structured, governed, and reused across data dictionaries and templates. Without a shared semantic framework and proper governance, the industry risk being populated with incomplete, inconsistent, or unusable data. The lack of governance—that is, clear roles and responsibilities for defining and maintaining data—remains a key obstacle. For manufacturers like Pipelife Norway, success has depended on early investment in structured data systems and active participation in standards development. However, for much of the industry, this remains an emerging and underdeveloped practice, and success is dependent of collaboration across the whole value chain.

5.2 Need for Further Development of Product Category Rules

PCRs provide the methodological framework for developing EPDs and are essential for ensuring consistency and comparability within product categories. In the European context, PCRs are typically divided into a general Part A (core rules) and a product-specific Part B (additional rules tailored to specific product groups). In practice, however, Part B documents are often missing, outdated, or simply not used. Many EPDs are developed based solely on the core PCR (Part A), without the additional detail and specificity that Part B is intended to provide. This limits the accuracy and comparability of environmental data, particularly for complex or non-standardized products. One recurring challenge is the definition of the declared unit, which must be scalable, meaningful, and appropriate for the product type. In many cases, standard units such as "per meter with a given diameter" are poorly suited to actual production processes. Environmental impacts may be relatively uniform across diameters, making diameter-based comparisons misleading. For non-linear components—such as bends, junctions, or fittings—linear units offer little relevance. In such cases, weight is often a more practical and scalable basis for comparison but may not be accepted under the PCR Part B specification. Products with multiple functions also present challenges. For example, a pipe might serve drinking water, sewer, and cable protection purposes simultaneously, but current PCR frameworks often lack the flexibility to accommodate this multifunctionality in a standardized way. This could result in a need to follow multiple PCR Part Bs and in turn lead to the development of multiple EPDs for each product. Finally, many product groups still lack a dedicated Part B, especially in cases of specialized or innovative construction products. This creates a barrier for manufacturers wishing to develop EPDs covering the whole life cycle of a product, and further restricts the potential for integrating environmental data into digital tools and procurement systems. To fully support digitalization, accurate benchmarking, and lifecycle assessment, PCRs need to evolve. This includes both expanding coverage across more product categories, ensuring functional descriptions are aligned with actual production practices and product usage, and adjusting declared units to be scalable from product groups to specific products.

An especially complex area within EPD methodology and implementation involves composite products—configurations made up of multiple components or materials, often from different product categories, which are designed to function together as a single unit. In most cases, no single EPD covers the entire composite product, requiring the manufacturer to combine data from multiple EPDs to calculate the overall environmental impact. To produce meaningful and scalable results, composite products may require EPD data to be distributed across several components, each with its own declaration. Alternatively, manufacturers would need to generate individual EPDs for each product variant. However, the current business models of most EPD program operators make this approach prohibitively expensive. Fees and administrative effort typically increase with the number of EPDs issued, discouraging high-resolution, article- or SKU-level declarations—despite their importance in digital construction workflows and product-specific climate assessments. Without digital tools that support the aggregation, harmonization, and comparison of EPD data across components, transparency and comparability are weakened. As a result, environmental performance calculations for composite products risk becoming incomplete or inconsistent—especially in contexts where accurate and project-specific data is most critical. There is an increasing need for new tools, standards, and business models that better reflect how products are designed, specified, and assembled in construction projects.

5.3   Traceability Between PDF and Digital Format

For digital EPDs to be reliable and verifiable, clear traceability back to the approved and verified PDF version is essential. Machine-readable EPD data should always include references to the EPD registration number and expiration date, ensuring alignment with the officially published declaration. In this way, the PDF version functions as a certificate of verification, while the structured data can be used directly in digital tools and automated processes—without the need to open or manually inspect the document itself. Maintaining this level of traceability requires systems that support digital linking, version control, and validation workflows, ensuring that users can trust the integrity and source of the environmental data used in calculations, comparisons, and procurement decisions.

6   Application and Value Across the Value Chain

Machine-readable environmental data offers value far beyond compliance and documentation—it enables practical integration into planning, design, procurement, and operations across the construction value chain. When structured through standardized data dictionaries, such as those based on ISO 23386 (ISO 23386:2020), 23387 (ISO 23387:2020), and 22057 (ISO 22057:2022), this data becomes usable from the earliest project phases. Planners and clients can use digital tools to define environmental performance targets during design, ensuring requirements are clear, interoperable, and aligned with shared industry standards.

Structured EPD data can be embedded in BIM models, integrated into LCA tools, and linked to procurement and comparison platforms, allowing environmental characteristics to directly inform specifications, tender documents, and material choices. This supports holistic decision-making where environmental and economic factors are assessed side by side—especially relevant in public procurement with multiple policy goals. Crucially, shared data dictionaries ensure consistent semantics and governance, preventing fragmentation and isolated property definitions.

Best practice demonstrates the power of combining structured product data with openBIM workflows and transparent environmental benchmarks. Tools like Define connect EPDs to digital product libraries, making environmental information both accessible and actionable. Forward-leaning manufacturers are already leveraging project-specific EPDs and full BIM integration to align technical performance with strategic environmental targets, setting a model for the broader industry.

7   Live Demo

The live demo consists of three parts:

(1) Definition in a data dictionary: Demonstrates how a property is defined within a structured data dictionary and grouped under a set of standardized properties.
(2) Application in early project phases: Shows how this property is used during the early design stage to define and set environmental-related requirements.
(3) Manufacturer data enrichment: Illustrates how the same property is utilized by a manufacturer to enrich a specific product with verified environmental performance data.

8   Conclusions

EPDs are essential tools for communicating the environmental performance of construction products in a standardized and transparent way. However, to make them truly applicable across the construction value chain, EPDs must move beyond the PDF format and become part of structured, machine-readable product data ecosystems. This enables integration into digital workflows—from early design and procurement to lifecycle analysis and regulatory compliance.

Pipelife Norway has demonstrated how both environmental and technical product data can be structured and governed according to international standards such as ISO 23386 (ISO 23386:2020), ISO 23387 (ISO 23387:2020), and ISO 22057 (ISO 22057:2022). This holistic approach ensures that all product data—not just EPD content—is interoperable, scalable, and usable across BIM, LCA tools, and procurement systems. It represents a model for how manufacturers can prepare for more data-driven, environmental construction practices.

Nevertheless, the industry still faces major challenges. Many companies rely on outdated IT systems and manual processes. The so-called "BIM gap" separates data-rich planning environments from product suppliers and manufacturers, limiting data flow and coordination. Additionally, many EPDs are developed without using Part B of the PCRs, and declared units are often poorly suited to real-world product characteristics—especially for non-linear or multifunctional products.

Composite products add further complexity, often requiring EPD data from multiple materials or components to calculate overall impacts. While it is possible to create one EPD per article, a more efficient and scalable approach is to assemble EPD data at the material level—provided that program operators and digital tools support this.

Looking ahead, the construction sector must address fragmentation in how product data is structured and exchanged. There are too many local product data initiatives that lack impact on the industry, and software platforms from global vendors often fail to support openBIM principles or interoperability. This fragmentation leads to inefficiencies, inconsistencies, and missed opportunities for better environmental outcomes.

DPPs may become the mechanism that finally aligns the industry around a common framework for structured, standardized, and interoperable product data. As DPP requirements become more concrete, they have the potential to unify stakeholders across the value chain—manufacturers, software vendors, specifiers, and regulators—around shared data standards and governance. With the right policy, tools, and collaboration in place, the transition to digital, transparent, and circular construction can accelerate meaningfully.

References

BuildingSMART Norway. https://www.buildingsmart.no (Accessed 15.07.2025)

Byggalliansen. https://byggalliansen.no/sertifisering (Accessed 15.07.2025)

Cobuilder. https://www.cobuilder.com/define (Accessed 15.07.2025)

European Commission. https://single-market-economy.ec.europa.eu (Accessed 15.07.2025)

European Committee for Standardization, EN 15804:2012+A2:2019—Sustainability of construction works—Environmental product declarations—Core rules for the product category of construction products, CEN (2019)

International Organization for Standardization, ISO 12006–3:2007—Building construction—Organization of information about construction works—Part 3: Framework for object-oriented information, ISO (2007)

International Organization for Standardization, ISO 14025:2010—Environmental labels and declarations—Type III environmental declarations—Principles and procedures, ISO (2010)

International Organization for Standardization, ISO 21930:2017—Sustainability in buildings and civil engineering works—Core rules for environmental product declarations of construction products and services, ISO (2017)

International Organization for Standardization, ISO 22057:2022—Sustainability in buildings and civil engineering works—Data templates for the use of environmental product declarations in building information modelling (BIM), ISO (2021)

International Organization for Standardization, ISO 23386:2020—Building information modelling and other digital processes used in construction—Methodology to describe, author and maintain properties in interconnected data dictionaries, ISO (2020)

International Organization for Standardization, ISO 23387:2020—Building information modelling (BIM) – Data templates for construction objects used in the life cycle of any built asset—Concepts and principles, ISO (2020)

LCA.no, https://www.lca.no (Accessed 15.07.2025)

Nye Veier. https://www.nyeveier.no (Accessed 15.07.2025)

Pipelife Norway. https://www.pipelife.no (Accessed 15.07.2025)

Statsbygg, https://www.statsbygg.no (Accessed 15.07.2025)

Wienerberger. https://www.wienerberger.com (Accessed 15.07.2025)

World Economic Forum. https://www.reuters.com/sustainability/climate-energy/comment-why-circular-built-environment-makes-economic-environmental-sense-2024-02-23 (Accessed 15.07.2025)

How Unsuitable Data and Data Choices Can Negatively Influence Reliability and Costs of Decarbonization Efforts, Supplier Choices, and Sustainability Approaches in Companies

Martin Baitz, Tim Becker, Lionel Thellier Mercier, Cecilia Makishi Colodel, Matthias Rudolf, and Magnus Piotrowski

Abstract Reliable Life Cycle Assessment (LCA) depends on high-quality, fit-for-purpose data. However, data may be unsuitable due to poor quality (a) or inappropriate user choices (b), especially in decarbonization contexts. Existing standards and initiatives lack clear data quality definitions, and new data concepts risk misuse—while relatively harmless in academia may have financial and legal implications in industry. This study analyzes various data concepts and key risks: reliance on statistical vs. specific data, poor regional/technological modeling, greenwashing via gap closing, and issues with AI-generated or outdated data. To mitigate risks under "(a)," data must be practically relevant and continuously confirmed; under "(b)," users must understand its background and suitability. Findings show not all data concepts meet industrial and regulatory LCA needs and concluding recommendations are given.

1 Introduction

Life cycle assessment (LCA) is established as a scientifically robust tool to assess the environmental impacts of products and organizations (Finkbeiner et al. 2025). This notion is supported by stakeholders from industry, policy, and academia over the world (Baitz et al. 2013). LCA is supportive or compulsory in policies and regulations like Packaging and Packaging Waste Regulation, Corporate Sustainability Reporting Directive, and Ecodesign for Sustainable Product Regulation. Therefore, reliable results are needed more than ever. Differences in results by poorly defined, misunderstood, or unsuitable data must be identified and avoided to minimize bad consequences and to achieve reliable LCA results. Certain data can be unsuitable due to fundamental issues: (a) data may not live up to the needs to support quality results, or (b) choices of data may not support the quality and future requirements

M. Baitz (✉) · T. Becker · L. T. Mercier · C. M. Colodel · M. Rudolf · M. Piotrowski
Sphera Solutions GmbH, Stuttgart, Germany
e-mail: mbaitz@sphera.com

© The Author(s) 2026

M. Traverso et al. (eds.), *Life Cycle Management from Global to Local*,
https://doi.org/10.1007/978-3-032-17987-6_8

in, e.g., decarbonization. Results of LCAs are influenced by multiple factors: Suitability and justifiability of the methodological options and their proper declaration in goal and scope; by the appropriateness of the choices and assumptions applied by users, by the availability of data or efforts to get appropriate data, and by the engineering quality of data as such (age, regional and technological specificity, and comprehensiveness). We will focus on data quality aspects in this article. International standards and sector initiatives lack proper data quality definitions. Immature data concepts are brought to market, imposing considerable risks of misuse or poor results; they may be harmless in academic applications, but with significant financial, reputational, and legal consequences when used in cooperations or for decision making; e.g., misinformation of customers and regulators, failure to meet targets, loss of trust, or ineffective investments. The research question is: Which attributes must be fulfilled by LCA data, which data development processes must be implemented, and which information must be at hand of the user to prevent negative influences on the reliability of results?

2   Methods

The study was conducted as a desktop analysis, combining experience from in-house data developments, 30 years of in-house LCA projects, work in standardization and LCA Initiatives, literature, as well as frequent internal and 3rd party interviews with sector experts. The evaluated criteria originate mainly from (a) the review of similarities and differences in published LCA method guides and standards (see examples below), (b) from LCM presentations of LCA frontrunners in industry (e.g., VW 2023), (c) from invitations to surveys (e.g., UBA 2025) and (d) invitations to special sessions and panel discussions at LCA conferences (like "LCA Databases of the World" at EcoBalance Conference 2024 in Sendai, Japan.

The findings aim to lead practitioners toward proper decisions and reliable results, based on suitable data quality in the respective goal and scope. The study was performed on the basis of thousands of LCA projects and initiative involvements of Sphera Solutions (and predecessors, thinkstep/PE International), including publications and developments in LCA standardization. Sphera is at the spearhead of the LCA method and data implementation, because MLC data serves in major sectors as data backbone. Therefore, quality and method issues, induced by immature, unsuitable, or unclear approaches, will appear in our daily LCA data work and are aligned for consistent models (details see Modeling Principles [Sphera 2025]). The findings are based on analysis of LCA-related ISO standards, industry and policy initiatives (e.g., ISO 2006a, b, 2019; PlasticsEurope (2022), Catena-X 2023; TfS 2024, European Commission (2010a), European Commission (2010b), European Commission 2020) as well as on comparisons with the Sphera Databases (Sphera 2025), undertaken for associations, companies, and own Sphera tool and services use. Dependencies of data suitability and quality in relation to result reliability were assessed with existing LCA methods and practices. Issues need to be identified and recommendations given to avoid unintended or unidentified risks of impaired results. This is important for companies operating LCA in accountable business environments beyond the freedom of science. The influence of data choices on reliability and associated costs of efforts is discussed on the aspects: product/process specific vs. input–output data; foreground vs. background data/primary vs secondary data; unsuitable regionalization and technology; treatment of data gaps and avoiding green washing; AI and digital solutions in LCA; misinterpretation of updates and data age. In focus are LCA data attributes and user choices relating to business-critical consequences on results in industrial decision and reporting applications.

3 Findings and Outcome

This work is not intended to judge data quality, but to support users to double-check the practice of choosing and trusting data or emission factors. Only if users are aware of the type, adequateness, and quality of data, related risks in result reliability can be identified, mitigated, accepted, or avoided. How to cope with result reliability stays with users, and is not affected by formal compliance. However, it affects the business responsibilities of a company and may have legal implications (e.g., data sources and result reliability are of concern, if an LCA for procurement is leading to supplier contract cancelations—in favor of a seemingly "better" supplier). The main data foundation used is the MLC LCA database (Sphera 2024), which includes data collected and developed by Sphera and data of 50+ industry associations (e.g., like Plastics Europe, Worldsteel, and many more; details see Sphera (2025)).

3.1 Process-Based Data Versus Market-Mediated Input—Output Data

The success of LCA is based on standardization of the method, but also on the use of physical, engineering-like, and process-specific data, without any "virtual," "non-physical," or even "creative" way of data generation, application, or modification. Due to its coarse, macroeconomic data, the sector-based input–output (IO)-LCA data must be separated from specific process-based LCA data (Finkbeiner et al. 2025, Sphera 2025). Deviations to product and process-specific results are caused, because it is assumed that economic factors can be translated into environmental figures with the necessary precision and differentiation; even if reference years are often aged, as often IO-tables of the last decade are used. Further, simplified matrix structures and missing technological specificness are causing deviations to product and process-specific results. Company or product-specific aspects and supply chain-related innovations cannot be addressed because of the sector-based scope. Therefore,

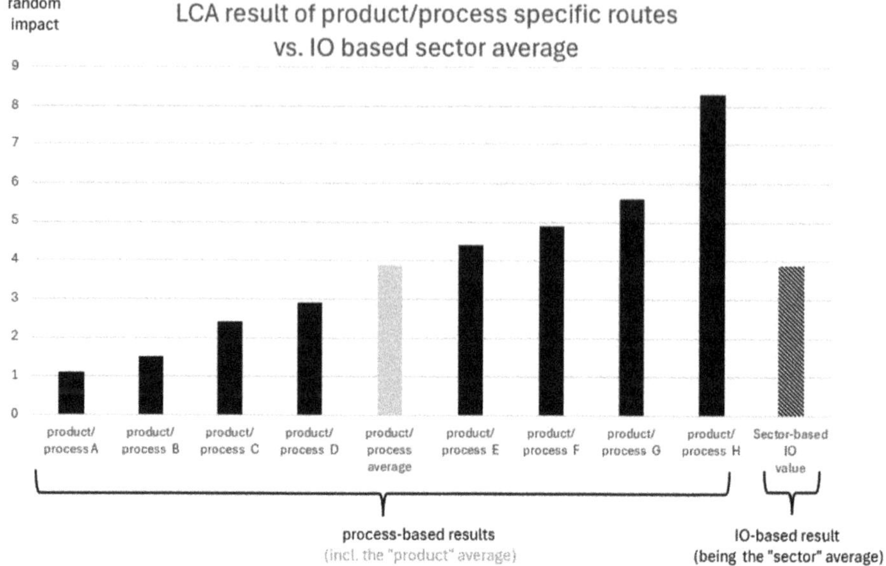

Fig. 1 Specific product and process-based data versus IO-based data

the variability of the results process based vs. IO based can be high, depending on the spread of technology and supply chain circumstances (See Fig. 1).

If IO-based data is used in a product or process-specific context (e.g., to compare product, process, or supplier options), the result would overestimate the real situation in cases A to D (imposing unnecessary burden or costs). In cases E to H, the real situation is underestimated (imposing a hidden burden and associated risks).

3.2 Foreground Versus Background or Primary Versus Secondary Data

The terms foreground/primary data as well as background/secondary data are partly used in different circumstances and meaning, being confusing and misleading. LCA primary data usually refers to process- or product-related data of the production process obtained from questionnaires, or enterprise resource planning (ERP) systems, as well as on-site measurements. Most guidelines encourage the use of primary data to ensure quality of the life-cycle system (e.g., PlasticsEurope 2022, TfS 2024, Baitz et al. 2013, and many more). Primary data typically define the foreground system of life cycle operations directly influenced by the operator, e.g., masses of materials, amounts of intermediates, utilities, energy used, and emissions of the production process. Background data typically refers to the data that the user or stakeholder has no direct influence, e.g., data on precursor production, energy supply, and emissions caused by suppliers and utilities. For these information, secondary data (e.g., from LCA databases) is commonly used. Secondary data can be a representative selection or average of certain released (direct or indirect) primary data pieces and supplemented by appropriate technical reference documents, manufacturer information, statistics, and appropriate literature sources. It is advantageous if at least the core foreground information is related to primary data. The energy, materials, and utilities needed can have a significant influence on the result: (a) via the influence on direct emissions and impacts of (foreground) operation on-site; plus (b) due to the determination of (background) upstream burden caused. Only suitable (primary) foreground data combined with consistent (secondary) background data leads to quality results. The data (collection) costs for "best available data" in a relevant decision context are far less than the opportunity cost based on a decision taken on data not fit for purpose.

3.3 Data in Relation to Regionalization and Technologies

Regionalization is important, as regions use different technologies having regional supply chains. However, technologies are easily adopted, sold, or licensed and applied to other parts of the world. Therefore, regionalization and technology specificity must be analyzed and modeled with care. Depending on the objective, averages derived from statistics, nameplate capacities, or other less relevant local information may not be sufficient. The "technological" situation in operation is important. For example, a steam cracker in the EU operates differently than in the US. While EU steam crackers operate primarily on naphtha, US steam crackers operate primarily on gas, resulting in differences in selectivity/severity of operations and a different product mix. User should be informed of regional differences in the upstream of the relevant process steps in the life cycle to reduce variability in the results. Supply chain data must respect the specific technological situation, production, and consumption conditions to maintain quality results.

Significant variations in results occur, even if the technology is known, but the region of production is unknown, or vice versa.

A whitepaper thinkstep (2012) analyzed more than 2000 production and synthesis processes. It showed that for Primary Energy Demand the variability in results at known technology but randomly chosen geography is -21% to $+27\%$, and for Global Warming it is -41% to $+70$ (concerning the 10 and 90% percentile). For other impacts like Eutrophication and Photosmog the variability of the results at known technology but randomly chosen geography is even higher.

The same whitepaper showed also that in the case of a known region of operation, but randomly chosen technology, the variability in results of Primary Energy Demand is -34% to $+52\%$ and for Global Warming -71% to $+248$ (concerning the 10 and 90% percentile). Other impacts like Eutrophication and Photosmog show a variability in the range of Global Warming.

Therefore, technology and geography of the supply chains must both be known for results of low "variability."

The results presented here were generated by an analysis in the GaBi-Master Database in 2012 (meanwhile called Managed Lifecycle Content (MLC) Database, Sphera (2024)), containing various datasets of the same kind of products, materials, and chemicals produced by different technologies in different regions.

3.4 Treatment of Data Gaps and Avoiding Green Washing

Data gaps are partly confused with cut-offs. Data gaps shall be "unavoidable" by definition, whereas cut-offs shall be made "purposeful" and must be justified. Data gaps pose a risk because of its "unknown relevance." Avoiding data gaps is crucial to avoid greenwashing or accusations (due to potentially relevant omissions). In LCA practice following approaches are usually used.

- Expert judgement or approximations (proxy)
- Worst case approach or precautionary principle approach

Figure 2 displays the potential differences of the approaches. If an expert judgement is used to close an LCA data gap, the expert (group) takes informal responsibility to make an "appropriate judgement" and the user informally trusts the expert (group) judgement. An approximation (or proxy) in the context of LCA data gap closing often is related to some physical, chemical, or engineering similarities. A rigorous way to treat gaps is the "worst case approach."

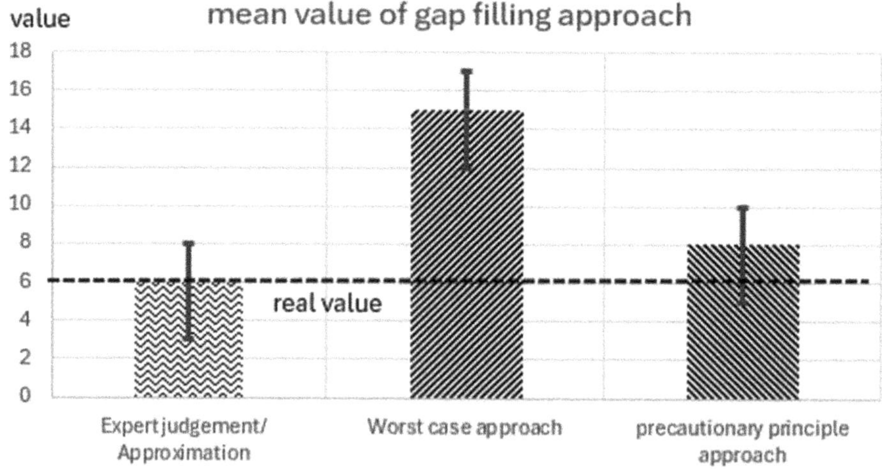

Fig. 2 Potential differences in gap closing approaches concerning the adequateness of the gap filling value in relation to the potential real value

It can have several levels of worst case assumptions (on LCI flow, on amount, and on characterization factor level), which can lead to overestimations. A suitable and justifiable way of closing data gaps is the so-called "precautionary principle approach" (mainly applied by Sphera). It can be understood as a combination of the approaches above: Industry-sector- and LCA expertise is connected to physical, chemical, or engineering similarities and in case of doubt rather a conservative (less optimistic) assumption is chosen. Expert judgment and approximation can have fuzzy distinctions, depending on the circumstances of the data gap. The assumptions may be interlinked between personal expert know-how and engineering similarities. However, if the assumptions are documented, the suitability of the gap closing can be assessed. If the "worst case approach" is applied on several levels, the result is likely to be overestimated. If, e.g., a process has an unknown chloro-organic emission, the "worst case approach" would call for the most impactful chloro-organic emission flow: R11 (trichlorofluoromethane), considering GWP as the major impact in this case. If further the emitted amount is unknown, the "worst case approach" would call on the highest known LCI emission level for chloro-organic emissions. This can be as high as 1E-04 kg emission or even 1E-03 kg emission per kg substance produced. This double effect has significant effects and may overestimate the impact by a factor of 20, let alone different GWP characterization factors. The "precautionary principle approach" aims to avoid such "systematic issues," but is still being on the conservative side of assumptions. The "precautionary principle approach" seems to be most suitable to avoid (unintended) greenwashing without unnecessary burden and costs in sustainability approaches and decarbonization efforts.

3.5 Artificial Intelligence (AI) or Digital Solutions in LCA Data

AI and automized digital solutions are not (yet) able to solve critical LCA data tasks; at least if "best available data quality" is in focus. If AI is used to predict data and is compared to industry data significant differences arise. AI is usually overestimating. AI can be useful for certain solutions around LCA; however, in the application of data generation AI faces several limitations. Selected search terms and literature, the lack of industry information for alignment, and missing background information are considered as main issues (Popowicz 2025). AI and other digital solutions can only focus on specific aspects, questions, or details. This means that fragmented details are combined into "LCA relevant information" without any "background check" of a sector or engineering expert. This imposes risks of inconsistencies due to different "sources". Applied in LCA projects AI can help to identify specific individual data pieces. Many specific examples are existing (see, e.g., Popowicz 2025). However, to generate or maintain LCA databases the specific information pieces must be combined systematically and consistently into LCA data. AI is not (yet) able to qualify the identified information for proper LCA usage. AI data solutions may be used for gap filling of less important data points. Caution is advised, if AI data is in the core of a data solution, just to avoid costly surprises due to lacking quality of data and results.

3.6 Data Update Cycles and Data Age

In changing societies and economies, the age of data is significant. The actuality of data has different components, and is often wrongly described or interpreted, having influence on quality results. Data age has at least three components:

- Underlying technical information pieces: Various years of the resource and energy usage, material and precursor consumption, waste generation, and emission release that form unit-processes involved.
- The reference year of the final dataset formed by the unit processes.
- The (re-)release date of the updated database including the final datasets.

Ignoring the statistical and market-mediated input–output data due to the reasons given above, the underlying technical information shall be of best available, different sources and compiled by sector experts, to ensure technical actuality and accuracy. To judge the time-related quality of data, solid sector expertise is needed. For example, the average age of a steam cracker in operation is 25–40 years, depending on the region. The technology is well optimized. Significant changes year over year are unlikely. This is in contrast to fast changing sectors like consumer goods and electronics, where materials and technologies have "innovation cycles" of 1 or 2 years, making constant monitoring and targeted update necessary. The reference year representing a supply chain is referring to data of different years (see, e.g., 12.2 Data quality aspects in European Commission (2010a)). The reference year must acknowledge the different actuality of underlying technical information and its relevance concerning the life-cycle and sectoral static or dynamic to represent a "weighted average" of the "most suitable" reference year. The date of the release of the data or database is just a formal time stamp, because it does not describe the time-related actuality of data. However, the (re-)release date of the databases is often documented as time-related quality indicators, which is wrong and frequently criticized in critical reviews as being inappropriate. The underlying technical information (please consult documentation and sector experts of your LCA database) and the adequate reference year of data are key time-related quality indicators.

4 Summary, Limitations, and Recommendations

We acknowledge that LCA is used in various contexts of academia, policy, consulting, and industry. Therefore, limitations apply. Specific situations apply, where some findings may not play a major role for the results quality. However, the specific situation should be described in goal and scope and reflected against the findings to transparently refute the relevancy of aspects mentioned above. Another limitation may be that not all aspects are covered, which influence the quality between data and results in business context and industrial applications. Companies are sometimes not willing, not able, or even not allowed to disclose all specific data (due to the competitive edge and competition laws). However, the findings cover the majority of cases in business and industry. The recommendations are meant to support using LCA, data, and results in business context of regulatory requirements and research and development.

- Quality LCA data needs to be collected, compiled, and provided not only with LCA know-how, but also with sector expertise and adequate data development, maintenance, review, and release procedures.
- Use a process-based LCA approach and related specific data.
- Use primary and complement with adequate secondary data.
- Acknowledgements adequate regionalization and technology specification.
- Avoid data gaps/greenwashing with a precautionary principle.
- Do not trust AI or auto-generated data without assurance of its quality.
- Check, assure, or choose data with adequate time-related quality.

It is an increasing error to only consider the (known or estimated) quantitative stochastic inventory data uncertainty and directly use this to demonstrate significance of differences in compared systems. The judgement of the overall data quality can ultimately only be done by expert judgement. Uncertainty calculations and qualitative or quantified accuracy assessment can substantially help but provide supporting, quantitative information only (European Commission (2010a)). Review and compare responsibly. This is particularly important when benchmarking, comparing, or testing different LCA software and databases, or if various published studies have different conclusions. To be protected from unjustified criticism and reporting inaccurate facts (with possible lawsuits), it is critical to thoroughly check the data quality, comparability, consistency, and the resulting conclusions.

References

Baitz M (2019) Responsibility in life cycle assessment practice. Int J Life Cycle Assess 24:179–180. https://doi.org/10.1007/s11367-018-1558-1(accessed31.05.2025)

Baitz M, Albrecht S, Brauner E et al (2013) LCA's theory and practice: like ebony and ivory living in perfect harmony. Int J Life Cycle Assess 18:5–13. https://doi.org/10.1007/s11367-012-0476-x(accessed22.05.2025)

Catena-X, Catena-X product carbon footprint rulebook CX-PCF rules, Version 2 (2023). https://catenax-ev.github.io/assets/files/CX-NFR-PCF-Rulebook_v.3.0-04874a80a6d27511df06e07ae3049278.pdf (accessed 22.05.2025)

European Commission, International Reference Life Cycle Data System (ILCD) Handbook—Specific guide for Life Cycle Inventory data sets, Joint Research Centre, 1st ed., 03/2010, Publications Office of the EU (2010). https://eplca.jrc.ec.europa.eu/uploads/ILCD-Handbook-Specific-guide-for-LCI-12March2010-ISBN-fin-v1.0-EN.pdf (accessed 23.05.2025)

European Commission, International Reference Life Cycle Data System (ILCD) Handbook—General guide for Life Cycle Assessment—Detailed guidance, Joint Research Centre—Institute for Environment and Sustainability, 1st ed., 03/2010, Publications Office of the EU (2010) https://eplca.jrc.ec.europa.eu/uploads/ILCD-Handbook-General-guide-for-LCA-detailed-guidance-12March2010-ISBN-fin-v1.0-EN.pdf (accessed 23.05.2025)

European Commission, Learning Materials, Webinar: Environmental Footprint (EF) Transition Phase, Impact Methods, Data Collection and Data Requirements (2020). https://green-business.ec.europa.eu/environmental-footprint-methods/learning-materials_en#paragraph_2381 (accessed 22.05.2025)

Finkbeiner M, Roche L, Holzapfel P (2025) From analysis-LCA to message-LCA: a lost cause? Int J Life Cycle Assess 30:803–810. https://doi.org/10.1007/s11367-025-02460-9(accessed22.05.2025)

ISO 14040 (2006) Environmental management—Life cycle assessment—Principles and framework. International Organization for Standardization, Geneva

ISO 14044 (2006) Environmental management—Life cycle assessment—Requirements and guidelines. International Organization for Standardization, Geneva

ISO 14067 (2019) Greenhouse gases—Carbon footprint of products—Requirements and guidelines for quantification. International Organization for Standardization, Geneva

PlasticsEurope, Eco-profiles program and methodology, Version 3.1 (2022). https://plasticseurope.org/wp-content/uploads/2024/03/PlasticsEurope-Ecoprofiles-program-and-methodology_V3.1.pdf (accessed 23.05.2025)

Popowicz M et al (2025) Digital technologies for life cycle assessment: a review and integrated combination framework. Int J Life Cycle Assess 30:405–428. https://doi.org/10.1007/s11367-024-02409-4(accessed28.05.2025)

Sphera (2024) LCA Master Database MLC (GaBi), CUP_2024_2, Sphera

Sphera, Managed LCA Content (MLC): LCA Databases Modeling Principles 2025 (2025). https://lcadatabase.sphera.com/dataset-documentation-download/?download=Download+Data (accessed 07.04.2025)

TfS, Together for Sustainability—The product carbon footprint guideline for the chemical industry (ver. 2.1). Specifications for suppliers' product carbon footprint calculation (2024). https://www.tfs-initiative.com/app/uploads/2024/03/TfS_PCF_guidelines_2024_EN_spreads-low.pdf (accessed 22.05.2025)

thinkstep, Koffler C, Baitz M, Koehler A (2012) Addressing uncertainty in LCI data with particular emphasis on variability in upstream supply chains, Whitepaper thinkstep AG. https://www.researchgate.net/publication/391901372_Addressing_uncertainty_in_LCI_data_with_particular_emphasis_on_variability_in_upstream_supply_chains_Whitepaper_Content_Content_Content_Content (accessed 20.05.2025)

UBA, German Environmental Protection Agency, Survey: Emission Factors for Carbon Footprinting in Companies/Erprobungsphase der Emissionsfaktorenliste für die organisationsbezogene THG-Bilanzierung (2025). https://www.soscisurvey.de/testphase_emissionsfaktoren/ (accessed 10.07.2025)

VW, Neef M, Fugger T, Dettmer T (2023) How to handle constantly changing life-cycle-based carbon calculation? Presentation of Volkswagen at LCM conference 2023, Lille, France

Toward Artificial Intelligence and Blockchain-Enabled Frameworks to Improve Critical Review Control and EPD Verification Process

Javier Martin Echazarreta and Claudia Peña

Abstract This paper explores the adoption of new technologies to improve consistent and reliable verification of environmental performance indicators of products and services. Life Cycle Assessment (LCA) Critical Review (CR) for Environmental Product Declaration (EPD) verification aimed at providing reliable information on environmental performance of products and services. There is a need to improve accuracy, facilitating comparability of results between EPDs, as well as ensure impartiality and transparency of the process. Using Artificial Intelligence (AI) and Machine Learning (ML) for auditing EPD verification processes can establish an effective framework for EPD Program Operators (POs) to validate the work of verifiers, improve the quality of EPDs, and promote continuous improvement and timesaving in the new era of massive EPD publication. AI and ML tools offer the ability to leverage vast datasets and address complex, multidimensional issues in LCA projects. These technologies can validate Life Cycle Inventory (LCI) data by identifying, isolating, and rectifying errors, inconsistencies, and outliers, as demonstrated in numerous case studies. Algorithms can also support the auditing of the CR process, correcting mistakes and preventing distortions by minimizing the risk of human error. The ML framework, aligned with ISO 14071 and ISO 17029 standards, enhances direct and efficient oversight of verifiers, maintaining the impartiality of the process and further reducing the potential for human mistakes. This integration not only improves accuracy but also streamlines the verification process, offering a more reliable and objective assessment of environmental performance. Furthermore, the paper emphasizes the necessity for POs to integrate these new tools to keep pace with emerging trends and future challenges, making EPDs remain effective and relevant.

J. M. Echazarreta (✉)
Instituto Nacional de Tecnología Industrial (INTI), Buenos Aires, Argentina
e-mail: jechazarreta@inti.gob.ar

C. Peña
PINDA LCT SpA, Santiago de Chile, Chile

M. Traverso et al. (eds.), *Life Cycle Management from Global to Local*,
https://doi.org/10.1007/978-3-032-17987-6_9

1 Introduction

An Environmental Product Declaration (EPD) is a standardized document that shows the environmental performance, e.g., emissions to the end user. An EPD shall be verified by third-party verifier in compliance with the ISO 14025 standard, General Programme Instructions (GPI), and a Product Category Rule (PCR) specified and published by an EPD Programme Operator (PO). In most cases, the EPDs are a pdf file; however, some operators modify the pdf file through an online platform. EPDs are published and registered on the website of the Programme Operator, who sets the rules for the verification and publication of EPDs (ISO 2006).

EPDs are critical instruments used to communicate a product's environmental impacts transparently. Traditionally static PDF documents, their verification requires compliance with international standards (e.g., ISO 14025), GPI, guided by Product Category Rules. However, the exponential growth in data and increasing demand for transparency have rendered manual verification processes inefficient and vulnerable to errors.

EPDs are required in the construction sector for certifying buildings, and as described by Brisson et al., they are recognized around the world as a means of assessing the environmental performance of buildings and facilitating product comparisons. The importance of using EPDs is therefore increasing (Stapel et al. 2024).

The digitization of the EPD format not only improves the effectiveness of EPDs, but also brings significant benefits in terms of accessibility, integrity, interoperability, standardization, and data exchange. It makes it possible to display them in a simple electronic format, the process launched in 2020 leading Eco-Portal displaying all digitized EPDs. The ILCD (International Life Cycle Data) + EPD format developed by the Indata network was considered the most suitable for digitization and included all the information contained in an EPD. Digital transformation—through standardized formats like ILCD + EPD and platforms like ECO Platform—has provided interoperability and data harmonization. Yet, challenges remain in ensuring consistent, high-quality, and scalable verification practices. Artificial Intelligence (AI), machine learning (ML), and blockchain offer a pathway to reimagine EPD verification, ensuring accuracy, transparency, and trust (Pannuti 2023).

Accompanying digitalization, Blanco et al. consider that ML will also tend to drive results toward expected values or ranges that are largely based on past data. Further, digitization is an essential step in the application of AI and ML to a life cycle assessment (LCA) project or verification process (Blanco et al. 2024).

In addition, the manual extraction of data from PDF format requires a resource-intensive task with the risk of introducing human error, confusion, and loss of environmental performance data. To avoid this, most of the EPD programme operators are contributing to a database such as ECO Platform and InData Network to drive digital transformation, reducing costs and human risk of data manipulation (Stapel et al. 2022).

One of the difficulties of digitalization is the different electronics formats no standardized, thus the ECO Platform has developed a web application programming interface (API) where each PO can upload the different results from each EPD to have the data harmonized and available for consultation in the same format. Anyone, who wishes to obtain information from the EPDs and work with the data, should download them from the Eco Platform website (Stapel et al. 2022).

While digitization alone is not enough to improve data quality, digitizing EPDs can help to improve data quality through AI in conjunction with the verification process. Furthermore, machine learning is a subset of AI and involves different mathematical and statistical algorithms linked with an existing data set, the statistical algorithms include, e.g., linear regression, cluster, and principal component as part of the analysis to help the verification process. Therefore, machine learning can be used not only in the LCA project as many papers described, but also it will be used in the critical reviews (Koyamparambath et al. 2022; Rao et al. 2017).

As Ibn-Mohamed et al. contemplated, the complex relationship between environmental factors, emissions, and impact categories, including global warming potential (GWP), acidification potential, eutrophication potential, and human toxicity, can be considered by AI in the environmental impact assessment. As Ibn-Mohammed et al. describe the integration of AI techniques into LCA can support data collection, modeling, analysis, monitoring, and presentation in the stages of a product's life cycle assessment (Ibn Mohammed et al. 2023). In addition, the new techniques can help identify potential errors and deviations in the LCA that may not have been previously identified by the verifier in the first step of the verification procedure.

As Pannuti (2023) considers, the environmental performance of building and infrastructure design can benefit from the EPD digitization process. However, the digitization process is necessary to start using AI in the EPD databases to support the verification process and transform it from a human activity into a hybrid process where human activities may be complemented by AI to improve the quality of data.

As Bogani et al. consider the Artificial Intelligence encompass in most aspect of our lives, however, ethical legal and safety and morals concerns should be integrated with our lives to improve the human productivity and the quality of the works (Bogani et al. 2022).

In addition to improving the quality of EPDs, Gailhofer et al. suggest that AI could be used to improve the detection of fraudulent ecolabels and product declarations. Such use would require third-party certified ecolabeling to start using AI in their procedures for tracking labeled products (Gailhofer et al. 2021).

The important role of AI in the sustainability approach could help address global environmental challenges. However, it will raise unique ethical, legal, and philosophical challenges that need to be addressed (Yigitcanlar 2021).

As Bogani et al. consider AI encompasses in most aspects of our lives, however, ethical legal and safety and morals concerns should be integrated with our lives to improve human productivity and the quality of the verification process. This working paper proposes a framework for the EPD PO to adopt, incorporating new technologies to ensure consistent and reliable product and service assessments. The hybrid verification process will ensure comparability and impartiality of results, reducing the time taken and improving the reliability of EPDs.

2 Methodological Framework

A hybrid framework integrates artificial intelligence, machine learning, and blockchain technologies to modernize and improve the EPD verification process, replacing the current human verification into a hybrid verification process.

The articles identified through Google Scholar were tracked and cited. Those published since 2014 that applied AI to LCA studies and the verification process were identified and screened based on their abstracts and full texts. The current EPD verification process of the International EPD System (IES) was selected as an example module to expand the hybrid process into a real process, based on GPI 5.0.1.

The studies from the LCA and AI perspectives were finally thoroughly evaluated, including those from the ML perspective. The study particularly examines the use of generative and discriminative AI to improve data quality, reduce human error, and strengthen auditability within the context of ISO 14025, ISO 14071, and ISO 17029 standards. It also addresses challenges in ML model development, data bias and proposes a robust methodology for deployment, governance, and sustainability. A deeper focus is given to the contextual variability in EPD results, the attributional nature of LCA models, and limitations to collaborate with the verification process to hybrid framework to reduce the human errors.

2.1 Scope and Design

The research adopts a mixed-method approach integrating:

- Literature review on AI/ML in LCA/EPD contexts from the last 10 years: a comprehensive literature survey of peer-reviewed articles was conducted to assess the current state of machine learning applications in life cycle assessment and their integration in EPD workflows, focusing on automation, accuracy, and standardization.
- Gap analysis of current EPD verification and the hybrid verification process against ISO frameworks: verification processes were benchmarked against ISO 14025, 14071, and 17029 to identify bottlenecks, human limitations, and areas where AI could enhance impartiality, reproducibility, and data transparency.

The focus of this paper is to derive the potential inclusion of AI in the current verification process through the development of a hybrid verification process to improve the current methodologies used by different operators such as IES.

2.2 Workflow Components

The different workflow components not only incorporate AI, but also include blockchain architecture for verification event logging: a blockchain-based system is proposed to capture every step in the EPD verification process, ensuring data provenance, immutability, and third-party validation compliance.

The principal workflow components are:

- The AI/ML model development life cycle should include the data acquisition from existing EPDs, training the model, and it should be validated from the same PCR and complementary Product Category Rule (c-PCR).
- Blockchain architecture for verification event logging.
- Validation frameworks for EPD databases include contextual harmonization methods.

3 Challenges in ML Model Development for EPD Verification

3.1 Data Quality and Annotation Bias

AI model performance in EPD verification is critically dependent on the accuracy and representativeness of the data used for model training. LCA and EPD datasets often originate from diverse and heterogeneous sources, such as regional databases, company-specific inventories, and industry-average data. This heterogeneity may lead to varying levels of granularity, conflicting assumptions, and divergent system boundaries.

The representation of the results from the AI will depend on the:

(1) Heterogeneous sources (e.g., national databases, industry reports): these introduce inconsistencies in functional units, time periods, and background data assumptions, making harmonized training difficult.
(2) Missing metadata or inconsistent units: gaps in metadata or poorly documented data quality indicators hinder effective normalization and model generalization across EPDs.
(3) Annotation bias, due to subjective human interpretation of product systems: This occurs when different practitioners interpret or define product systems differently, impacting the consistency of supervised learning outcomes.
(4) Compliance of the EPD with the GPI, PCR, and c-PCR, because it tends to harmonize the results of the data.
(5) Anomalies or outliers in data that are inconsistent with other data from the same product performed by the same GPI, PCR, or c-PCR.

Despite harmonized Product Category Rules (PCRs), EPDs show contextual differences rooted in regional practices, supply chain configurations, and local LCI datasets. These nuances highlight the critical role of domain-specific human oversight, as automation alone may not adequately reflect the real-world variability or the nuanced boundaries of attributional LCA, and the EPDs may show results aggregated (Köck et al. 2023; Parvatker and Eckelman 2019).

Elouariaghli et al. believe that deep learning and machine learning will be able to improve current databases and generate others for more or less direct use. Thus, improving the quality of databases through better Critical review (CR) is likely to improve EPDs (Elouariaghli et al. 2022).

The algorithms are designed to explicitly isolate anomalies rather than profile normal instances. Two quantitative properties are used: (i) they are in the minority and consist of a smaller number of cases, and (ii) they have attribute values that are very different from those of normal instances (Liu et al. 2008).

As most program operators have a large database where they can search for anomalies in new records or draft EPDs, when AI checks and compares them with the database, the results will be accepted or rejected if an anomaly appears. According to Samariya et al., an outlier is a data point that does not match the rest of the data. One option is to use anomaly detection algorithms based on isolation, such as the isolation forest algorithm (Samariya and Thakkar 2023).

3.2 Algorithm Selection

Choosing the appropriate machine learning algorithm is highly dependent on the type of data, the verification task at hand, and the stage within the EPD lifecycle being addressed. Different ML methods serve distinct roles and some of them in supporting the verification process, from anomaly detection and impact assessment validation to textual analysis of verifier comments. The following algorithms are among the most commonly used, with the following applications:

- Random Forests and SVMs perform well in structured tabular data (e.g., impact indicators): These algorithms are ideal for tasks such as classification of environmental performance levels or detection of outliers in GWP or energy use metrics, where input features are well-structured and labeled.
- Isolation Forest (iForest) can be applied to LCA databases, particularly for anomaly detection, such as identifying outlier processes, suspicious inventory data, or inconsistent environmental impact values.
- Neural networks (CNNs, LSTMs) are suitable for unstructured texts (e.g., reviewer comments): These models help in interpreting textual data from CR reports, enabling automated flagging of inconsistencies or omitted assumptions.
- Generative models (LLMs, GANs) can simulate missing data, reconstruct incomplete EPDs, and generate counterfactuals for review: Such models are especially useful when input data is sparse or when scenario testing is required to estimate ranges of impact.

However, as Neupane et al. (2025) conclude, the algorithm efficacy depends on the life cycle stage being analyzed (e.g., inventory vs. impact assessment) and contextual validity must always be preserved: Algorithms must be chosen to suit the data phase—e.g., inventory generation versus final indicator aggregation and must undergo strict contextual calibration to ensure methodological appropriateness; based on use-case (Neupane et al. 2025).

Meanwhile, Martínez et al. believe that machine learning models can efficiently handle large datasets and complex systems with rigorous model equations. iForest is distinguished from existing model-based distance-based and density-based methods in the following ways; (i) the isolation nature of iForest allows them to build partial models and exploit subsampling to an extent not possible with existing methods; (ii) iForest utilizes no distance or density measures to detect anomalies; (iii) iForest has a linear time complexity with a low constant and a low memory requirement; and (iv) iForest has the ability to scale up to handle extremely large amounts of data and high dimensional problems with a large number of irrelevant attributes (Martínez-Ramón et al. 2024; Liu et al. 2008).

4   Role of Generative AI in Verification Process

The verification process, defined by ISO 14025, has specific requirements to ensure the credibility and transparency of the environmental performance presented in an EPD. These requirements are essential to establish that the environmental information on a product is accurate, reliable, and compliant with the standard.

4.1   Traditional Framework—IES Current Verification Process (Human)

Here's a breakdown of the key requirements for the verification process in a traditional verification according to ISO 14025, 14071, and GPI 5.0.1 from IES:

(1)   Considerations from the independence of the Verifier

Third-party verification: The verifier shall be an independent third party with no direct involvement in the development of the LCA project or the preparation of the EPD and no conflict of interest with the organization submitting the LCA project & EPD for verification. The verifier shall act with impartiality and objectivity to ensure that the environmental claims are verified fairly and accurately in accordance with the principles of GPI 5.0.0 Sect. 8.

(2)   Competence of the Verifier

According to GPI 5.0.0 Sect. 5.10.1, verifiers must be independent and demonstrate a broad set of competencies. They should have general knowledge of environmental issues relevant to industry and products, along with specific process and product expertise, including familiarity with applicable standards in their verification sector. They must be well-versed in LCA and relevant standards, including ISO 14040, 14044, ISO 14020, ISO 14025, ISO 14071, and EN 15804. Additionally, they need knowledge of the International EPD System, its GPI, and any national or regional licensees involved. Practical experience in reviewing LCAs or verifying EPDs, or similar work, is essential. Finally, verifiers must be proficient in English to comprehend relevant documentation and to produce verification reports. There are also further requirements specific to the type of third-party verifier: individual verifiers and Accredited Certification Bodies.

(3)   Review of the Environmental Product Declaration and the methodology used

According to the main objectives of verification described in the GPI 5.0.1 Sect. 8.2.1. The EPD owner uploads the EPD and Excel file of the LCA results (if a machine-readable format is opted for) to the EPD Portal of the PO. The verifier shall do a verification of the accuracy of the life cycle data used to assess the environmental impacts (e.g., energy consumption, emissions, resource use) for both mandatory and optional indicators. The correct application of LCA methodology and calculations, ensuring that the LCA has been carried out in accordance with ISO 14040 and 14044, GPI, PCR, and c-PCR. The transparency of the EPD, ensuring that all data sources are clearly identified and that any assumptions or limitations in the data are disclosed, as well as compliance with environmental standards. In addition to the methodology, the verification includes checking that the company has used consistent functional units and appropriate impact categories (e.g., climate change, resource depletion, etc.), and the system boundaries of the product (what is included in the LCA) should be clearly defined.

(4)   Confirmation of Data Quality

The verifier shall assess the quality of the data used in the EPD, ensuring that the data are relevant to the product and system boundary. Reliable and obtained from trusted sources (e.g., peer-reviewed databases or industry-specific data). Any exclusions should be justified, and the verifiers should review the plausibility and accuracy of both the LCA and EPD by assessing precision, completeness, consistency, reproducibility, sources, and uncertainty of the presented information and data. The verifier should also confirm that the EPD owner has established internal follow-up procedures for EPD updates during its validity period, to reflect current environmental conditions and practices. The verifier should also check that the assumptions, estimates, or proxies used in the assessment are reasonable and adequately documented, including geographical coverage, period, and technology coverage.

(5)   Adequacy of Environmental Claims

The verifier shall ensure that all environmental claims made in the declaration are truthful, based on accurate data, do not mislead the public, and meet the requirements in ISO 14021 and national legislation.

(6)   Transparency and Clarity

The EPD needs to be written and presented in a way which is clear and understandable to all stakeholders (consumers, regulators, etc.).

(7)　EPD verification report

The verifier shall complete and upload the EPD verification report based on the EPD verification report template published by the PO. For construction products, the EPD verification report template is mandatory to use as it complies with ECO Platform verification checklist. Additionally, the dialogue between the verifier and EPD owner (or LCA practitioner) from the verification procedure, regarding, e.g., discrepancies or non-conformities discovered during the verification, shall be documented and included in the verification report.

The verifier shall upload the completed EPD verification report to the EPD Portal in the verification process.

A schema describes the sequence of the current verification in Fig. 1.

Besides the previous description of a traditional and current CR, the new framework is fully compliant with ISO standards and GPI of IES; however, it will improve and ensure the process by incorporating the tools of AI at different points transform traditional verification into hybrid verification process. Below we describe the AI tools to be applied in each of them.

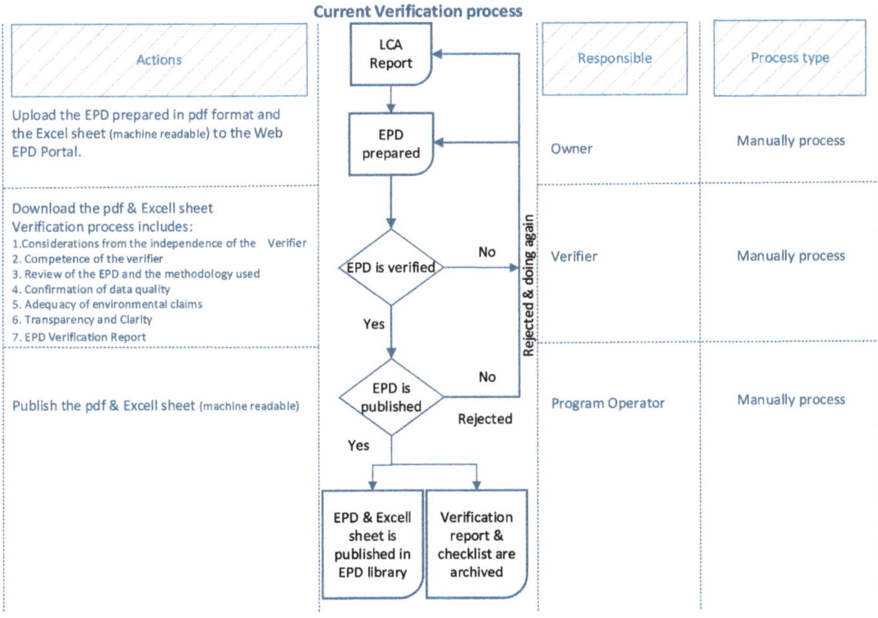

Fig. 1 Flowchart from current verification process

4.2   New Framework—Hybrid Verification Process

To adopt a hybrid process, the system shall be modified in the next steps of the verification process:

(1) The review process of Confirmation of Data Quality, the Review of the EPD, the methodology used, and the Adequacy of Environmental Claims

The verification including the confirmation, accuracy of the life cycle data used to assess the environmental impacts, and the adequacy of Environmental claims shall be done by the Verifier and the AI separately. The tool suggested to make a verification of accuracy of the life cycle data from the EPD, according to the bibliography, could be Isolation Forest to detect anomalies in the new EPD, and the results shall be incorporated as an AI Control Report, which the Verifier could access to validate it with the manual verification report. If a potential deviation is also called an anomaly detected, the Verifier rejects the EPD and asks EPD owner to adjust it and replace the wrong data or confirm it, justifying the deviation with documents to the verifier.

(2) Verification Checklist, Verification Report, and AI Control Report

Finally, the verification checklist shall be filled by the Verifier and the AI application separately; therefore, the PO shall control any potential deviation between the Verifier's report and the AI control report. Finally, if no differences are found between them, the EPD will be published. If some differences are detected, the EPD will be rejected in order to solve any potential deviations.

However, for any reason there is a potential deviation from the existing information in the different databases, and it was checked. The operator and verifier shall be notified by the Owner and justified in the EPD to be published. To safeguard the immutability of the information the AI Report and each step shall be registered in the Blockchain.

A schema describes the sequence of the hybrid verification process (See Fig. 2).

Thus, AI should support the identification of anomalies and data gaps, suggest likely corrections, and summarize documentation, while human experts must verify the alignment of LCA model structures with PCRs, validate regional assumptions, and ensure methodological appropriateness.

All the steps by a PO to adapt AI and ML to modify the current verification process to a hybrid verification process. For example, a flowchart adapted from Elsaid et al. has been designed to use the Isolation Forest for the hybrid verification process. It shows how a PO could use AI—ML to detect anomalies in the EPDs before they are published (Elsaid and Binbusayyis 2024) (See Fig. 3).

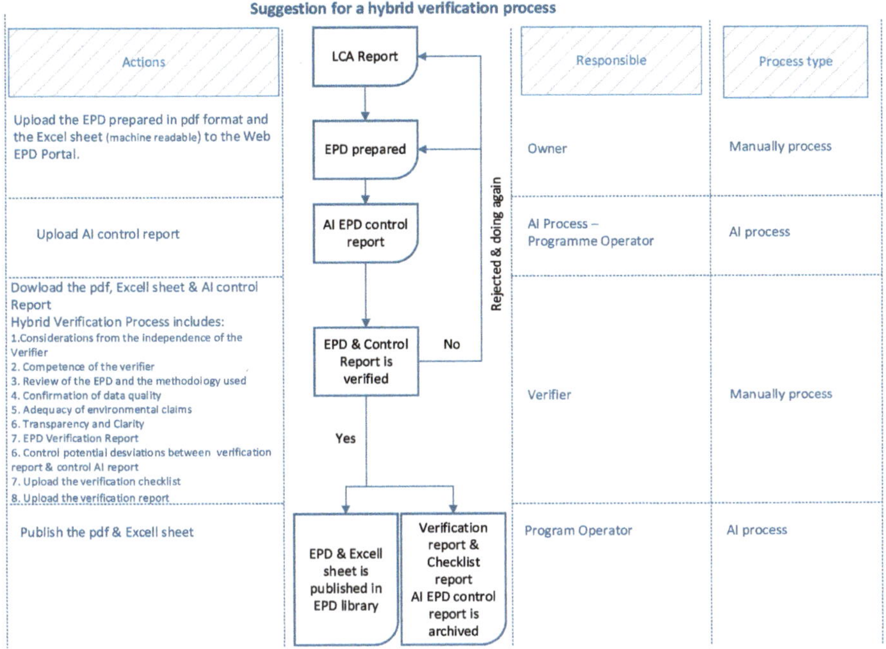

Fig. 2 Flowchart from current verification process

5　Discussion

The integration of artificial intelligence with Isolation Forest and blockchain technologies into the EPD verification process presents a transformative opportunity to improve transparency, reliability, and scalability, characteristics from blockchain described by Olanrewaju et al. AI can mitigate human error through rule-based automation and ML-driven logic checks, reducing the cognitive burden on critical reviewers and minimizing inconsistencies in assessments and detecting anomalies. This is especially critical when reviewing large-scale databases where manual validation is impractical (Olanrewaju et al. 2022).

Anomaly detection is another key benefit, as fairness-conscious algorithms can be trained on diverse, multi-source LCA datasets to identify anomalies and limit systemic bias introduced by regionally or sectorally biased data. However, these benefits are dependent on properly curated training data and robust oversight of the machine learning models applied.

In terms of scalability, AI facilitates automated parsing, normalization, and cross-verification of EPDs, even across different operators and geographies. This capability is essential for platforms such as InData and ECO Platform, which host thousands of digital EPDs and require consistent quality control to maintain public confidence. As a large database is required to use iForest, a PO, e.g., International EPD System

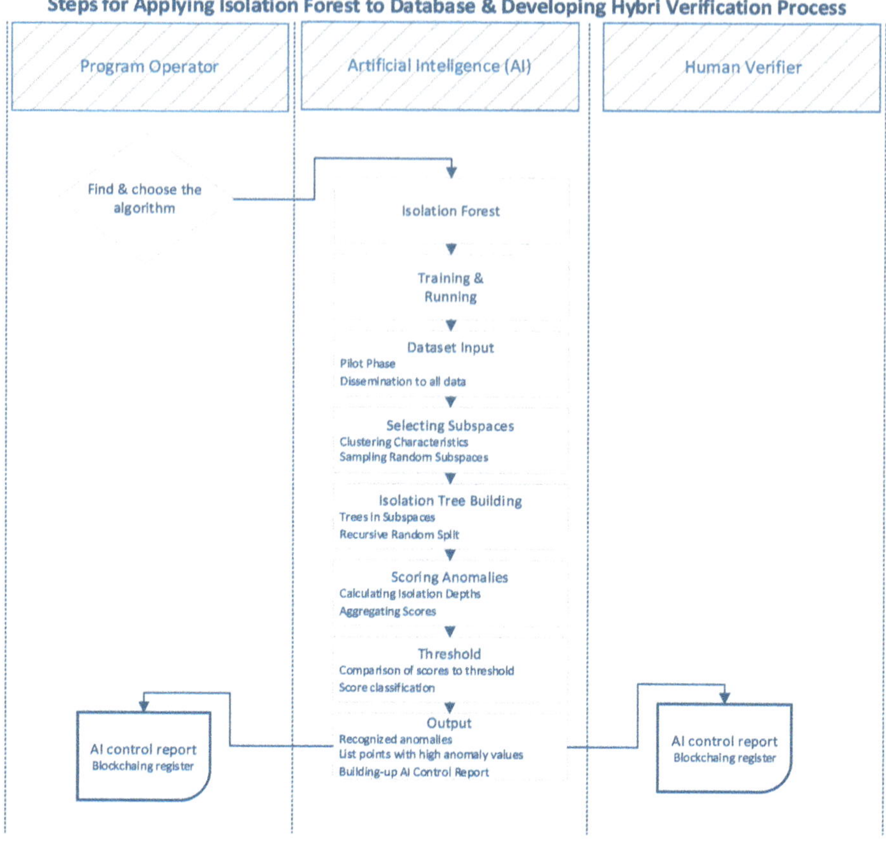

Fig. 3 Flowchart adapted from Elsaid et al. to isolating fores applied to hybrid verification process

with thousands of published EPDs, can use iForest to become an ongoing verification process in a hybrid verification process.

Blockchain complements these capabilities by embedding trust directly into the system architecture. Immutably logging verification events, smart contract execution, and verifier interventions, blockchain provides an auditable, tamper-proof ledger of the verification process. This reinforces impartiality and reproducibility, key tenets of ISO 17029 and ISO 14071 (ISO 17029, 2019; ISO 14071, 2014).

However, full automation is neither feasible nor advisable for all EPD verification tasks. As highlighted by Ghoroghi et al. (2022) and Romeiko et al. (2024), attributional LCAs often involve methodological judgements about system boundaries, attribution rules, and data quality—tasks that require human contextual interpretation and ethical reasoning. AI lacks semantic and domain-specific awareness to resolve these complexities with full autonomy. The verification process should be a hybrid activity to increase productivity and quality of the process (Ghoroghi et al. 2022; Romeiko et al. 2024).

Therefore, the optimal configuration is a hybrid framework where AI supports and augments expert reviewers without replacing the critical role of the human eye. Human reviewers must remain responsible for tasks involving subjective decisions, such as interpreting GPI, PCR, c-PCR, and specific requirements, or evaluating methodological assumptions in different regional contexts. AI, on the other hand, can be entrusted with repetitive, structured tasks such as data quality checks and anomaly detection.

In summary, while AI technologies such as IForest and blockchain greatly enhance the efficiency and auditability of EPD verification, the human element remains indispensable for ensuring contextual accuracy and upholding ethical standards.

6　Conclusions

A hybrid verification associated with the use of a blockchain framework that integrates AI with traditional verification processes can indeed strengthen the integrity and reliability of EPDs. Some of the benefits of using the new framework are: (i) Increased accuracy: AI-ML tools quickly analyze large datasets and identify potential anomalies or inconsistencies in the data reported in EPDs; (ii) When an anomaly is detected, the verifier can focus on more complex assessments or conduct an analysis of the cause of the anomaly; (iii) ML algorithms can quickly identify patterns and trends in environmental impacts that may not be visible through human analysis; (iv) as the demand for EPDs and the number of verifiers increases, AI can collaborate to provide more homogeneous verification and reduce the discrepancy between verifications of the same product EPDs by different experts, tending to harmonize the results from different verifications.

Combining AI with established verification practices, a hybrid model can enhance the overall credibility of EPDs, ultimately supporting sustainability efforts and informed decision-making to consumers and businesses like ensuring the impartiality from the process and improving the quality of data.

Incorporating new technologies (AI and blockchain) into the verification process, working together to ensure the accuracy, efficiency, and credibility of critical reviews in accordance with ISO 17029, 14071, and 14025 standards.

Integrating AI and blockchain technologies is not merely an upgrade but a paradigm shift in how environmental declarations are reviewed and trusted across supply chains. However, domain-specific human expertise, contextual analysis, and ethical governance must remain central to EPD verification.

References

Blanco C, Pauliks N, Donati F, Engberg N, Weber J (2024) Machine learning to support prospective life cycle assessment of emerging chemical technologies. Green Sustain Chem 50:1–8

Bogani R, Theodorou A, Arnaboldi L, Wortham R (2022) Garbage in, toxic data out: a proposal for ethical artificial intelligence sustainability impact statements. AI Ethics (Springer Nat) 3:1135–1142

Brisson Stapel E, Balouktsi M, Sørensen CG, Birgisdottir H (2024) Type III environmental product declarations—the perils and pitfalls of digitalization. In: IOP conference series: earth and environmental science, 1402, 1–10

Elouariaghli N, Kozderka M, Quaranta G, Pena D, Rose B, Hoarau Y (2022) Eco-design and life cycle management: consequential life cycle assessment, artificial intelligence and Green IT. IFAC PapersOnLine, 55–5, Elsevier, 49–53

Elsaid S, Binbusayyis A (2024) An optimized isolation forest based intrusion detection system for heterogeneous and streaming data in the industrial Internet of Things (IIoT) networks. Discov Appl Sci (Springer) 6(483):1–29

Gailhofer P, et al (2021) The role of artificial intelligence in the european green deal, policy department for economic, scientific and quality of life policies, European Parliament, L-2929, Luxembourg

Ghoroghi A, Rezgui Y, Petri I, Beach T (2022) Advances in application of machine learning to life cycle assessment: a literature review. Int J Life Cycle Assess (Springer) 27:433–456

Ibn Mohammed T, Mustapha KB, Abdulkareem M, Ucles Fuensanta A, Pecunia V, Dancer C (2023) Toward artificial intelligence and machine learning enabled frameworks for improved predictions of lifecycle environmental impacts of functional materials and devices. MRS Commun 13(5):795–811

ISO, ISO 14025 Environmental labels and declarations—Type III Environmental Declarations—Principles and procedures, July 2006, Zurich, The International Organization for Standardization

ISO, ISO 14071 Environmental management—Life cycle assessment—critical review processes and reviewer competencies: additional requirements and guidelines to ISO 14044:2006, 2014, Geneva, The International Organization for Standardization

ISO, ISO 17029 Conformity assessment—General principles and requirements for validation and verification bodies, October 2019, Zurich, The International Organization for Standardization

Köck B, Friedl A, Loaiza S, Wukovits W, Mihalyi-Schneider B (2023) Automation of life cycle assessment—a critical review of developments in the field of life cycle inventory analysis. Sustainability (MDPI) 15(5531):1–40

Koyamparambath A, Adibi N, Szablewski C, Adibi S, Sonnemann G (2022) Implementing artificial intelligence techniques to predict environmental impacts: case of construction products. Sustainability (MDPI) 14(3699):1–12

Liu FT, Ting KM, Zhou Z-H (2008) Isolation forest. Eighth IEEE Int Conf Data Min 08(1550–4786):413–422

Martínez-Ramón N, Calvo-Rodríguez F, Iribarren D, Dufour J (2024) Frameworks for the application of machine learning in life cycle assessment for process modeling. Clean Environ Syst (Elsevier) 14:1–11

Neupane B et al (2025) Machine learning algorithms for supporting life cycle assessment studies: an analytical review. Sustain Prod Consum (Elsevier) 56:37–53

Olanrewaju O, Enegbuma W, Donn M (2022) Data quality assurance in environmental product declaration electronic database: an integrated clark-wilson model, machine learning and blockchain conceptual framework. Arch Sci Assoc (ANZAScA) 201–210

Pannuti UR (2023) LCA and EPD need digitalization, life-cycle of structures and infrastructure systems (Biondini & Frangopol) 4086–4086

Parvatker A, Eckelman M (2019) Comparative evaluation of chemical life cycle inventory generation methods and implications for life cycle assessment results. ACS Sustain Chem Eng (Am Chem Soc) 350–367

Rao A, Kondaiah K, Chandra R, Kumar K (2017) A survey on machine learning: concept, algorithms and applications, international conference on innovative research in computer and communication engineering, ISSN (Print) 2302–9708, 1301–1309

Romeiko X et al (2024) A review of machine learning applications in life cycle assessment studies. Sci Total Environ (Elsevier) 912:1–43

Samariya D, Thakkar A (2023) A comprehensive survey of anomaly detection algorithms, annals of data science (Springer) 829–850

Stapel F, Tozan B, Sørensen C, Birgisdottir H (2022) Environmental product declarations—an extensive collection of availability, EN15804 revision and the ILCD+EPD format. In: IOP conference series: earth and environmental science, 1–10

Yigitcanlar T (2021) Greening the artificial intelligence for a sustainable planet: an editorial commentary. Sustainability (MDPI) 1–9

Carbon and Environmental Footprinting in Industry and Business

A Combined Standardization of LCA and Mass Balance: Allowing the Certification of PCFs for Mass Balanced Products

Jana Gerta Backes, Malina Nikolic, Patrick Ober, Stefan Majer, and Stefan Gärtner

Abstract The demand for standardized, comparable, and transparent methods to assess product-related environmental impacts—particularly the Product Carbon Footprint (PCF)—is increasing across industry, policy, and science. The mass balance (MB) is another assessment tool used primarily in the chemical industry to attribute sustainable feedstock to products. This paper examines the methodological and practical challenges as well as the potential of integrating MB in life cycle assessment (LCA) according to ISO 14040/44/67. While the MB method allows the book-keeping-based attribution of sustainable raw materials to products—without the need for physical segregation—LCA relies on actual material flows and correlating environmental impacts. This leads to differences in system logic and goals, making integration challenging. Using a fictional example from the chemical industry, several scenarios are defined and discussed to explore the methodological compatibility of both approaches (MB and LCA). The paper demonstrates that credible integration is particularly feasible when physical connectivity, functional equivalence, and site-specific system boundaries are respected. At the same time, there is further an urgent need for global standardization to ensure methodological consistency and verifiability, for example, with the help of certification frameworks. The findings contribute to the scientific advancement of combined sustainability assessment methods and the operationalization of PCF certification schemes that incorporate MB principles.

J. G. Backes (✉)
Junior Professor for Safety, Security and Sustainability Evaluations in Foresight Research, Aachen, Germany
e-mail: j.backes@saf.rwth-aachen.de

M. Nikolic · S. Gärtner
Meo Carbon Solutions GmbH, Cologne, Germany

P. Ober
ISCC System GmbH, Cologne, Germany

S. Majer
DBFZ GmbH, Leipzig, Germany

© The Author(s) 2026

M. Traverso et al. (eds.), *Life Cycle Management from Global to Local*,
https://doi.org/10.1007/978-3-032-17987-6_10

1   Introduction

Comparability, transparency, and this across years, national borders, and industries: a wish that is often expressed by industry and politics, partly also by scientists, with regard to the use of Life Cycle Assessment (LCA) and Product Carbon Footprints (PCF): A request for standardization (Backes, 2023).

Approaches to such standardization (outside of ISO standards 14040/44/67 (ISO 14040 2006; ISO 14044 2018; ISO 14067 2019)), can indeed be observed sector-dependent and cross-sectoral, such as Environmental Product Declaration (EPD) (ibu-epd 2021) and Product Environmental Footprint (PEF) (Del Borghi et al. 2020; European Commission, 2018), Together for Sustainability (TfS) (2025) or Catena-X (2025), as well as "peripheral" approaches such as the Greenhouse Gas (GHG) Protocol (2025) on company and reporting level.

Some of these standards make use of the mass balance (MB) approach (e.g., TfS) to identify and verify sustainability aspects—this approach is increasingly being used in complex product systems, such as in the chemical industry. The MB approach is a traceable accounting (on book-keeping level) for sustainable raw materials (defined here as of biogenic, bio-circular or circular origin), even if they are physically mixed with fossil or non-sustainable materials. MB is used when sustainable and conventional materials are mixed during processing, and their physical separation is not practicable or economically viable. Other approaches follow the LCA, which also have been established for years (e.g., ISO 14040/44/67, EPD & PEF), having a different logic than the MB does.

In addition to the desire for standardization, the demand for verification or even certification of emission values is rising. At this point, as one example, the International Sustainability and Carbon Certification (ISCC) (ISCC Systems GmbH 2025c) has started (besides other entities) to respond to the request of their stakeholders and has been developing a certification scheme since 2023, which deals with the carbon footprint of products and explicitly focuses on standardization and certification of PCFs—named ISCC Carbon Footprint Certification (CFC) (ISCC Systems GmbH 2025b). ISCC is a globally recognized certification system that has existed since 2010 and guarantees sustainable, deforestation-free, and traceable supply chains for biomass, biomaterials, and renewable energies (ISCC Systems GmbH 2025c). It is used both in regulated markets, e.g., the (bio-)fuel market within the framework of the EU Renewable Energy Directive (RED) (European Comission 2025), and in the voluntary market, e.g., food and feed as well as the (petro-) chemical sector. The primary objective of ISCC is to ensure environmental and social sustainability along the entire value chain and respond at the same time to European legislative-based requirements on sustainability (ISCC Systems GmbH 2025c).

Considering the different backgrounds of implemented standards for mass balancing in sustainability certification and LCA and PCFs, this article focusses on the potential combination of both instruments to answer market demand while ensuring scientifically sound and robust results, and at the same time further provide one standardized approach within the ISCC CFC certification scheme.

2 State of the Art and Current Challenges

In order to meet the industry's desire for a unified and standardized approach on the one hand and to develop a scientifically sound basic approach to the combination of LCA and MB on the other, reference is made below to the industry. A fictitious case study from the chemical industry is used as an illustrative example.

2.1 Mass Balance Approach: Definition and Motivation

The following example illustrates the general concept of the mass balancing approach: A company buys/produces a certain amount of sustainable raw materials (e.g., bio or circular (recycled)). These sustainable raw materials are mixed and processed with conventional materials. The sustainable raw material quantity is attributed on a book-keeping level to certain final products, even if these are not physically made exclusively from sustainable material, as the fossil and sustainable materials have the same properties and functions in the production process, but have different attributes (bio, circular vs. fossil). In the context of MB and also under ISCC PLUS (ISCC Systems GmbH 2025c), it is possible for companies to report as many 100%-sustainable final products as they have purchased/produced certified sustainable raw materials under appropriate consideration of production losses—even if the product would not have these sustainable attributes on sole physical basis. A complete physical separation of sustainable raw materials would often be inefficient or technologically unfeasible. Sustainable raw materials can be integrated more quickly into the existing infrastructure of production systems, and companies can report certified sustainable content on an MB basis in cases where physical segregation is not appropriate (ISO 22095 2020; ISO/DIS 13662 2025).

2.2 Life Cycle Assessment: Definition and Application

The LCA is a methodically standardized analysis of the potential environmental impacts of a product, process, or service, typically considering the entire life cycle—from cradle to grave. LCA is standardized in the ISO 14040 and ISO 14044 (ISO 14040 2006; ISO 14044 2018). One of the most frequently used indicators of the LCA (one of many potential environmental impacts which can be considered with an LCA) explicitly focuses only on the effects on climate change and thus considers the global warming potential—expressed in kg CO_2e (ISO 14040 2006; ISO 14044 2018). This indicator represents the PCF or the carbon intensity (CI).

2.3 *Challenges in Combining Mass Balance and PCF/LCA*

With increasing interest in sustainability (focus on environmental sustainability) and its accounting, more and more approaches are emerging that shed light on different aspects of sustainability assessment, such as the aforementioned MB and LCA. The challenge: both approaches focus on different calculation approaches, which makes a combination difficult. Nevertheless, there are increasing calls from the industry to establish a solution for the credible PCF of mass-balanced products and have it tested, in best case certified.

To enable the combination of MB and LCA—in other words: to develop or apply a PCF concept for MB products—the attributes of bio, circular or bio-circular, as well as conventional, must be provided with emissions (emission factors), so that in addition to the characteristics of the raw materials, the level of the emissions of these raw materials is considered. A bio raw material or a circular raw material can contribute to a change in the corresponding PCF.

The biggest challenges to be considered when combining MB (e.g., ISCC PLUS) and LCA/PCF (e.g., ISCC CFC) can be summarized as follows.

2.3.1 Conceptual Differences

The MB is an accounting method that mathematically assigns sustainable raw materials to specific products, even if fossil and sustainable raw materials are physically mixed (ISCC Systems GmbH 2025a, b, c). The LCA, on the other hand, analyzes the actual physical material flows and assesses environmental impacts.

The LCA is therefore based on real material flows, while the MB allows a mathematical attribution over specific timeframes. As a result, LCAs based on the MB might not necessarily reflect the physical reality of the analyzed products—a contradiction.

2.3.2 Challenges in the Distribution of Environmental Burdens

As the MB approach distributes sustainable raw materials across different products, the question arises which product receives which share of the environmental benefits (e.g., lower CO_2 emissions or reduced resource consumption)? Following, one is faced with the challenge of whether the allocation of emissions follows the attribution of (bio or circular) masses, which corresponds to the aforementioned aspects. If further the allocation of emissions is not transparent, this could lead to greenwashing or suggest excessive environmental relief. Furthermore, there is a risk of double counting. Double counting may occur when, for example, two MB batches are sold (one with sustainable attributes and one residual batch), and the PCF of the residual batch is calculated with actual physical material properties, leading to a double counting of sustainable attributes (Finkbeiner et al. 2025).

2.3.3   Methodological Incompatibilities with Existing Standards

ISO 14040/44/67 (LCA) requires objective and verifiable data, as done by ISCC CFC, which follows the mentioned norms. ISO 22095 (mass balance) allows a mathematical attribution of sustainable raw materials (ISO 22095 2020). Compliance with the LCA standards could be compromised if MB deviates too much from physical reality. Consequently, standardization is needed to specify how far MB may be integrated into LCAs. The challenges mentioned above will be addressed below in Chap. 3.

3   Combination of Mass Balance and PCF/LCA Under a Certification Scheme

The greatest challenge for the combination of MB and LCA is that for an LCA a detailed statement of actual physical flows including physical contents of raw materials in products, which is not necessarily given in the context of the MB. Consequently, it is not possible to determine in detail whether the final product under consideration actually contains exactly this one specific proportion of, e.g., bio raw materials, or whether there is more or less of this sustainable attribute (see Fig. 1).

Let's imagine a fictive company A and its suppliers A–D. Company A produces a final product built of different raw materials (feedstocks), namely triangles and stars (see Fig. 1, system flow chart). The raw materials are either purchased from (different) suppliers or (partially) produced by company A.

Circles, as fictitious raw material, are required to produce triangles and can be of bio/circular or conventional origin. The upside-down triangle represents the purchased triangle material needed for the production process of company A, whereas the standard triangle represents the triangle material produced on site with purchased circles. The final product at company A therefore requires circles to produce triangles, which then form the final product together with stars (see Fig. 1). Based on the fictitious production of company A and its supplier described, a wide range of scenarios would be possible under the MB approach, which, however, represent challenges in the LCA due to the differences of MB and LCA mentioned above (see Fig. 1, lower box: PCF calculation of final product (designed by authors)). The potential scenarios shown in Fig. 1 will be further described in the following subsections.

3.1   Physical Reality

Scenario 1 (see Fig. 1, lower box) represents the calculation of the PCF, based on the raw materials shares in the physical reality of the production including any (partial) exchange of conventional with sustainable raw material. The PCF of the final product is therefore calculated by considering three stars, two standard triangles, of which

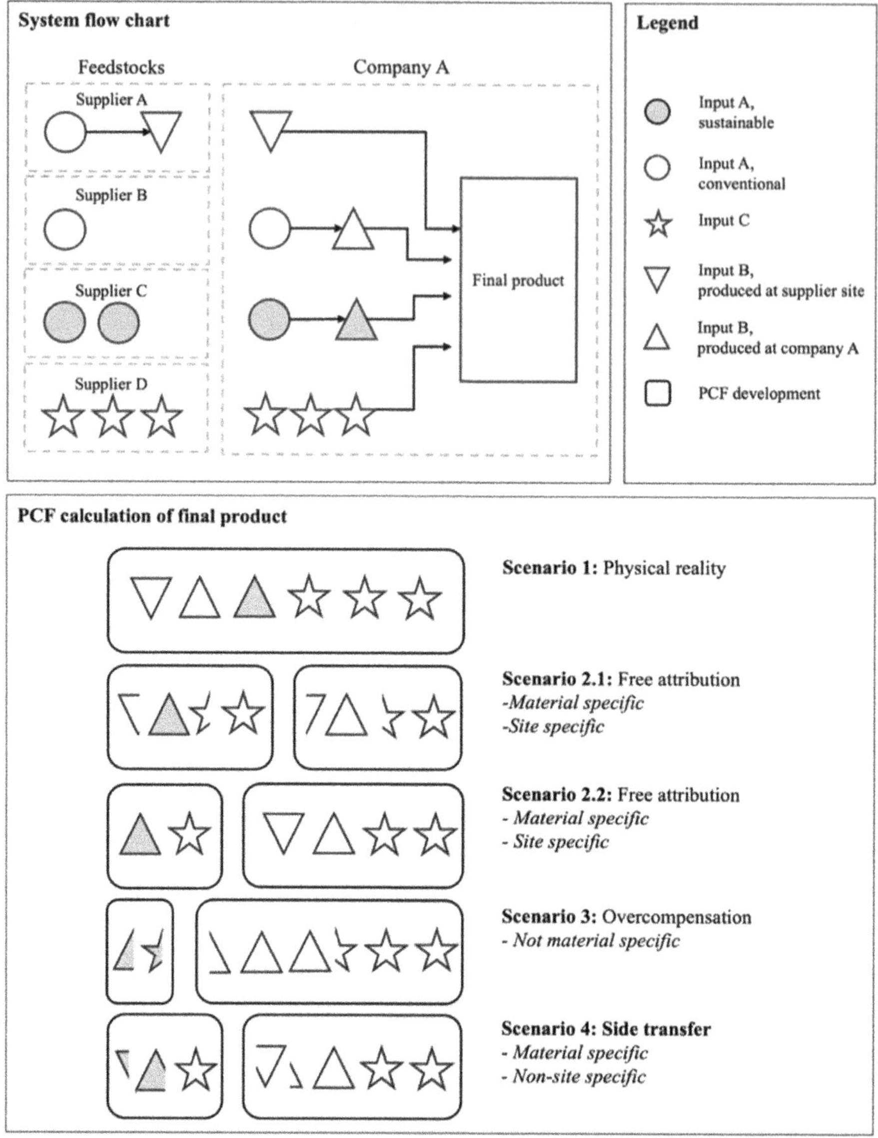

Fig. 1 Exemplary possible combinations of mass balance and LCA (own figure)

one is gray/sustainable, and one upside-down triangle. In scenario 1, the PCF is the same for every final product (see Fig. 1). The circles represent the raw material of the triangle and are therefore not recognizable as such in the final product.

In the context of physical segregation, the calculation of the PCF follows the same logic, as sustainable and conventional inputs and production streams are physically separated, so that the physical reality is then represented in the PCFs. Unlike in this example, there is the possibility to include a system separation and implement two potential PCFs in this case, one for the conventional and one for the sustainable production line.

3.2 Free Attribution

Free attribution (as second example in Fig. 1) is defined within this publication as a virtual segregation allowing a split of PCFs on an accounting level as in physical segregation. This requires that a white triangle has the same function (i.e., same material, same chemical molecule) as a gray triangle but different (sustainability) attributes and therefore different origins. A real-world example could be the exchange of fossil ethylene by bio-ethylene. Such an exchange does not only change the attribute of the feedstock to the final product that contains this feedstock but likely also the PCF, because a bio or circular material typically has a different CI/PCF than the original conventional/fossil material. At this point, it is therefore a matter of material exchange based on the raw material mass, for example, in kg per kg of product. MB in this context works like a book-keeping system. This means that when a product is sold, it doesn't necessarily have to physically contain the exact amount of bio-based or circular material (just in the books). Instead, the system ensures that over a set period of time—say, three months—the total amount of products sold with bio or circular claims does not exceed the amount of certified input materials that entered the system (the book-keeping needs to be equal, even though the real material is not yet used or already fully used within just one month) (gray circles and respective triangles, Fig. 1). At the end of the fixed time period, it must be ensured that the MB is completely balanced. If attributes/feedstocks of bio or circular are still available in surplus, these, under ISCC, could be carried over to the next three-month period, including their emission performance. The MB will need to be reviewed on an annual basis by an external auditor to ensure the correct attribution and allocation.

Generally, if there is an opportunity or the need to replace a portion of conventional/fossil raw material with a sustainable bio or circular raw material as part of a production process, the following logic is applied: If 100% of the conventional/fossil raw material is replaced with 100% of the bio raw material, the mass fractions of the product remain the same—provided the sustainable substitute has the same properties/functions as the conventional raw material—but the CI or PCFs of this raw material changes. If it is possible to replace only a proportion of the raw material with that of the bio raw material, for example, 50%, the methodology allows to be applied as follows: 50% of the conventional mass proportion remains identical and thus also retains the CI/PCF of the conventional raw material. The other 50% of the mass may be replaced with biomaterial analogous to the 100% case mentioned above and thus also have the corresponding bio CI/PCF for 50% of the mass required to produce the final product. All possible combinations between 100% fossil and 100% sustainable raw materials would be possible to assess and to be certified (e.g., 20/80,

10/90, 60/40) if the sum of the raw materials (non-sustainable + sustainable) equals 100% of the required mass fraction and the mass of the sustainable is not exceeded.

This approach reflects the requirements of ISO 14040/44/67, which require the real material flows—even if this can be stretched over a period of, e.g., three months in accounting. Under ISCC CFC this approach can only be applied within the same site—therefore a site-specific approach and the physical connection of the material would be required and given.

When it comes to the PCF calculation based on the inventory data in the example, two scenarios can be considered under free attribution according to Fig. 1.

Scenario 2.1 still reflects that the processed mixture contains triangles that are purchased (upside-down) or produced by company A. However, it is permissible to calculate two PCFs: One PCF based on 1.5 stars, 1 white standard triangle, and 0.5 white upside-down triangle, and the other PCF where the white standard triangle is replaced by the gray one. The product with the second PCF has also sustainable characteristics attributed to it, in the mass share of the gray triangle within the product.

Scenario 2.2 only reflects that a triangle is combined with a star. It is permissible to calculate one PCF based on 1 star and 1 gray triangle. This product has also the sustainable characteristics attributed to it. Another PCF can be calculated based on the residual mix as shown in Fig. 1. However, one could also consider another split within the residual mix into a PCF based on the white upside-down triangle and the other one based on the white standard triangle.

From the authors' point of view these scenarios are defendable against the existing ISO 14040/44/67, particularly if a certification of the mass balance is in place and the attribution/allocation is transparently explained.

3.3 Overcompensation

Another option (scenario 3) in Fig. 1 is named as overcompensation. It has the goal to show an increased sustainable share in the final product based on only one sustainable input. Let's stay with the example mentioned above: If the conventional raw material C (e.g., star) as part of the final product receives characteristics from the sustainable raw material B (e.g., gray triangle) in the aforementioned production, this is no longer an exchange with an identical function of the raw material. Depending on the details of the chosen MB certification scheme, such an overcompensation could be possible from an MB perspective. However, if this transfer of characteristics is included in a PCF calculation, it would no longer be in line with the initial idea of an LCA, according to the authors' understanding.

3.4 Side Transfer

The last option shown in Fig. 1 (scenario 4) represents a side transfer, including free attribution and overcompensation. The special feature of this scenario, compared to the previously described combination options, is that a transfer of sustainable characteristics occurs between materials with the same function but different PCFs and across companies/production pathways. Consequently, our named company A accepts production aspects or raw material aspects from its suppliers without knowing them in detail or even knowing their LCA/PCF approach. Furthermore, PCFs might become optimized. To the authors understanding this exceeds defined system boundaries in the context of the LCA and the same site principle of ISCC and is therefore not considered an optimal combination of MB and LCA by the authors.

4 Conclusion

The increasing industrial and regulatory demand for credible, transparent, and standardized methods for assessing and communicating product-related carbon emissions has brought the convergence of LCA and MB approaches into the spotlight. This paper critically examined the conceptual and methodological tensions that arise when attempting to combine these two different approaches, particularly in the context of emerging certification frameworks.

The MB enables the attribution of environmental sustainability characteristics (see above: bio, circular) to products based on accounting principles, even when such attributes are not physically segregated within the supply chain. In contrast, LCA, grounded in ISO 14040/44/67, requires that emissions and environmental impacts be derived from physical actual, verifiable material flows and process data. These divergent logics create considerable challenges when attempting to integrate both approaches within a single methodological and operational framework.

The analysis presented in this paper, including a structured scenario typology (see Fig. 1), has shown that not all integration pathways are methodologically sound or certifiable under established LCA principles, according to the authors' understanding of the existing norms (ISO 14040/44/67). Specifically, scenarios that involve physical connection and site-specific mass balancing offer the greatest methodological coherence and certification potential (scenarios 1–2). These cases allow for the substitution of functionally equivalent raw materials (e.g., fossil versus bio) with appropriately adjusted emission factors, ensuring that the PCF reflects material reality over a defined accounting period. Conversely, approaches such as overcompensation, which rely on functionally non-equivalent substitutions, pose significant risks to methodological integrity, traceability, and ultimately the credibility of PCF calculations.

From a certification perspective, this analysis underscores the importance of harmonizing the MB logic with the rigor of LCA. It also reveals a pressing need for (further) formal standardization and the development of unambiguous methodological guidance to ensure consistency, auditability, and comparability across companies and industries. In particular, questions regarding emission allocation, system boundaries, substitution rules, and timeframes must be addressed in a way that bridges the operational flexibility of MB with the environmental accountability required by LCA.

Strategic foresight and policy-oriented research are needed to explore the long-term implications of MB integration in the context of evolving carbon pricing, product labeling, and supply chain due diligence regulations. This includes scenario modeling of how different balancing regimes might affect corporate carbon disclosures, green finance eligibility, and cross-border sustainability reporting (e.g., CSRD, CBAM, SEC climate rules), which is a limitation of this publication.

In sum, the methodological and governance challenges identified in this paper represent not only barriers but also opportunities for advancing a more integrated, scalable, and credible environmental sustainability assessment architecture. Cross-disciplinary collaboration among LCA practitioners, certification bodies, standardization agencies, and policymakers will be key to translating these insights into workable, future-proof solutions.

In conclusion, while the integration of LCA and MB is both conceptually feasible and practically necessary to scale PCF accounting in complex, mixed-material supply chains, it demands careful methodological framing and robust governance through certification schemes and norms. The future development of clear, transparent, and science-based standards will be critical to enabling industry-wide application while avoiding the pitfalls of greenwashing or unverifiable claims. The insights from this study contribute to the foundational work needed to align sustainability certification with advanced environmental assessment practices—ultimately supporting more credible and actionable climate strategies across sectors.

References

Backes JG (2023) Opportunities, challenges and future development of life cycle sustainability assessment in the construction sector with focus on carbon reinforced concrete. [Dissertation], RWTH Aachen University

Catena-X. https://catena-x.net. Accessed 25 Apr 2025

Del Borghi A, Moreschi L, Gallo M (2020) Communication through ecolabels: how discrepancies between the EU PEF and EPD schemes could affect outcome consistency. Int J Life Cycle Assess 25(5):905–920

EPD Programm. https://ibu-epd.com/epd-programm/. Accessed 25 Apr 2025

European Commission, PEFCR Guidance document—guidance for Product Environmental Footprint Category Rules (PEFCRs), https://ec.europa.eu/environment/eussd/smgp/pdf/PEFCR_guidance_v6.3.pd. Accessed 25 Apr 2025

European Commission, Renewable Energy Directive. https://energy.ec.europa.eu/topics/renewable-energy/renewable-energy-directive-targets-and-rules/renewable-energy-directive_en. Accessed 25 Apr 2025

Finkbeiner M, Roche L, Holzapfel P (2025) From analysis-LCA to message-LCA: a lost cause? Int J Life Cycle Assess 30(5):803–810

GHG Protocol. https://ghgprotocol.org/standards-guidance. Accessed 25 Apr 2025

ISCC Systems GmbH, International Sustainability & Carbon Certification. https://www.iscc-system.org. Accessed 25 Apr 2025a

ISCC Systems GmbH, ISCC Carbon Footprint Certification. https://www.iscc-system.org/wp-content/uploads/2025/03/ISCC-CFC_1.3_March25.pdf. Accessed 25 Apr 2025b

ISCC Systems GmbH, ISCC PLUS. https://www.iscc-system.org/certification/iscc-certification-schemes/iscc-plus/. Accessed 25 Apr 2025c

ISO, ISO 14040—Environmental management—life cycle assessment—principles and framework (2006)

ISO, ISO 14044—Environmental management—life cycle assessment—requirements and guidelines (2018)

ISO, ISO 14067:2019-02—Greenhouse gases—carbon footprint of products—requirements and guidelines for quantification (2019)

ISO, ISO 22095:2020—Chain of custody—general terminology and models (2020)

ISO, ISO/DIS 13662—Chain of custody—mass balance—requirements and guidelines (2025)

Together for sustainability. https://www.tfs-initiative.com. Accessed 25 Apr 2025

Scaling-Up Life Cycle Thinking into Corporate Footprinting: The Electrolux Group Strategy

Stefano Zuin, Vsevolod Dengin, Natalia Lopez Marin, Vincenzo Lariccia, and Monica Celotto

Abstract Electrolux is a leading global appliance company that adopts a science-based approach for its sustainability strategy and initiatives. The Life Cycle Assessment (LCA) methodology is well established within the company, with continued efforts to further broaden its scope and application. LCA has been used for years for identifying the main environmental hotspots for products and for guiding ecodesign developments. These results have also been leveraged to strengthen marketing claims and brand reputation, besides nudging consumers toward more sustainable use of appliances. In order to support a consistent methodology and credible communication of the environmental impacts of our products, we've developed and adopted an internal Appliance LCA Guideline, which was verified by a third-party according to the ISO 14040 and 14044. LCA methodology is also playing a key role in supporting corporate strategy and climate targets. LCA is indeed used in the context of corporate footprint to provide valuable insights in understanding Scope 3 greenhouse gases (GHG) emissions, particularly for main contributors such as for Category 1 (purchase goods and services) and 11 (use of sold products). LCA methodology will continue to play an important role in Electrolux's strategy, with particular focus in the management of complex data and implementation of future scenarios in the analysis of product portfolio to support decision-makers.

1 Introduction

The human-caused climate change is a global challenge that requires effective strategies to reduce Greenhouse Gases (GHG) emissions from various sources. According to The Intergovernmental Panel on Climate Change (IPCC), limiting warming to around 1.5 °C requires global GHG emissions to peak before 2025 at the latest, and

S. Zuin (✉) · N. L. Marin · V. Lariccia · M. Celotto
Electrolux Italia SpA, Porcia, PN, Italy
e-mail: stefano.zuin@electrolux.com

V. Dengin
Electrolux AB, Stockholm, Sweden

© The Author(s) 2026

M. Traverso et al. (eds.), *Life Cycle Management from Global to Local*,
https://doi.org/10.1007/978-3-032-17987-6_11

be reduced by 43% by 2030 (IPCC 2023). The residential sector accounts for 30% of global final electricity consumption (International Energy Agency/IEA 2022), and the appliances category is responsible for around 39% of energy-related CO_2 emissions because tens of billions of appliances are in use today, all of which consume electricity or natural gas (e.g., gas hobs used for cooking) during their lifetime (CLASP 2023).

As a leading global appliance company, the Electrolux Group is one of the first 100 companies globally to have science-based climate targets approved by the Science Based Targets (SBT) initiative in 2018. The Group has recently set a new expanded SBT target to be achieved by 2030, such as to reduce Scope 1 and 2 emissions by 85% by 2030 compared to 2021 levels, and to reduce Scope 3 emissions by 42% in the same period (Electrolux 2025). This carbon emission reduction pathway is a fundamental pillar of the Electrolux Group's sustainability framework—For the Better 2030. The majority of GHG emissions, approximately 85%, occur during the use phase of our appliances; therefore, our most important contribution to reduce climate change is to create more energy-efficient appliances. Additionally, home appliance manufacturers are committed to designing appliances that are not only energy but also resource efficient, especially in Europe, where the new Ecodesign for Sustainable Product Regulation (ESPR) is challenging the industry to reduce the environmental impacts of products and to improve their circularity, recyclability, and durability (EU 2024). This requires a more comprehensive perspective by considering the entire appliance life cycle and by implementing the Life Cycle Assessment (LCA) methodology.

LCA has become a mature and widely used methodology for assessing environmental impacts and is increasingly applied by the Electrolux Group firstly at product level then at organization level because it addresses both upstream (e.g., purchased goods and services) and downstream (e.g., use of products) impacts, therefore it is an ideal tool to effectively measure Scope 3 emissions and account for the full impact of Electrolux's products. However, when it comes to a product portfolio composed of hundreds of appliance categories which are used worldwide, the integration of LCA results into Scope 3 accounting is challenging, as well as time and resource-consuming. Therefore, an effective approach to analyze the product portfolio to meet the corporate value chain Scope 3 accounting scheme is needed.

This paper aims to share how LCA has been used at Electrolux Group so far, and how we are using it at the corporate level to provide valuable insights in understanding Scope 3 emissions, and to address aspects of consistent GHG emissions accounting and assessment. In detail, we will describe how LCA is supporting the GHG inventory of Electrolux Group for the Scope 3 Category 1 (purchase goods and services) and 11 (use of sold products).

2   The Life Cycle Approach at Electrolux Group

The Electrolux Group has been performing LCAs aimed at identifying the main hotspots, and at optimizing its product designs, since the late 1990s. An example is the certified Environmental Product Declaration (EPD) of the Electrolux ER 8199 B refrigerator model issued in 2000. More recent examples concern the Carbon Footprint (CF) analysis of an oven, induction hob, hood, refrigerator and dishwasher, analysis reviewed by third-party according to the ISO 14067:2018 and aimed at supporting the commercial launch of these new AEG kitchen range appliances at the IFA 2024, one of the world's leading trade shows for consumer electronics and home appliances (Electrolux 2024a, b). These AEG appliances are characterized by the highest energy labels and were analyzed by considering their new energy-saving features. Indeed, for energy-using products the use phase is the most relevant life cycle phase, especially in terms of GHG emissions. The use phase contributes with more than three-quarters of the overall CF for a major appliance, followed by raw materials and their processing as shown in Fig. 1. The remaining life cycle stages have an insignificant CF impact (<5%).

Also, to support a consistent and credible communication of the environmental impacts of our products, we developed an Appliance LCA Guideline. This guide proposes a global framework and harmonized rules to determine the environmental impacts of Electrolux appliances based on the state of the art for LCA developments and methodologies. It is based on ISO 14040 and ISO 14044 and was also verified by TUV Rheinland certification body (Certificate No. CO C01-2024-09-21262582).

Here, particular attention has been dedicated to defining the most suitable functional units and to the key processes that should be included in the LCA of appliances and their multi-material components. With respect to inventory analysis, the recommendations and default scenarios have been provided according to the production, transport, use, and end-of-life phases, since each of these phases is characterized by challenges.

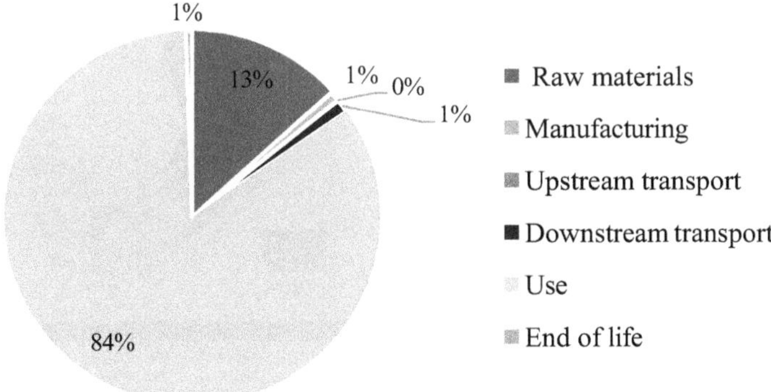

Fig. 1   Relative CF impact of the life cycle of an Electrolux dishwasher used in EU

The flexibility of the LCA is one of its strengths; indeed, we have exploited this feature over the last years to support the Electrolux sustainability strategy. In detail, within the Electrolux Better Living initiative, LCA has been used to guide our customers to minimize environmental impacts during the use of appliances. In 2021, we published the results of an LCA study performed on an Electrolux washing machine, comparing the Global Warming Potential (GWP) of a 40 °C versus a 30 °C wash within the Electrolux Better Living initiative (Electrolux 2021). The findings were clear: reducing the wash duration and temperature from 40 to 30 °C can significantly reduce the GWP by approximately 20–25%. Building upon this, we have developed the Electrolux Care LCA model to quantify the environmental impact of different washing treatments on a garment's overall footprint (Zuin et al. 2025). The Electrolux Care LCA may provide reliable information that can guide consumer behavior to reduce the environmental impact through Electrolux's innovative wash cycles and technologies, and this methodology was also reviewed by TUV Rheinland based on ISO 14040 and 14044 (Certificate No. CO C01-2024-10-21262584).

LCA requires significant time and expertise, taking several months to complete the analysis, based on the complexity of the product. For this reason, we developed a Power BI LCA dashboard (Fig. 2), a product analytic dashboard which helps to display and export key LCA data of the Electrolux appliances, and then to inform decision-making processes.

Although there are several data and informations available from Electrolux LCA studies done so far (e.g., product compositions, use scenarios, environmental impacts, etc.), these data and information have never been organized into a single tool that can be easily used by non-LCA experts. Now with this dashboard, Electrolux users (e.g., business users, R&D engineers, designers) may extrapolate useful data from master models already analyzed by LCA experts. The dashboard provides a user-friendly

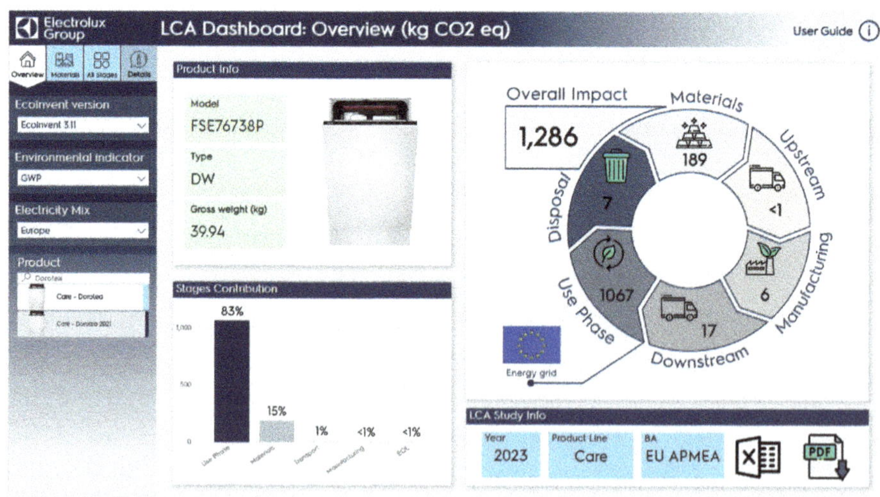

Fig. 2 Electrolux LCA dashboard

interface and an immediate view of environmental hotspots, key appliance features, and Life Cycle Inventory (LCI) data. Also, the dashboard simplifies the scenario analysis because defined key parameters (e.g., the country where the appliance is used) can be varied by users to display results in real-time.

Another significant example of the use of the LCA in the Electrolux Group has been the use of product footprinting as an integral element of our GHG accounting scheme. This will be discussed in the next chapter.

3   The Use of LCA in the Electrolux Corporate Footprinting

The Electrolux Group accounts for its GHG emissions in accordance with the Greenhouse Gas Protocol (GHG Protocol). The GHG Protocol is an international accounting scheme for quantifying, managing, and reporting GHG emissions (World Resources Institute & World Business Council for Sustainable Development 2004).

The GHG Protocol categorizes a company's GHG emissions into three scopes: Scope 1, Scope 2, and Scope 3. Scope 1 covers direct emissions from a production facility and offices, plus transport by company owned passenger cars and trucks. Scope 2 covers indirect emissions from the generation of purchased electricity, heat, and steam. Scope 3 includes all other GHG indirect emissions that occur in a company's value chain, such as emissions from purchased materials and goods, from transportation of purchased materials and sold products, from the use of sold product and from end-of-life treatment of sold products. The Scope 3 GHG emissions account for around 85% of Electrolux Group's CF. This makes Scope 3 emissions a critical decarbonization target for the Electrolux Group; in detail, the category 1—purchased goods and materials, and category 11—use of sold products, are the two key Scope 3 emissions categories for our decarbonization strategy.

However, for an efficient GHG assessment of Scope 3 emissions a broader view of the company's entire product portfolio should be done. Since large companies like Electrolux have a product portfolio composed of different categories with various variants (e.g., for the washing machine product category there are several variants such as top-load washer, front-load washer, slim washer, etc.), it would be necessary to model many LCAs for assessing the entire product portfolio. Due to the high effort to run LCAs individually for each product at Electrolux Group, we decided to focus on the calculation of a manageable number of products, which are representative of the product portfolio. The selection of representative products is mainly guided by the annual sales of products, by the coverage of all product categories, and by the commercial launch of new models.

3.1 The Use of LCA for Scope 3 Category 1

As explained in the GHG Protocol, GHG emissions are quantified using either direct measurement or calculation methods. As direct measurement data for Scope 3 emissions are difficult to obtain, usually such information is estimated using calculation methods, making use of activity data and emission factors. An "activity data" is a quantitative measure of the level of activity that results in GHG emissions, e.g., tons of purchased materials. An Emission Factor (EF) is a factor that converts activity data into GHG emissions (i.e., kg CO_2 emitted per ton) (World Resources Institute & World Business Council for Sustainable Development 2011). Activity data used by Electrolux for calculating Scope 3.1 emissions are basically the tons of procured raw materials (i.e., metals, glass, plastics, etc.) which are used to manufacture our product portfolio. Therefore, understanding the detailed material composition of our product portfolio is essential for calculating the Scope 3.1 emissions.

To get deeper insights on the amount and type of materials required for the Electrolux product portfolio along their entire supply chain, and their environmental impacts, we've developed a dedicated tool, an internal Material Footprint (MF) dashboard. The MF dashboard is based on the Bills of Materials (BOMs) of representative products analyzed through LCA. These BOMs provide a detailed product composition because they were prepared for performing LCA. The approach behind the MF calculation is shown in Fig. 3.

A "representative product" is basically a product representative of part of the product portfolio sold by the Electrolux Group. We use these reference products to simplify the data required for the analysis, while ensuring high coverage. The concept of "representative product" was also recently introduced by the European Commission (EC) within the Product Environmental Footprint (PEF) initiative (Zampori and Pant 2019). The most time-consuming part in our LCI phase is indeed the BOM elaboration, as far as material and weight information is concerned. A BOM is a list of components, materials, and sub-assemblies needed to build a product. It provides a detailed breakdown of the parts, quantities, and relationships between components.

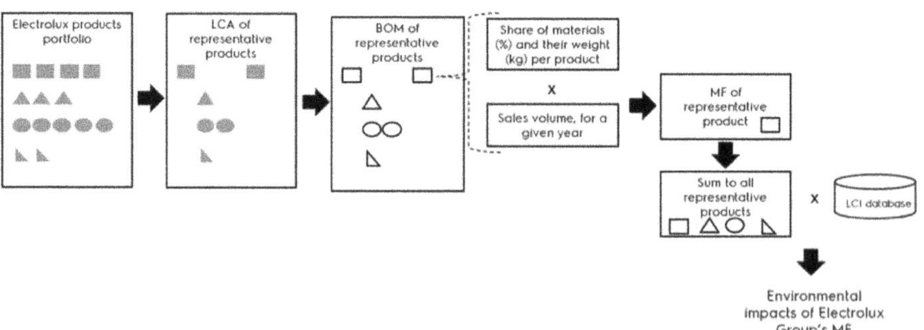

Fig. 3 Approach used for calculating the MF of the Electrolux product portfolio

Our BOMs must be reworked for LCA scope, because they include several multi-material components (e.g., compressors, motors, pumps, etc.) which must be broken down into individual materials, and irrelevant entries (e.g., logistic instructions) must be removed. This step is very time-consuming because we need to use various data sources (i.e., technical drawings, supplier's datasheet, disassembled components, etc.), which have different material nomenclature (e.g., the stainless steel is denoted as AISI 304 or AISI 316 according to the American Iron and Steel Institute nomenclature). An example of a BOM prepared for LCA scope and then used for the MF is reported in Table 1. Then, the materials and associated weights per representative product are multiplied by the quantity sold of that representative product for a given year to calculate its MF (in metric tons). The approach shown in Fig. 3 can then be applied to all representative products to obtain the Electrolux MF, such as the mass of materials and their share, embedded in the Electrolux Group's appliances, for a given year.

As we do not have the LCA for all groups of appliances, we've applied a proxy in some cases: for some representative products, we used the available BOM from the LCA of similar product categories. According to this approach, we were able to reach a coverage of around 90% with fewer than 100 groups of appliances. Concerning the system boundary, the material acquisition and its processing are included here. Such as, it starts when resources are extracted from nature and ends when the product components enter the Electrolux factory gate.

The EF used to translate the metric ton of materials into GWP impact comes from the Ecoinvent database. Each raw material is then mapped to an Ecoinvent reference associated with processing (e.g., injection molding, plastic extrusion, metal working, etc.) to determine the impact of upstream (supplier) processing. The Ecoinvent database provides a rich pool of secondary data for Scope 3 GHG reporting. Here, Global (GLO) or Rest of World (ROW) datasets are used since the origin of the materials is not easily determined, given the many intermediate suppliers. The system model "Allocation, cut-off by classification" and the "market" datasets of Ecoinvent were used because they represent the consumption mix of a specific material within a defined geographical region, including average transport and losses in trade.

Table 1 Example of a full-size dishwasher composition

Representative product	Material	Share (%)
Full size dishwasher	Ferrous metals	40.8
	Nonferrous metals	10.6
	Plastics	25.2
	Rubbers	3
	Electronics	0.7
	Others (dumping and insulation materials, adhesive, and foams)	19.7

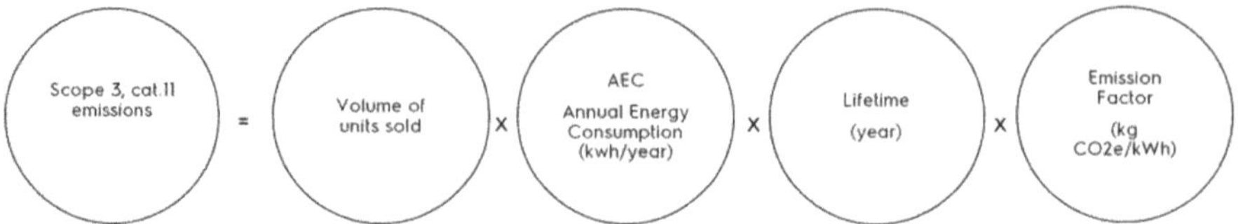

Fig. 4 Calculation method applied for Scope 3.11 emissions

3.2 The Use of LCA for Scope 3 Category 11

This category includes emissions from the use of goods and services sold by the reporting company in the reporting year, such as the domestic appliances sold by the Electrolux Group. To calculate these emissions (in ton CO_2-eq) we consider: the amount of products sold annually, their annual energy consumption (domestic appliances consume during their usage electricity or natural gas from the grid), their lifetime in years, and the EF of electricity or natural gas. Figure 4 shows more in depth the calculation method.

The Annual Energy Consumption (AEC; kWh/year) comes from energy labels or standards wherever possible (e.g., Energy Star in the USA, the EU Energy label, etc.). An energy label is an instrument for conveying information to consumers about the energy performance of appliances. The appliance lifetime was also extrapolated from up-to-date sectoral studies, as the EU Preparatory Studies within the Ecodesign Directive 2009/125/EC, which has been recently replaced by the Ecodesign for Sustainable Products Regulation (EU 2024). The use scenario defined by the AEC and lifetime has a significant impact on the result, especially for appliances which have large sales volumes, long lifetimes, and high energy consumption (e.g., refrigerators). The AEC and expected lifetime are parameters included also in the use scenario definition in the Electrolux Appliance LCA Guideline (there for appliances that use water and detergents during the use stage, the latter are additionally considered to model the use stage).

The most interesting elements from an LCA angle are the EFs for the electricity and natural gas consumed during the lifetime of appliances. According to GHG protocol, there are two methods for calculating the emissions linked to electricity use: the location-based method (i.e., emissions from electricity consumption based on the average regional electricity mix from the grid) and the market-based method (emissions from the electricity we have purchased from a selected supplier who guarantees the origin). The GHG Protocol advises the use of regional or national grid EFs and allows the use of global average EFs, which are consistent with location-based method. However, the GHG Protocol does not specify which EFs shall be used, and it suggests available life cycle databases and sources to be used. Neither the LCA standard ISO 14044:2006 specifies which data on electricity should be selected in which case (ISO 14044, 2006). Indeed, one of the challenges for determining a grid-average EF lies in selecting suitable data sources, because these databases may have different methodological aspects, system boundaries, etc. (Schäfer et al. 2024; Bastos et al. 2024).

We measure the impact of the electricity consumption of our products based on the specific energy mix of the region where the products are sold. These region-specific EFs are derived from the International Energy Agency (IEA) database. The IEA publishes EF for electricity grids in various countries and regions, since 2015. The IEA database includes the direct combustion of fuels at the point of electricity and heat generation (default values) without emissions associated with transmission and distribution losses, and with electricity trade between countries. The EFs for transmission and distribution losses and electricity trade have been recently added in the IEA database (from 2023 version), but they are not available for all countries (IEA 2023). Based on this approach, in the 2021 base year, Scope 3 emissions were 39.5 Mt CO_2 eq., which were reduced to around 31% by the end of 2024 (Electrolux 2025).

4 Conclusions

In this paper, we have presented how LCA is implemented in Electrolux's sustainability strategy, with a focus on corporate footprinting, considering the entire product portfolio. Based on this, we have identified representative products and calculated the Scope 3 emissions starting from a uniform data structure. Our future research will focus on further extending the use of LCA to cover a broader and more granular range of product categories for the Group. Also, one of the challenges in defining the use scenario necessary for Scope 3.11 emissions is the absence of comprehensive regulations and harmonized standards tailored to both AEC and lifetime in different regions/countries where appliances are sold, to represent realistic product usage patterns.

From a regulatory angle, the last EU regulations and directives incite companies to consider circular economy approaches; therefore, new criteria and methodology are needed to integrate these circularity criteria into corporate footprinting, and to ensure that reported values from organizations are comparable across sectors or product categories.

Finally, further research is needed to analyze future scenarios related to product portfolio and use patterns evolution to make the future impact of reduction measures visible for all stakeholders.

References

Bastos J, Monforti-Ferrario F, Melica G (2024) Covenant of mayors for climate and energy: greenhouse gas emission factors for local emission inventories. Publications Office of the European Union, Luxembourg, JRC136272

CLASP (2023) Net zero heroes: scaling efficient appliances for climate change mitigation, adaptation & resilience, mountain view, pp 1–126

Electrolux (2021) The truth about the laundry, findings of a European study into laundry, caring for clothes and environmental impacts. Stockholm. https://admin.betterlivingprogram.com/wp-content/uploads/2021/02/Electrolux_TheTruthAboutLaundry_WhitePaper-1.pdf. Accessed 10 June 2025

Electrolux (2024) IFA 2024: premium brand AEG unveils best-ever kitchen range with AI-assisted cooking and striking design. Press release. Stockholm. https://www.electroluxgroup.com/en/ifa-2024-premium-brand-aeg-unveilsbestever-kitchen-range-with-ai-assisted-cooking-and-striking-design-42444/. Accessed 03 Sep 2024

Electrolux (2025) Shape living for the better. Annual Report 2024. Stockholm. https://www.electroluxgroup.com/en/electrolux-sustainability-statement-2024-43534/. Accessed 20 Feb 2025

European Union (2024) Regulation (EU) 2024/1781 of the European Parliament and the Council of 13 June 2024 establishing a framework for the setting of ecodesign requirements for sustainable products, amending Directive (EU) 2020/1828 and Regulation (EU) 2023/1542 and repealing Directive 2009/125/EC, Brussels

IEA (2024) Emission factors 2023. Database documentation, Paris

IEA (2022) Technology and innovation pathways for zero-carbon-ready Buildings by 2030, Paris

International Organization for Standardization (2006a) ISO 14040:2006–environmental management–life cycle assessment–principles and framework. Geneva

International Organization for Standardization (2006b) ISO 14044:2006–Environmental management–life cycle assessment–requirements and guidelines. Geneva

International Organization for Standardization (2018) ISO 14067:2018–Greenhouse gases–carbon footprint of products–requirements and guidelines for quantification. Geneva

IPCC (2023) Climate change 2023. Synthesis Report, Summary for Policymakers, Geneva, pp 1–42

Schäfer M, Cerdas F, Herrmann C (2024) Towards standardized grid emission factors: methodological insights and best practices. Energy Environ Sci 17:2776–2786

World Resources Institute, World Business Council for Sustainable Development (2004) Greenhouse gas protocol: a corporate accounting and reporting standard. Washington, pp 1–116

World Resources Institute, World Business Council for Sustainable Development (2011) Greenhouse gas protocol product life cycle accounting and reporting standard (Product Standard). Washington, pp 1–148

Zampori L, Pant R (2019) Suggestions for updating the Product Environmental Footprint (PEF) method. Publications Office of the European Union, Luxembourg, JRC115959

Zuin S, Dengin V, Perzolla V, Bisaro F, Pipita M, Azzano A, Garzena F, Stabon E (2025) The laundry care LCA project, 6th PLATE product lifetimes and the environment conference. Denmark 6:1030–1034

Applying Matrix-Based Modeling for GHG Emission Forecasting and Corporate Decarbonization Strategy Evaluation: A Steel Manufacturing Case Study

Alessandro Marson and Alessandro Manzardo

Abstract The escalating climate crisis necessitates robust corporate decarbonization, yet businesses often struggle with the complexities of greenhouse gas (GHG) emission forecasting and evaluating mitigation strategies. This paper introduces a novel matrix-based computational model designed to simplify these challenges, enabling companies to simulate future emissions trajectories and assess the combined impacts of various reduction initiatives. The model systematically structures the GHG inventory, projects emissions under Business-as-Usual scenarios by accounting for internal growth and evolving external factors, and quantifies the effects of decarbonization actions described through activity data modifications. A steel manufacturing case study demonstrates its practical application.

1 Introduction

The escalating climate crisis presents an unprecedented global challenge, demanding urgent, concerted action. Central to this response is rapid economic decarbonization, where businesses are critical agents. As significant contributors to greenhouse gas (GHG) emissions through their operations and value chains, companies are increasingly recognized for their responsibility and capacity to drive transformative change.

Effectively addressing this requires integrating decarbonization strategies deeply into core business policies, strategic planning, and industrial operations, rather than treating them as peripheral initiatives. This integration is crucial for aligning climate action with business objectives, fostering innovation, and ensuring a just transition.

A. Marson (✉) · A. Manzardo
CESQA (Quality and Environmental Research Centre), Department of Civil, Environmental and Architectural Engineering, University of Padova, Padova, Italy
e-mail: alessandro.marson@unipd.it

A. Marson
Bluegreen Ecoinnovations Srl Società Benefit, Trieste, Italy

M. Traverso et al. (eds.), *Life Cycle Management from Global to Local*,
https://doi.org/10.1007/978-3-032-17987-6_12

However, the corporate path to decarbonization is fraught with complexity. Accurately measuring an organization's carbon footprint demands significant expertise and robust data management. Furthermore, identifying effective decarbonization interventions and rigorously evaluating their individual and combined impacts is a multidisciplinary task. This inherent complexity can be a barrier for decision-makers, diverting resources from strategic action toward data collection.

Consequently, there is a pressing need for approaches that simplify the computational burden of emissions accounting and intervention analysis. Such tools should also provide a comprehensive understanding of the complexities, potential positive externalities, and synergistic effects of combined reduction actions (e.g., reduction in energy consumption, change of material). This would empower business leaders to focus on strategic decision-making, identifying and implementing the most effective decarbonization pathways.

Crucially, evaluating decarbonization efforts must transcend static assessments, employing dynamic simulations of future effects that account for evolving internal factors (e.g., growth trajectories) and external shifts (e.g., energy mix decarbonization, technological advancements, regulatory changes). Accurate forecasting and scenario planning are thus essential for developing resilient, effective long-term strategies.

It is in this context that the present research introduces a novel mathematical structure for a computational model to support corporate decarbonization. This model aims to provide businesses with a powerful yet accessible tool for simulating future emissions trajectories. It also facilitates the application of pre-configured reduction actions that automatically interact with the company's GHG inventory, importantly evaluating their combined effects. This integrated approach simplifies emissions forecasting and intervention analysis, enabling decision-makers to explore various decarbonization scenarios with greater ease and confidence. Beyond its theoretical presentation, this paper will illustrate the model's practical application through a detailed simulation using Bluegreen (www.bluegreen.eco), demonstrating its potential to inform strategic planning and accelerate corporate climate action.

2 Materials and Methods

This section outlines the definitions and calculation methods for three principal elements of the model: the latest reported GHG inventory Sect. 2.1, projections of future inventories under Business-as-Usual (BaU) conditions Sect. 2.2, and the inventory trajectory considering the combined effects of an organization's decarbonization plan (conceived as a set of initiatives) Sect. 2.3.

To further support this theoretical description and enhance comprehension, a simplified case study will be presented.

Collectively, these elements offer insight into an organization's emissions trend without intervention and an estimate of results attainable via its strategic measures.

Notation used in this paper: italic letters (e.g., x) indicate scalars, roman bold lowercase letters indicate vectors (e.g., $\mathbf{x}$), and roman bold uppercase letters indicate matrices (e.g., $\mathbf{X}$). The superscript T indicates the transpose of a vector.

2.1 The Computational Structure of GHG Inventory for the Most-Recent Year

The most common approach for quantifying GHG emissions (expressed in tCO_2e) is to apply an emission factor to activity data (WIR and WBCSD 2004). This study uses a specific structure and model that refer solely to this quantification methodology.

In this context, an organization's GHG inventory for its most recent year (I_j for the j-th year) can be expressed as

$$I_j = \mathbf{d}_j^T \mathbf{f}_j = \begin{pmatrix} d_{j,1} & \dots & d_{j,N} \end{pmatrix} \begin{pmatrix} f_{j,1} \\ \dots \\ f_{j,n} \end{pmatrix} = \begin{pmatrix} d_{j,1} \cdot f_{j,1} + \dots + d_{j,N} \cdot f_{j,N} \end{pmatrix} \tag{1}$$

where $\mathbf{d}_j^T$ is the transpose of the activity data vector (each row of the vector displays the activity data for a specific emission source, expressed in its own unit), $\mathbf{f}_j$ is the emission factors vector for the j-th year (tCO2e/unit) and N is the number of activity data/emission factors. Therefore, the scalar I_j represents the sum of emissions from the organization's n emission sources.

2.2 The Computational Structure of Business-As-Usual GHG Inventory for Subsequent Years

This section details the mathematical structure used for projecting GHG inventories for future years beyond the latest available inventory.

The first element of this model is the BaU projection, which quantifies the trajectory of corporate emissions if the company implements no direct reduction measures.

Nevertheless, this scenario accounts for fluctuations in the company's economic activity, thereby influencing activity data, as well as shifts in the external environment that impact emission factors.

For year j + 1, Eq. (1) develops as follows:

$$I_{j+1}^{BaU} = \mathbf{d}_{j+1}^{\mathrm{BaU}\,T} \mathbf{f}_{j+1} = \left(\mathbf{c}_{j,j+1} \mathbf{d}_j \right)^T \cdot \mathbf{f}_{j+1} \tag{2}$$

where I_{j+1}^{BaU} is the BaU GHG inventory projection and the vector of coefficients $\mathbf{c}_{j,j+1}$ represents the change in activity data between year j and year j + 1 (in terms of increase or decrease), while the vector of emission factors $\mathbf{f}_{j+1}$ may differ from that of the previous year due to changes in the external context.

Therefore, Eq. (2) can be extended to a generic year $j + y$:

$$I_{j+y}^{\mathrm{BaU}} = \mathbf{d}_{j+y}^{\mathrm{BaU}\,T}\mathbf{f}_{j+y} = \left(\left(\prod_{l=0}^{y-1}\mathbf{c}_{j+l,j+l+1}\right)\mathbf{d}_j\right)^{T} \cdot \mathbf{f}_{j+y} \tag{3}$$

where $\prod_{l=0}^{n-1}\mathbf{c}_{j+l,j+l+1}$ represents the product of the vectors of the coefficients of variation of the activity data. Please, refer to the properties of empty products and products with equal indices to demonstrate that in the case of $y = 0$ and $y = 1$, (3) coincides with (1) and (2), respectively.

2.3 The Computational Structure of the Reduction Initiatives and the Modified Inventory

In this model, reduction initiatives are represented as modifications to activity data, effective from a specific year. This approach assumes that implemented reduction measures are fully operational and sustained in subsequent years. Defining these measures through activity data, rather than emission factors, does not limit the model's scope, as new emission factors can be introduced into the time series with a corresponding reallocation of relevant activity data.

A reduction initiative, denoted as $\mathbf{A}$, is formalized as an $N \times N$ matrix (where N is the dimension of vectors $\mathbf{d}$ and $\mathbf{f}$). Its non-zero coefficients identify the activity data components influenced by the initiative.

In a generic year $j + y$, the GHG inventory modified by R reduction initiatives $\left(I_{j+y}^{\mathrm{Mod}}\right)$ can therefore be expressed as:

$$I_{j+y}^{\mathrm{Mod}} = \mathbf{d}_{j+y}^{\mathrm{Mod}\,T}\mathbf{f}_{j+y} \tag{4}$$

With:

$$\mathbf{d}_{j+y}^{\mathrm{Mod}} = \left(\mathbf{c}_{j+y-1,j+y}\mathbf{d}_{j+y-1}^{\mathrm{Mod}}\right) + \left(\left(\mathbf{c}_{j+y-1,j+y}\mathbf{d}_{j+y-1}^{\mathrm{Mod}}\right)^{T} \cdot \sum_{l=1}^{R} i_{l,j+y}\mathbf{A}_l\right)^{T} \tag{5}$$

where $i_{l,j+y}$ is an intensity coefficient that is useful for generalizing matrices $\mathbf{A}$ and enabling the organization to accurately configure the level of implementation of the initiative. For example, matrix $\mathbf{A}_l$ could represent an initiative that describes the replacement of a material with a recycled alternative, and coefficient $i_{l,j+y}$ could describe the proportion of the material intended for replacement in year $j + y$, e.g., 20% of the total.

Please note that, unlike the BaU calculation, modified inventories are calculated from the previous year onwards, so it is not possible to carry out calculations from the starting year. Therefore, if a new initiative is included in the reduction action plan, the calculation algorithm must also be re-applied to all subsequent years. This ensures that the simulation model considers the combination of actions implemented in different years.

2.4 Solutions for Implementing the Model and Related Sources of Uncertainty

The theoretical model description introduced information, coefficients, and mathematical structures extending beyond conventional activity data and emission factors. This section, therefore, aims to explore the meaning of these components in greater depth and propose methodologies for their quantification (See Fig. 1).

The primary element introduced for the definition of the BaU evolutionary scenario is the year-on-year variation coefficient of activity data $c_{j,j+1}$. This information, typically derived from corporate strategic or industrial plans, aims to represent the organization's temporal evolution. This coefficient can be defined for each individual activity datum (and thus for each emission source). However, simplified approaches, such as assigning a single coefficient to a group of emission sources, are also pragmatically justifiable. For instance, in a manufacturing context, emissions sources directly linked to production—such as energy consumption in production units, raw material inputs, upstream and downstream transportation, and product use

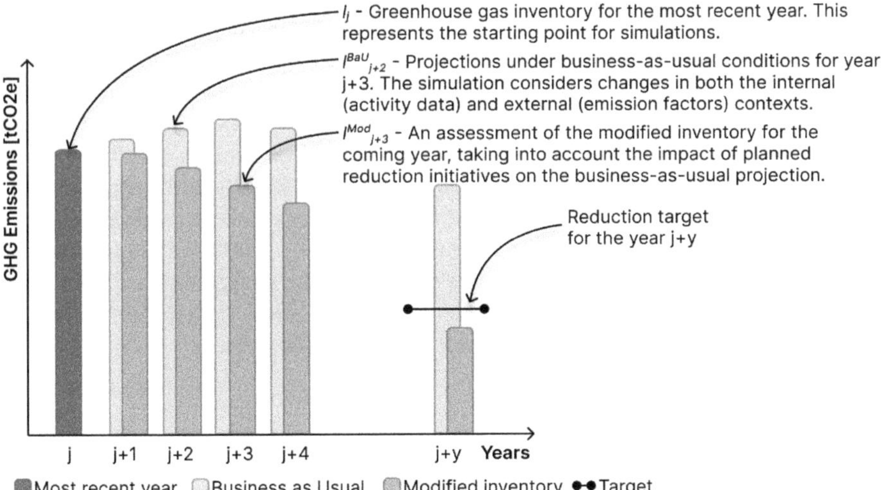

Fig. 1 Main elements of the model

and end-of-life phases—can be assumed to scale proportionally with sales volumes, which are typically forecasted in multi-year corporate plans. Concurrently, sources predominantly associated with personnel—including office energy consumption, employee commuting, and business travel—can be correlated with projected hiring plans.

These coefficients undoubtedly represent a primary source of model uncertainty, particularly over the long term. Organizations often faced with such uncertainty through periodic comparisons of budgeted versus actual performance data. A common approach to address this uncertainty involves computing multiple scenarios, thereby identifying potential best-case and worst-case outcomes (Igos et al. 2019).

The evolution of activity data is subsequently integrated with changes in the external context, which are represented by future emission factors ($\mathbf{f}_{j+y}$). The derivation of these emission factors can vary depending on the databases employed or the use of primary data. Generally, their estimation can draw upon prospective Life Cycle Assessment methodologies, leveraging contextual evolutions described by Integrated Assessment Models (IAMs) (Sacchi et al. 2004).

This information can thus be directly incorporated into the model, minimizing additional user effort, as the decarbonization rate (or multiple rates, if managing various scenarios) can be entered as an inherent property of each emission factor. Here too, the recommended approach for managing uncertainty involves evaluating a set of evolutionary scenarios, which can range from more to less conservative. It is therefore crucial that this complexity is managed in the background by the model. While this may increase computational demand, it should not impose significant additional burdens on the user.

Once the BaU scenario is established, quantifying the emissions trajectory that incorporates the effects of a reduction plan (I^{Mod}) requires no additional general information beyond the specific description of individual initiatives. A reduction initiative can be comprehensively described by three key elements: the year of implementation, a matrix detailing the modification of activity data ($\mathbf{A}$), and an intensity coefficient (i). Theoretically, all matrices describing any given reduction initiative can be pre-generated and made available to the user, as they do not require any organization-specific information. Indeed, once the reportable emission sources within the model are defined, all such matrices can be constructed and subsequently applied to the GHG inventory. Consequently, given that these matrices can be provided via a dedicated library, the sole input required from the user is to specify the implementation year for an initiative and to determine its intensity. This straightforward coefficient (ranging from 0 to 1), when applied to the organization's activity data vector, enables the quantification of the initiative's effects in the specific context.

3   Illustrative Case Study

This section presents a simplified case study to further elucidate and facilitate comprehension of the application of the previously described model. The case study concerns a steel mill that produces semi-finished products using Electric Arc Furnace (EAF) technology. The facility conducted its most recent inventory in 2025 and intends to evaluate the effects of a reduction plan extending to 2030. An attributional approach is used. The reduction strategy encompasses three interventions: the partial substitution of Direct Reduced Iron (DRI) with steel scrap (in 2026), the partial substitution of natural gas with biogas (2027), and a decrease in electricity consumption achievable through enhancements in machinery efficiency (2027). Adhering to the procedural steps outlined in §2, the various stages will be presented, and the methodologies employed, and data sources utilized will be described in detail. The calculations and reduction matrices were obtained using Bluegreen software. For transparency, the main steps of the calculations are shown below. The initial step involves quantifying the inventory for the most recent year, 2025. For the sake of simplicity, only activity data pertinent to this case study are presented below (and are considered in the quantification. The values of the activity data ($\mathbf{d}_{2025}$) and emission factors ($\mathbf{f}_{2025}$) used refer to the production of 1 million tons of steel and are shown below (and reported in Table 1 with all the other data).

$$
\mathbf{d}_{2025} = \begin{pmatrix} 216.000 \\ 0 \\ 39.604 \\ 400.000 \\ 322.581 \\ 752.688 \end{pmatrix} \begin{matrix} \text{GJ} \\ \text{GJ} \\ \text{t} \\ \text{MWh} \\ \text{t} \\ \text{t} \end{matrix} \quad \begin{matrix} \text{Natural gas} \\ \text{Biogas} \\ \text{Other } CO_2 \text{ EAF direct emission} \\ \text{Electricity (Italy)} \\ \text{DRI (coal} - \text{based, produced in India)} \\ \text{Scraps} \end{matrix} \tag{6}
$$

$$
\mathbf{f}_{2025} = \begin{pmatrix} 0{,}057 \\ 0{,}000 \\ 1{,}000 \\ 0{,}374 \\ 2{,}011 \\ 0{,}0000 \end{pmatrix} \begin{matrix} tCO_2e/GJ \\ tCO_2e/GJ \\ tCO_2e/t \\ tCO_2e/MWh \\ tCO_2e/t \\ tCO_2e/t \end{matrix} \quad \begin{matrix} \text{DEFRA, 2024} \\ \text{DEFRA, 2024} \\ - \\ \text{Ecoinvent v3.10} \\ \text{Ecoinvent v3.10} \\ - \end{matrix}
$$

It should be noted that the CO_2 emissions from the EAF furnace do not include contributions from natural gas combustion, which are calculated separately. Instead, they refer to contributions from graphite electrodes, injected carbon, and ferroalloys. The emission factors used were obtained from DEFRA (2024) and Ecoinvent v3.10 (Wernet et al. 2016). Specifically, the scrap is classified as waste, with no associated impact in accordance with the cut-off allocation principle.

Applying (1) the GHG inventory for 2025 is obtained:

Table 1 The evolution of activity data, emission factors, and total greenhouse gas (GHG) emissions during the examined period

Source	Units	2025	2026	2027	2028	2029	2030
Activity data							
Natural gas	GJ	216.000	219.240	222.529	225.867	229.255	232.693
Biogas	GJ	0	0	0	0	0	0
Other CO_2	t	39.604	40.198	40.801	41.413	42.034	42.665
Electricity	MWh	400.000	406.000	412.090	418.271	424.545	430.914
DRI	t	322.581	327.420	332.331	337.316	342.376	347.511
Scraps	t	752.688	763.978	775.438	787.070	798.876	810.859
Emission factors							
Natural gas	$tCO_2e/$ GJ	0,0568	0,057	0,057	0,057	0,057	0,057
Biogas	$tCO_2e/$ GJ	0,000	0,000	0,000	0,000	0,000	0,000
Other CO_2	$tCO_2e/$ t	1,000	1,000	1,000	1,000	1,000	1,000
Electricity	$tCO_2e/$ MWh	0,374	0,365	0,356	0,348	0,340	0,331
DRI	$tCO_2e/$ t	2,011	2,005	2,000	1,995	1,990	1,985
Scraps	$tCO_2e/$ t	0,000	0,000	0,000	0,000	0,000	0,000
Modified activity data							
Natural gas	GJ	216.000	215.311	174.833	177.455	180.117	182.819
Biogas	GJ	0	0	43.708	44.364	45.029	45.705
Other CO_2	t	39.604	38.234	38.807	39.389	39.980	40.580
Electricity	MWh	400.000	376.532	370.715	376.276	381.920	387.648
DRI	t	322.581	130.968	132.932	134.926	136.950	139.005
Scraps	t	752.688	960.430	974.837	989.459	1.004.301	1.019.366
Total emissions							
I^{BaU}	tCO_2e	850.075	857.541	865.132	872.849	880.694	888.668
I^{Mod}	tCO_2e	850.075	245.934	234.920	234.128	233.370	232.644
$I^{Mod} - I^{BaU}$	tCO_2e	0	−607.383	−621.710	−625.885	−630.098	−634.350

$$I_{2025} = \mathbf{d}^{T}_{2025}\mathbf{f}_{2025} = 850.075 tCO2e$$

Once the inventory for the most recent year has been obtained, the BaU projections can be constructed. To represent company growth, a growth factor of 1.5% was used for all activity data, with this figure applied on a year-on-year basis until 2030.

The assessment of the evolution of emission factors focused on Italian electricity and DRI (direct reduced iron) consumption. Annual decarbonization factors were then calculated using Premise and an SSP2 IMAGE scenario (Sacchi et al. 2022) (assuming a linear trend): -2.4% for electricity and -0.25% for DRI (sponge iron from India).

Once the BaU projection has been established, the subsequent phase involves detailing the planned reduction initiative. For each action, the provided information includes a description of the specific intervention (including the $\mathbf{A}$ matrix), its effective year of implementation, and the applied intensity coefficient.

The first reduction measure to be implemented in 2026 is replacing DRI with metal scraps. This action directly reduces the consumption of virgin materials in favor of materials obtained from the waste chain. However, it also leads to significant changes in terms of electricity consumption (DRI requires around $0{,}120$ MWh/t more electricity to melt than scrap) and less significant effects on natural gas consumption ($-0{,}02$ GJ/t) and direct CO_2 emissions from the reaction of electrodes and carbons ($-0{,}01$ tCO_2/t) (these data were obtained by consulting industry experts).

These effects can all be summarized in the reduction action matrix $\mathbf{A}_1$ (the order of emission sources is the same as proposed in (6)).

$$
\mathbf{A}_1 = \begin{pmatrix}
0 & 0 & 0 & 0 & 0 & 0 \\
0 & 0 & 0 & 0 & 0 & 0 \\
0 & 0 & 0 & 0 & 0 & 0 \\
0 & 0 & 0 & 0 & 0 & 0 \\
-0{,}02 & 0 & -0{,}01 & -0{,}15 & -1 & 1 \\
0 & 0 & 0 & 0 & 0 & 0
\end{pmatrix}
$$

Therefore, considering an intensity of $i_{1,2026} = 0{,}6$ (60% of the DRI will be replaced by scraps), and considering that $\mathbf{c}_{2025,26}\mathbf{d}^{\mathrm{Mod}}_{2025} = \mathbf{d}^{\mathrm{BaU}}_{2026}$, applying (5) we obtain:

$$
\mathbf{d}^{\mathrm{Mod}}_{2026} = \mathbf{d}^{\mathrm{BaU}}_{2026} + \left(\mathbf{d}^{\mathrm{BaU}\,T}_{2026} \cdot \sum_{l=1}^{1} i_{l,2026}\mathbf{A}_l \right)^{T} =
$$

$$
= \mathbf{d}^{\mathrm{BaU}}_{2026} + \left(\left(\begin{pmatrix} 219.240 \\ 0 \\ 40.198 \\ 406.000 \\ 327.420 \\ 763.978 \end{pmatrix}^{T} \cdot 0{,}6 \begin{pmatrix} 0 & 0 & 0 & 0 & 0 & 0 \\ 0 & 0 & 0 & 0 & 0 & 0 \\ 0 & 0 & 0 & 0 & 0 & 0 \\ 0 & 0 & 0 & 0 & 0 & 0 \\ -0{,}02 & 0 & -0{,}01 & -0{,}15 & -1 & 1 \\ 0 & 0 & 0 & 0 & 0 & 0 \end{pmatrix} \right)^{T} \right) =
$$

$$= \begin{pmatrix} 219.240 \\ 0 \\ 40.198 \\ 406.000 \\ 327.420 \\ 763.978 \end{pmatrix} + 0{,}6 \begin{pmatrix} -327.420 \cdot 0{,}02 \\ 0 \\ -327.420 \cdot 0{,}01 \\ -327.420 \cdot 0{,}15 \\ -327.420 \cdot 1 \\ +327.420 \cdot 1 \end{pmatrix} = \begin{pmatrix} 215.311 \\ 0 \\ 38.234 \\ 376.532 \\ 130.968 \\ 960.430 \end{pmatrix}$$

Then applying (4): $I_{2026}^{\mathrm{Mod}} = \mathbf{d}_{2026}^{\mathrm{Mod}\,T} \mathbf{f}_{2026} = 450.615\,\mathrm{tCO_2e}$.

The remaining two reduction initiatives ($\mathbf{A}_2$ and $\mathbf{A}_3$) are both scheduled for implementation in 2027. Initiative $\mathbf{A}_2$ involves replacing some of the natural gas with biogas, while $\mathbf{A}_3$ consists of improving the plant's overall electricity consumption efficiency.

$$\mathbf{A}_2 = \begin{pmatrix} -1 & 1 & 0 & 0 & 0 & 0 \\ 0 & 0 & 0 & 0 & 0 & 0 \\ 0 & 0 & 0 & 0 & 0 & 0 \\ 0 & 0 & 0 & 0 & 0 & 0 \\ 0 & 0 & 0 & 0 & 0 & 0 \\ 0 & 0 & 0 & 0 & 0 & 0 \end{pmatrix} \quad \mathbf{A}_3 = \begin{pmatrix} 0 & 0 & 0 & 0 & 0 & 0 \\ 0 & 0 & 0 & 0 & 0 & 0 \\ 0 & 0 & 0 & 0 & 0 & 0 \\ 0 & 0 & 0 & -1 & 0 & 0 \\ 0 & 0 & 0 & 0 & 0 & 0 \\ 0 & 0 & 0 & 0 & 0 & 0 \end{pmatrix}$$

Therefore, it is possible to proceed with the contextual calculation of $\mathbf{d}_{2027}^{\mathrm{Mod}}$ by applying Eq. (4). In this case, however, it is not possible to refer to the BaU of the reference year, since changes were made in 2026. Intensities of $i_{2,2027} = 0{,}2$ and $i_{3,2027} = 0{,}03$ were assumed for the two actions, respectively (please note that $\mathbf{c}$ is equal to a 1,5% increase across all sources).

$$\mathbf{d}_{2027}^{\mathrm{Mod}} = \left(\mathbf{c}_{2026,27}\mathbf{d}_{2026}^{\mathrm{Mod}} \right) + \left(\left(\mathbf{c}_{2026,27}\mathbf{d}_{2026}^{\mathrm{Mod}} \right)^{T} \cdot \sum_{l=2}^{3} i_{l,2027}\mathbf{A}_l \right)^{T} =$$

$$\begin{pmatrix} 218.541 \\ 0 \\ 38.807 \\ 382.180 \\ 132.932 \\ 974.837 \end{pmatrix} + \left(\begin{pmatrix} 218.541 \\ 0 \\ 38.807 \\ 382.180 \\ 132.932 \\ 974.837 \end{pmatrix}^{T} \cdot (0{,}2\mathbf{A}_2 + 0{,}03\mathbf{A}_3) \right)^{T} = \begin{pmatrix} 174.883 \\ 43.708 \\ 38.807 \\ 370.715 \\ 132.932 \\ 974.837 \end{pmatrix}$$

Using this information, it is feasible to calculate the modified inventory for 2027 and projections for all subsequent years, as no additional initiatives are currently planned. The data and emission factors can be found in Table 1. To evaluate more cautious and optimistic scenarios, the parameters for initiative intensity and company growth have been modified (worst case: $c = 2{,}5\%$, $i_{1,2026} = 0{,}6$, $i_{2,2027} = 0{,}05$, $i_{3,2027} = 0$; best case: $= 1\%$, $i_{1,2026} = 0{,}9$, $i_{2,2027} = 0{,}3$, $i_{3,2027} = 0{,}05$) (Fig. 2).

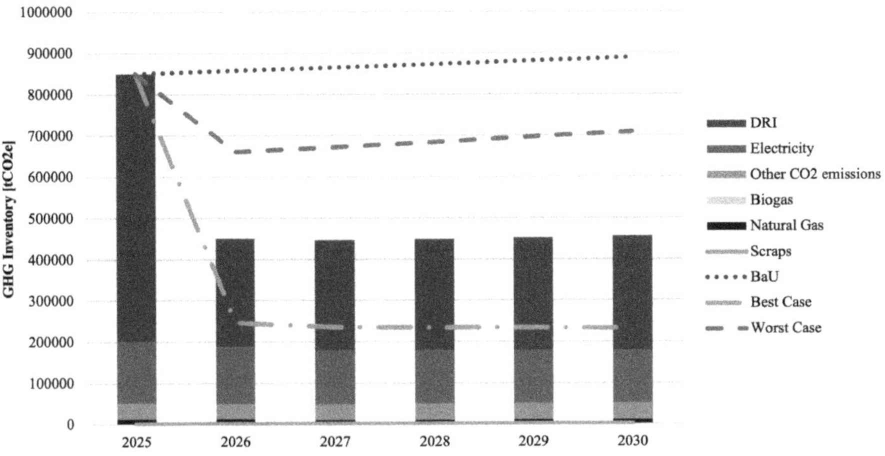

Fig. 2 The graphical results of the case study (also including worst- and best-case scenarios)

The case study findings highlight the model's potential as a powerful decision-support tool. Its capability for systematic evaluation of various decarbonization levers—individually and in combination—enables businesses to identify the most impactful pathways to achieve emission reduction targets. A key strength is the framework's dynamic simulation of future effects, accounting for both internal corporate developments (e.g., production growth) and external contextual shifts (e.g., grid decarbonization). This dynamic approach surpasses static carbon accounting, providing more realistic, forward-looking assessments crucial for long-term strategic planning.

The model's architecture, which represents reduction initiatives as modifications to activity data via predefined matrices and user-defined intensity coefficients, simplifies the computational burden for end-users. This allows companies to focus on strategic aspects, such as the timing and scale of interventions, rather than complex underlying calculations. A significant advantage of this clear computational structure is its facilitation of the digitalization of decarbonization assessment processes and straightforward implementation into software tools, enhancing efficiency, scalability, consistency, and promoting wider adoption of systematic planning. The potential to integrate a library of pre-configured reduction actions further boosts its accessibility and practical utility.

However, the model's outputs are inherently dependent on input data quality and accuracy. Uncertainties in forecasting future business activity, the evolution of emission factors, the intensity of implementation, and the precise impact of reduction measures represent limitations. The model attempts to mitigate this by enabling scenario analysis (e.g., best-case/worst-case outcomes based on different growth trajectories or external decarbonization rates), a critical feature for robust decision-making under uncertainty. The model's reliance on the activity data and emission factor approach also defines its operational scope.

4 Conclusions

This paper has presented a novel matrix-based mathematical model for forecasting corporate GHG emissions and evaluating the efficacy of decarbonization strategies. The core methodology involves structuring the latest GHG inventory, projecting future emissions under a BaU scenario, and simulating a modified emissions trajectory that incorporates a portfolio of reduction initiatives. The proposed model has been successfully applied to an illustrative case study involving a steel manufacturer. Future research could focus on the integration of economic analyses, allowing for the evaluation of decarbonization strategies not only in terms of GHG reduction potential but also cost-effectiveness. Enhancements could also include the integration of artificial intelligence in the generation of reduction initiatives and optimization of reduction strategies. In conclusion, the matrix-based modeling approach detailed in this paper offers a structured, adaptable, and comprehensive framework for corporate decarbonization strategy evaluation. By simplifying complex emissions forecasting and intervention analysis, it empowers businesses to make more informed decisions, effectively integrate climate action into their core operations, and accelerate their transition toward sustainable, low-carbon business models, ultimately contributing to the broader goals of mitigating climate change.

References

Gov.UK (2024). https://www.gov.uk/government/publications/greenhouse-gas-reporting-conversion-factors-2024. Accessed 04 June 2025

Igos E, Benetto E, Meyer R, Baustert P, Othoniel B (2019) How to treat uncertainties in life cycle assessment studies? Int J Life Cycle Assess 24:794–807

Sacchi R, Terlouw T, Siala K, Dirnaichner A, Bauer C, Cox B, Mutel C, Daioglou V, Luderer G (2022) PRospective environmental impact asSEment (premise): a streamlined approach to producing databases for prospective life cycle assessment using integrated assessment models. Renew Sustain Energy Rev 160:112311

Wernet G, Bauer C, Steubing B, Reinhard J, Moreno-Ruiz E, Weidema B (2016) The ecoinvent database version 3 (part I): overview and methodology. Int J Life Cycle Assess 21:1218–1230

World Resources Institute (2004) World business council for sustainable development, GHG protocol corporate accounting and reporting standard, Revised edition

Estimating Embodied Impacts of UCL New Student Center Using Construction KPI Data

Daniel Anyanya, Charnett Chau, Andrea Paulillo, Silvia Fiorini, and Paola Lettieri

Abstract The construction sector's environmental footprint necessitates robust methodologies to quantify and mitigate construction phase impacts. This study advances Life Cycle Assessment (LCA) methodology by integrating real-time construction phase Key Performance Indicator (KPI) data into a dynamic framework, applied to the University College London (UCL) New Student Center—a BREEAM "Outstanding" institutional building. Through detailed construction phase analysis, the research identifies mechanical, electrical, and plumbing equipment (MEP) installation as the dominant contributor to environmental impacts (60–90% of construction burdens), followed by superstructure works. The integration of KPI datasets enabled unprecedented granularity in identifying construction hotspots. The findings advocate for targeted construction phase interventions, enhanced MEP component environmental product declarations (EPDs), and real-time monitoring integration.

1 Introduction

The unprecedented pace of global urbanization presents significant challenges for sustainable development, with urban population growth projected to reach 83% in high-income countries by 2030 from 76% in 2000 (Nations et al. 2018). The construction sector's environmental footprint is particularly significant, contributing approximately 40% of global final energy consumption and GHG emissions, while consuming half of all extracted natural resources (UK Green Building Council n.d). In the UK alone, the construction industry contributes 7% to gross domestic product (GDP) yet generates nearly 25% of national carbon emissions (Department for Business 2013).

D. Anyanya · C. Chau · A. Paulillo · P. Lettieri (✉)
University College London, London, UK
e-mail: p.lettieri@ucl.ac.uk

S. Fiorini
Mace, London, UK

© The Author(s) 2026

M. Traverso et al. (eds.), *Life Cycle Management from Global to Local*,
https://doi.org/10.1007/978-3-032-17987-6_13

Life Cycle Assessment (LCA) has evolved into a crucial methodology for evaluating environmental impacts of buildings (Guinée 2002; Anyanya et al. 2025). However, traditional construction phase assessments have predominantly relied on industry-average data and simplified assumptions, leading to limited accuracy in assessing real-world construction impacts (Crawford 2011). Recent advances in building energy efficiency have shifted the relative importance of construction phase impacts, as embodied energy from materials and construction processes increasingly represents a larger proportion of total building lifecycle impacts in high-performance buildings (Ramesh et al. 2010; Sartori and Hestnes 2007). Conventional assessments are typically retrospective and fail to account for the dynamic nature of construction activities, where numerous stakeholders, complex material flows, and logistical challenges introduce significant variability (Khasreen et al. 2009).

The construction phase encompasses multiple distinct activities that contribute differently to environmental impacts. Material production and manufacturing (A1-A3 stages) typically represent the largest component of construction phase impacts, while transportation to site (A4) and construction installation processes (A5) contribute additional burdens (Jaemoon et al. 2023). The relative significance of these stages varies considerably depending on material choices, transportation distances, construction methods, and project-specific factors.

To address the challenges resulting from the dynamic nature of construction and LCA data limitations, the construction sector has increasingly adopted innovative tools for real-time environmental sustainability management during construction. Mace's proprietary Optimize platform facilitates comprehensive tracking of key performance indicators (KPIs) related to environmental metrics during construction, including carbon emissions, material consumption, waste generation, energy use, and transportation activities.

The potential of integrated construction sustainability management systems is exemplified by University College London's (UCL) New Student Centre (NSC). Completed in 2018, this flagship project achieved BREEAM "Outstanding" certification and incorporated cutting-edge construction strategies including advanced material selection, comprehensive waste management protocols, and detailed environmental monitoring throughout the construction process.

This study addresses the challenges of implementing construction LCA by developing a systematic approach to integrating construction phase KPI datasets into LCA frameworks. Key objectives include enhancing construction phase LCA accuracy through project-specific construction KPI data; enabling phase allocation of construction impacts across different construction phases; and conducting comprehensive construction hotspot identification to support targeted improvement strategies.

2 Methods

This study implements a comprehensive cradle-to-construction Life Cycle Assessment (LCA) spanning all construction activities from material production through on-site installation and commissioning. The construction phase LCA methodology adheres to ISO 14040/14044 and follows the EN15978:2011 framework modules A1-A5 for assessing construction phase environmental performance (ISO 2020; CEN/TC 350 2011).

The goal is to conduct a construction phase LCA on the UCL NSC that provides actionable insights for construction phase improvements. The functional unit was defined as the whole building construction, with impacts also reported per square meter. The system boundaries capture all construction phase impacts, encompassing material production, transportation to the site, and on-site construction activities. The construction activities are subdivided across phases as described in Table 1 and Fig. 1, and as follows.

Phase I—Pre-Construction: Preliminary works including site preparation, clearing areas, setting up facilities, and minimal demolition work. Phase II—Substructure Work: Earthworks, excavation, piling, and underpinning to establish foundations and structural support. Phase III—Superstructure Work: Constructing main structural elements including concrete framing, steelwork, roof installation, and facades. Phase IV—External Works: Landscaping and infrastructure changes surrounding the building. Phase V—Fit outs/Internal Finishes: Making spaces operational through plaster boarding, flooring, painting, and other finishes. Phase VI—MEP Installation: Installing HVAC systems, electrical networks, plumbing fixtures, and operational infrastructure. Phase VII—Commissioning: Ensuring building and systems comply with safety and performance standards.

The system boundary includes Product Stage (A1-A3) covering raw material extraction, processing, and manufacturing of construction materials; Transportation stage (A4) encompasses transportation of materials from manufacturing facilities to the construction site; and Construction Process Stage (A5) covers on-site and off-site assembly, installation, and commissioning activities.

The foundation of the construction phase life cycle inventory (LCI) is primary data collected during construction, supplemented by secondary data from industry databases. Primary construction data sources include:

- Energy consumption from the construction stage data: Extracted from Mace's Optimize platform, logging detailed records of electricity consumption (15-min intervals), fuel usage (daily), and water consumption (hourly).
- Material Delivery Data: Comprehensive tracking of material deliveries, quantities, and specifications.
- Construction Waste Generation: Detailed records by construction phase, material type, and end-of-life method. End-of-life treatment methods for construction waste were incorporated into the A5 stage based on actual waste management

Table 1 Phase allocation approaches

Inventory	Data provided	Allocation approach used
Site fuel (diesel, petrol, and CNG) usage	L/month	The MEP phase was assumed to contribute 5% of the total fuel usage during construction. After which, where only one construction phase occurred over a period, the average fuel used per month per phase was calculated and extrapolated to the full duration of the phase. Otherwise, usage per month was allocated equally among the phases occurring in parallel
Site water usage	m^3/month (per meter)	Where only one construction phase occurred over a period, an average water use per month per phase was calculated and extrapolated to the full duration of the phase. Otherwise, usage per month was allocated equally among the phases occurring in parallel No water was assumed to be used for the installation of MEPs, as water was instead used at the testing and commissioning stage of the construction
Construction materials[a]	Tons/delivery (dated)	Phase allocation was based on sub- contractor's role in construction. Certain materials were assumed to be used in specific phases only; for instance, gypsum and paint were allocated to internal finishes
MEP[a]	Item/delivery (dated)	Mass was calculated from transport data, where the contractor and vehicle loading were provided
Timber[a]	m^3/delivery (dated)	Descriptions of timber that were provided for each delivery were used in conjunction to sub-contractor's role in construction to assist phase allocation
Waste outputs[a]	Tons/collection (dated)	Phase allocation was based on sub- contractor's role in construction, otherwise it was assumed that certain phases only dispose of certain materials Waste reports were used to understand the end-of-life of different waste types
Transport[b]	Mileage/vehicle type and % load	Phase allocation was based on sub- contractor's role in construction. Delivery loads were also correlated against materials and waste mass inventories

[a] Contractor/sub-contractor specified
[b] Contractor/sub-contractor specified, whether the transport was for incoming materials or outgoing waste specified

practices documented during construction. The treatment allocations were determined from contractor waste reports that specified the actual treatment method for each waste stream type.

- Transportation Logistics: Captured via DataScope's system, recording vehicle movements, load factors, and routing information.

Secondary data sources include Sphera LCA Database (version 10.0.1) and EcoInvent Database (version 3.8, cut-off system model). The differences between the

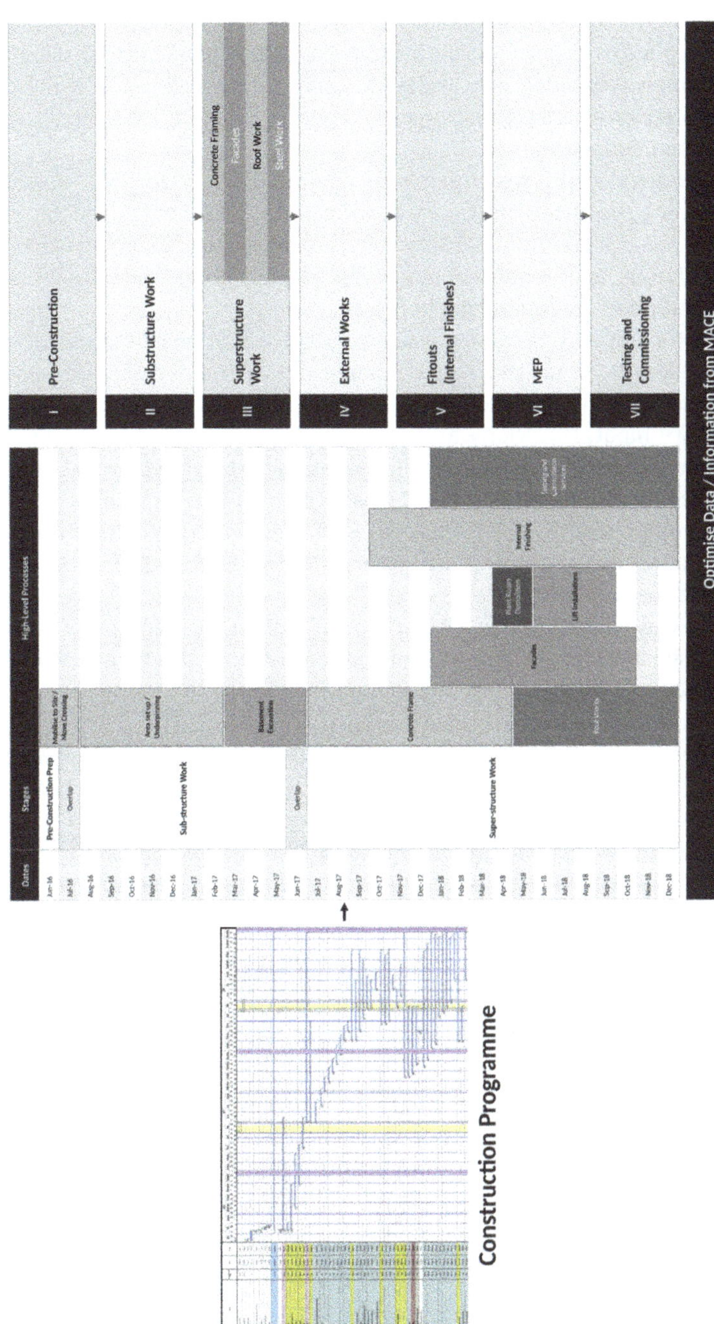

Fig. 1 High-level processes for defining construction phases

Sphera LCA and EcoInvent databases have been considered in this assessment. Where possible, consistent system boundaries were applied, and temporal representativeness was aligned by prioritizing the most recent data available in both databases. Database selection was based on data availability and regional relevance, with EcoInvent preferred for European-specific processes and the Sphera LCA database for industry-specific materials.

Material quantities were cross-referenced against delivery records, contractor reports, and building specifications. Construction activities were organized into seven distinct phases: pre-construction, substructure, superstructure, external works, fit-out, MEP installation, and commissioning. This segmentation enabled systematic analysis of construction environmental hotspots.

Several inventory inputs (e.g., water or site fuel) had to be allocated across different construction phases due to the overlapping nature of construction activities and the format of collected data. Table 1 includes a summary of the allocation approach used for each inventory input. Electricity and fuel usage were allocated based on phase duration and parallel activities. Material allocation was based on sub-contractor roles and material-specific usage patterns. MEP mass was calculated from transport data due to limited component-level information, with material composition estimated through expert consultation: 40% plastic piping, 15% aluminum, 35% steel, and 10% polystyrene foam.

Construction waste allocation was based on sub-contractor roles and waste type relationships to specific phases. Transport allocation considered sub-contractor roles and correlated delivery loads against material inventories.

The construction phase inventory data is reported in Table 2. This reveals substantial resource consumption. Diesel consumption totalled 36,532 L, with substructure work consuming the largest portion (33,168 L). Electricity consumption reached 298,662 kWh, with superstructure work representing the highest demand (122,562 kWh). Construction water consumption totalled 3,128 m^3, with superstructure and substructure phases accounting for the majority. Material consumption reached 39,415 tons, dominated by concrete representing 77% of total construction materials by mass.

The construction phase LCA was conducted using LCA for Experts Software, incorporating Sphera LCA and EcoInvent v3.8 databases. The study adopts the Environmental Footprint 3.0 (EF 3.0) method to evaluate construction phase environmental impacts across multiple categories including climate change, acidification, eutrophication, human toxicity, resource depletion, and land use impacts. A systematic construction hotspot analysis was conducted across three levels: construction life cycle stage level (A1-A3 vs A4-A5), construction phase level (seven phases), and within-phase level (individual materials and activities contributing more than 5% to any construction impact category).

Table 2 Construction materials allocated by phase (tons)

Phase	Brick	Rebar	Structural steel	Concrete (in situ)	Concrete (pre-cast)	Cement and mortar	Insulation	Plasterboard	Glass	Other metal	Bituminous materials	Stone and gravel
I	3.2	–	–	–	–	–	–	–	–	–	–	–
II	0.0	320.0	0.0	4520.9	0.0	0.0	0.0	0.0	0.0	0.0	0.0	0.0
III	224.0	381.1	203.1	8115.2	17,571.4	47.0	50.7	0.0	72.6	65.3	5234.0	2010.0
IV	0.0	8.2	0.0	83.0	0.0	0.0	0.0	0.0	0.0	0.0	0.0	0.0
V	1.1	0.0	0.0	0.0	0.0	0.0	0.3	150.1	0.0	293.2	0.0	4.0
Total	228.3	709.3	203.1	12,719.2	17,571.4	47.0	51.1	150.1	72.6	358.5	5234.0	2014.0
Percentage	0.6	1.8	0.5	32.3	44.6	0.1	0.1	0.4	0.2	0.9	13.3	5.1

3 Results and Discussion

The environmental impact assessment reveals distinct patterns across impact categories as reported in Table 3, raw material extraction, transportation of raw material, and manufacturing (A1-A3) emerging as the dominant contributor to construction environmental burdens, followed by transportation to construction site (A4) and construction stage (A5).

Cradle to gate contributions consistently dominate most environmental indicators, typically constituting 70–85% of total construction environmental impact assessment results. This predominance underscores the significant environmental burden associated with material extraction, processing, and manufacturing prior to construction activities. The construction process stages (A4-A5), encompassing transportation and on-site construction activities, contribute the remaining 15–30% of construction environmental impacts. The GWP100 category within construction shows material production phases contributing approximately 80% of construction impacts, with transportation and construction installation accounting for roughly 20%. Construction transportation (A4) demonstrates higher contributions in impact categories related to combustion emissions, such as acidification and eutrophication, while construction installation processes (A5) show greater influence in categories related to energy consumption and direct construction activities.

3.1 Construction Stage Analysis

The detailed breakdown across seven construction phases is reported in Fig. 2 and reveals distinct patterns in environmental contributions, with variations in impact magnitude.

The results show that Phase VI (MEP installation) dominates most construction impact categories, typically contributing 60–90% of the total construction environmental burden. This striking dominance indicates that MEP systems installation constitutes the primary environmental hotspot across nearly all categories within the construction phase. Phase III (superstructure construction) emerges as the second most significant contributor across numerous construction impact categories, particularly prominent in construction-related eutrophication categories, where it accounts for approximately 40–60% of construction impacts.

Phase II (substructure construction) shows notable contributions in several construction categories, particularly construction-related climate change impacts and certain toxicity measures, reflecting the environmental impacts of foundation systems and below-grade construction activities. Phase V (internal finishes) demonstrates particularly high contributions in construction human toxicity categories, due to chemical compounds in finishing materials used during construction activities. The remaining phases (pre-construction, external works, and commissioning) contribute minimal proportions across most construction impact categories.

Table 3 Construction phase environmental impact per m^2

LCIA method	Raw material extraction, transportation, and manufacturing			Transportation to site and construction stage	
	A1	A2	A3	A4	A5
EF v3.0 LCIA method	Raw material supply	Transport	Manufacturing	Transport	Construction installation process
Acidification (A)—accumulated exceedance [Mole of H + eq.]	1.09E+01			1.33E+00	4.51E−01
Climate change (CC)—radiative forcing as global warming potential (GWP100) [kg CO$_2$ eq.]	2.66E+03			2.14E+02	7.89E+01
Climate change, biogenic—radiative forcing as global warming potential (GWP100) [kg CO$_2$ eq.]	5.65E+00			7.29E−01	8.10E+00
Ecotoxicity, Freshwater—comparative toxic unit for ecosystems [CTUe]	4.53E+04			1.97E+03	4.72E+03
Eutrophication freshwater—fraction of nutrients reaching freshwater end compartment [kg P eq.]	5.55E−01			8.83E−04	9.83E−04
Eutrophication marine—fraction of nutrients reaching marine end compartment [kg N eq.]	2.19E+00			6.56E−01	9.59E−02
Eutrophication terrestrial—accumulated exceedance [Mole of N eq.]	2.28E+01			7.28E+00	8.07E−01
Human toxicity cancer—comparative toxic unit for human [CTUh]	5.62E−06			4.00E−08	5.43E−08

(continued)

Table 3 (continued)

LCIA method	Raw material extraction, transportation, and manufacturing			Transportation to site and construction stage	
	A1	A2	A3	A4	A5
EF v3.0 LCIA method	Raw material supply	Transport	Manufacturing	Transport	Construction installation process
Human toxicity non-cancer—comparative toxic unit for human [CTUh]	4.66E−05			2.51E−06	9.56E−07
Ionizing radiation, human health—human exposure efficiency relative to U235 [kBq U235 eq.]	1.14E+02			4.88E−01	6.51E+00
Land Use—soil quality index [Pt]	1.71E+04			1.34E+03	2.77E+02
Ozone depletion—ozone depletion potential [kg CFC-11 eq.]	7.39E−05			2.08E−11	6.08E−07
EF particulate matter—impact on human health [Disease incidences]	1.61E−04			8.64E−06	5.38E−06
Photochemical ozone formation—human health—tropospheric ozone concentration increase [kg NMVOC eq.]	8.14E+00			1.26E+00	2.49E−01
Resource use, fossils—abiotic resource depletion fossils fuels (ADP-fossil) [MJ]	4.06E+04			2.70E+03	1.05E+03
Resource use, minerals and metals—abiotic resource depletion (ADP ultimate reserve) [kg Sb eq.]	2.17E−02			1.76E−05	1.51E−05
Water use—user deprivation potential (deprivation-weighted water consumption) [m^3 world equiv.]	1.01E+03			3.08E+00	2.05E+01

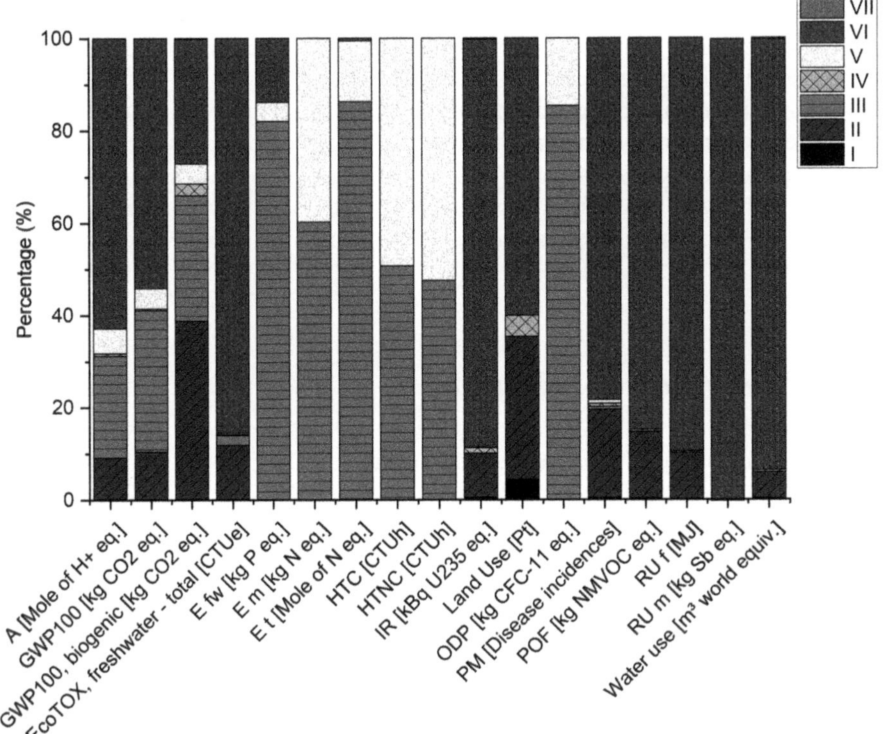

Fig. 2 Construction phase impacts

Regarding material-specific hotspots, steel rebar emerges as a dominant contributor in substructure work, generating 56.3% of particulate matter emissions attributed to dust generation during metallurgical processes and contributing substantially to construction human toxicity (48.5% of cancer impacts) through heavy metal releases. Timber demonstrates concentrated impacts, accounting for 77.1% of land use impacts in superstructure construction and 77.6% of ozone depletion in substructure work, linked to forestry operations and chemical preservation treatments.

Construction waste transport activities constitute significant contributors to environmental impacts, particularly regarding acidification (42.4%) from diesel combustion in heavy vehicles during substructure construction. These transport operations drive terrestrial eutrophication (59.5%) and marine eutrophication (57.8%) through nitrogen oxide emissions. Mixed waste streams display distinctive environmental profiles, particularly regarding biogenic climate change (82.2%), indicating substantial organic carbon content resulting in methane emissions during decomposition.

Precast concrete demonstrates significant contributions to climate change impacts during superstructure construction, accounting for 37.8% of total greenhouse gas emissions primarily from carbon dioxide released during cement production. Other metals used in construction applications show concentrated impacts in mineral and metal resource depletion (37.3%) reflecting intensive mining operations and consumption of limited mineral resources.

Across all construction phases, material production stages (A1-A3) consistently dominate environmental impacts, contributing 70–85% of total construction impacts, while construction process stages (A4–A5) contribute the remaining 15–30%. This pattern emphasizes the critical importance of material selection decisions during construction planning phases rather than solely focusing on construction process optimization.

3.2 Implications and Limitations

The environmental performance analysis reveals critical insights for construction industry sustainability practices. The dominance of MEP installation impacts (60–90% of construction burdens) identifies this construction activity as the primary target for environmental improvement efforts. These findings challenge conventional construction sustainability approaches that often focus primarily on structural materials. The construction phase analysis demonstrates that material production stages (A1-A3) contribute 70–85% of total construction impacts, emphasizing the critical importance of material selection decisions made during construction planning and design phases. Construction waste transport impacts, contributing 42.4% to acidification during substructure construction, identify construction logistics optimization as a significant opportunity for environmental improvement. This study's integration of construction phase KPI data with LCA reveals methodological advances that can transform construction environmental assessment practices. The systematic allocation of construction impacts across seven construction phases enables unprecedented granularity in construction environmental hotspot identification, supporting targeted construction phase improvement strategies.

The construction phase analysis methodology addresses significant limitations in conventional construction LCA approaches that rely on industry-average data and simplified construction assumptions. The integration of real-time construction monitoring data through platforms like Optimize demonstrates the potential for more accurate and actionable construction environmental assessments.

However, several methodological limitations merit acknowledgment. Geographic specificity gaps emerged from using pan-European datasets for construction materials like precast concrete and steel, potentially affecting the accuracy of construction impact assessments for UK-specific construction practices. The NSC's status as a BREEAM "Outstanding" project limits generalisability to conventional construction projects due to featuring advanced strategies like GGBS concrete and passive cooling.

Construction energy consumption data allocation across overlapping construction phases required assumptions about phase-specific consumption patterns. While validated through stakeholder consultation, these allocation methods introduce uncertainties in construction phase impact attribution that could affect the precision of construction hotspot identification.

4 Conclusion

This construction phase study demonstrates the transformative potential of integrating construction KPI data with LCA specifically focused on construction activities. By leveraging detailed primary construction data from Mace's Optimize platform, this research overcomes longstanding challenges in construction sector LCAs, notably the reliance on generic data and simplified assumptions about construction processes. The construction phase analysis reveals MEP installation as the dominant contributor to construction environmental impacts across nearly all categories, accounting for 60–90% of construction burdens. This finding fundamentally challenges conventional construction sustainability approaches and underscores the urgent need for focused interventions in MEP construction practices, material selection, and installation processes.

The integration of detailed construction KPI datasets enabled systematic identification of construction environmental hotspots often overlooked in conventional construction LCAs. Internal finishes contributing significantly to toxicity impacts during construction, polymeric piping systems accounting for substantial fossil resource depletion during MEP construction, and the complex environmental profile of construction waste streams demonstrate the value of granular construction phase analysis.

This research suggests several pathways for improving construction phase sustainability: industry-wide initiatives to standardize Environmental Product Declarations for MEP components to address critical data gaps, enhanced construction phase monitoring and data collection protocols for more accurate environmental impact assessments, development of construction phase-specific improvement strategies targeting identified hotspots, and policy frameworks that support sustainable construction material selection and construction process optimization.

References

Anyanya D, Paulillo A, Fiorini S, Lettieri P (2025) Evaluating sustainable building assessment systems: a comparative analysis of GBRS and WBLCA. Front Built Environ 11. https://doi.org/10.3389/fbuil.2025.1550733

CEN / TC 350 (2011) EN 15978:2011 Sustainability of construction works-assessment of environmental performance of buildings-calculation method

Crawford R (2011) Life cycle assessment in the built environment. Routledge. https://doi.org/10.4324/9780203868171

UK Green Building Council (n.d) Net zero carbon buildings: a framework definition. https://www.ukgbc.org/ukgbc-work/net-zero-carbon-buildings-framework/. Accessed 09 Aug 2025

Guinée JB (2002) Handbook on life cycle assessment: operational guide to the ISO standards. Kluwer Academic Publishers

Department for Business (2013) I. & S., Construction 2025, Industrial Strategy: government and industry in partnership

ISO (2020) Environmental management—life cycle assessment —principles and framework

Jaemoon K, Duhwan L, Seunghoon N (2023) Potential for environmental impact reduction through building LCA (Life Cycle Assessment) of school facilities in material production stage. Build Environ 238:110329. https://doi.org/10.1016/J.BUILDENV.2023.110329

Khasreen MM, Banfill PFG, Menzies GF (2009) Life-cycle assessment and the environmental impact of buildings: a review. Sustainability 1(3):674–701. https://doi.org/10.3390/su1030674

Ramesh T, Prakash R, Shukla KK (2010) Life cycle energy analysis of buildings: an overview. Energy Build 42(10):1592–1600. https://doi.org/10.1016/J.ENBUILD.2010.05.007

Sartori I, Hestnes AG (2007) Energy use in the life cycle of conventional and low-energy buildings: a review article. Energy Build 39(3):249–257. https://doi.org/10.1016/J.ENBUILD.2006.07.001

United Nations (2018) Department of economic and social affairs, population division, world urbanization prospects: the 2018 revision

Integrating LCA in the Design Process. A BIM-Based Tool to Quantify the Environmental Footprint of Railway Infrastructure Projects

Francesca Reale, Carla Di Biccari, Christian Fontana, and Mattia Mangia

Abstract The enhanced circularity required to the construction sector, together with the need to reduce the environmental impacts ask designers and contracting authorities to integrate environmental considerations in the process of design and construction. New "integrated" tools are needed, especially for infrastructure whose design process is complex. In the context of the BIM 4 Rail LCA project, a WebApp was developed able to provide LCA-based information and data for railway infrastructures in an integrated way, interoperable with Building Information Modeling (BIM). The article describes the approach behind the WebApp development, including the description of the context and the main achievements.

1 Introduction

The new concept of infrastructure development cannot be decoupled anymore from an overall evaluation of works under the environmental, social, and economic dimensions. The economic viability of infrastructures has to match social progress, environment, and health protection. In this context, new "systems" should avoid negative impacts on health and ecosystems, while minimizing the impact on climate change and increasing efficiency in terms of material reuse, so as to contribute in the implementation of a circular and sustainable economy. Such need is enhanced by several European policies, among all the New Circular Economy Action Plan (EC 2020) on one hand and the directive for the public procurement (UE 2023) on the other hand, which are putting higher attention on key aspects such as waste prevention and recycling or are pushing toward the use of awarding criteria that consider social and/or environmental aspects addressing any life cycle stage of the works, from raw material extraction to the disposal (UE 2023, art. 64).

F. Reale (✉) · C. Fontana
Ecoinnovazione srl, Bologna, Italy
e-mail: f.reale@ecoinnovazione.it

C. Di Biccari · M. Mangia
Università degli Studi del Salento, Lecce, Italy

© The Author(s) 2026
M. Traverso et al. (eds.), *Life Cycle Management from Global to Local*,
https://doi.org/10.1007/978-3-032-17987-6_14

The project BIM 4 Rail LCA[1] was born to provide railway infrastructures with tools able to address the environmental sustainability and a more efficient management of the design process, construction, and operation of infrastructures. To this aim, the project combines the Life Cycle Assessment (LCA) and the Building Information Modeling (BIM), i.e., the two approaches that better answer, from a scientific and standardization perspective, the two issues. More in detail the WebApp "Rail LCA" was developed, which provides LCA-based information and data for railway infrastructures, in an integrated way, interoperable with BIM. The final goal is to provide environmental sustainability evaluations to be combined with other evaluation types at the different stages of the design, construction, and operation of the railway infrastructure.

The present article explains the methodology applied in the project, including the description of the context, and the main achievements from the point of view of both the final result (WebApp) and main outputs (e.g., the LCA/LCI database).

1.1 The Railways Infrastructure and Design

Each railway infrastructure is made of a civil infrastructure, a railway superstructure, and technology systems. The civil infrastructure gathers alternative types of projects, namely embankment, cutting, bridge (also named viaducts), and tunnel, which can be natural or artificial. The embankment is used to bring up to a defined height the railway infrastructure, and it is basically implemented with soils from a quarry (natural aggregates), which can be mixed with recycled aggregates. On the contrary, the cutting is done to lower down the railway infrastructure level. Viaducts are designed to maintain the infrastructure level higher compared to that one of the surrounding ground, whereas tunnels are designed to allow the transit of railway lines through natural obstacles like mountains. The superstructure includes the ballast and all other components and processes needed to build the track systems. Technology systems include electric traction, signaling systems, and communication systems, mechanical and safety systems (e.g., fire prevention, heating, cooling).

The design process of railway infrastructures starts with a preliminary design, which is the first phase. It is known in Italian as PFTE—Technical and Economic Feasibility Design Project. The following phase is named PD—Detailed design Project, which implies a more technical design. The output of this phase is used as a basis for the public procurement of the construction phase (see also Sect. 2.2). The BIM is used since the preliminary design.

According to the standards adopted in Italferr, the comprehensive BIM model of a railway project is organized into a series of families—that is, spatial structures corresponding to the principal families of the railway infrastructure (Embankments,

---

[1] The project, implemented in 2020–2023, was funded by the Ministry of Economic Development within the National Operative Programme "Business and Competitiveness 2014–2020-Funds for sustainable growth: Smart Factory, Agrifood - Prog. n. F/190007/01-03/X44.

Cuttings, Viaducts, etc.). Such families are reflected in the BIM models as entities that can be traced back to Industry Foundation Classes (IFC) such as IfcFacility (bridge) or IfcFacilityPart or IfcFacilityPartTypeEnum (buildingSMART 2024). In the BIM authoring environment, each object is assigned two high-level key parameters: Work Breakdown Structure (WBS), which defines the macro-family to which the object belongs (for instance, "Embankment" or "Viaduct") and ObjectName, which specifies the exact type of element (for example, "subgrade," "channelDrain," "subballast").

During the PFTE, models are developed at a lower Level of Development (LOD) than those produced in the PD. As the project advances, additional detail is progressively introduced to each BIM object's geometry and parameters. To streamline cost estimation and speed up the design stage, the company employs a standardized set of Bill of Quantities (BoQs), each of them identified by a "GOR" code (Representative Homogeneous Group). Each GOR code basically describes the composition of the unitary quantity of a specific BIM's object, i.e., of a specific project component. Such standardized BoQs are based on design requirements valid at Italferr. BIM specialists populate custom parameters within each object so as to enable automatic matching with the appropriate GOR. For example, a foundational pile in a Viaduct might employ parameters "Diameter," "BentoniteSlurries," and "FormTube"—set to "1500," "False," and "True" respectively, to identify BoQ/GOR VI PL 03.

Different disciplines (civil engineers, structural specialists, technical system designers, etc.) each develop their own BIM models using tools suited to their scope (e.g., Autodesk Revit for discrete elements, Bentley OpenRail for linear infrastructure). These discipline-specific files are exported to IFC and federated into a single, coordinated model. All BIM files and revisions are stored in a Common Data Environment (CDE), where version control and change tracking ensure data consistency throughout the project lifecycle. This integrated approach allows Italferr to maintain a direct link between the graphical model, BoQ cost schedules, and GOR definitions, thereby optimizing both design coordination and cost management.

1.2 Tools for Design and Barriers in the Integration of Environmental Considerations in the Design Process of Infrastructure

Generally speaking, when designing railway infrastructure excluding buildings such as train stations, spare parts warehouses, etc., the types of works involved correspond to the categories identified by Italferr, namely, civil infrastructure, railway superstructure, and technology systems.

Within the BIM methodology, there are several software tools specifically designed for railway infrastructure design (Zhi et al. 2018; Pasetto et al. 2020), including Bentley OpenRail Designer, Autodesk Civil 3D, Midas CIM, and Sierra-Soft Infra Design Studio.

Compared to the BIM design of discrete structures such as buildings, in the infrastructure sector the starting point is the railway path and its integration within the territory. The path is both a conceptual and graphical element composed of lines, curves, and clothoids that must meet both the safety and navigability requirements of the route as well as the constraints of the surrounding terrain, to ensure that the infrastructure is environmentally and economically sustainable.

Therefore, the foundational basis for evaluating the design aspects of a railway line consists of the terrain represented by a Digital Elevation Model (DEM) and the path, which includes two key components: one is the XY planimetric layout of the route, as if seen from above on a 2D map; the other is the associated elevation profiles that represent the Z-coordinate (height) of each point along the 2D patch. This path entity is used as the reference for the 3D BIM design of the railway. From this, using parametric profiles, the different sections—such as elevated segments, cuttings, embankments, and tunnels—are generated. Where the creation of 3D models of bridges/viaducts is required, given their specific structural characteristics, the design is integrated using software such as Autodesk Revit, Tekla Structures, or other BIM tools developed primarily for vertical or discrete works.

Today, a significant amount of environmental footprint data is available covering construction products and/or components. A relevant part of them comes from EPD Programmes which gather and make public Environmental Product Declarations (EPDs), and where, in the last 10 years, a significant increasing of EPD was registered. In some cases, national benchmarks have been developed; this is the case of the thermal insulation in Italy.[2] However, most of the available data concerns products and components related to the building sector (e.g., windows, doors, paints for indoor, etc.). On the contrary, when it comes to the infrastructures, few data are available (e.g., for raw materials like cement or concrete), and initiatives are scarce. Indeed, an ISO standard to quantify the environmental footprint of infrastructures does not exist, and environmental data and information on products and components used in infrastructures or environmental footprint studies of infrastructures themselves are available only to a limited extent (Reale et al. 2024). In turn, such a gap prevents the integration of environmental considerations in the design process of infrastructures, where the potential for reducing the environmental impacts is the highest.

Indeed, today, designers and/or engineering firms are not able to integrate environmental footprint data in their design process because of missing data also combined with unsustainable efforts in terms of time needed for that integration. Contracting authorities cannot identify life cycle-based environmental criteria or a benchmark against which designers can measure their offers. In addition, contracting authorities might have difficulties in the procurement procedure, due to the lack of a well defined and recognized methodology to measure the environmental footprint of infrastructure and evaluate the offers.

---

[2] This was done in the context of the Green Public Procurement. An excel tool can be accessed and downloaded at https://gpp.mase.gov.it/strumenti.

2 Setting an Integrated Design Process for Railways Infrastructure

The approach applied in the project to deliver the WebApp included: the systemic collection and qualification of data related to systems and components used in the railway families, the development of methodology requirements for the quantification of the (Environmental Footprint) EF of the railway infrastructures within the WebApp, the definition of the requirements of the WebApp, the development of the LCI/LCA data, the development of the WebApp itself, the testing and validation of the WebApp in case studies.

The new tools and methodologies introduced by means of the activities of the BIM 4 Rail LCA project reflected a change in the design process. This change has been addressed and documented with a business process mapping activity in which the traditional design process, known as the AS-IS Process, shifted toward the TO-BE Process. In the context of the professional discipline of Business Analysis, the term "AS-IS process" refers to a detailed analysis of how a business process is performed at the current state, while the term "TO-BE process" depicts the future version of the process, which is intended to enhance it and to address gaps and issues detected within the "AS-IS process" (Vanwersch et al. 2016). The two versions of the process have been reported by means of the BPMN standard (Business Process Modeling Notation) (Allweyer 2016). The most evident novelty introduced in the process consists in the introduction of a dedicated LCA Specialist as a new actor, which integrates its workflow with the other design offices for the railway path, tunnel, embankments, etc. The BPMN allows to understand how the process changed and how the developed tool impacts on activities providing new capabilities and opening to new interactions and sub-processes between the actors.

2.1 Data Collection and Completion

In order to support the development of LCA/LCI data, an analysis of railway infrastructures has been performed by Italferr. Systems and components used in the railway for each type of railway infrastructure have been analyzed and BoQs defined based on information already available at Italferr, complemented with additional information based on internal expertise (e.g., interview to own designers). The strategy put in place within the project was to describe the composition of systems and components following the same approach used in Italferr to quantify the cost of projects. Thus, information at the core of the resulting BoQ for systems and components are those used by Italferr in the analysis of RFI (Italian Railways Network) tariffs and addressing the materials used, the processing, and the transport.

Such a strategy was adopted for several reasons. One of such reasons is to ensure the consistency with the other tools already put in place in Italferr to answer to other environmental commitments or internal policies targeting the carbon footprint, (ii) optimizing the efforts by Italferr to collect data and future efforts to maintain data.

2.2 LCI/LCA Development and the Interface Requirements for the Tool

The development of the LCI/LCA data for systems and components used in the infrastructure projects required, firstly, the definition of a methodology to quantify the environmental footprint of the railway infrastructure. This methodology is built upon standard for LCA, intersectoral and specific for construction products (ISO 14040, ISO 14044 and EN 15804 + A2) and on existing literature specific for railways, namely the International EPD® System Product Category Rules (PCRs) 2013:19 for railways (version 2.11), LCA case studies from previous Italferr projects and the Italferr Guidelines for LCA of Railway Infrastructure (2018).

The methodology defines, among other aspects, the type of data to be used for environmental footprint quantification based on the design stage and the quality of available environmental data (e.g., generic datasets). Infrastructure elements are described at the component level, and for each component (and sub-component, where applicable), the material and energy input/output flows are specified and quantified, per unitary amount of component. This approach enabled the creation of an LCI/LCA dataset for each component, potentially usable in the overall set of railway projects (embankment, tunnel, etc.). All datasets are gathered within the BIM Rail LCA tool database, where they are referenced (named) using the same GOR code in use at Italferr. For example, when referencing the GOR VI PL 03, the dataset represents the LCI/LCA data of 1 m^3 of a foundational pile of 1500, in concrete C25/30 N/mm^2, implemented with temporary form-tubes.

This tailor-made structure allows the BIM Rail LCA tool to easily assign the appropriate dataset to each BIM's object (project component), streamlining the bill of materials quantification process and reducing the level of LCA expertise required for the assessment.

Starting from standardized BoQ provided by Italferr (see Sect. 2.1), each component has been quantified by Ecoinnovazione which performed the LCA analysis using a system boundaries "cradle to gate (modules A1-A3)," with construction phase (modules A4-A5), use phase (B1-B5) limited to regular maintenance, end-of-life (modules C1-C4) and optional module D. Ecoinnovazione also generated the datasets using the EPD-Editor v.5. In fact, the format adopted for the datasets is the ILCD + EPD, which is typically used to vehiculate in a structured way information included in the EPD of construction products and/or simply compliant to EN15804 (Reale et al. 2024).

The database currently includes several datasets, namely 22 for embankments' components, 17 for cuttings' components, 16 for bridges' components, 10 for tunnels' components, 5 for railway superstructure' components, and 11 for technology systems components.

In the context of the activities aimed to develop the methodology for quantifying the environmental footprint of railway infrastructures, the datasets and the database, also computational and interface requirements, were drafted to support the tool development. The computational requirements address the impact calculation, the results calculation (e.g., per km and per whole project length), and the data gap treatment. The interface requirements concern the input and output interface and identify minimum content to be visualized (e.g., the length of the project under assessment, the amount of each GOR code involved in the project) and the minimum interactions to be provided (e.g., the amount validation, the button for launching the assessment, the button to export project data and LCA results, etc.).

2.3 The WebApp Development and Validation

To enable effective interaction between BIM data and LCA workflows, particularly within infrastructure projects such as railways, a web-based application was developed with the goal of being independent from commercial BIM authoring tools or LCA software. The choice to implement the solution as a WebApp was led by a literature review performed in the context of the project, in which several BIM/LCA integration methods were assessed and analyzed by authors of previous reviews (Crippa et al. 2020; Potrč Obrecht et al. 2020; Xue et al. 2021). The choice of the WebApp approach addressed several gaps, such as the scarcity of integration methods based on BIM open non-proprietary software and formats, and the infrastructural sector, which is less addressed compared to buildings.

The application was designed to be accessible to professionals from both the BIM and LCA domains, offering a shared environment where users can view and verify model data, assign classification parameters, and retrieve impact assessment results. The chosen approach relied on a microservice architecture, which allows the system to be composed of independent modules—each responsible for a specific task—while remaining flexible, scalable, and easy to maintain. This architectural choice enabled the integration of various functionalities such as querying IFC files, performing classification procedures, and calculating LCA results based on a rail-specific dataset. The web application is installable on local servers and was developed using open-source libraries and frameworks, promoting adaptability and transparency.

From a functional standpoint, the system comprises a set of interactive modules—such as the IFC Checker, IFC Viewer, and LCA Results Report—which allow users to navigate the model, validate classification attributes (GORs), and visualize environmental impact results. Other background services, such as the GOR Identifier and BIM-LCA Calculation modules, automate the data mapping process between BIM objects and LCA records. These services interpret the attributes of BIM elements and match them with predefined environmental profiles (GORs), enabling the quantification of impacts based on official Bill of Quantities (BoQ) entries.

The WebApp offers an intuitive React-based interface, underpinned by Python and JavaScript frameworks, with data stored in a PostgreSQL database and a BIM viewer based on the IfcOpenshell Library (ifcOpenShell n.d). This architecture enables users to iteratively manage input and regenerate LCA results without relying on specialized software, enhancing accessibility and interoperability. Users can select lifecycle stages according to EN 15804 (e.g., A1–C4, D) and apply a domain-specific LCA dataset for railway infrastructure, derived from the ILCD database via semi-automated conversion using openLCA.

To cover maintenance phase assessments (stage B), a custom script calculates required quantities directly in the web environment, avoiding the need to enrich BIM models. For instance, Italferr's declared units for tunnel wall maintenance were used by recognizing BIM elements through coded names. A second script aligns LCA profiles with site-related tasks and BoQ items by leveraging parametric BIM data.

Results are expressed using EN 15804 indicators (e.g., Global Warming Potential, Acidification, Resource Use, etc.), and following the requirements mentioned in the methodology defined in the context of the project, supporting benchmarking and reporting needs. Validation against commercial LCA software, in collaboration with Ecoinnovazione, showed relative errors below 0.5%, confirming the tool's accuracy and suitability for design and sustainability workflows.

3 Discussions

The developed WebApp represents an example of BIM-LCA integration within a specific sector of civil engineering, namely railway infrastructure, while adopting a generalizable and transferable approach that could be extended to other infrastructural and building contexts. As with the integration of other data sources, the combination of BIM and LCA requires careful attention to the development, maintenance, and management of the underlying data model. In the case of Italferr, the adoption of an existing Work Breakdown Structure derived from the company's cost tariff system demonstrates how a single data model can effectively serve multiple interrelated purposes—specifically, economic feasibility analysis and environmental sustainability assessment. Another key design criterion was the platform's ease of use. Designed for both BIM and LCA specialists, the WebApp features a simplified user interface and a user manual that clearly explains its functionalities. Reporting tools, the ability to modify the design categories of building elements (i.e., GORs), and the option to save temporary changes in .csv format without altering the original IFC file enable users to explore design alternatives even with limited expertise in either domain—for example, a BIM specialist with minimal LCA knowledge, or vice versa.

Compared to other approaches found in the literature, which often rely on proprietary plugins embedded in BIM authoring software, the strategy adopted in this research emphasizes an open-source architecture. While this may limit the implementation of certain user-friendly features—such as attribute-based color overlays, which are natively supported in tools like Autodesk Revit—it aligns more closely with openBIM principles and the requirements set out in the Italian Public Procurement Code, especially the new one (D.Lgs. 36/2023 2023), which mandates the use of non-proprietary methods and tools for information management. Moreover, in the case of Italferr—and potentially in other similar contexts—BIM models may be developed by different providers and with different BIM authoring tools, which would have required a BIM/LCA integration effort for each of adopted software. Therefore, ensuring independence from specific software ecosystems contributes to a more interoperable and scalable solution across various phases and stakeholders in infrastructure development.

4 Conclusions and Outlook

This work has presented the results of a BIM-LCA integration approach for the early design stages of railway infrastructure. The research highlighted how such integration entails a shift not only in the technologies to be adopted but also in the need to incorporate appropriate resources and expertise within processes and methods that themselves require transformation. This change represents the introduction of LCA within a workflow already undergoing transformation due to the mandatory adoption of BIM in the Italian context. However, this additional layer (i.e., LCA), if properly supported by suitable integration methods, builds upon the information management logic introduced by BIM, enabling the management of LCA-relevant data in early design stages through attributes embedded in parametric BIM models.

In this research, the integration was achieved using open, non-proprietary formats, a key objective for any kind of data integration that extends beyond the core design aspects (e.g., structural capacities, bill of materials, etc.). Through the creation of a dedicated LCA dataset and a WebApp designed to manage this dataset in conjunction with user-uploaded IFC models, the platform provides users with a clear and entry-level overview for assessing design solutions that incorporate sustainability considerations from the outset of the project.

References

Allweyer T (2016) BPMN 2.0: introduction to the standard for business process modeling, 2nd updated and extended ed. BOD-Books on Demand

buildingSMART, IfcRailwayPartTypeEnum. https://ifc43-docs.standards.buildingsmart.org/IFC/RELEASE/IFC4x3/HTML/lexical/IfcRailwayPartTypeEnum.htm. Accessed 15 July 2025

Crippa J, Araujo AMF, Bem D, Ugaya CML, Scheer S (2020) A systematic review of BIM usage for life cycle impact assessment. Built Environ Proj Asset Manag 10(4):603–618

D.Lgs. 36/2023 (2023) Decreto Legislativo 31 marzo 2023, n 36, 2023. https://www.normattiva.it/uri-res/N2Ls?urn:nir:stato:decreto.legislativo:2023-03-31;36. Accessed 15 July 2025

EN 15804:2012+A2:2019, Sustainability of construction works–environmental product declarations–core rules for the product category of construction products

EN ISO 14040:2006 (2006) Environmental management–life cycle assessment–principles and framework

EN ISO 14044:2006 (2006) Environmental management–life cycle assessment–requirements and guidelines

European Commission (EC) (2020) Communication from the commission: a new circular economy action plan for a cleaner and more competitive Europe. COM(2020)98, 11.03.2020

European Union (EU) (2014) Directive 2014/23/EU of the European parliament and of the council on the award of concession contracts, 26.02.2014

Italferr (2018) Linee guida per la valutazione LCA di infrastrutture ferroviarie, Report

ifcOpenShell. https://ifcopenshell.org/

Pasetto M, Giordano A, Borin P, Giacomello G (2020) Integrated railway design using infrastructure-building information modeling: the case study of the Port of Venice. Transp Res Procedia 45:850–857

Potrč OT, Röck M, Hoxha E, Passer A (2020) BIM and LCA integration: a systematic literature review. Sustainability 12(14):5534

Reale F, Buonamici R, Fontana C, Genovesi S, Iuliano D, Khodadadi J, Masoni P, Parisi L, Salamanno G, Zamagni A (2024) Sustainability of construction sector: development of an LCA database for the quantification of the environmental footprint of railways infrastructures. Proc Ital LCA Netw Conf, Pescara

The International EPD® System, Product Category Rules (PCRs) 2013:19 for railways, v. 2.11

Vanwersch RJB, Shahzad K, Vanderfeesten I, et al (2016) A critical evaluation and framework of business process improvement methods. Bus Inf Syst Eng 58(1):43–53

Xue K, Hossain MU, Liu M, Ma M, Zhang Y, Hu M, Chen XY, Cao G (2021) BIM integrated LCA for promoting circular economy towards sustainable construction: an analytical review. Sustainability 13(3):1310

Zhi P, Shi T, Wang W, Wang H (2018) Application of BIM technology in the construction management of shield tunnel. Proc 4th Int Conf Inf Manag (ICIM), IEEE, 284–289

Trenchless Technologies for Modern and Sustainable Infrastructures

Alexandra Kalnev, Stefan Schmitz, Monu George Varghese, and Zaid Hashash

Abstract This article evaluates the environmental performance of trenchless construction technologies, focusing on TRACTO's GRUNDOMAT 45 soil displacement hammer, in comparison to traditional open-cut excavation. Part of an ongoing study also covering Horizontal Directional Drilling (HDD) and pipe bursting methods, the analysis applies a cradle-to-gate Life Cycle Assessment (LCA) in line with ISO 14040 and ISO 14044 standards. Environmental impacts are assessed across eight categories. Findings show that trenchless methods, especially the GRUNDOMAT 45, result in notably lower impacts—particularly in global warming potential (GWP) and resource use—due to reduced excavation and material needs. These outcomes align with existing LCA research and support the integration of life cycle thinking in infrastructure planning. The study also reflects how sustainable construction practices can support broader goals such as the European Green Deal and the UN Sustainable Development Goals.

1 Introduction

The global economy is facing major changes. Challenges such as climate change, urban growth, and increasing public awareness of environmental issues are putting pressure on governments, companies, and societies to act. These global trends require new ideas and long-term solutions to support sustainable development.

At the international level, several frameworks shape the way we understand and implement sustainability. The United Nations Agenda 2030 set out 17 Sustainable Development Goals (SDGs) to guide global transformation across environmental, social, and economic dimensions (UN 2015). The Paris Agreement focuses on limiting global warming to below 2 °C, with efforts to stay within 1.5 °C (UNFCCC

A. Kalnev (✉) · S. Schmitz
TRACTO-Technik GmbH & Co. KG, Lennestadt, Germany
e-mail: alexandra.kalnev@tracto.com

M. G. Varghese · Z. Hashash
Institute of Sustainability in Civil Engineering, RWTH Aachen University, Aachen, Germany

M. Traverso et al. (eds.), *Life Cycle Management from Global to Local*,
https://doi.org/10.1007/978-3-032-17987-6_15

2015). In Europe, the European Green Deal outlines a pathway to make the continent climate-neutral by 2050, including strategies for clean energy, circular economy, and biodiversity (European Commission 2019).

For businesses, sustainability has become a core strategic concern. Integrating Environmental, Social, and Governance (ESG) factors is essential not just for meeting regulations, but for staying competitive and gaining social acceptance (OECD 2020).

In this context, sustainability is no longer just a responsibility—it is a strategic opportunity for companies to lead and adapt in a changing world, particularly within the construction and infrastructure sector, which plays a vital role in advancing sustainable development. Trenchless methods for installing pipes and cables illustrate this potential by offering significant environmental, economic, and social benefits — including reduced emissions, lower material use, minimized surface disruption, and improved safety — yet they remain underutilized in many projects. The study aims to explore the potential of trenchless construction technologies, particularly soil displacement methods, and assess their environmental sustainability impacts.

1.1 The Role of Life Cycle Thinking in Global and Political Context

In recent years, sustainability regulations in Europe have become increasingly strict and binding. Following the adoption of the European Green Deal, the European Union (EU) has introduced a series of legal measures that go beyond reporting duties, requiring companies to actively incorporate sustainability into their operations (European Commission 2019). These policies aim to ensure that environmental and social impacts are addressed along a product life cycle.

The financial sector plays a key role in this transformation. As investment decisions increasingly consider ESG criteria (few of them include life cycle assessment), capital is being directed toward sustainable technologies and projects. This reinforces the importance for companies to innovate and develop solutions aligned with climate and sustainability goals (OECD 2020).

The United Nations Agenda 2030 further underlines the need for systemic change. With 17 SDGs, it encourages private sector engagement in tackling global challenges such as poverty, inequality, and environmental degradation (UN 2015). Similarly, the Paris Agreement promotes a global framework to limit climate change, with each country setting its own emissions targets (UNFCCC 2015).

In this political context, life cycle thinking (LCT) has emerged as a practical and science-based approach to evaluate the long-term impacts of products and infrastructure (ISO 2022a, b; Finkbeiner et al. 2010). Life Cycle Assessment (LCA), as defined by international standards (ISO 14040:2022; ISO 14044:2022), provides a consistent methodology for assessing environmental performance throughout a system's life cycle. When extended into Life Cycle Sustainability Assessment (LCSA), this approach also integrates economic and social dimensions, enabling more comprehensive evaluations of sustainability performance (Finkbeiner et al. 2010).

Trenchless construction methods are increasingly assessed using LCA to compare their environmental performance with conventional open-cut techniques. Studies indicate that these methods may result in lower energy use and reduced greenhouse gas emissions due to minimized excavation and material needs. Key impact categories include resource consumption and emissions (Chadalawada 2024).

Incorporating life cycle-based approaches into infrastructure planning and industrial strategy helps align technology choices with regulatory goals and societal expectations (Del Rosario and Traverso 2023). It also provides a solid foundation for comparing the long-term effects of different technical solutions and for promoting resource-efficient, socially responsible, and economically viable outcomes.

1.2 *TRACTO Strategy for Sustainable Pipeline Construction*

TRACTO-Technik GmbH & Co. KG (hereafter referred to as TRACTO) is a pioneer in the field of trenchless technology. The company has long focused on enabling underground infrastructure development and pipeline renewal without the need for traditional open-cut excavation. Instead of using large-scale earthmoving equipment to dig open trenches, TRACTO applies "closed" or trenchless construction methods, offering more sustainable alternatives in terms of environmental performance, cost efficiency, and social impact.

From an early stage, TRACTO recognized that innovation and sustainability are inseparable when it comes to minimizing environmental impacts in construction. This principle has guided TRACTO for formulating a clear strategic approach to environmentally sustainable pipeline construction, which forms also the basis for the study.

The goal is to transform global infrastructure toward more sustainable, efficient, and resilient utility systems. This can only be achieved by continuously innovating advanced trenchless technologies and ensuring they are technically mature and compliant with applicable industry and regulatory standards.

This strategic framework takes a life cycle approach, which underpins not only the company's technological development but also its broader corporate responsibility efforts. The present LCA was initiated to evaluate and validate the environmental performance of trenchless technology as part of the company's commitment to sustainable infrastructure and strategic orientation.

2 Methodological Framework—Life Cycle Assessment Applied to Trenchless Construction

To evaluate the environmental impacts associated with modern underground construction methods, a comparative LCA was conducted. The study focused on analyzing the differences between a conventional open-cut trenching technique and a trenchless installation method,—specifically, a soil displacement hammer—implemented using the GRUNDOMAT 45 system developed by TRACTO. In addition to this method, two further trenchless techniques are included in the broader study: the HDD and bursting.

The soil displacement hammer method works by propelling a pneumatic hammer through the ground, displacing the soil to create a borehole into which cables or pipes can be inserted. Small entry and exit pits are excavated at both ends of the borehole, while the ground surface in between remains untouched. Figure 1 illustrates this process.

The methodology follows the internationally recognized LCA framework outlined in ISO 14040:2022 and ISO 14044:2022, along with guidance from ISO 14075:2024 for civil engineering applications. The study is further informed by the Life Cycle Sustainability Assessment (LCSA) framework (Finkbeiner et al. 2010), though the current analysis focuses solely on environmental indicators.

The goal of the study is to quantify and compare the environmental burdens associated with the underground installation of a 10-m optical fiber cable using the aforementioned construction techniques. The functional unit is defined as the complete installation of this 10-m segment. The assessment adopts a cradle-to-gate system boundary, including raw material extraction, energy and equipment used during construction, and site-level impacts. End-of-life processes and use-stage considerations are excluded.

Primary data for the trenchless scenario were obtained through TRACTO's collaboration with Nosconnect GmbH, capturing field-specific operational data for the GRUNDOMAT 45 (Fig. 1). The conventional open-cut trenching scenario was

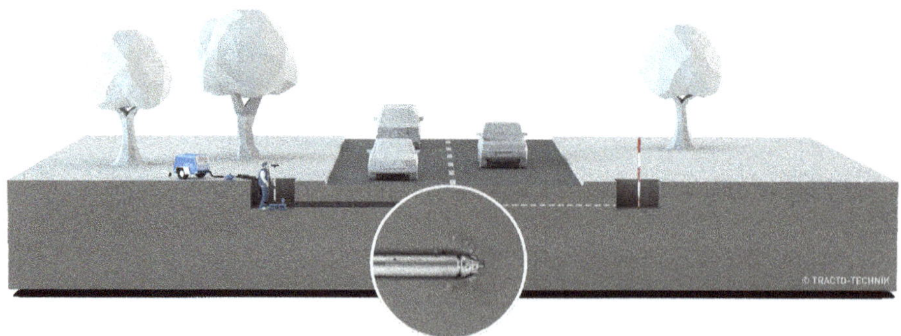

Fig. 1 Soil displacement hammer method using the GRUNDOMAT 45

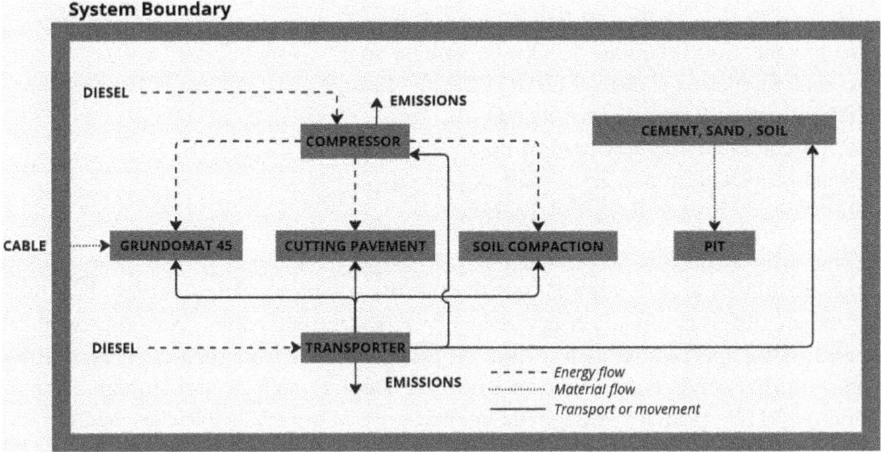

Fig. 2 System boundary for the LCA of the soil displacement hammer method using the GRUNDOMAT 45 (cradle-to-gate)

modeled using representative secondary data from the Ecoinvent v3.7 database and verified industry assumptions.

The system boundary (Fig. 2) covers all major processes associated with the soil displacement hammer method, including the operation of key equipment such as the GRUNDOMAT 45, compressor, transporter, and compacter. A cutting tool, such as a Hilti demolition hammer, is used to break concrete surfaces and remove pavement layers prior to pit excavation. The optical fiber cable itself is excluded from the system boundary, as it is identical in both the trenchless and open-cut scenarios. The life cycle inventory (LCI) includes both primary and secondary data. As previously noted, primary data was obtained from TRACTO's field use of the GRUNDOMAT 45 system, while secondary background data was sourced from the ecoinvent v3.7 database. Inventory inputs include diesel fuel for machinery (compressor, transporter), backfill materials (cement, sand, soil), and the use of a Hilti demolition tool for initial access. Emission factors related to fuel combustion and material processing were included in the analysis based on established LCI datasets.

3 Results and Interpretation

To comprehensively evaluate the environmental impacts of the two scenarios, a life cycle impact assessment was conducted using the CML (Centre for Environmental Science) baseline method, developed by the Institute of Environmental Sciences at the University of Leiden in the Netherlands (Ecoinvent 2023). The assessment follows ISO 14044 guidelines and includes the following midpoint impact categories:

- Global Warming Potential (GWP, kg CO_2 eq)

- Human Toxicity Potential (HTP, kg DCB eq)
- Eutrophication Potential (EP, kg Phosphate eq)
- Ozone Depletion Potential (ODP, kg CFC-11 eq)
- Acidification Potential (AP, kg SO_2 eq)
- Photochemical Ozone Creation Potential (POCP, kg C_2H_4 eq)
- Resource Depletion (Fossil Fuels, kg oil eq)
- Ecotoxicity Potential (kg 1.4-DCB eq)

The results (Fig. 3) show that the conventional open-cut method consistently exhibited higher environmental impacts across all selected indicators. The most significant difference was observed in GWP, where emissions from the open-cut scenario were approximately five times higher than those from the trenchless method were. Similar trends were found in Resource Depletion (RD) and Human Toxicity Potential (HTP), primarily due to the intensive use of materials like cement and sand, and energy-intensive excavation equipment.

The main contributors to the elevated impacts in the open-cut scenario were the substantial use of cement and sand, along with emissions from extensive soil excavation and backfilling using heavy construction equipment. In contrast, the soil displacement hammer method, implemented with the GRUNDOMAT 45, required only localized excavation, significantly reducing the need for material inputs and fuel consumption, and thereby limiting overall environmental disturbance.

These findings are supported by LCAs in civil infrastructure contexts, where trenchless techniques have been shown to reduce CO_2 emissions by up to 59% and overall ecological impact by 60–70% (Chorazy et al. 2024; Kaushal and Najafi 2020). Additional research has confirmed that avoiding large-scale excavation and material transport contributes to both lower fuel consumption and reduced emissions

Fig. 3 Comparative environmental impacts of open-cut and trenchless methods across selected LCA categories

that affect human health (Loss et al. 2018; British Columbia Climate Action Guidance 2024). Together, these studies reinforce the potential of trenchless technologies to support more sustainable infrastructure development when guided by life cycle thinking. While previous studies such as Chorazy et al. (2024) and Kaushal and Najafi (2020) provided valuable insights into the environmental performance of trenchless versus open-cut techniques—primarily focusing on sewer and water infrastructure—this study contributes a novel perspective by evaluating the soil displacement hammer method specifically for optical fiber cable installation. It further distinguishes itself by relying on recent primary field data from TRACTO's operations, applying a cradle-to-gate boundary, and concentrating solely on environmental indicators relevant to small-scale, high-precision utility projects.

4 Discussion

The results of the study indicate that the soil displacement hammer method significantly outperforms the conventional open-cut trenching approach in terms of environmental performance across all assessed midpoint impact categories. This advantage stems from the inherently less invasive nature of the trenchless technique, which minimizes the need for extensive excavation, reduces fuel consumption, and eliminates the high material demands associated with backfilling and surface restoration. Notably, the most pronounced benefits were seen in global warming potential, where emissions from open-cut excavation were found to be approximately five times higher. These trends suggest that environmental impacts in underground construction are strongly influenced by material intensity and energy demand, rather than solely by the length or scale of the installation. Furthermore, the results reflect how even relatively small-scale interventions—such as localizing excavation—can yield substantial environmental benefits. These insights reinforce the importance of early design-stage decisions in infrastructure planning and support the broader integration of life cycle thinking in construction practices.

While this article focuses exclusively on environmental indicators, the broader aim of the study is to contribute to a comprehensive sustainability assessment that also considers economic and social impacts.

It is important to note that the current focus on environmental impacts reflects a deliberate first step in the broader research framework. The exclusion of social and economic dimensions in this paper represents a defined scope limitation, rather than a shift in the overall research direction. Using life cycle thinking in this context helped clarify where environmental improvements occur. By looking at the entire construction process—from material supply to installation—LCA provided a more holistic view of the impacts. This approach supports better decision-making for infrastructure planning and highlights how companies like TRACTO are aligning technology development with sustainability goals.

5 Conclusion

The study applied LCA to compare two underground installation methods: conventional open-cut trenching and the soil displacement hammer method using TRACTO's GRUNDOMAT 45. The analysis showed that the trenchless approach resulted in lower environmental impacts in key categories, particularly global warming potential and resource use.

These reductions were mainly due to more efficient construction processes and less material consumption. While results are case-specific, they point to the potential for innovative construction technologies, such as those developed by TRACTO, to contribute to environmental improvement in infrastructure projects.

Overall, the study shows how LCA can support a more sustainable planning in civil engineering by revealing the impacts of different construction strategies. Consequently, sustainable construction projects can be initiated and sustainable infrastructure can be developed.

Further research can expand on this work by including economic and social criteria, as well as additional project types and scales.

6 Future Work and Outlook

While this study focused on environmental impacts using a cradle-to-gate LCA of trenchless versus open-cut cable installation, several areas remain for future exploration. Expanding the system boundary to a cradle-to-grave approach—accounting for operation, maintenance, and end-of-life stages—would provide a more comprehensive view of long-term environmental sustainability performance.

Moreover, the current analysis considers only environmental indicators. A more holistic LCSA could incorporate social and economic dimensions, including worker safety, construction time, traffic disruptions, and cost efficiency. These aspects are not only critical in urban environments but also in other sensitive contexts—such as residential neighborhoods, industrial zones, public facilities, and ecologically protected areas—where minimizing surface disruption, noise, and construction time is essential. In such settings, the societal and economic impacts of infrastructure work can be as important as the environmental footprint and should be systematically considered in future studies.

Further investigations may also assess the trenchless method's performance under varying geographic, soil, and climatic conditions to test its adaptability. The integration of digital innovations—such as AI-based route planning and real-time monitoring—could further improve the precision and resource efficiency of trenchless operations.

As a next step in the broader research effort, a comparative LCSA study is planned to evaluate additional trenchless techniques, including HDD and pipe bursting, in contrast to traditional open-cut methods. This will support a deeper understanding of the sustainability trade-offs across different construction technologies.

As infrastructure sustainability becomes increasingly prioritized in both public and private sectors, methods that combine technical efficiency with minimized environmental and social disruption—such as those applied by TRACTO—will play an essential role in building resilient, future-ready utility systems.

References

British Columbia Ministry of Environment and Climate Change Strategy (2024) Project profile: trenchless technology capital projects (quantifying greenhouse gas emissions). Government of British Columbia, March 1, 2024. https://www2.gov.bc.ca/assets/gov/environment/climate-change/cng/guidance-documents/project_profile_trenchless_technology_guidance.pdf

Chadalawada R (2024) Innovative trenchless technologies for installing underground fiber optic cables are improving efficiency while minimizing environmental impact. Eur J Adv Eng Technol 11(10):85–98

Chorazy T, Hlavínek P, Raček J, Pietrucha-Urbanik K, Tchórzewska-Cieślak B, Keprdová Š, Dufek Z (2024) Comparison of trenchless and excavation technologies in the restoration of a sewage network and their carbon footprints. Resources 13(1):12

Del Rosario P, Traverso M (2023) Towards sustainable roads: a systematic review of triple-bottom-line-based assessment methods. Sustainability 15:11

Ecoinvent (2023) Ecoinvent database v3.7. Swiss Centre for Life Cycle Inventories

European Commission (2019) The European Green Deal, European Commission

Finkbeiner M, Schau EM, Lehmann A, Traverso M (2010) Towards life cycle sustainability assessment: drawing on the triple bottom line. Int J Life Cycle Assess 15(5):373–393

ISO (2022a) ISO 14040:2022–Environmental management–life cycle assessment–principles and framework. International Organization for Standardization, Geneva

ISO (2022b) ISO 14044:2022–Environmental management–life cycle assessment–requirements and guidelines. International Organization for Standardization, Geneva

ISO (2024) ISO 14075:2024–Environmental performance evaluation–principles and guidance for verifiable claims. International Organization for Standardization, Geneva

Kaushal P, Najafi M (2020) Environmental life-cycle assessment for pipe bursting and open-cut pipeline replacement techniques. Tunn Undergr Space Technol 99:103376. https://doi.org/10.1016/j.tust.2020.103376

Loss A, Gagnon GA, Bissonnette B (2018) Comparative LCA of trenchless and open-cut pipe replacement techniques for water infrastructure. J Clean Prod 180:75–83. https://doi.org/10.1016/j.jclepro.2018.01.132

OECD (2020) ESG investing and climate transition: market practices, issues and policy considerations. OECD Publishing

Tracto-Technik. https://en.tracto.com/company. Accessed 28 May 2025

United Nations (2015) Transforming our world: the 2030 agenda for sustainable development. United Nations

United Nations Framework Convention on Climate Change (UNFCCC) (2015) The Paris agreement. UNFCCC Secretariat

Life Cycle Evaluations of Energy Systems and Technologies

Life Cycle Assessment of Thermal Energy Storage in Buildings

Haoyang Dong, Maryna Henrysson, Luka Smajila,
Saman Nimali Gunasekara, and Justin Ningwei Chiu

Abstract Thermal energy storage (TES) plays an important role in enhancing energy efficiency and flexibility in building systems. This study develops a method for conducting life cycle assessment (LCA) studies for evaluating the environmental performance of TES in different configurations. A case study is carried out, comparing phase change material (PCM)-TES, water-TES, and borehole-TES. Cradle-to-grave life cycle inventory is analyzed over a 25-year lifetime, and ReCiPe 2016 midpoint (H) is used for midpoint impact categories quantification. Results show that TES integrations yield the lowest life cycle impact in global warming potential (GWP), surplus ore potential, marine eutrophication and water consumption among others against non-TES reference during operational phase. It is also shown that the predominant GWP impact of TES integration comes from production, installation, and end-of-life phases.

1   Introduction

Building operations are responsible for 30% of global final energy use and 26% of energy-related emissions worldwide (IEA 2023). A global transition of the building sector requires increasing use of renewable energy and energy efficient buildings design and operation. Thermal energy storage (TES) systems play a crucial role in balancing the energy supply and demand and enabling tariff-based operation strategy and peak shaving for cost saving. Technical and economic studies on TES in buildings have been substantially developed, whereas life cycle assessment (LCA) is a relatively new topic that has emerged over the past decade through the performed literature survey presented in the current work. TES can evolve from a purely technical component to a policy-aligned solution that contributes to both emissions reduction and circular resource use. Designing TES with recyclable or bio-based materials,

H. Dong (✉) · M. Henrysson · L. Smajila · S. N. Gunasekara · J. N. Chiu
Department of Energy Technology, School of Industrial Engineering and Management, KTH Royal Institute of Technology, Stockholm, Sweden
e-mail: haoyangd@kth.se

M. Traverso et al. (eds.), *Life Cycle Management from Global to Local*,
https://doi.org/10.1007/978-3-032-17987-6_16

disassembly potential, and robust end-of-life recovery is no longer optional; it is essential for regulatory compliance and long-term sustainability.

This paper addresses a timely and underexplored topic: the integration of energy storage into building systems as a pathway to more sustainable energy solutions. The aim of this study is to evaluate the environmental impact of TES by benchmarking the life cycle performance of a TES-integrated building against that of a conventional reference building without TES. The assessment focuses on key environmental indicators, including energy consumption and greenhouse gas (GHG) emissions. A case study is conducted on a 3,000 m^2 office building in Bergen, Norway, comparing phase change material (PCM) TES, water-TES, and borehole TES configurations over a 25-year life cycle. Realization of TES's full potential as a policy aligned, circular resource solution hinges on involvement beyond the construction sector. Norway's circular buildings roadmap highlights the need for clearer legislation, stronger value chain collaboration, and integrated research infrastructure actors across the entire TES value chain should therefore be included in circularity analyses (Knoth et al. 2022).

## 2	Methodology

### 2.1	Proposed Method for LCA Study of TES in Building Application

Components of the building heating and cooling system, including heat pumps, TES, electric heaters and chillers, circulation pumps, and fans, adhere to Product Category Rules (PCR) 2019:14 "Construction products (EN 15804 + A2)" v2.0.1 (Swedish Environmental Research Institute 2025). According to this PCR, the life-cycle information modules are specified as follows: A1–A3 product stage; A4–A5 construction process stage; B1–B7 use stage; C1–C4 end-of-life stage; D benefit and load beyond system boundary.

While interest in TES within building systems is gaining momentum, LCAs of different TES solutions in building applications remain limited in the existing litera-ture. This study proposes a systematic method for conducting LCA of TES solutions in building heating and cooling systems, in accordance with the LCA standard ISO 14040/44 (International Organization for Standardization 2020) and covering the full life cycle from cradle to grave, as shown in Fig. 1. The goal of the study is to bench-mark the environmental impact of different TES integrations and support early-stage design decisions.

The functional unit is defined as the annual heating and cooling provided by the heating and cooling system, in kWh/yr. Thus, all key components in the system that are involved in the energy and material flow are included in the life cycle inventory analysis. ReCiPe 2016 midpoint (H) method is used to translate inventory flows into environmental indicators.

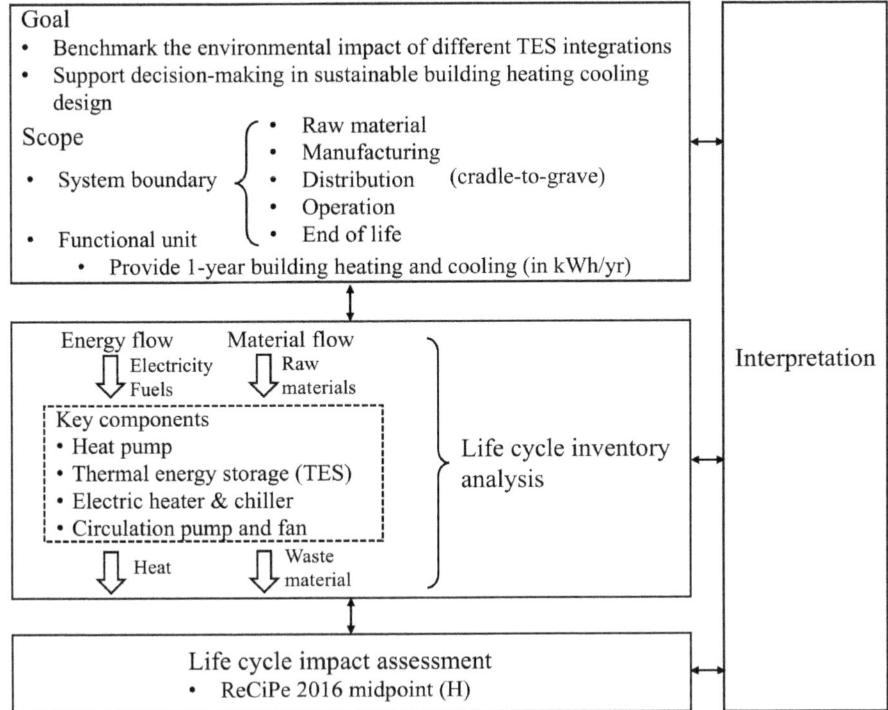

Fig. 1 Proposed framework of LCA study for TES in building heating and cooling application based on ISO 14040

2.2 Case Study

2.2.1 System Description

The case study is a heating and cooling system of an office building in Bergen called Sigba, Norway, which utilizes a ground source heat pump (GSHP) system coupled with boreholes thermal energy storage (BTES) of 1250 m length. The system provides space heating, domestic hot water (DHW), and comfort cooling across a 3,000 m^2 area. Heating is primarily delivered via an 11–44 kW heat pump (Thermia Mega M, R410A refrigerant) supported by a 90 kW electric boiler as peak backup. For cooling, the system combines direct free cooling from the borehole loop, active cooling via the heat pump, and a 146 kWh PCM-TES using Axiotherm 15 (organic and encapsulated), with the PCM charged at night. Thermal response testing confirmed a ground thermal conductivity of 3.0 W/m·K and borehole thermal resistance of 0.10 m·K/W. The system diagram is displayed in Fig. 2.

Figure 3 displays the yearly heating and cooling demand profile of the Sigba office building, based on monitoring from March 2023 to March 2024. A heating load is consistently present most of the year due to DHW demand, with peak demand

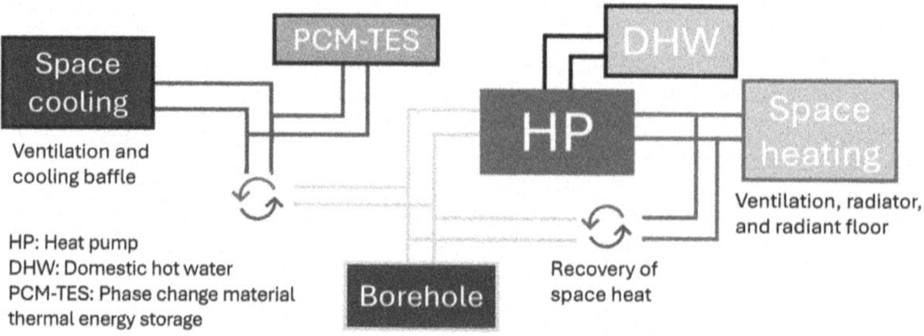

Fig. 2 System diagram of the case study

occurring during winter months due to peak space heating needs. Cooling demand is concentrated in summer, particularly in July. The system reaches a maximum heating demand of 118.9 kW and a maximum cooling demand of 78.2 kW. The total annual heating demand amounts to 186 MWh, while the total cooling demand is 13 MWh. PCM-TES went through 9.1 cycles in one year of operation with a thermal efficiency of 76%. The annual electricity consumption by the pumps of PCM-TES and borehole TES is 158 kWh and 15,250 kWh, respectively.

The case study evaluates four TES configurations, summarized in Table 1. The base case (nr. 0) uses PCM-TES for night charging and daytime discharge. The sensible TES scenario (nr. 1) replaces PCM-TES with a chilled water TES to assess material and energy trade-offs. The extended BTES scenario (nr. 2) doubles the total borehole depth to 2000 m to enhance seasonal TES without additional cooling peak backup devices. The district heating scenario (nr. 3) eliminates on-site TES and BTES, relying instead on a chiller as a cooling device.

The functional unit is defined as annual heating (186MWh/yr) and cooling (13 MWh/yr) provided by the system. All scenarios assume identical building thermal loads and operational schedules. A 25-year system lifetime is assumed in order to take into account the replacement of mechanical equipment such as heat pump and

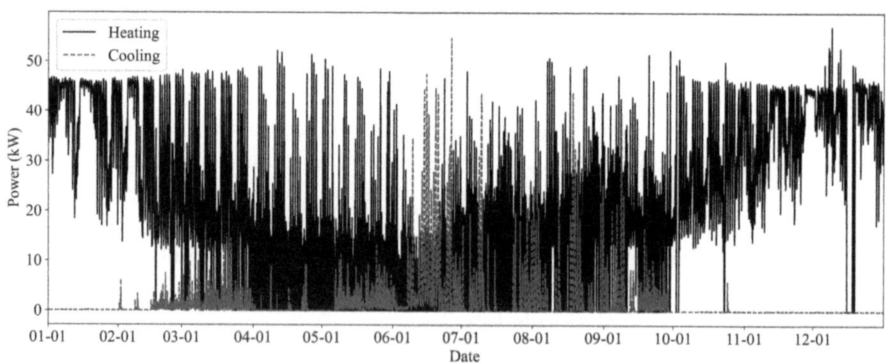

Fig. 3 Annual heating and cooling demand profile of Sigba office building (hourly average)

Table 1 Scenario of different TES solutions

Scenario	Cooling		Heating		BTES length (m)
	Base load	Peak load	Base load	Peak load	
0	BTES	PCM-TES	GSHP (44 kW)	Electric boiler	1250
1	BTES	Water TES	GSHP (44 kW)	Electric boiler	1250
2	BTES	BTES	GSHP (70 kW)	Electric boiler	2000
3	Chiller	Chiller	District heating (90 kW)	Electric boiler	–

chiller, which commonly have a lifetime less than 20 years. The geographical location is Europe by default, and is chosen as global when the specific data for Europe are not available. Reuse, recovery, and recycling potential (module D as defined in the PCRs for construction products) at the end of life are not considered.

2.2.2 Life Cycle Inventory (LCI)

The life cycle inventory (LCI) is generated based on each TES and related equipment in the production of heating and cooling. Environmental product declarations (EPD) and Ecoinvent v3.11 (Ecoinvent 2025) of technical equivalent products are selected covering all raw material, process and energy used in the stage of production, distribution, and end of life. The geographical focus is centered on Europe, with global data integrated where regional sources are lacking. Transportation emissions are included in the analysis. An extrapolation factor is utilized to scale the product to the demand side of the case study in terms of size of the tank, power capacity, length of the borehole for instance, i.e., an extrapolation factor of 2.8 is used for scaling 5600 L of steel tank from available data on heat storage tank of 2000 L. Table 2 summarizes the inventory of the key components. Electric heaters are assumed to be installed in all scenarios, which are only considered during the operation stage as the backup and peak device. The column "Inventory data" refers to the exact activity name from Ecoinvent, as well as the product name from the Norwegian EPD Foundation (2024) and EPD International AB (2023).

Operational LCI accounts for electricity used by circulation pumps, heat loss through storage systems, and replacement of mechanical equipment according to estimated service life. LCI during operational stages from the real case studies, projected over the 25-year lifetime, is summarized in Table 3. Here, the inventory of operational phase only considers electricity and heating.

2.2.3 Life Cycle Impact Assessment (LCIA)

ReCiPe 2016 midpoint (hierarchist) is selected in this study. To transfer different impact category units used in EPDs, conversion factors suggested by Dong et al.

Table 2 Inventory of the key components in the stages of production, distribution, and end of life

Item	Product	Quantity	Unit	Stage	Inventory data	EF
PCM-TES	Paraffin PCM	2562	kg	P, D	Market for paraffin	1
	Paraffin PCM	2562	kg	EoL	Treatment of waste plastic, mixture, municipal incineration	1
	Steel tank, 5600 L	1	–	P, D, EoL	Market for heat storage, 2000 l	2.8
	Circulation pump, 0.5 kW	1	–	P, D	Market for water pump, 22 kW	0.05
	Circulation pump, 0.5 kW	6.8	kg	Eol	Market for waste steel	1
Water TES	Steel tank, 12,300 L	1	–	P, D, EoL	Market for heat storage, 2000 l	6.15
	Circulation pump, 0.5 kW	1	–	P, D	Market for water pump, 22 kW	0.05
	Circulation pump, 0.5 kW	6.8	kg	EoL	Market for waste steel	1
BTES	Borehole 1250 m	1	–	P, D	Market for borehole heat exchanger, 150 m	8.33
	PE100 collector	661	kg	EoL	Treatment of waste plastic, mixture, municipal incineration	1
	Brine, 35 wt% ethanol solution	1500	kg	EoL	Treatment of spent anti-freeze liquid, hazardous waste incineration	1
	Circulation pump, 1.4 kW	2	–	P&S	Market for water pump, 22 kW	0.05
	Circulation pumps, 1.4 kW	38.2	kg	EoL	Market for waste steel	1
Heat pump	Brine-water heat pump 44 kW	1	–	P&S	Market for heat pump, brine-water, 10 kW	4.4
	Refrigerant	4.4	kg	EoL	Treatment of used refrigerant R134a, reclamation	1
	Steel	201	kg	EoL	Market for waste steel	1
	Copper	69	kg	EoL	Market for waste copper	1

(continued)

Table 2 (continued)

Item	Product	Quantity	Unit	Stage	Inventory data	EF
	Plastic insulation	1.1	kg	EoL	Treatment of waste plastic, mixture, municipal incineration	1
Chiller	Chiller 100 kW	1	–	P, D, EoL	Chiller and heat pump VLS 160–315 kW (EPD International AB 2025)	0.4
District heating	Heat	1	kWh	P, D, EoL	District heating in Bergen area (The Norwegian EPD foundation 2024)	1

* Note: P, production; D, distribution; EoL, end-of-life; EF, extrapolation factor

Table 3 Life cycle inventory of the scenarios during operational phase

Scenario	Electricity (MWh)	District heat (MWh)	Equipment replacement	Number of replacements
0	2005.8	0	Heat pump	1
1	2005.8	0	Heat pump	1
2	1634.2	0	Heat pump	1
3	118.2	4650	Chiller	1

2021 are applied. For instance, 1 MJ is converted to 41.9 kg oil-eq for fossil fuel depletion. The electricity emission factor of Norway is 0.018 g/kWh in 2024 (IEA 2024).

3   Result Interpretation and Discussion

Table 4 summarizes the estimated midpoint impacts of all four scenarios. Scenarios 0 (PCM-TES) and 1 (water-TES) exhibit comparable overall impacts, although the PCM-TES system, due to its reliance on paraffin by-products, yields higher fossil-fuel potential and GWP, while the water-TES system has higher impacts across most categories due to its greater steel-related activity. Scenario 2's expanded BTES field substantially increases impacts during borehole installation (a highly fuel-intensive process) and heat pump installation with higher capacity, which drives up acidification, ecotoxicity, and fossil-fuel impacts. Scenario 3, which is based on conventional chillers and district heating, shows the highest impacts overall, particularly in terrestrial acidification, marine eutrophication, and water-consumption potentials.

Table 4 Life cycle impact in the midpoint impact categories

Impact category	Scenario				Unit
	0	1	2	3	
Terrestrial acidification potential (TAP)	205	207	296	632	kg SO_2-Eq
Global warming potential (GWP) 100	92,731	88,530	102,500	120,689	kg CO_2-Eq
Freshwater ecotoxicity potential (FETP)	25,570	26,159	38,517	–	kg 1.4-DCB-Eq
Marine ecotoxicity potential (METP)	31,267	32,061	46,886	–	kg 1.4-DCB-Eq
Terrestrial ecotoxicity potential (TETP)	272,852	299,069	378,455	–	kg 1.4-DCB-Eq
Fossil fuel potential (FFP)	14,886	12,694	17,695	29,476	kg oil-Eq
Freshwater eutrophication potential (FEP)	15	17	21	38	kg P-Eq
Marine eutrophication potential (MEP)	1,2	1,4	1,4	292	kg N-Eq
Human toxicity potential carcinogenicity (HTPc)	28,438	43,748	23,333	–	kg 1.4-DCB-Eq
Human toxicity potential non-carcinogenicity (HTPnc)	167,598	188,887	217,549	–	kg 1.4-DCB-Eq
Ionizing radiation potential (IRP)	892	1004	1151	–	kBq Co-60-Eq
Agricultural land occupation (LOP)	770	1039	753	–	m^2*a crop-Eq
Surplus ore potential (SOP)	3351	4457	3781	4899	kg Cu-Eq
Ozone depletion potential (ODPinfinite)	0.033	0.031	0.046	0.063	kg CFC-11-Eq
Particulate matter formation potential (PMFP)	90	94	129	–	kg PM2.5-Eq
Photochemical oxidant formation potential: Humans (HOFP)	228	228	339	–	kg NOx-Eq
Photochemical oxidant Formation potential: ecosystems (EOFP)	236	237	352	–	kg NOx-Eq
Water consumption potential (WCP)	204	225	251	44,122	m^3

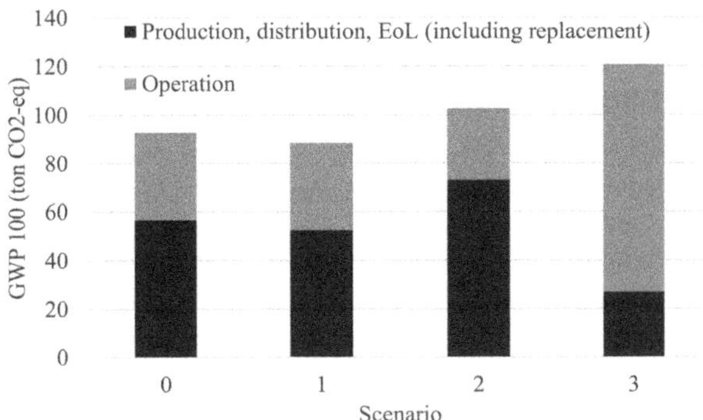

Fig. 4 Life cycle GWP of different cooling and TES solutions

Figure 4 shows the GWP of the scenario in the non-operation and operation phases. Systems with BTES (Scenarios 0–2) shift most of the CO_2 emissions into the non-operation phase by installing boreholes plus either PCM- or water-tank TES with roughly 50–60 t CO_2-Eq. Scenario 2 (only with BTES and heat pump) reaches around 73 t due to its larger BTES field and larger heat pump. Operational emissions for scenarios with BTES and heat pump remain comparatively low, since the ground-source heat pumps are efficient in heating production. The lowest, 29 t, is achieved in scenario 2, where the electric heater is used the least. In contrast, the conventional chiller with district-heating layout (Scenario 3) has the minimal non-operation carbon emissions (27 t) but has the highest carbon emission during operation (94 t) contributed by district heating (77%) and chiller (23%).

4 Conclusions

The study presents a methodology for system-level cradle-to-grave LCA for TES in buildings. A case study is carried out based on a 3,000 m^2 office building in Norway yielding the following key findings:

- As a peak shaving and load shifting component, PCM-TES has higher impacts on GWP and fossil fuel depletion due to the use of a paraffin product (PCM) compared to water-tank TES, whereas the latter shows a higher impact in other categories due to higher usage of steel.
- In a 25-year lifetime, TES integration reduces the environmental impacts compared to a system without TES, mainly due to the much lower life cycle impact during operational phase.
- In the long term, GSHP (heat pump with BTES) is a more sustainable solution compared to district heating and chiller configuration, either with or without TES. This is so because, even though the production, distribution and EoL generate higher impacts due to, e.g., drilling work, the operation phase impact is significantly lower than district heating.

The present case study is constrained by the use of a single year of operational data and projected future impacts. Future research should encompass dynamic operational modeling that captures the interaction between TES performance, variable renewable energy supply, and real-time grid conditions. It should also integrate end-of-life considerations and life-span extension strategies such as reuse, recycling, repurpose, and components and material recovery to fully reflect the circular economy model potential of TES systems. Moreover, expanding the scope to encompass a wider range of building typologies and climatic contexts, including hotter, drier, and tropical regions, would enable more comprehensive validation.

References

Dong Y, Hossain MU, Li H, Liu P (2021) Developing conversion factors of Lcia methods for comparison of Lca results in the construction sector. Sustainability (Switzerland) 13(16):9016. https://doi.org/10.3390/SU13169016/S1

Ecoinvent, Ecoinvent v3.11 (2025). https://ecoinvent.org/

EPD International AB, Environmental Product Declaration: CHILLER AND HEAT PUMP VLS—VLS 160—315 KW, 2025. https://www.environdec.com/

IEA (2024) Emissions factors 2024—data product—IEA. https://www.iea.org/data-and-statistics/data-product/emissions-factors-2024

IEA (2023) Tracking clean energy progress 2023. https://www.iea.org/reports/tracking-clean-energy-progress-2023#overview

ISO (2020) ISO 14040:2006 environmental management—life cycle assessment—principles and framework, Geneva, Switzerland

Knoth K, Fufa SM, Seilskjær E (2022) Barriers, success factors, and perspectives for the reuse of construction products in Norway. J Clean Prod 337:130494. https://doi.org/10.1016/J.JCLEPRO.2022.130494

Swedish environmental research institute, PCR 2019:14 construction products (EN 15804+A2) (2.0.1) (2025)

The Norwegian EPD foundation, environmental product declaration, district heating in Bergen Eviny Termo (2024). https://epd-global.no/epder/energi/fjernvarme

Biomethane Splitting: The Hindered Carbon Removal Potential of Biohydrogen

Isabella Bulfaro, Anna Sánchez, Mikel Fadul, Gabriela Benveniste, Beatriz Amante, and Victor José Ferreira

Abstract Biomethane splitting for hydrogen and solid carbon production is an emerging carbon removal technology. It sequesters biogenic CO_2 from biomass into solid carbon, enabling negative GHG emissions. Reported GHG emissions vary depending on the accounting framework used. This is the first study to systematically assess MS emissions across the EU Renewable Energy Directive (Red III), the UK Low Carbon Hydrogen Standard, and ISO LCA, addressing the implications of the inclusion and exclusion of biogenic CO_2 and carbon credits. GHG emissions are 2.11, -7.41, and 5.32 $kgCO_2/kgH_2$, respectively. Red III underestimates the removal potential by excluding both biogenic CO_2 and carbon credits. LCA including biogenic CO_2 results vary based on biomethane origin, creating an unfavorable scenario for biowaste biomethane. The UK standard offers the most balanced approach, recognizing credits for the solid carbon sequestration.

1 Introduction

Low emissions hydrogen is recognized by the International Energy Agency (IEA) as a strategic technology projected to contribute approximately 4% to cumulative global greenhouse gas (GHG) emissions reductions by 2050 (IEA 2024a). Its application varies from heavy goods traffic, energy balance, to hard-to-abate industries such as cement, steel, and fertilizer (IEA 2024b).

Focusing on hydrogen production today, it is based on unabated fossil fuel processes, emitting 920 Mt CO_2 annually (IEA 2024b). Low emission hydrogen accounts for only 1% of the global hydrogen produced. Specifically, in Europe, the reforming process represents 68.8% of the total production, while reforming

I. Bulfaro (✉) · A. Sánchez · M. Fadul · G. Benveniste · V. J. Ferreira
Catalonia Institute for Energy Research, Sant Adrià de Besòs, Spain
e-mail: ibulfaro@irec.cat

I. Bulfaro · B. Amante
ENMA (Environmental Engineering), ESEIAAT (The School of Industrial, Aerospace and Audiovisual Engineering of Terrassa), Universitat Politècnica de Catalunya (UPC), Terrassa, Spain

© The Author(s) 2026
M. Traverso et al. (eds.), *Life Cycle Management from Global to Local*,
https://doi.org/10.1007/978-3-032-17987-6_17

with carbon capture and storage (CCS) and water electrolysis account for 0.5% and 0.4%, respectively (Hydrogen Europe 2024a). Both reforming with CCS and water electrolysis technologies present specific global and regional constraints.

Reforming with CCS is deeply related to the natural gas resources, water availability, presence of CO_2 storage sites, CO_2 transport infrastructure and policy development (Griffiths et al. 2021; IEA 2024b); while high electricity consumption, water availability, land use supply, and depletion of critical materials pose challenges for electrolysis development.

In a hydrogen economy, Methane Splitting (MS) is a possible bridging technology for climate-friendly and cost-effective hydrogen production. The reaction does not produce direct CO_2 emissions, avoiding the need for a transport and storage CO_2 system. Meanwhile, the electricity and water demands are lower than those of water electrolysis and reforming processes. MS technologies reach, today, a high-technology readiness level, ranging from 5 to 8 (Hydrogen Europe 2024; IEA 2024b). The EU and the US play a central role in the development of MS technology, with the majority of leading companies in this field headquartered in these regions (Hydrogen Europe 2024b). The EU Commission incentivizes the MS technologies development through a first and second round of research funding (Clean Hydrogen Partnership 2025). The last call for proposals specifically targets biomass and biofuel as feedstock, excluding the use of fossil fuel. When biomethane is consumed as feedstock, the MS shows carbon removal potential. The Life Cycle Assessment (LCA) study of Diab et al. 2022 calculates GHG emissions of -5.22 kgCO_2/kgH_2, accounting for biogenic CO_2 in the calculation. However, different results are obtained when changing the applied biofuel. The same process, consuming soybean biofuel, emits 0.52 kgCO_2/kgH_2, including the biogenic CO_2 (Lawson et al. 2024). The values increase to 1.41 when biogenic CO_2 is excluded. In both studies, any carbon credits are attributed to the solid carbon.

This research focuses on the possible approach for GHG calculation on biomethane splitting, with a focus on EU regulation. This is the first study to systematically assess the GHG emissions of biomethane-based MS using multiple regulatory and LCA calculation frameworks. It uniquely compares the Renewable Energy Directive (Red III) (The European Parliament and the Council of the European Union 2018), UK Low Carbon Hydrogen Standard (UK Department for Energy Security and Net Zero 2024), and ISO LCA methodologies (ISO 2006a, ISO 2006b), explicitly addressing the implications of including or excluding biogenic CO_2 and carbon credits. The analysis also considers different biomethane origins and performs a sensitivity analysis on carbon allocation, providing an unprecedented regulatory perspective tailored to the EU context.

2 Methodology

2.1 Technology Description and Study Assumptions

Methane Splitting (MS), also known as methane pyrolysis or methane decomposition, is a thermochemical process that splits methane into hydrogen and solid carbon (Eq. 1).

$$CH_{4(g)} \leftrightarrows 2H_{2(g)} + C_{(s)} \; \Delta H_0 = +74.9\text{kJ} \tag{1}$$

The primary process steps of MS comprise (1) the splitting of methane into gaseous hydrogen and solid carbon, (2) the removal of the solid carbon from the gas stream through a filter or cyclone system, and (3) the purification of the gas stream, mainly using pressure swing adsorption (PSA) or membranes. The plasma reactor has been selected for this work, as the reactor system with the highest TRL (Monolith 2025). The technical parameters of the process are indicated in Table 1. The electricity consumption is fixed at 15 kWh per kg of hydrogen in alignment with the target defined by the research call (Clean Hydrogen Partnership 2025). This is a conservative assumption, reflecting the average electricity consumption for the actual thermal plasma operation (Timmerberg et al. 2020). However, it can be optimized by lowering the reactor temperature, as demonstrated in non-thermal plasma systems. Biomethane consumption is defined as 4 kg per kg of hydrogen, in alignment with the stoichiometric limit and with the parameters identified in the literature (Postels et al. 2016). The electricity consumption from purification is sourced from Postels et al. 2016. Hydrogen delivered at atmospheric pressure is subsequently compressed to 3 bar, and the energy required for this compression is calculated in accordance with the UK GHG Emissions Calculation Guidelines (UK Department for Energy Security and Net Zero 2024).

Table 1 Biomethane splitting technical performance

	Value
Electricity consumption MS reaction (kWh/kg H_2)	15.00
Biomethane consumption (kg/kg H_2)	4.00
Electricity consumption (purification) (kWh/kg H_2)	1.67
Electricity consumption (compression 3 bar) (kWh/kg H_2)	0.76

2.2 GHG Emissions Calculation

The GHG emissions of the biomethane splitting process are calculated following three methodological approaches identified respectively in: (1) the EU Renewable Energy Directive (The European Parliament and the Council of the European Union 2018); (2) the UK Low Carbon Hydrogen Standard (UK Department for Energy Security and Net Zero 2024); and (3) LCA including biogenic CO_2. The methodologies are summarized in Table 2 and will be discussed in the following paragraphs.

A functional unit (FU) of 1 kg of hydrogen is adopted as the basis for comparison across all approaches. The final hydrogen pressure varies according to the requirements of the selected methodology. The system boundary of the study is defined based on the methodological requirements and is illustrated in Fig. 1.

For the MS process, no direct emissions are produced during the splitting process. In this case study, the process is located in Norway, and it is powered by renewable

Table 2 GHG emissions methodologies applied

Methodology	System boundaries	Threshold	Biogenic CO_2	Allocation solid carbon	Solid carbon sequestration credits	Renewable electricity GHG emissions
EU Red III	Well-to-wheel	3.4	Excluded	Energy allocation	Excluded	Excluded
UK Low Hydrogen Standard	Well-to-gate	2.4	Excluded	No allocation	Included	Excluded
LCA including biogenic CO_2	Well-to-gate	/	Included	Energy allocation	Excluded	Included

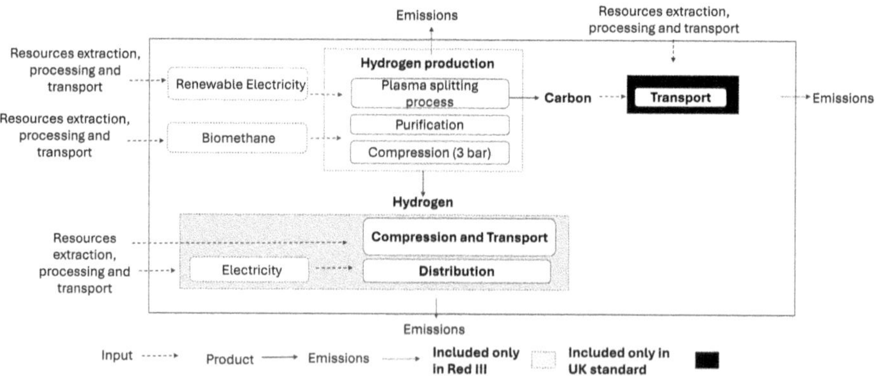

Fig. 1 System boundaries

energy. The upstream emissions of biomethane, electricity, and transport are sourced from Ecoinvent 3.11 cutoff approach by classification (Wernet et al. 2016), and Activity Browser software is used (Steubing et al. 2020). Norwegian (NO) data are selected as the primary option; when unavailable, European (ReR) data are used, and as a final alternative, Rest of the World (RoW) data are applied. Two types of biomethane are considered in the scenario analysis: (1) generic biomethane from RoW and (2) wood-derived biomethane from RoW. The biomethane data selection has been defined according to the Ecoinvent data availability. The impact assessment method applied is the IPCC 2021 GWP100 (IPCC 2021). For the EU and UK regulatory approaches, total GWP emissions, excluding biogenic CO_2, are considered. In contrast, for the LCA approach, total GWP emissions, including biogenic CO_2, are used. In this study, the carbon is assumed to be destined for incorporation into concrete or cement.

2.2.1 Renewable Energy Directive

By 2050, the European Union targets renewable hydrogen to supply approximately 10% of its total energy demand (European Commission 2025). The EU regulation for sustainable hydrogen is the (The European Parliament and the Council of the European Union 2018). It includes hydrogen from biomass origin and "renewable fuel of non-biological origin" (RFNBO). To be defined as renewable biomass has to meet the sustainability criteria, while the hydrogen GHG emissions need to stay below the GHG emissions of 3.4 $kgCO_2eq/kgH_2$. In the Red III GHG emissions methodology, the defined system boundaries are well-to-wheel. The biogenic CO_2 and renewable electricity upstream emissions are counted as 0. This approach assumes that all the biogenic CO_2 uptake from the air and sequestered in the biomass will be released, counting at the end as 0. The energy allocation approach is applied, assigning a portion of the emissions to the solid carbon by-product based on its lower heating value (LHV). The regulation does not include solid carbon as a CO_2 storage system, and any credits can be associated with it. The system boundaries are well-to-wheel, including the hydrogen distribution and storage.

In this study, the final pressure of hydrogen is fixed at 800 bar, as required in the fuel station, and the hydrogen transport and distribution stages are modeled in alignment with the study of Wulf and Kaltschmitt 2018. An average transport distance of 100 km between the plant and the fueling station is assumed, based on the range of values identified by Wulf and Kaltschmitt 2018. The GHG emissions of the electricity consumption at the fuel station are calculated using the IPCC average Norway electricity GHG emissions, excluding emissions from renewable production (Bastos et al. 2024). The upstream emissions from the electricity consumption at the hydrogen production plant are 0 because powered by renewable energy.

2.2.2 UK Low Carbon Hydrogen Standard

The UK Low Carbon Hydrogen Standard (UK Department for Energy Security and Net Zero 2024) sets a maximum threshold for the amount of GHG allowed in the production process for hydrogen to be considered "low-carbon hydrogen." It includes all the hydrogen production processes from fossil-based processes, bio-based and water electrolysis production pathways. The GHG target is fixed at 2.4 $kgCO_2eq/kgH_2$ and adopts a well-to-gate approach. The biogenic CO_2 and renewable electricity upstream emissions are counted as 0. The regulation includes solid carbon sequestration as a CO_2 capture and storage system, considering 3.664 kg of sequestered CO_2 per kg H_2. The solid carbon can be included as a sequestration system only if incorporated into concrete or cement for construction or kept in inert underground storage (e.g., disused mines and bunkers, inert landfill, and spent oil and gas wells). Emissions from solid carbon distribution have to be included in the calculation. No allocation is applied to the solid carbon. In this study, the solid carbon transport distribution is modeled with the EURO 6 transport (freight, lorry, >32 metric ton, diesel). The transport distance is fixed at 130 km, in alignment with the Product Environmental Footprint (PEF) value for suppliers located inside Europe (Zampori et al. 2019). Hydrogen transport and distribution are excluded from the system boundaries, and the final pressure is defined at 3 bar.

2.2.3 LCA Including Biogenic CO_2

The Life Cycle Assessment (LCA) approach including biogenic CO_2, is performed in alignment with the ISO 14040–14044 (ISO 2006a; ISO 2006b), carbon footprint ISO 14067 (ISO 2018) and H_2 LCA guidelines (Bargiacchi et al. 2022). System boundaries are well-to-gate. The final pressure of the hydrogen is fixed at 3 bar. In this case, the CO_2 biogenic upstream emission of biomethane is included, attributing a characterization factor of $+1$ for the CO_2 release and -1 for the CO_2 biogenic uptake during the biomass growth. While biogenic CO_2 is excluded from the H_2 LCA guideline (Bargiacchi et al. 2022), its reporting is required by ISO 14067 (ISO 2018), though is not included in the final results. This approach is selected to cover all the different methodological approaches and to compare results with the studies of Diab et al. 2022 and Lawson et al. 2024. No credits are provided to the solid carbon, excluding double counting. In alignment with the H_2 guideline (Bargiacchi et al. 2022), the electricity emissions from renewable energy production are included.

3 Results and Discussion

In this study, the GHG emissions of the biomethane splitting have been calculated following the methodological approaches of: (1) EU Renewable Energy Directive (European Parliament and of the Council 2023); (2) the UK Low Carbon Hydrogen

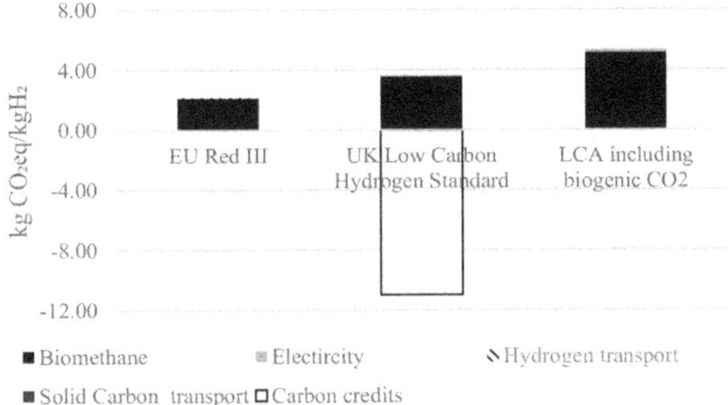

Fig. 2 Hydrogen GHG emissions (kgCO$_2$eq/kgH$_2$) from the biomethane splitting process following the EU Red III, UK Low Carbon Hydrogen Standard, and LCA including biogenic CO$_2$ emissions methodologies

Standard (UK Department for Energy Security and Net Zero 2024); and (3) LCA including biogenic CO$_2$. The obtained results are presented in Fig. 2, showing a range of -7.41 to 5.32 kgCO$_2$/kgH$_2$.

The lowest emissions of -7.41 kgCO$_2$eq/kgH$_2$ are observed when applying the UK low-carbon emissions approach. The biogenic CO$_2$ embedded in the biomass is not released but sequestered in the carbon, destined for cement and concrete products. This allows for long-term CO$_2$ storage, typically exceeding 100 years, and its credits can be subtracted as defined by ISO 14067 (ISO 2018). Applying this methodology, the biomethane splitting process not only is classified as low-carbon hydrogen, but also shows its carbon removal potential. However, excluding the carbon by-product from GHG emissions allocation may overestimate hydrogen's carbon removal potential, which would otherwise be partially attributed to the solid by-product.

Applying the Red III guideline, the GHG emissions are equal to 2.11 kgCO$_2$eq/kgH$_2$. Despite the emissions remaining below the defined threshold of 3.4 kgCO$_2$eq/kgH$_2$, the emissions results increased compared to the results obtained by the UK guideline. The higher results are related to the absence of carbon credits associated with CO$_2$ sequestration into solid carbon. Indeed, the EU guideline includes the credits only when the CO$_2$ is sequestered in gaseous form. The identified gap in Red III disadvantages the biomethane splitting process compared to other biohydrogen processes with a gaseous CCS system. Finally, the highest value of 5.32 kgCO$_2$/kgH$_2$ is observed when the GHG emissions are calculated including the biogenic CO$_2$. In this case, the highest emission value is attributed to biogenic CO$_2$ released during the biomethane production process. However, the total biogenic CO$_2$ uptake by biomass is not fully allocated to the biomethane, as a fraction of the feedstock originates from waste, for which no biogenic carbon uptake is assigned according to the cutoff by application standard.

Across all methodologies, upstream emissions from biomethane production contribute more than 95% of total GHG emissions, even when emissions from

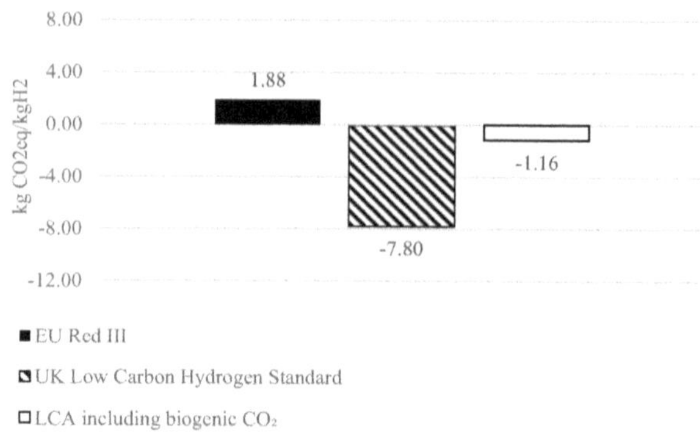

Fig. 3 Hydrogen GHG emissions (kgCO$_2$eq/kgH$_2$) for wood biomethane scenario

renewable electricity are included, as in the LCA, including biogenic CO$_2$. Considering the biomethane contribution to the hydrogen GHG emissions, a scenario analysis is performed varying the feedstock from generic biomethane to wood-derived biomethane, and results are presented in Fig. 3.

The widest variation is observed in the LCA, including biogenic CO$_2$, achieving a negative value of -1.16 kgCO$_2$/kgH$_2$. In contrast to generic biomethane, which includes the use of biowaste, the wood-based biomethane scenario allocates the entire CO$_2$ biogenic uptake to biomass and consequently to the biomethane. The embedded CO$_2$ is only partially released during the biomethanation process, whereas no biogenic CO$_2$ is emitted during plasma splitting, resulting in a net negative GHG outcome. Thie cut-off approach by classification in the LCA methodology disadvantages the use of biowaste compared to other biomass sources, creating a distorted scenario in which the circularity and waste valorization are undervalued from the GHG emissions point of view. Finally, the GHG in Red III and UK Low Carbon Hydrogen Standard methodologies decrease by 11% and 5%, respectively, reflecting their sensitivities to the biomethane origin.

The splitting process is characterized by two main products: hydrogen and carbon.

However, the two products present different markets in terms of volume today, and in future predictions (Dagle et al. 2017). The carbon market, although developing, is considerably smaller than the hydrogen one, risking its rapid saturation. Therefore, a sensitivity analysis is conducted on the percentage of solid carbon identified as carbon by-products and thus eligible for allocation. The remaining fraction is classified as waste and is not subject to allocation. Results are presented in Fig. 4.

In LCA, including biogenic CO$_2$ and EU methodology, the GHG emissions range from 5.32 to 6.97 and 2.11 to 3.57 kgCO$_2$eq/kgH$_2$, respectively, when the percentage of solid by-product eligible for allocation increases from 0 to 100%. A limit of 6.38% of carbon by-product is identified for the Eu Red III methodology to not exceed the regulation threshold. However, this value could change according to the upstream emissions of biomethane. In the UK Low Carbon Hydrogen Standard, the GHG emissions are unchanged to the % of eligible carbon by-product, considering no allocation applies to it. In this case, the methodology does not differentiate between carbon waste and carbon by-product.

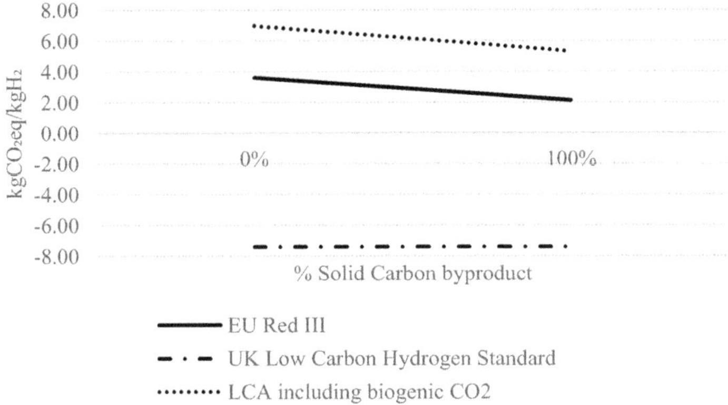

Fig. 4 Sensitivity analysis of GHG emissions ($kgCO_2eq/kgH_2$) to the % of solid carbon byproducts (eligible for allocation)

4 Conclusions

This study investigates the GHG emissions associated with the biomethane splitting process for hydrogen and carbon production. The analysis focuses on EU regulation and its comparison with the UK Low Carbon Hydrogen Standard, and ISO LCA approaches. The methodologies are selected to provide a comprehensive overview and assess the implications of including or excluding biogenic CO_2 and carbon credits. The hydrogen GHG emissions are equivalent to 2.11 (Red III), -7.41 (UK Low Carbon Standard), 5.32 (LCA including biogenic CO_2) $kgCO_2/kgH_2$. In the EU regulation, the biogenic CO_2 emissions are accounted as 0, considering that all the carbon uptake of biomass will be released, giving a final sum of 0. The solid carbon sequestered is not accounted as CCS system and no credits are attributed to it. In the UK regulations, the biogenic CO_2 is excluded, no GHG emissions are allocated to the carbon, and the solid carbon sequestration accounts for credits when its storage is assumed for more than 100 years. Specifically, when the solid carbon is destined for concrete or cement for construction or kept in inert underground storage. Finally, in the LCA including biogenic CO_2, the uptake of CO_2 and CO_2 emissions are considered in the calculation, with a characterization factor of $+1$ and -1. The simultaneous exclusion of both biogenic CO_2 and carbon credits in the EU regulation places MS on par with fossil-based hydrogen and at a disadvantage compared to biohydrogen pathways with gaseous CCS. Moreover, it fails to differentiate between carbon use in long-lived products (e.g., concrete) and combustion. The UK standard includes the credits from solid carbon sequestration, solving the identified gap in the Red III. However, the exclusion of solid carbon from allocation could overestimate the hydrogen carbon removal potential, which could be otherwise partly allocated to the solid by-product. The LCA methodology, including biogenic CO_2, shows the highest value among the methodologies analyzed. It is related to the biogenic CO_2 emissions of biomethane and its biomass origin. When the biomethane is sourced from biowaste, the uptake emissions will not be allocated to the biowaste, and therefore, the GHG emitted during the biomethanation is not compensated by the CO_2 uptake. However, when the biomethane source is from wood, the hydrogen shows emissions of $-1.16\ kgCO_2eq/kgH_2$. This creates an unfavorable outcome for biowaste-derived biomethane. While including biogenic emissions with cut-off approach by classification can represent an interesting approach for LCA studies with the highest level of accuracy, its application to regulation would create a distorted scenario in regulation where the use of biowaste is disadvantaged, discouraging circular economy practices. In conclusion, this study reveals a regulatory gap in the EU framework regarding the accounting of solid carbon sequestration. This omission may lead to competitive distortions in the hydrogen market by underrepresenting the climate benefits of biomethane splitting. Additionally, it highlights the critical importance of considering the final use of carbon, whether for long-term storage or short-lived applications, in both allocation methodology and emission crediting.

References

Bargiacchi E, Puis-Samper G, Campos Carriedo F, et al (2022) D2.2 definition of FCH-LCA guidelines

Bastos J, Monforti-Ferrario F, Melica G (2024) GHG Emission factors for electricity consumption. European Commission, JRC136340

Clean hydrogen partnership, strategic research and innovation agenda—clean hydrogen partnership (2025). https://www.clean-hydrogen.europa.eu/about-us/key-documents/strategic-research-and-innovation-agenda_en (Accessed 13.06.2025)

Dagle RA, Dagle V, Bearden MD, Holladay JD, Krause TR, Ahmed S (2017) An overview of natural gas conversion technologies for co-production of hydrogen and value-added solid carbon products, pacific northwest national lab. (PNNL), Richland, WA; Argonne National Lab. (ANL), Argonne, IL

Diab J, Fulcheri L, Hessel V, Rohani V, Frenklach M (2022) Why turquoise hydrogen will be a game changer for the energy transition. Int J Hydrogen Energy 47(61):25831–25848. https://doi.org/10.1016/j.ijhydene.2022.05.299

European Commission (2025) Hydrogen. https://energy.ec.europa.eu/topics/eus-energy-system/hydrogen_en (Accessed 30.06.2025)

Griffiths S, Sovacool BK, Kim J, Bazilian M, Uratani JM (2021) Industrial decarbonization via hydrogen: a critical and systematic review of developments, socio-technical systems and policy options. Energy Res Soc Sci 80:102208. https://doi.org/10.1016/j.erss.2021.102208

Hydrogen Europe (2024) Clean hydrogen monitor 2024

IEA (2024b) Global hydrogen review 2024

IEA (2024a) Net Zero Emissions by 2050 Scenario (NZE)—global energy and climate model—analysis. https://www.iea.org/reports/global-energy-and-climate-model/net-zero-emissions-by-2050-scenario-nze (Accessed 10.02.2025)

ISO (2006a) ISO 14040:2006(en), Environmental management—Life cycle assessment—Principles and framework

ISO (2006b) ISO 14044:2006(en), Environmental management—Life cycle assessment—Requirements and guidelines

ISO (2018) ISO 14067:2018 Greenhouse gases—Carbon footprint of products—Requirements and guidelines for quantification

Intergovernmental Panel on Climate Change (IPCC) (2023) The earth's energy budget, climate feedbacks and climate sensitivity. In: Climate change 2021 – The physical science basis: working group I contribution to the sixth assessment report of the intergovernmental panel on climate change. Cambridge University Press, pp 923–1054

Lawson R, Dasappa S, Diab J, McCormick M, Wyse E, Hardman N, Fulcheri L, Dames E, Biodiesel surrogate and ethane evaluation for green carbon black and turquoise hydrogen synthesis via thermal plasma. Energy Convers Manag 302:118149. https://doi.org/10.1016/j.enconman.2024.118149

Monolith (2025) Homepage. Monolith. https://monolith-corp.com. Accessed 1 Jul 2025

Postels S, Abánades A, von der Assen N, Rathnam RK, Stückrad S, Bardow A (2016) Life cycle assessment of hydrogen production by thermal cracking of methane based on liquid-metal technology. Int J Hydrogen Energy 41(48):23204–23212. https://doi.org/10.1016/j.ijhydene.2016.09.167

Steubing B, de Koning D, Haas A, Mutel CL (2020) The activity browser—an open source LCA software building on top of the brightway framework. Softw Impacts 3:100012. https://doi.org/10.1016/j.simpa.2019.100012

The European Parliament and the Council of the European Union (2018) Directive (EU) 2018/2001 of the European Parliament and of the council of 11 December 2018 on the promotion of the use of energy from renewable sources (recast) (Text with EEA relevance)

Timmerberg S, Kaltschmitt M, Finkbeiner M (2020) Hydrogen and hydrogen-derived fuels through methane decomposition of natural gas—GHG emissions and costs. Energy Convers Manag X(7):100043. KEIPI

UK Department for Energy Security and Net Zero, UK low carbon hydrogen standard: greenhouse gas emissions methodology and conditions of standard compliance (2024)

Wernet G, Bauer C, Steubing B, Reinhard J, Moreno-Ruiz E, Weidema B (2016) The ecoinvent database version 3 (part I): overview and methodology. Int J Life Cycle Assess 21(9):1218–1230. https://doi.org/10.1007/s11367-016-1087-8

Wulf C, Kaltschmitt M (2018) Hydrogen supply chains for mobility—environmental and economic assessment. Sustainability 10(6):1699. https://doi.org/10.3390/su10061699

Zampori L, Pant R, European Commission (eds) (2019) Suggestions for updating the organisation environmental footprint (OEF) method, Publications Office, Luxembourg

Flying with Liquid Hydrogen: An Assessment of CFRP Fuel Tank Manufacturing in the Aviation Industry and Its Challenges

Karina Kroos, Jens Bachmann, Christian Bülow, Sabrina Diniz, Sebastian Freund, Markus Kleineberg, Antonia Rahn, and Steffen Opitz

Abstract The use of liquid hydrogen in aircraft is a promising option for a more sustainable aviation. Concepts for hydrogen-powered aircraft explore the use of lightweight carbon fibre-reinforced polymer tanks without liners. This study aims to support decision makers in the aviation industry by assessing the potential environmental impact of the manufacturing of these tanks. To this end, a life cycle assessment of hydrogen tanks for a defined use case is conducted using primary data from the manufacture of a half-tank, with the parameters defined by aircraft predesign using the common parametric aircraft configuration schema, an aircraft description language. Contribution analysis identifies the main contributors, while a Monte Carlo analysis reveals the model's uncertainties. Finally, a scenario analysis illustrates the differences in datasets from various carbon fibre-reinforced polymer datasets.

1 Introduction

Environmental impacts of aviation in general and especially the effects on global warming are gaining in importance in the research and development of new airplane concepts and their propulsion. Today, the aircraft use phase is responsible for a majority of climate change impacts from a combination of direct CO_2 emissions and non-CO_2 effects like contrails. The impacts depend on a variety of parameters, particularly during the use phase. One of these is flight distance. In 2023, short-haul flights accounted for 54% of CO_2 emissions emitted by the aviation industry (HAV 2024).

K. Kroos (✉) · J. Bachmann · C. Bülow · S. Diniz · S. Freund · M. Kleineberg · S. Opitz
Institute of Lightweight Systems (SY), DLR German Aerospace Center, Braunschweig, Germany
e-mail: Karina.Kroos@dlr.de

A. Rahn
Institute of Maintenance, Repair and Overhaul (MO), DLR German Aerospace Center, Hamburg, Germany

© The Author(s) 2026

M. Traverso et al. (eds.), *Life Cycle Management from Global to Local*,
https://doi.org/10.1007/978-3-032-17987-6_18

This number demonstrates that short-haul flights are a highly energy- and cost-sensitive segment. This is further demonstrated by comparing it with ground-based transportation options (Ritchie 2023).

To counteract these high impacts, fuel alternatives are taken into focus. Sustainable Aviation Fuels (SAF) and liquid hydrogen (LH2) are promising alternatives to petrol-based kerosene to reduce the environmental impact (Baharozu et al. 2017; Bergero et al. 2023).

From an economic viewpoint, due to the steady growth of the hydrogen sector, the price of sustainably produced hydrogen is expected to drop significantly by 2050, rendering it more cost-effective than synthetic aviation fuel (Gallucci 2022).

Other life phases like raw material extraction, production, Maintenance, Repair and Overhaul (MRO) and End of Life (EoL) just have a minor impact over the life time of an aircraft in comparison to the use phase (Mignanelli et al. 2024). With the upcoming introduction of innovative technologies that aim to emit considerably less greenhouse gases, the relative importance of other phases increases. However, the environmental impacts of SAF and hydrogen production need to be evaluated for better understanding as they depend considerably on the raw materials and production methods. Beside that the indirect climate impact of hydrogen needs to be understood better. Also, high-performance materials like carbon fibres have a considerable environmental impact on their own (Bachmann et al. 2017; Benitez et al. 2021; Kamali et al. 2025), on the other hand, they enable the production of components that fulfil ever rising requirements of future aircraft. Here, Life Cycle Assessment (LCA) can be a precious support for choosing the right materials and processes.

This study is based on the use case of an aircraft configuration taken from the ALICIA (Aviation Life Cycle and Impact Assessment) project. The aircraft concept is a short-range aircraft that uses LH2 as energy carrier and has a mild hybrid electric propulsion (MHEP) system for 250 passengers.

Researchers and industry are currently investigating two different tank arrangements for the aft fuselage tanks (see Fig. 1). The 'side-by-side' approach is advantageous from the perspective of tank manufacturing and certification because only one type of tank with a smaller diameter needs to be produced and certified (FlightGlobal 2024). However, the 'tandem' tank arrangement utilises the given fuselage shape more efficiently.

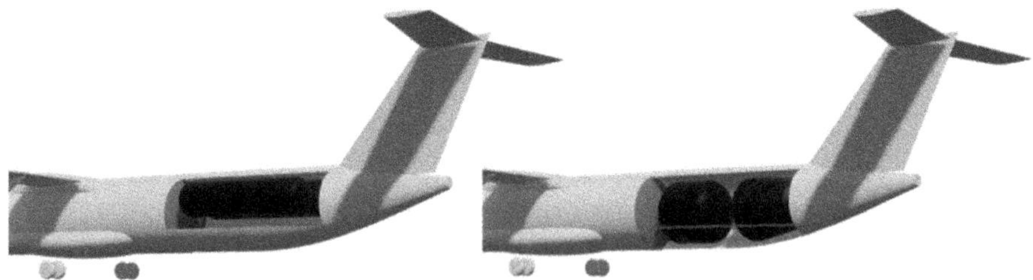

Fig. 1 Aft fuselage mounted 'side by side' (left) and 'tandem' (right) tank

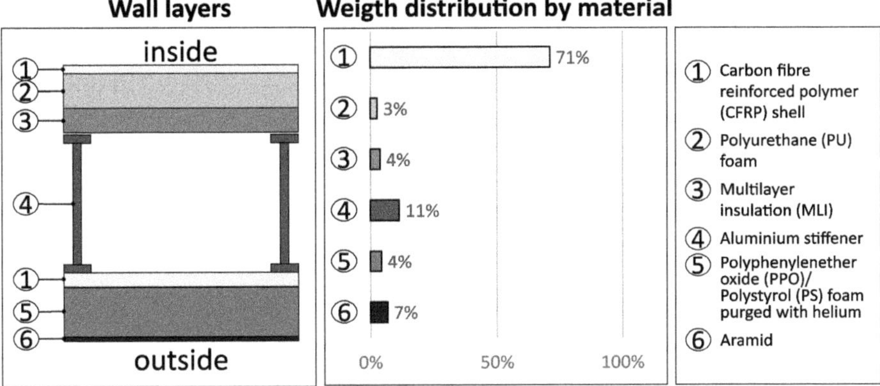

Fig. 2 Exemplary tank wall with layers as defined in the CPACS file and the calculated mass distribution

For this use case, the hydrogen is stored under cryogenic conditions in two tanks in the 'tandem' configuration located behind the rear bulkhead as shown in Fig. 1, right. Figure 2 presents a principle cross section of the tank shells made of an inner and an outer Carbon Fibre-Reinforced Polymer (CFRP) shell including insulation.

The parametrizations of the whole aircraft are defined in the Common Parametric Aircraft Configuration Schema (CPACS) (Alder et al. 2020). More information about the use case can be found in (DLR Institute of System Architectures in Aeronautics 2025).

LCAs of composite hydrogen tanks are scarce in the literature. Most of the assessments in this area relate to the automotive sector and the tanks are mainly included in the assessment of a fuel cell electric car as a component. The main focus is on Type III and IV tanks for 350 and 700 bar gaseous states, manufactured using filament winding technology (Benitez et al. 2021; Evangelisti et al. 2017; Mignanelli et al. 2024). LCAs on Type V pressure vessels for use under cryogenic conditions, especially for aircrafts, are underrepresented.

2 Materials and Methods

This study carries out a cradle-to-gate life cycle assessment of a CFRP LH2 tank in accordance with ISO standards 14040 and 14044, ensuring a systematic approach to quantify potential environmental impacts. This methodology consists of four phases: Goal and Scope Definition; Life Cycle Inventory (LCI); Life Cycle Impact Assessment (LCIA); and Interpretation. All of these phases are covered in the following sections.

2.1 Goal and Scope

The main purpose of this study is to assess the environmental impact of producing hydrogen tanks for the use case (see Chap. 1), based on primary data from an industrial-scale production process involving half of the inner CFRP shell, scaled to the dimensions of the use case. Additionally, a comparison is conducted between the CFRP inner shell and a comparable aluminium shell.

2.1.1 Tank Layers of the Use Case

The manufacturing process of the CFRP shells (carbon fibre + epoxy resin) is based on Automated Fibre Placement (AFP). All manufacturing steps from preparation until dismantling are included as well as the raw material extraction and the assembly with the insulation (see Fig. 3). Uncertainties in the manufacturing process are considered, and a scenario analysis will contribute to provide decision-makers in the field of sustainable aviation with a transparent and detailed overview of the results.

The product system itself consists primarily of the inner and outer composite shells, as well as the various insulation layers (see Fig. 2).

The total mass of the tanks is calculated as 1839 kg, with the cylindrical tank accounting for 55% and the conical tank for 45%. This estimation is based on geometric calculations using data from CPACS and literature. For the purposes of this study, the Functional Unit (FU) is defined as the LH2 tanks of the CPACS aircraft configuration, consisting of a conical tank and a cylindrical tank with insulation layers, as previously described. The two tanks together have a volume of 59.72 m³.

Figure 3 illustrates the system boundary for the cradle-to-gate study. The study is conducted using the brightway2 package, calculating the 16 impact categories

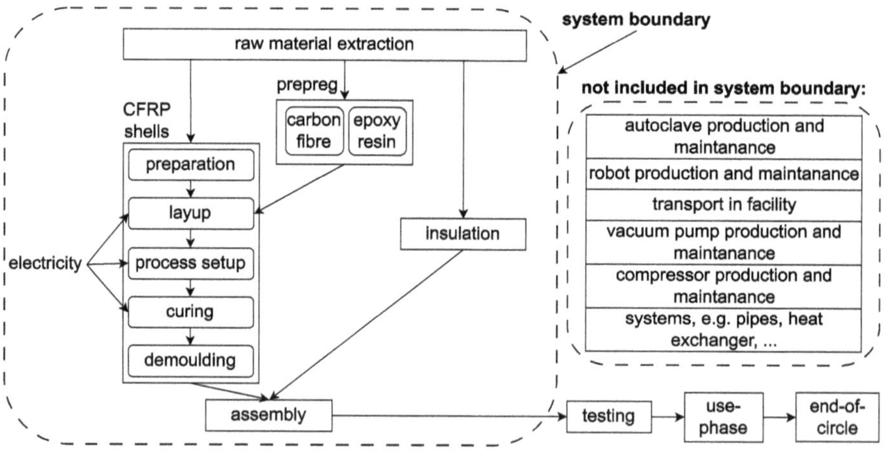

Fig. 3 System boundary

with the related indicators of the Environmental Footprint (EF) v3.1 method. The background database used is the ecoinvent 3.9.1 cut-off version.

The different manufacturing processes are assumed to be located in Europe. It is also important to note that the manufacturing process of the inner shell is assumed to be scalable for the tank diameter of the use case, and that the outer shell is assumed to have the same manufacturing process as the inner shell.

2.1.2 Comparison of CFRP with Metal Inner Vessels

This study includes a preliminary, high-level comparison of metal and composite shells. For this comparison, an inner metal shell based on the same technical characteristics has been designed.

Both vessels use a safety factor of 2.0 as analogy to pressurized fuselages in EASA/CS 25.365 d that requires a factor of 1.33 to the load carrying capability besides the ultimate load factor of 1.5 (CS 25.303). The operational pressure is set to 2 bars and an operational cycling from 1 bar to operational pressure is assumed for each of the 50,000 flight cycles as fatigue load spectrum.

The aluminium mass assessment is based on Barlow's formula as described in the AD2000 (Arbeitsgemeinschaft Druckbehälter e.V. 2016) standard and extended by a fatigue calculation from MMPDS-07 (Rice et al. 2012). Here aluminium 6061 T6 is used. CFRP vessels are calculated using the commercial structural solver μWind from MeFeX (Meyer 2016) in combination with DLRs tank optimization tool tankoh2 (Freund et al. 2025) as described in (Freund & Franzoni 2022) using pucks inter fibre failure criterion. Additionally, internal thermomechanical stresses due to the cryogenic temperatures using classical lamination theory are considered as well as in situ strengths (Camanho et al. 2006).

2.2 Life Cycle Inventory (LCI)

All the collected data is scaled up to the CPACS data for the use case. The primary data is supplemented with uncertainty distributions.

2.2.1 Data Collection for the Tank Layers of the Use Case

The foreground data for manufacturing the CFRP shells is collected from an industrial-scale AFP process used to manufacture a half-tank. This data is either measured or estimated by experts. This includes the various activities shown in Fig. 3, such as preparing the tool, laying up the material (prepreg) and setting up the vacuum build-up, as well as curing in the autoclave. Data for the curing step also includes the use of nitrogen for inert gas filling of the autoclave and the electricity

Table 1 Structural characteristics of the aluminium and CFRP inner shells for comparison

	Unit	Tank 1		Tank 2	
		Aluminium	CFRP	Aluminium	CFRP
Diameter	m	3.592		2.977	
Axial length	m	4.197		3.829	
LH2 volume	m^3	36.518		23.2	
LH2 mass	kg	2225		1414	
Shell mass	kg	395	316	244	220

consumed by the autoclave. Scaling the primary data to the use case dimensions is based on parameters such as surface area and layer volume.

Due to the absence of carbon fibre data in the background database, the detailed virgin carbon fibre (Toray T700 G) production dataset from Benitez et al. (2021) is employed instead.

All layers of insulation are taken from the CPACS file and supported by literature data. Although there is no dataset for polyphenylenether oxide (PPO), it is modelled based on stoichiometric calculations, following the ecoinvent methodology (Hischier et al. 2005). The Multilayer Insulation (MLI) material is assumed to consist of 30 layers of reflector and separator each. It is modelled based on (Finckenor & Dooling 1999).

2.2.2 Data Collection for the Comparison of CFRP with Metal Inner Vessels

The aluminium layer described in Chapter 2.1.2 is modelled based on ecoinvent datasets for aluminium and metal working as well as welding.

The structural masses of the inner vessel shells used are listed in Table 1.

3 Life Cycle Impact Assessment (LCIA)

The LCIA is divided into sections. First, a contribution analysis is conducted to identify the main contributors. Secondly, a scenario analysis illustrates how results can differ when using datasets from different databases for the same product before finally a comparison of the environmental impacts of an inner tank shell made out of CFRP is compared to one out of aluminium.

3.1 Contribution Analysis

The first part of the LCIA is the contribution analysis. As the bottom graph in Fig. 4 shows, the impact of the CFRP shells is consistently greater than 75%. Taking a closer look at the individual activities behind the CFRP shells reveals that the prepreg is the main contributor to impacts in all categories except Ozone Depletion Potential (ODP). Here, the release film from the process set-up has the highest contribution at 95.5%.

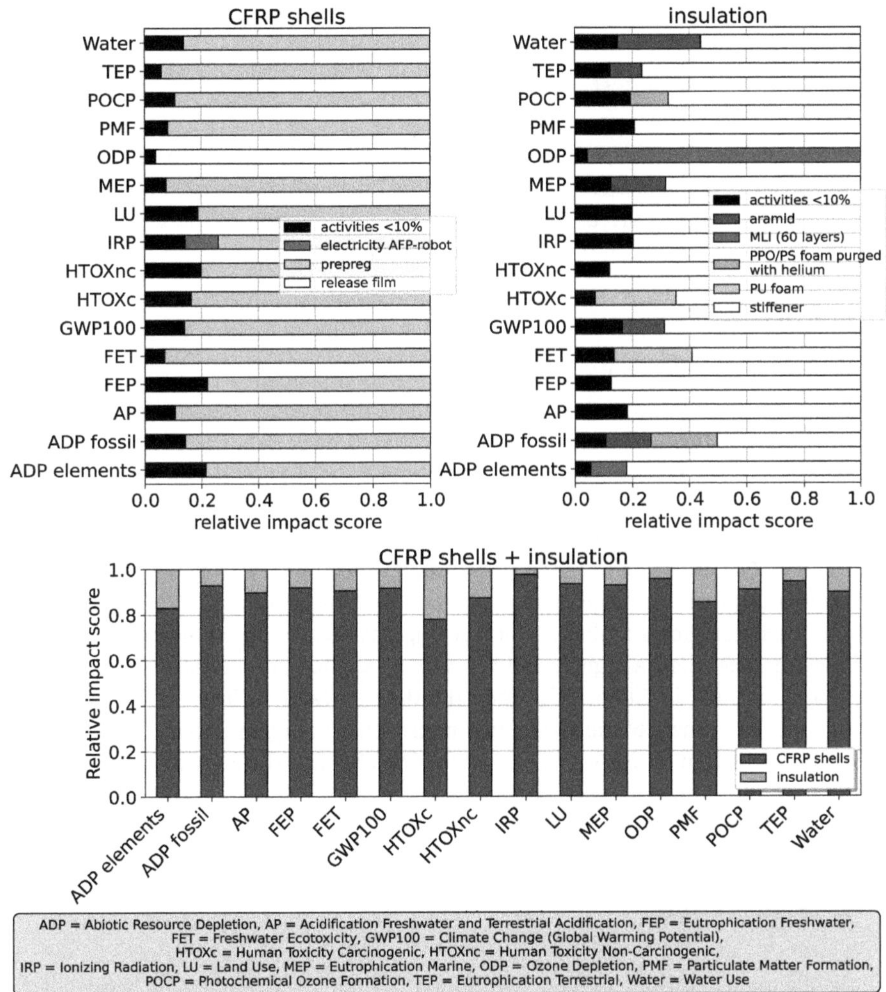

Fig. 4 Relative contributions of the CFRP shells and the insulation to the impacts of all EF v3.1 indicators

In contrast, no obvious element stands out as having the greatest impact on all insulation layer categories. However, the aluminium stiffener layer is identified as the main contributor in most categories, accounting for over 50% of the total in all but three indicators: Ecotoxicity Freshwater (FET), Human Toxicity Non-carcinogenic (HTOXnc) and ODP. For FET and HTOXnc, the Polyurethane (PU) foam is the main contributor. The MLI stands out in the Ozone Depletion category, accounting for 96% of the total impact.

The Global Warming Potential (GWP100) divides into 80,934 $kgCO_2eq$ for the CFRP shells and 7,505 $kgCO_2eq$ for the insulation.

3.2 *Scenario Analysis*

As shown in Chap. 3.1, the prepreg is identified as the main contributor. The scenarios are chosen based on this. Specifically, alongside the model based on foreground data, two additional datasets are used to represent the production of the CFRP shells as best fits from two LCA databases:

(1) the ecoinvent dataset: carbon fibre-reinforced plastic, injection moulded, GLO
(2) the esa LCA database dataset: CFRP, cured by autoclave process {RER} | production | Cut-off, U.

These two datasets are then linearly scaled by the product weight. For the MHEP use case based on primary data, the scaling is based on certain parameters described in Chap. 2.2.1. The differences between the scenarios are displayed in Fig. 5 for all EF v3.1 categories as deviation from the reference which is defined as the results based on the model with primary data. Existing datasets for CFRP from the ecoinvent 3.9.1 and esa LCA databases show deviations from the model using primary data. While environmental impacts are generally lower using the esa LCA dataset, the ecoinvent dataset shows higher and lower impacts depending on the category.

Based on these data, a high-level LCA was conducted in accordance with the goal and scope outlined in Chap. 2.1, using the data in Table 1. The results show that limiting the assessment to the production phase leads to higher impact of inner shells made from CFRP in all EF v3.1 categories. The absolute difference for GWP100 is 32 t CO_2 eq more for the CFRP shell.

4 Interpretation

Firstly, the main contributors were identified in Chap. 3.1. Coupling the results of the contribution analysis with the scenario analysis shows that, especially for the CFRP shells, the results depend heavily on the data source and the scaling approach.

It should be noted that data on composite production cannot be scaled based solely on weight. This underlines the need for parameterized and representative LCIs for

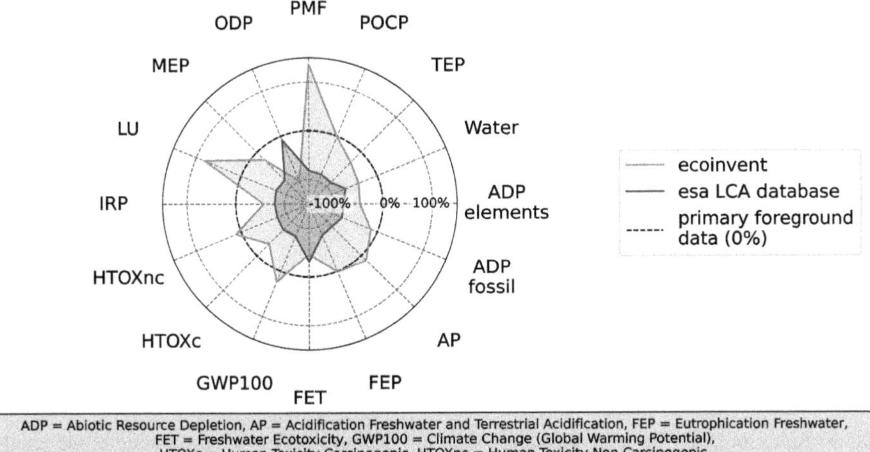

ADP = Abiotic Resource Depletion, AP = Acidification Freshwater and Terrestrial Acidification, FEP = Eutrophication Freshwater, FET = Freshwater Ecotoxicity, GWP100 = Climate Change (Global Warming Potential), HTOXc = Human Toxicity Carcinogenic, HTOXnc = Human Toxicity Non-Carcinogenic, IRP = Ionizing Radiation, LU = Land Use, MEP = Eutrophication Marine, ODP = Ozone Depletion, PMF = Particulate Matter Formation, POCP = Photochemical Ozone Formation, TEP = Eutrophication Terrestrial, Water = Water Use

Fig. 5 Relative deviation from using the model with primary data as reference for all impact indicators in EF v3.1

composite production processes that enable a more realistic modelling of real-world processes. The comparison of the environmental impact of manufacturing metal and CFRP inner shells shows the need to consider the entire life cycle of an aircraft, because the CFRP shell's lower weight means that its environmental impact during the use phase will be potentially lower than that of the aluminium one.

The contribution analysis step does not include the characterization of uncertainties. Therefore, a Monte Carlo analysis involving 10,000 runs was conducted. Uncertainties from the background data (lognormal distributions) and for the CFRP shell manufacturing (uniform distributions) are included. Figure 6 shows the distributions of each category using box plots. The boxes represent the central 50% of the values (interquartile), while the median is marked with a horizontal line in the box. The results are normalized to represent the deviation from the mean in % which is set as 0%. The data behind all boxes show that the second quartile is for all smaller than the third which can be characterized by the location of the median in the box. The Human Toxicity (HTOX) and the FET categories show the least concentrated distributions (Fig. 6, right). Also interesting is the indicator Abiotic Resource Depletion (ADP) elements which is the only distribution where the mean is not located in the interquartile. Furthermore, the median is always lower than the mean.

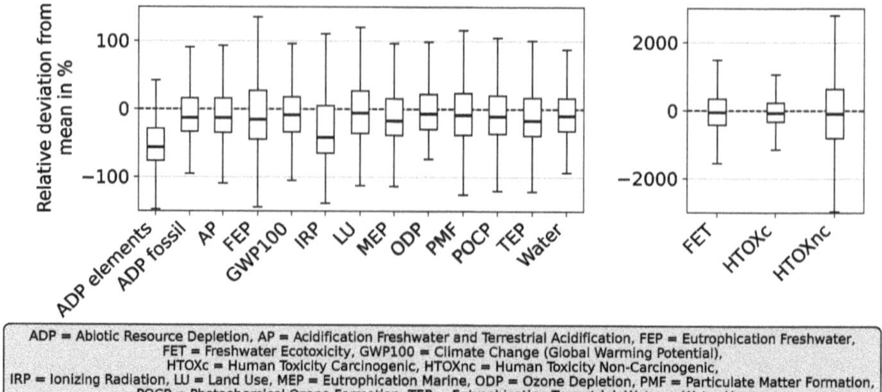

Fig. 6 Box plot of all EF v3.1 indicators showing the distributions of the Monte Carlo results separated into two plots due to the differences in the scale

5 Conclusion and Outlook

Assessing the environmental impacts for the defined functional unit shows that more primary data is needed. However, it is important to bear in mind that even primary data is subject to uncertainty and should always be challenged. The need for parameterized LCIs is also evident, as is the need to consider uncertainties in order to gain a broader understanding of the results. Further work should also include a sensitivity analysis of the identified parameters in order to best interpret the results with their uncertainties.

Also, a full comparison of metal and CFRP tank variants needs to be extended to the full life cycle including weight impacts on use phase fuel consumption, maintenance schedule and end-of-life solutions.

References

Alder M, Moerland E, Jepsen J, Nahel B (2020) Recent advances in establishing a common language for aircraft design with CPACS. In: Aerospace Europe conference 2020, Bordeaux, France, 2020, 1–14

Bachmann J, Hidalgo C, Bricout S (2017) Environmental analysis of innovative sustainable composites with potential use in aviation sector—a life cycle assessment review. SCIENCE CHINA Technol Sci 60:1301–1317

Baharozu E, Soykan G, Ozerdem MB (2017) Future aircraft concepts in terms of energy efficiency and environmental factors. Energy 140:1368–1377

Benitez A, Wulf C, de Palmenaer A, Lengersdorf M, Röding T, Grube T, Robinius M, Stolten D, Kuckshinrichs W (2021) Ecological assessment of fuel cell electric vehicles with special focus on type IV carbon fiber hydrogen tank. J Clean Prod 278:1–13

Bergero C, Gosnell G, Gielen D, Kang S, Bazilian M, Davis SJ (2023) Pathways to net-zero emissions from aviation. Nat Sustain 6:404–414

Camanho PP, Dávila CG, Pinho ST, Iannucci L, Robinson P (2006) Prediction of in situ strengths and matrix cracking in composites under transverse tension and in-plane shear. Compos A Appl Sci Manuf 37(2):165–176

DLR Institute of System Architectures in Aeronautics, D250-TFLH2-MHEP-2040. https://www.digital-hangar.de/portfolio/d250-tflh2-mhep-2040/ (Accessed 04.06.2025)

Arbeitsgemeinschaft Druckbehälter e.V. (2016) AD 2000-Merkblatt S 3/0 (2016–09): Materialien für Druckbehälter, Beuth Verlag GmbH, Berlin

Evangelisti S, Tagliaferri C, Brett DJL, Lettieri P (2017) Life cycle assessment of a polymer electrolyte membrane fuel cell system for passenger vehicles. J Clean Prod 142:4339–4355

Finckenor MM, Dooling D (1999) Multilayer insulation material guidelines, National Aeronautics and Space Administration, George C. Marshall Space Flight Center, Huntsville, AL

FlightGlobal, Harnessing hydrogen will fuel growth for UK plc. https://www.flightglobal.com/paid-content/harnessing-hydrogen-will-fuelgrowth-for-uk-plc/159055.article (Accessed 27.05.2025)

Freund S, Jacobsen L, Rickert C (2025) tankoh2: design and optimization of H_2 tanks (2.7.0), Zenodo

Freund S, Franzoni F (2022) Automated liquid hydrogen tank design optimization using filament winding simulation and subsequent comparison with aluminium vessels. DLRK, Dresden, Germany

Gallucci M (n.d.) Major airlines are getting serious about hydrogen-powered planes. https://www.canarymedia.com/articles/air-travel/major-airlines-are-getting-serious-about-hydrogen-powered-planes (Accessed 27.05.2025)

HAV, Growing connections, not emissions across the regional air network. https://www.hybridairvehicles.com/news/overview/blog/growing-connections-not-emissions-across-the-regional-air-network/ (Accessed 27.05.2025)

Hischier R, Hellweg S, Capello C, Primas A (2005) Establishing life cycle inventories of chemicals based on differing data availability. Int J Life Cycle Assess 10:59–67

Kamali AK, Isayev J, Laratte B, Sonnemann G (2025) Harmonizing life cycle assessment studies of emerging technologies: the case of virgin and recycled carbon fibers. Resour, Conserv Recycl 220

Meyer N (2016) Entwicklung einer virtuellen Simulationskette zur Auslegung und Optimierung von gewickelten Faser-Kunststoff-Verbund-Druckbehältern unter Berücksichtigung des nichtlinearen Werkstoffverhaltens. Shaker Verlag, Aachen

Mignanelli C, Bianchi I, Forcellese A, Simoncini M (2024) Life cycle assessment of pressure vessels realized with thermoplastic and thermosetting matrix composites. Int J Adv Manuf Technol 135:4491–4509

Rahn A, Wicke K, Wende G (2022) Using discrete-event simulation for a holistic aircraft life cycle assessment. Sustainability 14(17)

Rice RC, Jackson JL, Bakuckas J, Thompson S (2012) MMPDS-07: metallic materials properties development and standardization (MMPDS). Federal Aviation Administration, Washington, D.C.

Ritchie H (n.d.) Travel carbon footprint. https://ourworldindata.org/travel-carbon-footprint (Accessed 27.05.2025)

Life Cycle Assessment and Levelized Cost of Storage of Secondary- and Primary-Use LFP Batteries for Energy Storage Systems

Wenxin Yang, Xinyuan Wang, and Maximilian Schüler

Abstract This study presents a comparative life cycle assessment (LCA) and levelized cost of storage (LCOS) analysis of primary-use and secondary-use lithium iron phosphate (LFP) batteries in grid-scale energy storage systems (ESS), using California as a representative case. A cradle-to-grave scope is adopted to evaluate environmental impacts, including global warming potential (GWP) and water scarcity footprint (WF), alongside economic performance. Both LCA and LCOS assessments are harmonized to a common functional unit (FU) of 1 kWh of electricity delivered to the grid over the system lifetime. Results show that while secondary-use batteries exhibit lower energy delivery efficiency and higher GWP per kWh delivered, they outperform primary batteries in water footprint and remain cost-competitive, with an LCOS of 0.422 \$/kWhe compared to 0.438 \$/kWhe for new batteries. Uncertainty and sensitivity analyses showed that round-trip efficiency (RTE) and depth of discharge (DoD) are the most influential parameters for energy output and losses, while rated storage duration is the most influential factor for LCOS. Grid emissions dominate GWP, while precursor production governs WF results. These findings demonstrate that ESS is an economically promising second-life application for repurposed LFP batteries, highlighting the importance of regional electricity mix in shaping sustainability outcomes.

W. Yang
University of California Berkeley, Berkeley, USA

X. Wang
University of Illinois Urbana-Champaign, Champaign, USA

M. Schüler (✉)
Technische Hochschule Luebeck, Luebeck, Germany
e-mail: maximilian.schueler@th-luebeck.de

© The Author(s) 2026
M. Traverso et al. (eds.), *Life Cycle Management from Global to Local*,
https://doi.org/10.1007/978-3-032-17987-6_19

1 Introduction

The global shift toward decarbonization has accelerated the integration of solar and wind power into electricity grids, but their intermittency demands large-scale energy storage systems (ESS) to ensure grid stability (Tan et al. 2021). Lithium-ion batteries (LIBs), particularly lithium iron phosphate (LFP) types, are widely used in grid-scale storage due to their multiple advantages.

Meanwhile, the electric vehicle (EV) boom that began a decade ago means many LFP packs will soon retire from vehicles (Alfaro-Algaba and Ramírez 2020). Since most packs retain ~80% of their capacity, repurposing them for energy storage offers a promising approach to extend lifespan and promote a circular economy (Fan et al. 2023).

California exemplifies the parallel rise: it accounts for 35% of all registered EVs in the United States, and its grid faces an increasingly steep "duck curve" as renewable penetration grows (Raugei et al. 2020). Anticipating the large volume of retired batteries, companies are scaling up repurposing and recycling capacity across the state (Machala et al. 2025). Collectively, these factors make California a key state for assessing the economic and environmental potential of second-life LFP batteries.

LCA is the environmental impacts of products from a whole life cycle perspective. LCOS represents the levelized cost of all capital and operation and maintenance costs per unit of energy storage. This study conducts a cradle-to-grave LCA and LCOS analysis of primary- and secondary-use LFP batteries deployed in grid-scale ESS in California. By harmonizing environmental and economic metrics, this study aims to evaluate whether repurposed batteries can deliver competitive performance compared to new batteries in order to provide decision-relevant insight into the circular economy.

2 Literature Review

The environmental impacts and commercial potential of LIBs had been quantitatively assessed in several LCA studies. Fan et al. (2023) demonstrated that reusing secondary LIBs in ESS outperformed lead–acid batteries in environmental terms through the LCA model. Mostert et al. (2018) differentiated secondary-use LIBs from new LIBs based on varying cycle numbers and utilized LCA to determine that the GWP of secondary-use LIBs is 2 g CO_2-eq/kWh lower than the new one. Carvalho et al. (2021) utilized primary data on battery production from an Italian manufacturer, emphasizing the significance of grid mix and manufacturing energy intensity.

On the economic front, LCOS-based evaluations vary significantly. Yang et al. (2023) revealed that LFP new batteries for ESS have lower costs compared to other types of LIBs. Steckel et al. (2021) pointed out that second-life systems can be more expensive than new ones due to reduced cycle life and performance uncertainties and showed that LCOS is highly sensitive to discharge rate, depth of discharge, and module-repurposing costs.

However, few studies have jointly assessed both LCA and LCOS analyses specifically for secondary use versus new LFP batteries in energy storage systems. Moreover, assumptions regarding usage scenarios, repurposing procedures, and end-of-life (EoL) treatment vary widely, limiting the comparability of results.

3 Methodology

3.1 Life Cycle Assessment

3.1.1 Goal and Scope Definition

LCA is a methodology of evaluating the life cycle environmental impacts of a product. By conducting LCA, the environmental loads of LFP in different scenarios can be assessed. LCA is carried out under ISO 14040 and ISO 14044 (ISO 14040, 2022; ISO 14044, 2022). Two scenarios are taken into comparison in the study: (1) scenario 1: primary-use LFP batteries manufactured and deployed directly in stationary ESS; (2) scenario 2: retired EV LFP batteries repurposed for secondary use in ESS. Introducing LCA enables us to assess the environmental impacts of two scenarios within the same system boundary and to compare the environmental benefits of two circumstances.

3.1.2 System Boundaries and Functional Unit (FU)

The assessment of both product systems adopts a "cradle-to-grave" life cycle perspective, encompassing all relevant unit processes. Figure 1 below indicates the research boundary, covering precursor production, battery manufacturing, usage stage, and end-of-life (EoL) treatment, where:

(1) For secondary-use batteries, only the EV stage's energy losses are carried over as embedded emissions; its delivered electricity is ignored. In the ESS stage, both the delivered energy and losses are accounted for.
(2) In the testing and repackaging phase, only package components (housing, BMS, wiring) are replaced; cells and modules remain intact.
(3) LFP batteries are assumed to be recycled by hydrometallurgy, which is the predominant industry approach (Stegemann and Gutsch 2025).
(4) The infrastructure construction is excluded, and the electricity mix in California is incorporated to reflect region-specific impacts.

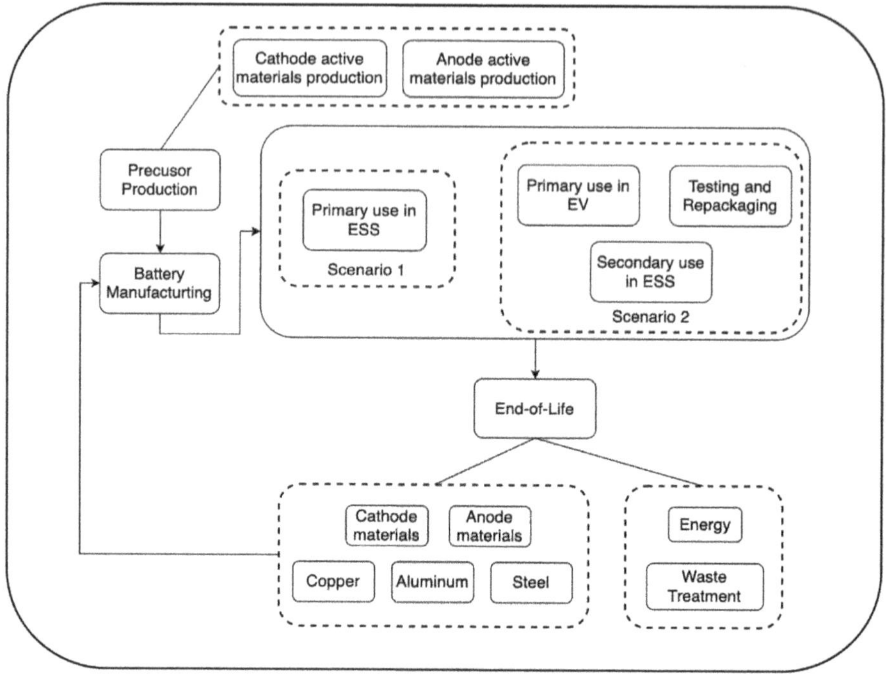

Fig. 1 System boundary of LFP batteries under two usage scenarios

To align the system-level environmental performance and cost metrics in both LCA and the following LCOS analyses, FU of 1 kWh of electricity delivered over the system lifetime is adopted. The reference flow corresponds to the total electricity delivered over the system lifetime per 1 kWh nominal capacity battery: 2,213 kWhe/kWh for primary-use battery, 1,416 kWhe/kWh for secondary-use battery. Note that these values may change depending on RTE and DoD.

3.1.3   Life Cycle Inventory

The data used in this study are categorized into background and foreground data. Background data are primarily obtained from the Ecoinvent database and BatPaC model to retrieve energy and emission factors. In addition, published literature is referenced to improve data accuracy, such as the GWP of graphite production (Engels et al. 2022). Foreground data are collected based on a combination of sources, including BatPaC and Environmental Product Declarations (EPD) for the mass distribution of battery materials. Additional data on processes are compiled from literatures, obtaining primary data in related industries.

To address both the study's geographic scope and the actual supply chain distribution, all processes are modeled as occurring in California (USA) except for precursor production, which is located in China.

Precursor production phase includes the production of both cathode and anode precursors, specifically LFP and battery-grade graphite. International shipping is included in the inventory.

Adjustments in battery manufacturing are made to the pack-level components based on an Environmental Product Declaration (EPD) from Huawei (2022), specifically to reflect the differences in casing materials used for stationary ESS. Another conservative adjustment is made for pack assembly in scenario 1: 0.03 kWh/kg of battery pack produced is used in the life cycle inventory.

For primary-use batteries, only the application in ESS as an energy storage battery is considered. Electricity losses occurring throughout the battery's lifetime are calculated. Previous studies had demonstrated that the capacity loss of LIBs is related to the cycle life (Arora et al. 1998).

To estimate electricity losses and energy delivered, we adopted established semi-empirical degradation and efficiency models (Han et al. 2014; Quan et al. 2022). These models account for capacity fade across cycles, and calculate total energy delivered based on nominal capacity, DoD, RTE, and projected cycle life.

For the secondary-use battery, losses from the initial EV phase are approximated using Tesla Model 3 and mass-based energy consumption ratios (Zackrisson et al. 2010). The repurposing stage is modeled with added material and energy inputs, and the second-use phase is assumed to operate at 80% of initial capacity.

Once retired from EV use, battery modules are tested and repackaged for ESS applications. The retired modules would be sorted by performance characteristics and age, then reconfigured into standardized battery packs with additional casing materials. Material inputs and electricity consumption during this process are adopted from Quan et al. (2022).

After testing and repackaging, the battery enters secondary use with 80% of its original nominal capacity. Reused batteries can have the longest possible second life with a DoD of 70%. This allows 4500 cycles can be reached throughout their second life (Saez de Bikuña et al. 2025).

In the hydrometallurgical recycling pathway modeled by Dai et al. (2019), cathode and anode active materials are partially recycled. Valuable metals are also recovered through the chemical leaching process, while residual materials are either partially combusted for energy recovery or assumed to be disposed of in the landfill. For simplification, all recovered materials are treated as functionally equivalent to primary materials in the battery manufacturing phase.

3.1.4 Life Cycle Impact Assessment

Two environmental impact indicators are considered in the analysis: GWP and WF. GWP is assessed based on the IPCC 100-year time horizon (GWP-100) method, while WF, obtaining and processing from SimaPro and GREET (2024), is evaluated using the AWARE method, which quantifies the potential to deprive other users of water.

3.2 Levelized Cost of Storage

3.2.1 Goal and Scope Definition

LCOS is applied to evaluate the economic performance of ESS under two scenarios with harmonized inputs and settings. The configuration of 60 MW capacity and 4 h storage duration are selected to reflect an increasingly popular and representative ESS design in the current market. Based on the same storage characteristics assumed in Sect. 3.2. One cycle per day and 320 cycles per year are assumed, resulting in an operation lifespan of 15 years for primary-use and 14 years for secondary-use systems.

3.2.2 Methodology

The methodology used is based on the LCOS calculator developed by Pacific Northwest National Laboratory (PNNL 2023). Capital cost inputs for utility-scale battery storage are adjusted to 2025 USD using a 2.5% annual inflation rate, consistent with long-term escalation rates recommended by the National Renewable Energy Laboratory (NREL.a). Similarly, fixed O&M costs are adjusted accordingly. For the secondary-use scenario, repurposing costs are derived using the Battery Second Use (B2U) Cost Calculator developed by NREL (NREL.b).

3.3 Sensitivity and Uncertainty Analysis

Piecewise polynomial response-surface regression is used to approximate two computational sub-modules: (1) the LCOS calculator developed by PNNL and (2) the iteration-based $E_{delivered}$ simulation discussed in Sect. 3.1.3. The fitted models capture key nonlinear interactions with $R^2 > 0.99$. 5000 input vectors are generated by independent uniform sampling within the specified RTE and DoD.

The surrogates are combined in a master Microsoft Excel workbook that reproduces the complete LCOS and LCA workflow for both scenarios. 5,000 input vectors are generated independent uniform sampling each variable across the following ranges: storage duration = 2–10 h; RTE = 85–95% for primary batteries and 70–90% for secondary; DoD = 80–90% for primary and 60–90% for secondary; and $\pm 10\%$ around the nominal values for all GWP and WF impact factors.

Deterministic calculations are subsequently executed using Excel macro. Output distributions are visualized with box-and-whisker plots, showing medians, interquartile ranges, and extreme values. Pearson correlation coefficients between each stochastic input and each output are computed across the 5,000 runs; these coefficients serve as first-order sensitivity indices, indicating both the direction and relative strength of each input's influence on the results.

4 Results and Discussions

4.1 Life Cycle Impact Assessments

Figure 2 presents the life cycle GWP results for both LFP battery scenarios. Overall, the total GWP for primary- and secondary-use battery are 177.32 gCO_2/kWhe and 214.53 gCO_2/kWhe, respectively. The usage phase contributes the largest share in both cases, accounting for over 60% of the total GWP. The GWP of secondary-use batteries is 37.21 gCO_2/kWhe higher than that of primary-use batteries, primarily due to the energy delivery accounting assumption.

A breakdown of the usage phase emissions in the secondary-use scenario reveals that the majority of GWP accounts for the ESS operation, and the emissions from testing and repackaging are minimal and even can be negligible. This reinforces the environmental feasibility for repurposing retired batteries. Meanwhile, the importance of battery performance in second ESS application is pointed out.

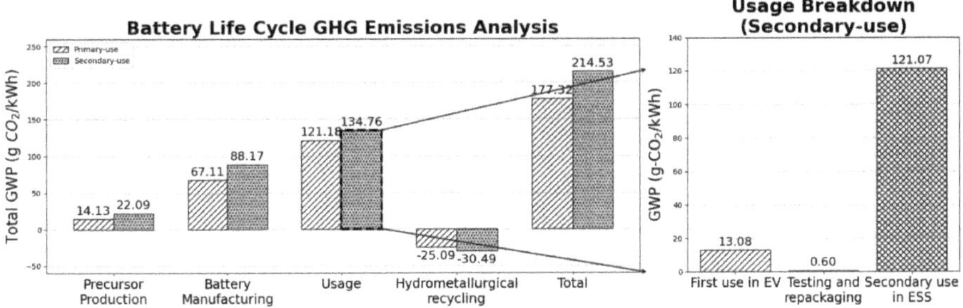

Fig. 2 The total GWP of primary- and secondary-use LFP batteries

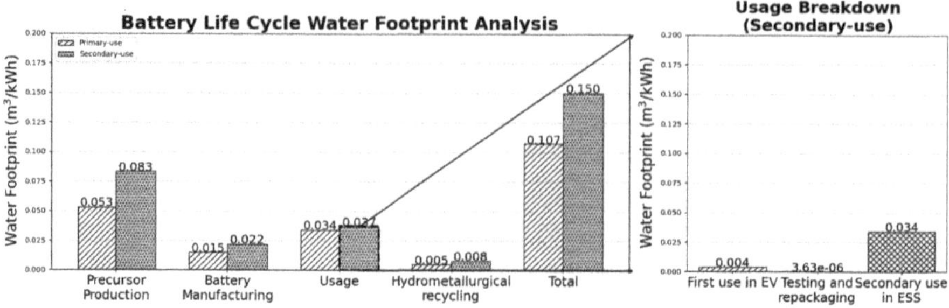

Fig. 3 The total WF of primary- and secondary-use LFP batteries

Figure 3 illustrates the life cycle WF of two LFP batteries. Across both scenarios, the precursor production stage also dominates the overall water footprint, contributing nearly half of the total net impact. The total water footprint for primary and secondary uses is calculated to be 0.107 m^3/kWhe and 0.150 m^3/kWhe, respectively. When disaggregating the usage stage of the secondary-use battery, the ESS application remains the largest contributor within this phase, 0.034 m^3/kWhe,, while the first use in EVs and testing & repackaging processes contribute marginally. Overall, these findings emphasize that the upstream material synthesis (precursor production) rather than battery operation governs the total water footprint. Thus, reducing water consumption in cathode precursor and material processing would yield the most substantial improvements, whereas regional electricity profiles continue to shape the relative impact of the usage stage.

4.2 Levelized Cost of Storage Results

The primary-use LFP battery exhibits an LCOS of 0.438 $/*kWhe*, while the secondary-use shows a slightly lower LCOS of 0.422 $/*kWhe*. The cost advantage of secondary-use scenario is primarily driven by the significant lower upfront battery cost. In the study, the cost of buying and repurposing retired LFP batteries is assumed to be 24 $/*kWh* according to the second-life battery pricing calculator from B2U Storage solutions, indicating the reduced capital costs offset the diminished performance. The results support the economic viability of repurposed batteries for subsequent ESS application.

4.3 *Interpretation*

As shown in Fig. 4, primary-use batteries deliver nearly twice as much energy per unit of nominal capacity compared to secondary-use batteries, reflecting the latter's lower RTE and remaining cycle life. Figure 5 further confirms that RTE and DoD are the principal determinants of $E_{delivered}$. DoD has a larger impact on the primary-use battery, whereas RTE plays a more significant role for the secondary-use battery. In both cases, energy losses are primarily driven by RTE assumptions, which also account for most of the observed uncertainty. Figures 5 also shows that storage duration appears to exert the greatest influence on the LCOS outcomes. The secondary-use battery again displays a marginally higher mode and substantially wider spread. Coupling batteries with longer-duration applications is economically favorable on battery capacity (in kWh) basis. Future research should focus on improving the RTE of secondary batteries during the repurposing process, thereby reducing both financial uncertainty and LCOS.

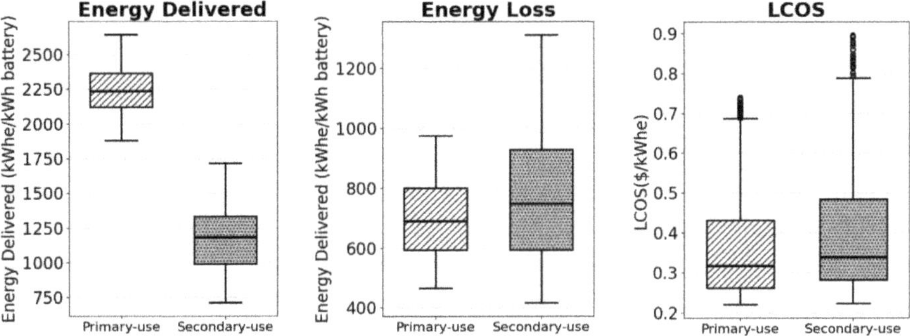

Fig. 4 Uncertainty analysis of **a** energy delivered, **b** energy loss, **c** LCOS

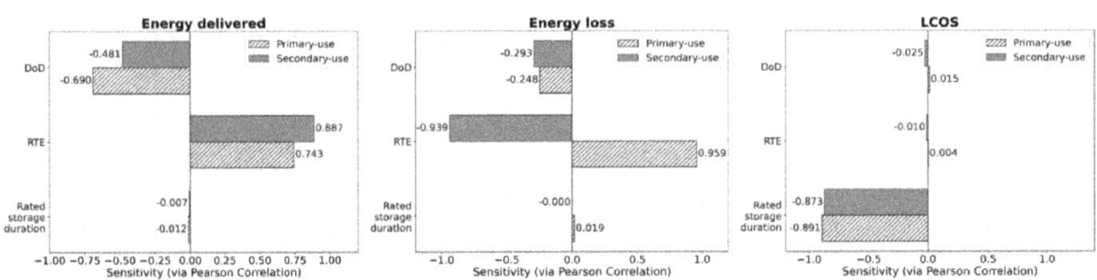

Fig. 5 Sensitivity analysis of **a** energy delivered, **b** energy loss, **c** LCOS

Figure 6 summarizes the uncertainties of environmental impacts. Secondary-use batteries exhibit higher median GWP and WF per kWh electricity delivered. This is mainly due to their lower RTE mode, which result in greater cumulative energy losses over the life cycle. These losses, when multiplied by the grid's GWP, lead to substantially higher total emissions. However, when assessed per kWh of battery capacity, the discrepancy narrow, and the secondary-use battery shows a comparable or even slightly lower water footprint. This observation underscores that both performance degradation and functional metrics should be considered.

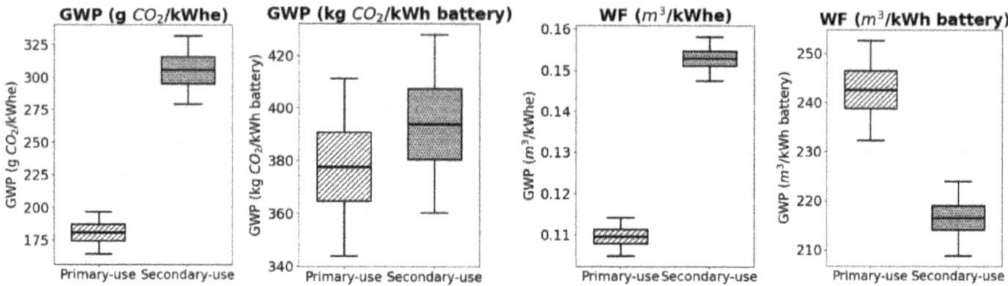

Fig. 6 Uncertainty analysis of **a** GWP per energy delivered, **b** GWP per 1 kWh nominal capacity of battery, **c** WE per energy delivered, **d** WF per 1 kWh nominal capacity of battery

The California grid is the most dominant factor influencing GWP, while WF is jointly influenced by both grid and non-grid parameters as shown in Fig. 7. In contrast, non-grid parameters such as battery materials and settings exhibit only minor influence. These results suggest that optimizing grid carbon and water intensity can substantially improve environmental outcomes for both primary and secondary battery deployments. Further research could examine how variations in grid carbon intensity and WF affect ESS across different regions or future grid scenarios.

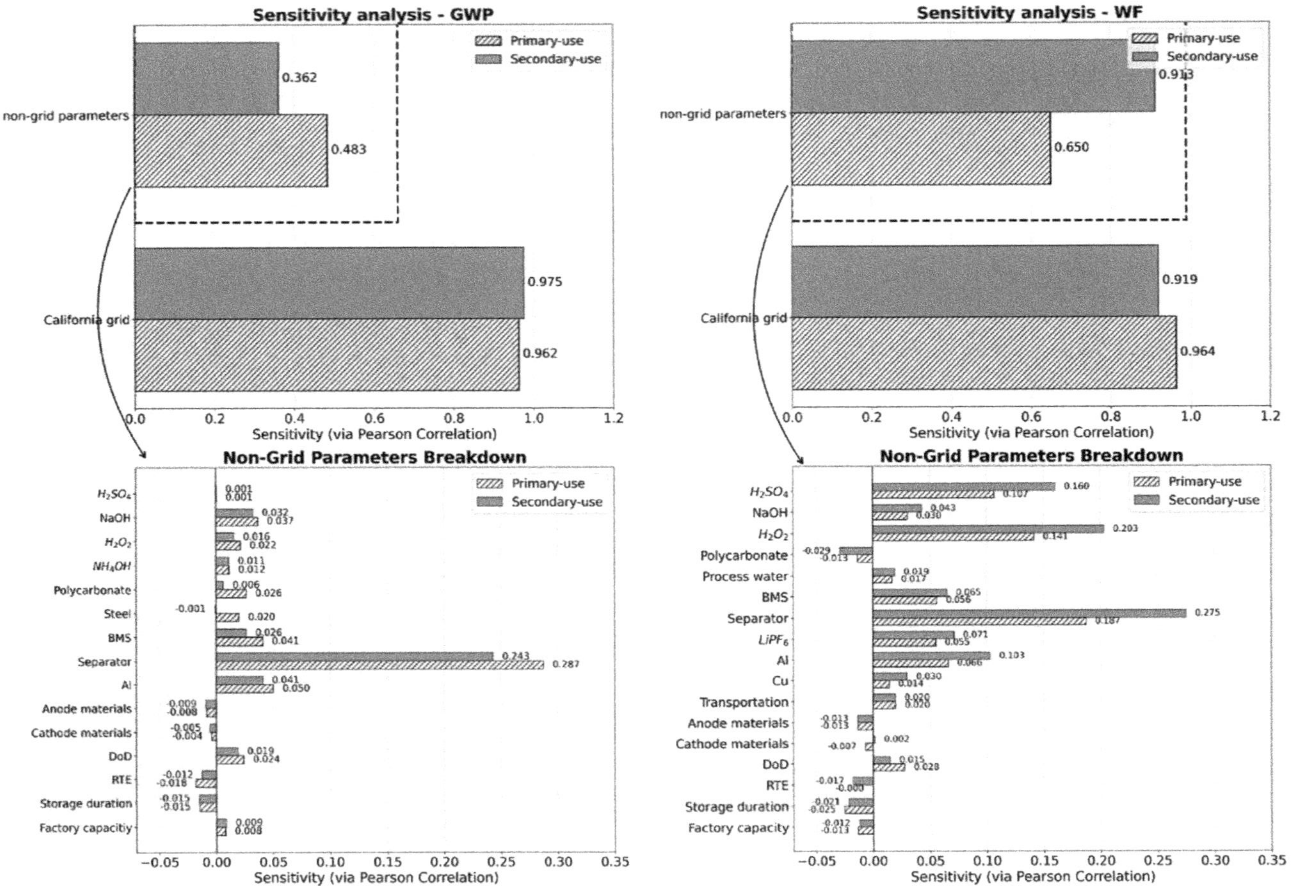

Fig. 7 Sensitivity analysis of GWP and WF per energy delivered and non-grid parameters breakdown

5 Conclusion

This study presented a comparative LCA and LCOS analysis for primary- and secondary-use LFP batteries in ESS applications, using California as a representative. Results show that secondary-use batteries, while exhibiting lower round-trip efficiency and higher GWP per unit of electricity delivered, offer significant advantages in WF and cost competitiveness.

Sensitivity and uncertainty analyses confirmed that RTE and DoD are the primary determinants of performance-related outcomes, while grid electricity emissions dominate environmental impacts. These findings highlight the importance of aligning repurposed batteries with low-carbon, low-water-intensity grids and improving round-trip efficiency during second-life preparation to maximize environmental and economic benefits.

However, it should be noted that these findings are based on California-specific conditions. Therefore, careful reassessment is needed before generalizing the results to other regions. Future research should focus on exploring region-specific deployment strategies and advancing repurposing technologies to improve the adaptability of second-life battery systems.

References

Alfaro-Algaba M, Ramirez FJ (2020) Techno-economic and environmental disassembly planning of lithium-ion electric vehicle battery packs for remanufacturing. Resour Conserv Recycl 154:104461

Argonne National Laboratory. https://www.anl.gov/partnerships/batpac-battery-manufacturing-cost-estimation (Accessed 01.06.2025)

Argonne National Laboratory. https://greet.anl.gov/ (Accessed 10.10.2025)

Arora P, White RE, Doyle M (1998) Capacity fade mechanisms and side reactions in lithium-ion batteries. J Electrochem Soc 145(10):3647

Carvalho ML, Temporelli A, Girardi P (2021) Life cycle assessment of stationary storage systems within the Italian electric network. Energies 14(8):Article 8

Dai Q, Spangenberger J, Ahmed S, Gaines L, Kelly JC, Wang M (2019) EverBatt: a closed-loop battery recycling cost and environmental impacts model, Argonne national laboratory (ANL), Report No. ANL-19/16

Ellingsen LAW, Hung CR, Strømman AH (2017) Identifying key assumptions and differences in life cycle assessment studies of lithium-ion traction batteries with focus on greenhouse gas emissions. Transp Res Part d: Transp Environ 55:82–90

Engels P, Cerdas F, Dettmer T, Frey C, Hentschel J, Herrmann C, Mirfabrikikar T, Schueler M (2022) Life cycle assessment of natural graphite production for lithium-ion battery anodes based on industrial primary data. J Clean Prod 336:130474

EPD Italy. https://www.epditaly.it/en/epd/battery-modules-of-the-stationary-energy-storage-system/ (Accessed 07.06.2025)

Fan T, Liang W, Guo W, Feng T, Li W (2023) Life cycle assessment of electric vehicles' lithium-ion batteries reused for energy storage. J Energy Storage 71:108126

ISO, ISO 14040. https://www.iso.org/standard/37456.html (Accessed 01.06.2025)

ISO, ISO 14044. https://www.iso.org/standard/38498.html (Accessed 01.06.2025)

Machala ML, Chen X, Bunke SP, Forbes G, Yegizbay A, de Chalendar JA, Azevedo IL, Benson S, Tarpeh WA (2025) Life cycle comparison of industrial-scale lithium-ion battery recycling and mining supply chains. Nat Commun 16(1):988

Mostert C, Ostrander B, Bringezu S, Kneiske TM (2018) Comparing electrical energy storage technologies regarding their material and carbon footprint. Energies 11(12):Article 12

NREL,a. https://atb.nrel.gov/electricity/2024/definitions (Accessed 07.06.2025)

NREL,b. https://www.nrel.gov/transportation/b2u-calculator?NREL (Accessed 07.06.2025)

PNNL. https://www.pnnl.gov/projects/esgc-cost-performance/lcos-estimates (Accessed 07.06.2025)

Quan J, Zhao S, Song D, Wang T, He W, Li G (2022) Comparative life cycle assessment of LFP and NCM batteries including the secondary use and different recycling technologies. Sci Total Environ 819:153105

Raugei M, Peluso A, Leccisi E, Fthenakis V (2020) Life-cycle carbon emissions and energy return on investment for 80% domestic renewable electricity with battery storage in California (U.S.A.). Energies 13(15):Article 15

Saez de Bikuña K, Pierobon M, Soldati C, Vale M, Picone N (2025) repurposing of electric vehicle batteries for second life stationary applications in residential photovoltaic systems: an environmental and economic sustainability assessment. Int J Environ Res 19(4):119

Steckel T, Kendall A, Ambrose H (2021) Applying levelized cost of storage methodology to utility-scale second-life lithium-ion battery energy storage systems. Appl Energy 300:117309

Stegemann L, Moritz Gutsch M (2025) Environmental impacts of pyro- and hydrometallurgical recycling for lithium-ion batteries—a review. Bus Chem

Tan KM, Babu TS, Ramachandaramurthy VK, Kasinathan P, Solanki SG, Raveendran SK (2021) Empowering smart grid: a comprehensive review of energy storage technology and application with renewable energy integration. J Energy Storage 39:102591

Yang Y, Ye Y, Cheng Z, Ruan G, Lu Q, Wang X, Zhong H (2023) Life cycle economic viability analysis of battery storage in electricity market. J Energy Storage 70:107800

Zackrisson M, Avellán L, Orlenius J (2010) Life cycle assessment of lithium-ion batteries for plug-in hybrid electric vehicles—critical issues. J Clean Prod 18(15):1519

Life Cycle Assessment of Fluff Residue Gasification as a Substitute for Natural Gas

Stefane Caldeira, Luciana Magalhães, Wesley Cavalcante, Julio Rivera, Leonardo Ribeiro, and Patrícia Rezende

Abstract The steel industry's transition to Electric Arc Furnace (EAF) technology is key to decarbonization, as it enables increased scrap recycling. However, energy use in downstream processes—especially billet reheating furnaces—remains a challenge due to their high reliance on fossil fuels. Replacing natural gas with cleaner alternatives is therefore essential. This study evaluates the use of fluff, a residue from automotive scrap processing, as feedstock for syngas production via gasification. The syngas is then used as fuel in billet reheating. A Life Cycle Assessment was performed on SimaPro software 9.6.0.1 and the ReCiPe 2016 Midpoint method using pilot-scale data to compare environmental impacts between syngas and natural gas use as well as the role of sourcing renewable electricity from wind power in the production of syngas. Results show lower impacts for syngas in all categories except water consumption.

1 Introduction

The global movement to address climate change seeks to promote various public and private initiatives aimed at reducing carbon emissions. The steel industry, for instance, is a significant source of greenhouse gas (GHG) emissions, accounting for approximately 7–9% of global emissions (ResponsibleSteel 2022). A challenge facing the industry is how to reduce global emissions despite the growing demand for per capita steel consumption, as steel is essential across a wide range of applications. In this context, steel production through scrap recycling in the electric

S. Caldeira (✉) · P. Rezende
Postgraduate Program in Product and Process Technology, Federal Center for Technological Education of Minas Gerais State, Belo Horizonte, Brazil
e-mail: stefane.caldeira@arcelormittal.com.br

S. Caldeira · L. Magalhães · W. Cavalcante
Department Decarbonization and Circularity, ArcelorMittal Long Carbon Steel, Belo Horizonte, Brazil

J. Rivera · L. Ribeiro
ArcelorMittal Maizieres Research SA, Maizières-Lès-Metz, France

M. Traverso et al. (eds.), *Life Cycle Management from Global to Local*,
https://doi.org/10.1007/978-3-032-17987-6_20

239

arc furnace (EAF) has emerged as a key strategy in the sector's decarbonization process, given that it emits fewer GHGs when compared to the integrated production route—involving the reduction of iron ore in blast furnaces (ArcelorMittal Brasil 2021).

A significant portion of the scrap used in steelmaking originates from end-of-life vehicles (ELVs), which undergo dismantling and shredding processes to separate their constituent materials. Shredder equipment enables the recovery of ferrous materials—mainly steel—which are subsequently directed to EAFs. Non-ferrous fractions, such as aluminium and copper, are also recovered and redirected to their respective recycling chains. (Caetano et al. 2020).

However, a substantial portion of the material—typically 15–25% of the vehicle's original mass—remains as a heterogeneous mix of non-metallic residues, commonly referred to as automotive shredder residue (ASR) or fluff. This fraction includes plastics, expanded polystyrene, rubber, textiles, wood, fines, and other complex materials. Fluff is generally sent to landfill due to technical and economic challenges associated with its treatment, which contributes to environmental impacts and represents a loss of potentially valuable energy content. Nevertheless, due to its high calorific value, fluff has attracted attention for its potential use as an alternative fuel in energy recovery technologies such as gasification (Caetano et al. 2020; Manente et al. 2022).

Some studies have demonstrated the potential of fluff gasification for the production of alternative fuels (Ciacci et al. 2010; Manente et al. 2022), as well as evidence of reduced environmental impacts compared to incineration (Ciacci et al. 2010). However, studies specifically applied to the steel sector remain limited.

Manente et al. (2022) conducted a literature review on the pyrolysis and gasification processes of ASR, aiming at the production of fuels and the recovery of valuable chemicals. The authors listed studies employing different thermochemical conversion routes, evaluating product yields, composition, and energy efficiency. Studies on pyrolysis highlighted the production of char, bio-oil, and gases, which can be used as fuels for energy recovery. Additionally, the potential for metal recovery from char via leaching and the upgrading of bio-oil into industrially relevant organic compounds—such as aromatic hydrocarbons and olefins—was noted. Gasification, in turn, favoured the formation of H_2 and CO gases. It was observed that higher temperatures and oxidising atmospheres enhance gasification, increasing gas yield and syngas quality while reducing tar formation. Finally, the authors underscored the potential of ASR as an alternative energy source, aligning with sustainable automotive waste management and circular economy goals.

In order to make more assertive decisions in terms of reducing environmental impacts, the application of the Life Cycle Assessment (LCA) tool has been growing in recent years. LCA considers all environmental impacts associated with products and services throughout their life cycle, ranging from the extraction of raw materials, through production and distribution, to final use and disposal. By integrating this systemic view, it is possible to identify and implement solutions that reduce environmental impacts and optimize the use of resources (ISO14040 2006). Ciacci et al. (2010) applied LCA to compare five ASR treatment scenarios—landfilling, non-ferrous recovery, waste-to-energy incineration, post-shredder technologies (PST), and a PST-incineration hybrid. Landfilling had the highest impacts due to land use and lack of energy recovery, while the combined PST with incineration scenario showed the best environmental performance by maximizing material recovery and reducing emissions.

This study presents an LCA comparing two scenarios for thermal energy generation at a Brazilian steelmaking site: one scenario involving energy recovery through fluff gasification and syngas production, and another based on the use of natural gas (status-quo). The aim is to evaluate the environmental performance of both scenarios and support more sustainable decision-making through improved waste and energy management in the steel industry.

2 Methodology

To support the environmental assessment carried out in this study, a pilot-scale gasification test was performed to characterise the energy efficiency and environmental profile of fluff derived from automotive metal scrap processing. The tests took place in a semi-industrial facility equipped with a fixed-bed, downdraft gasifier operating with patented technology. Technical specifications were not disclosed due to industrial confidentiality.

Originally designed for biomass, the system was adapted to process pre-treated fluff waste—screened, shredded, dried, and metal-separated—classified as industrial Residue-Derived Fuel (RDF). The pilot test involved three consecutive days of operation, during which samples of the feedstock and gasification by-products, including syngas, ashes, and condensed oily residues, were collected.

Laboratory analyses comprised elemental characterisation (C, H, N, S, Cl), ash content and waste classification in accordance with ASTM standards and ABNT NBR 10004. The syngas composition was monitored using portable gas analysers and stack sampling techniques based on CETESB, ABNT and USEPA protocols, allowing the quantification of emissions as HCl, Cl_2, VOCs, CO_2, CO, and CH_4.

The interchangeability between natural gas and syngas was validated through performance testing, demonstrating that syngas can be used effectively.

The material composition of the fluff was determined through gravimetric analysis, with carbon content measured in the lab. Biogenic and fossil carbon shares were assigned based on literature values (Table 1).

Regarding the LCA, SimaPro software 9.6.0.1 and the ReCiPe 2016 Midpoint method were used to carry out the LCA for this study.

Table 1 Gravimetric results, non-fossil and fossil carbon content, and the calculation of respective biogenic and fossil emissions, originating from fluff

Gravimetric of the fluff	Proportions in the fluff this work (%)	Total carbon in each material (IPCC 2006) (%)	Total carbon fossil in each material (Giegrich 2021) (%)	Total carbon biogenic in each material (Giegrich 2021) (%)
Minerals (stones)	2	0	0	0
Metals	15	0	0	0
Plastics	29	75	100	0
Rubber	11	56	20	36
Textiles	3	40	20	20
Wood	7	43	0	100
Paper	1	41	1	1
Fines (soil)	9	0	0	0
Others (unidentified)[1]	23	–	100	0

[1] For unidentified materials, the most conservative assumption was applied, considering that 100% of the emissions are from fossil carbon

2.1 Life Cycle Assessment

2.1.1 Goal and Sope Definition

In accordance with ISO 14040 and ISO 14044 standards, this LCA study was conducted to compare the environmental impacts of replacing natural gas with synthesis gas (syngas), produced through the gasification of fluff, as an energy source in the billet reheating process at a Brazilian steelmaking facility. The LCA aims to support the environmental feasibility analysis of implementing an industrial-scale gasification system, providing technical justification for investment decisions and innovation driven by sustainability.

Functional Unit

The functional unit was defined as the generation of 1 MJ of useful thermal energy delivered to the industrial reheating furnace. This definition ensures a consistent comparison by requiring both alternatives—natural gas and syngas produced from fluff—to supply 1 MJ of useful energy with a lower heating value (LHV) comparable to that of natural gas.

System Boundaries

The system boundaries for the syngas scenario extend from the point where fluff exits the shredder facility to the combustion of syngas in the reheating furnace. A cut-off criterion was applied to exclude all upstream flows prior to the generation of fluff, in accordance with the principle of omitting equivalent processes in comparative LCA.

For the natural gas scenario, the system boundaries include the impacts associated with the production and transportation of market-supplied natural gas, up to its combustion in the industrial furnace.

In the landfill scenario, the system boundaries include transportation and final disposal, assuming no energy recovery.

Data Quality Assessment

The data used in this study were obtained from three main sources: (i) primary data from a pilot-scale gasification test conducted specifically for this study; (ii) literature-based references for classifying the origin of carbon; and (iii) background data from the Ecoinvent v3.10 database for standard processes such as natural gas production and landfill operations.

A qualitative assessment of data quality was carried out using the pedigree matrix method. This method evaluates data across five dimensions: reliability, representativeness, temporal correlation, geographical correlation, and technological correlation (Weidema and Wesnaes 1996). The inventory data for syngas production based on measurements from pilot-scale testing results in high scores for temporal, geographical, and technological relevance.

The datasets used for natural gas supply and landfilling of fluff were sourced from the Ecoinvent v3.10 database, using the "cut-off by classification" system model.

2.1.2 Inventory Phase

Scenario 1—Syngas Production and Use

In this scenario, the modelling begins with fluff, which is considered to carry zero environmental burden since its impacts are already accounted for in the upstream steel recycling system (i.e. previous product system boundary). Fluff is converted into RDF, and this transformation includes pre-treatment processes such as screening, drying, and metallic separation. Non-ferrous metals separated during this stage are directed to recycling, and fine inert residues (soil) is sent to landfill. From 1 kg of fluff processed, 0.61951 kg of RDF is obtained. Additionally, 0.10928 kg of non-ferrous metals are recovered and directed to recycling, while 0.18550 kg of inert soil are separated and sent to landfill. Transforming fluff into RDF involves minor evaporation losses, estimated at 0.1134 kg of water released into the air. Electricity consumption for the drying stage at this phase was estimated at 34.62 kWh per ton of fluff processed.

In the gasification step RDF is converted into syngas. where 0.034 kg of RDF is required to produce 1 MJ of syngas, equivalent to 0.0348 Nm^3. The gasification process consumes 0.01027 kWh of electricity and 0.01553 kg of oxygen per MJ of syngas generated. The system also includes the transportation of RDF over 45 km from the preparation site to the gasifier.

Residues from gasification include approximately 0.00596 kg of ash, which is sent to a sanitary landfill. The scenario also incorporates transportation of this residue over 80 km (assumed based on process layout). Air emissions per MJ of syngas were characterised based on pilot test results and include fossil CO_2 (0.0551 kg), biogenic CO_2 (0.0138 kg), HCl (0.00014591 kg), and Cl_2 (0.00010816 kg). Direct monitoring was performed for HCl and Cl_2 emissions, while CO_2 values were estimated from the measured chemical composition of the syngas. Post-combustion CO_2 emissions were calculated assuming full oxidation of the main carbon-containing components (CO, CH_4, and CO_2). The origin of the carbon—biogenic or fossil—was determined using gravimetric analysis of fluff composition and literature-based carbon content values.

All flows were modelled under the "cut-off by classification" system, with electricity sourced from wind power under the Brazilian northeastern grid. Residual materials, such as non-condensable oils or secondary by-products, were not included in the inventory due to their marginal contribution to the system's energy function.

Scenario 2–Natural Gas Production and Use (Status-Quo)

In this scenario, the generation of thermal energy from natural gas was modelled using background data available in the Ecoinvent v3.10 database, under the "cut-off by classification" system model. The reference flow was defined as the amount of natural gas required to generate 1 MJ of useful thermal energy, calculated based on its lower heating value (LHV). The dataset selected includes the full life cycle of natural gas supply, encompassing extraction, processing, high-pressure transmission, and final combustion.

All inputs, emissions, and infrastructure requirements associated with the upstream stages of natural gas production and distribution are included in the dataset, representing an average global market supply. The modelling assumes that natural gas is delivered directly to the steel plant and combusted in the same type of reheating furnace used in the syngas scenario.

Since the inventory was based entirely on secondary data from a recognised and widely adopted LCI database, no further disaggregation or customisation was applied. The Ecoinvent v3.10 dataset used was validated for the time period relevant to this study and was deemed representative of large-scale industrial consumption in Brazil.

Scenario 3—Landfilling of Fluff (Avoided Impact)

A complementary scenario was modelled to evaluate the environmental impact of landfilling of the fluff. Based on pilot data, 0.035 Nm^3 of syngas are needed per MJ, produced from 0.055 kg of fluff. Thus, this mass of fluff was used as the reference flow for the landfill scenario. The aim is to quantify the impact that could be avoided by gasifying fluff instead of landfilling it.

In this scenario, the environmental impact of disposing of fluff in landfill was modelled as an alternative to its valorisation through gasification. The reference flow adopted was the same mass of fluff required to produce 1 MJ of thermal energy via gasification: 0.055 kg. Therefore, the landfill scenario considers the disposal of 0.055 kg of fluff, including its transportation and treatment by waste type.

As in Scenario 1, the modelling begins with fluff, considered to carry zero environmental burden since its impacts are already accounted for in the upstream steel recycling system.

Each material fraction was assigned to an appropriate waste treatment process, using landfill datasets for residual municipal solid waste or inert materials, depending on its classification. All fractions were modelled as sent to sanitary landfill, without energy recovery or gas capture, in line with conservative assumptions and cut-off modelling rules.

2.1.3 Limits and Assumptions of the Study

This study is based on a set of assumptions and simplifications inherent to the modelling approach, which should be acknowledged and understood before interpreting the results.

Regarding CO_2 emissions, values were estimated based on the syngas composition measured during the pilot-scale gasification test. However, these results may vary depending on the characteristics of the fluff, which is a heterogeneous material. Furthermore, the biogenic carbon fraction of the fluff was estimated based on literature data rather than direct analysis, which may introduce some uncertainty into the carbon accounting results.

The monitoring of pollutants was limited to HCl and Cl_2 due to their relevance in chlorine-rich waste. Other compounds, such as PAHs and dioxins/furans were not measured, but literature indicates that their formation is negligible in gasification systems operated above 800 °C under reducing conditions (Manente et al. 2022).

A sensitivity analysis was not performed due to the high specificity of the study. Parameters such as transport distances and electricity mix reflect actual conditions at the industrial site and could not be meaningfully varied. Likewise, oxygen and electricity consumption were directly measured during the pilot test and treated as fixed values. Although this limits extrapolation to other contexts, the use of primary data enhances the robustness of the results within the scope of this case study.

3   Results and Discussion

Environmental impacts for the three modelled scenarios were assessed using the ReCiPe 2016 method at the midpoint level. Eighteen environmental impact categories were evaluated for each scenario to provide a comprehensive understanding of their environmental performance.

Figure 1 presents the normalised results comparing Scenario 1 (syngas production and use) and Scenario 2 (natural gas production and use). Normalisation was applied using the highest impact value observed in each category as the baseline (100%). The syngas scenario demonstrated lower impacts across the majority of categories.

Scenario 3, representing the landfilling of fluff, was also modelled to estimate the avoided impacts resulting from diverting this waste to energy recovery via gasification. The results of this simulation are detailed in Table 2, which lists the avoided impacts across all midpoint categories.

The results reflect a consistent reduction in environmental impacts across most categories due to the replacement of a fossil-based energy source with syngas derived from waste containing a biogenic carbon fraction. The use of electricity from wind power throughout the syngas production system also contributed to lowering upstream emissions, enhancing the overall environmental performance of this alternative.

Among all categories analysed, water consumption was the only one in which the syngas scenario presented higher impacts compared to natural gas. This outcome is mainly attributed to the water requirements associated with oxygen production through cryogenic air separation, a step present in the syngas route but absent in the natural gas supply chain. This result emphasises the need for a comprehensive assessment of trade-offs, particularly in categories related to resource use.

One of the most relevant differences observed between scenarios relates to the climate change impact. The syngas scenario showed a considerable reduction in global warming potential compared to natural gas, due primarily to two factors: the partial biogenic origin of carbon in fluff and the avoidance of fossil fuel extraction and combustion.

The carbon origin was estimated based on the composition shown in Table 1, using the methodology proposed by Giegrich (2021) for waste streams in Brazil.

However, as the exact origin of biogenic carbon in waste is often uncertain, the assumption of carbon neutrality should be interpreted with caution. While it aligns with common LCA practices, it represents a limitation when claiming climate benefits from waste-derived fuels.

Fossil resource scarcity was significantly reduced in the syngas scenario due to the replacement of natural gas with syngas derived from waste. This directly contributes to the conservation of non-renewable resources, aligning with circular economy strategies. Although the syngas system consumes energy (notably electricity), its fossil fuel savings outweigh the upstream impacts.

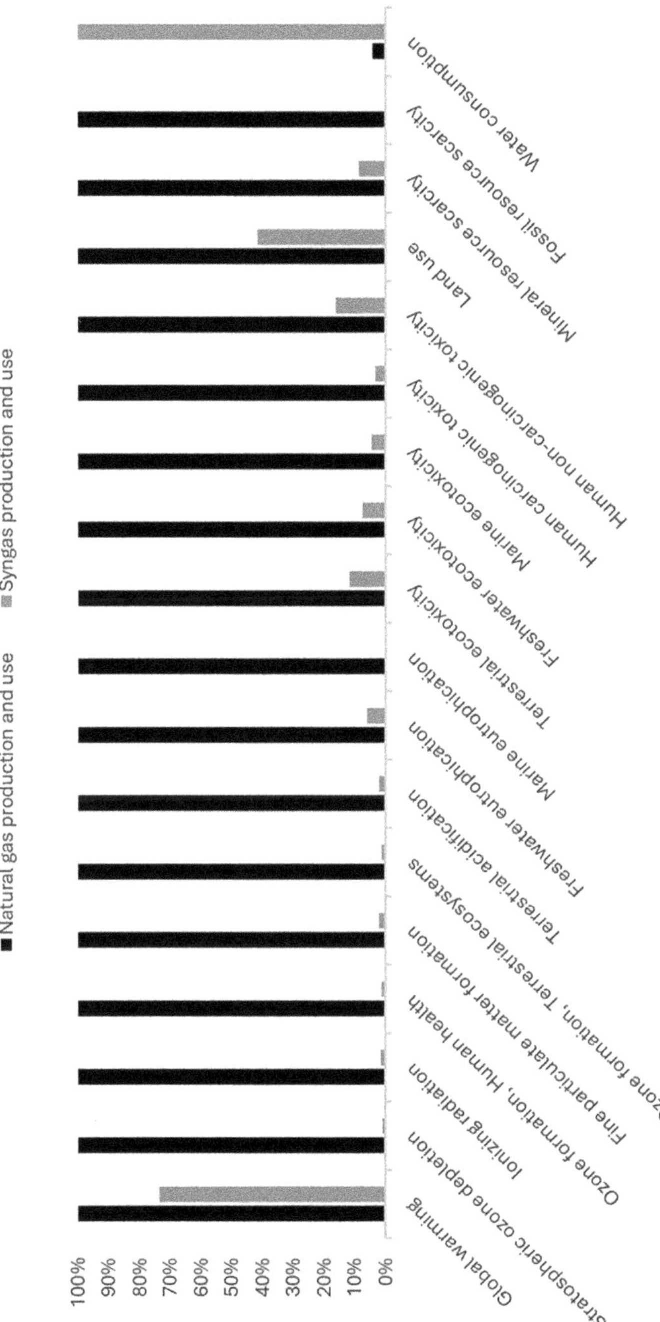

Fig. 1 Comparative bar chart of midpoint environmental impacts for syngas and natural gas scenarios (ReCiPe 2016 Endpoint (H), World 2010)

Table 2 Environmental impacts avoided by diverting fluff from landfill to syngas production (ReCiPe 2016 Midpoint, H)

Impact category	Unit	Landfilling of fluff (impact avoided)
Global warming	kg CO_2 eq	1.42E-02
Stratospheric ozone depletion	kg CFC11 eq	1.09E-09
Ionizing radiation	kBq Co-60 eq	9.36E-06
Ozone formation, Human health	kg NOx eq	3.41E-05
Fine particulate matter formation	kg PM2.5 eq	1.08E-05
Ozone formation, Terrestrial ecosystems	kg NOx eq	3.48E-05
Terrestrial acidification	kg SO_2 eq	2.60E-05
Freshwater eutrophication	kg P eq	2.51E-07
Marine eutrophication	kg N eq	1.19E-06
Terrestrial ecotoxicity	kg 1.4-DCB	2.50E-02
Freshwater ecotoxicity	kg 1.4-DCB	1.12E-02
Marine ecotoxicity	kg 1.4-DCB	1.53E-02
Human carcinogenic toxicity	kg 1.4-DCB	1.31E-02
Human non-carcinogenic toxicity	kg 1.4-DCB	1.36E + 00
Land use	m^2a crop eq	1.44E-04
Mineral resource scarcity	kg Cu eq	4.52E-05
Fossil resource scarcity	kg oil eq	1.71E-03
Water consumption	m^3	8.45E-06

By avoiding landfill disposal, environmental benefits were observed across all midpoint categories, with significant avoided burdens in terrestrial ecotoxicity, human non-carcinogenic toxicity and climate change (global warming).

The environmental performance observed in this study is consistent with the findings of Ciacci et al. (2010), who evaluated multiple end-of-life scenarios for Automotive Shredder Residue (ASR), including landfilling, incineration, and gasification. In their analysis, the feedstock recycling scenario, based on syngas production, demonstrated substantial environmental benefits in terms of climate change mitigation and resource conservation when compared to conventional disposal routes. Despite some trade-offs related to ecotoxicity, largely associated with gaseous emissions from thermal processes, the overall performance of gasification was superior when appropriate gas cleaning was applied.

Similarly, a Danish study on the life cycle impacts of fluff management (Danish Environmental Protection Agency 2015) compared incineration, landfilling, and co-processing in cement kilns. It concluded that energy recovery scenarios consistently outperformed landfill disposal, particularly in categories such as global warming, photochemical ozone formation, and resource use. However, the study also highlighted that impacts related to toxicity—especially human non-carcinogenic toxicity and freshwater ecotoxicity—could increase if potentially hazardous elements and halogenated compounds are not properly managed prior to thermal treatment.

In our case, the syngas scenario demonstrated consistently lower impacts across nearly all categories, reinforcing its potential as a low-emission alternative to natural gas. The inventory for syngas was considered to have a high-quality ranking because it was based on primary measurements obtained from the pilot-scale test, ensuring high temporal, geographical, and technological representativeness. The natural gas and landfill scenarios relied on background datasets from Ecoinvent v3.10, which, although consistent and valid for the period of interest, were rated as medium quality due to regional and technological generalisations, especially regarding upstream fuel burdens and landfill operations. Overall, the data quality was deemed sufficient to support the study's comparative objectives. Nonetheless, potential uncertainties associated with background datasets were considered during result interpretation.

It is important to note that not all combustion by-products were monitored in the pilot test—such as polycyclic aromatic hydrocarbons (PAHs) and dioxins—limiting a more comprehensive assessment of toxicity-related categories. These limitations highlight the need for continued research and improved monitoring of gasification emissions to ensure that environmental benefits are not offset by unaccounted localised effects.

4 Conclusions

This study evaluated the environmental performance of replacing natural gas with syngas—produced from automotive fluff via gasification—for billet reheating in a Brazilian steel plant. Based on pilot-scale data, the LCA indicated lower environmental impacts across most categories for the syngas scenario, except for water consumption. The use of waste as fuel supports circular economy objectives, reduces fossil dependency, and mitigates toxicity-related impacts by diverting fluff from landfilling. Although electricity is required for syngas production, the use of wind energy helps to preserve environmental benefits.

The study highlights the importance of conducting LCA during early technological stages. Applying LCA at the pilot scale allowed for the identification of benefits and trade-offs before scaling up, supporting more sustainable development. Despite some limitations, the findings offer valuable insights for integrating waste-derived syngas into decarbonisation strategies in the steel sector.

References

ArcelorMittal. https://corporate.arcelormittal.com/sustainability/climate-action-reports (Accessed 20.10.2024)

Caetano JA, Schalch V, Pablos JM (2020) Characterization and recycling of the fine fraction of automotive shredder residue (ASR) for concrete paving blocks production. Clean Technol Environ Policy 22:835–847. https://doi.org/10.1007/s10098-020-01825-y

Ciacci L, Morselli L, Passarini F, Santini A, Vassura I (2010) A comparison among different automotive shredder residue treatment processes. Int J Life Cycle Assess 15:896–906. https://doi.org/10.1007/s11367-010-0222-1

Ecoinvent Association (2024) Ecoinvent database version 3.10. Swiss centre for life cycle inventories, Zurich

Giegrich J. Manual da calculadora de emissões de GEE para resíduos. Vol. 1. IFEU – Institut für Energie und Umweltforschung Heidelberg GmbH; 2021. p. 134.

Intergovernmental Panel on Climate Change (IPCC). Guidelines for national greenhouse gas inventories. Institute for Global Environmental Strategies (IGES)

ISO (2006) ISO 14040: environmental management—life cycle assessment—principles and framework. Geneva, Switzerland

Manente G, Martignano S, Ficarella A, Cavaliere P (2022) The pyrolysis and gasification pathways of automotive shredder residue targeting the production of fuels and chemicals. J Phys: Conf Ser 2385:012003. https://doi.org/10.1088/1742-6596/2385/1/012003

ResponsibleSteel. https://www.responsiblesteel.org/about/steelzero/ (Accessed 20.09.2024)

Weidema BP, Wesnaes M (1996) Data quality management for life cycle inventories—an example for using data quality indicators. J Clean Prod 4(3–4):167–174

Embedding Circularity in LCM

Life Cycle Assessment of a Novel Chemical Recycling Technology for Conversion of Waste to Olefins

Aurora Cavaliere, Massimiliano Materazzi, Andrea Paulillo, Paola Lettieri, Semra Bakkaloglu, and Nilay Shah

Abstract This paper summarises results from a large UK project focused on a novel olefin synthesis process specifically tailored for operating with waste feedstock and medium scale plants (~100,000 tons/year), as in the case of syngas obtained from the gasification of waste. A comprehensive Life Cycle Assessment (LCA) analysis unveils the environmental sustainability but also the bottlenecks of the process, comparing with conventional olefin synthesis processes, analysing different impact categories. Overall, this work presents a promising pathway for sustainable olefin production from waste biomass, offering substantial environmental advantages over fossil-based technologies.

1 Introduction

The olefin industry is an important part of the UK economy, but it is also fundamentally reliant on fossil feedstock (Fsadni et al. 2025). Biomass represents the only readily available source of renewable carbon from which carbon-based materials can be produced, and its use in the chemicals and materials sector is essential if we are to reduce the reliance of these sectors on fossil feedstocks (Zhu et al. 2025). In this context, waste and lignocellulosic biomass are the best placed to provide substantial emission reductions, thanks to their low cost and ubiquity. Although thermochemical technologies for waste biomass conversion, such as gasification, have been successfully used for decades with fossil feedstock, the implementation of new bio-based processes also requires solving new problems, both at the front-end (generation of clean syngas from biomass/waste) and back-end (CO_2 utilisation/separation, low selectivity and costly product upgrading unsuitable at small scale) (Yadav et al. 2024). The production of olefins from fossil (coal or natural gas) syngas occurs traditionally

A. Cavaliere · M. Materazzi (✉) · A. Paulillo · P. Lettieri
University College London, London, UK
e-mail: massimiliano.materazzi.09@ucl.ac.uk

S. Bakkaloglu · N. Shah
Department of Chemical Engineering, Imperial College London, London, UK

© The Author(s) 2026
M. Traverso et al. (eds.), *Life Cycle Management from Global to Local*,
https://doi.org/10.1007/978-3-032-17987-6_21

via Fischer Tropsch and steam cracking to olefins (FTO), or methanol synthesis and subsequent Methanol-to-Olefins (MTO). Bi-functional catalysts have been shown to potentially convert syngas to olefins directly in one step (Direct Syngas-to-Olefins, DSO), thus significantly simplifying the process layout and reducing waste by more than 30% (Mahmoudi et al. 2024). DSO relies on metal oxides and acidic zeolites working synergistically. However, in case of a CO_2 -rich feed gas, as obtained from biomass gasification, the synergy progressively fades, because of water production that makes the thermodynamic limitations more stringent. Moreover, the large amount of steam produced in the presence of CO_2 hinders the catalytic activity and reversibly deactivates the dehydration catalyst. In this project, we propose a novel Sorption-Enhanced Olefin Synthesis (SEOS) concept, based on integrated DSO with in-situ water removal, and cyclic regeneration. Thanks to the removal of thermodynamic barriers, SEOS performance is less sensitive to variations in CO/CO_2 ratio and can be operated at much lower pressures than typical refinery processes. This paper summarises results from a large UK project focusing on a SEOS system retrofitted to a waste biomass gasification plant with integrated carbon capture. A sophisticated set of models has been developed using Aspen Plus to simulate the performance of the fully integrated waste-to-olefins system, as well as to produce the mass and energy balance of a first 100,000 tonnes of waste per annum (tpa) of waste input plant. Finally, the optimised mass and energy balances are used to perform a comprehensive Life Cycle Assessment (LCA) analysis to unveil the environmental sustainability but also the bottlenecks of the process, comparing it with conventional olefin synthesis processes, analysing 9 impact categories, such as climate change, resource use, water consumption, using Environmental Footprint (EF) 3.1 method and a cradle-to-gate approach.

2 Methodology

2.1 *Waste Biomass Gasification Plant*

The plant scale and layout reflect those of a first medium scale plant, treating approximately 100,000 tpa of waste biomass feedstock. Upon collection, waste wood (Grade B and C industrial waste wood) is dried, separated, and recyclable/external materials are recovered. The obtained chipped feedstock is fed into a high-temperature process for steam-oxygen gasification, ash vitrification and tar-reforming (Syngas generation). This can take place in a single-stage system (e.g. former Thermoselect concept, now licensed by JFE) or in two-stage processes, such as those commercialised in UK by ABSL (with addition of electric power in arc furnaces for ash collection and melting). The former data were used to validate the gasification runs, as reported in (Amaya-Santos et al. 2021). Large incombustible material is removed from the gasifier (as bottom ash) and sent to inert landfill, while the vitrified ash is collected and then exported from site for use as an aggregate. Syngas stream is then cleaned using conventional cleaning technologies, including dry filters, and a combination of acid and alkaline scrubbers (Syngas cleaning). Effluents are discharged to drain as wastewater. The hydrogen fraction of the clean syngas is then increased via a single high-temperature water gas shift reactor, followed by additional guard beds for removal of trace impurities (sulphur mostly) down to ppb levels. Given the excess carbon present in the system, carbon dioxide stream is chemically absorbed in a commercial amine-based solvent with high CO_2 recovery (~95%) followed by liquefaction at 150 bar for storage and transportation (CO_2 separation; CO_2 liquefaction & storage). A detailed description of the gasification and gas conditioning steps is provided in other works from the authors (Amaya-Santos et al. 2021; Materazzi et al. 2024).

2.2 SEOS Concept and Process Integration

Sorption enhanced olefins synthesis (SEOS) is a novel process for the production of light olefins from synthesis gas via DSO, in which water is removed in situ by the use of a solid adsorbent, typically a 3A or 4A zeolite (Gabruś et al. 2015). The concept is based on Le Chatelier's principle stating that reactant conversion to products in an equilibrium limited reaction is increased by selectively removing reaction products, which is utilized for various processes and products mainly considering CO_2 separation (Fig. 1). By continuously removing water as the reaction product during methanol synthesis and MTO, the oxygen surplus of the feed no longer ends up in CO_2, as is the case for direct DSO synthesis, resulting in higher carbon utilization and better process efficiencies.

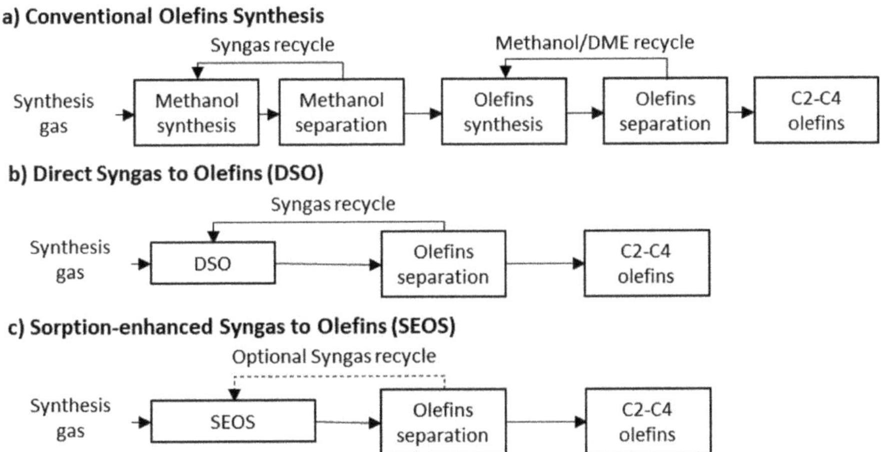

Fig. 1 Comparison of three syngas-to-olefins conversion technologies. **a** Conventional olefins synthesis via methanol synthesis and MTO **b** Direct syngas to olefins (DSO) **c** Sorption-enhanced syngas to olefins (SEOS)

In a sorption enhanced system, as the SEOS process, the periodic regeneration of the saturated adsorbent is typically done by pressure swing, temperature swing, purge/concentration swing or combinations of these methods. While the different regeneration procedures have their own typical duration (timings), the timing of adsorption and regeneration have to be carefully tuned. The simplest case would be a two-reactor column system, in which one column is producing the sorption enhanced product, while the other column is regenerating. This can be conveniently done in a dual fluidised bed system, in which the adsorbent is circulated between the two reactors, sized differently to allow for sufficient residence time for the two processes to go to completion. For zeolite-type adsorbents, a normal temperature swing regeneration cycle can be performed, by using hot dry air as fluidising agent in the regeneration reactor. This relatively simple two-column fluidised bed system is used as a base for the SEOS model study in this work.

The fully integrated process Waste-to-Olefins (WtO) through SEOS was modelled in ASPEN Plus V14, with the various process packages validated from demonstration (gasification, gas cleaning and conditioning) and industrial (methanol synthesis, MTO, water adsorption) data. The SEOS modelling package includes kinetic reactors for the simulation of chemical reactions (e.g. methanol synthesis, methanol dehydration, etc.), GSTA (Generalized Statistical Thermodynamic Adsorption) model for water removal and adsorbent regeneration, and equilibrium-based phase separators for products upgrading. Sorbent hydration and regeneration kinetics were implemented in MATLAB and integrated into the ASPEN model via Calculator Blocks.

2.3 LCA Methodology

The LCA was conducted following the steps outlined in ISO 14040/14044. The software LCA for Experts (formerly GaBi) V10.9 was used to build the LCA models (Sphera 2025; ISO 2006).

2.3.1 Goal and Scope

This study aims at quantifying the environmental impacts of converting wood biomass waste into olefins based on the SEOS concept. It compares the environmental performance against alternative pathways, including waste to olefin and coal to olefins with conventional MTO process. Although real plant data are not available yet to validate the SEOS models, a preliminary analysis based on model projections would provide an insight on major process deviation and potential benefits. In line with literature LCA studies on olefin production, the functional unit is set to 1 kg of olefin products, allowing direct comparison of the three alternative pathways (Cuevas-Castillo et al. 2024; Xiang et al. 2015). The system boundaries (Fig. 2) are "cradle to gate", extending from the feedstock input to the point of product output.

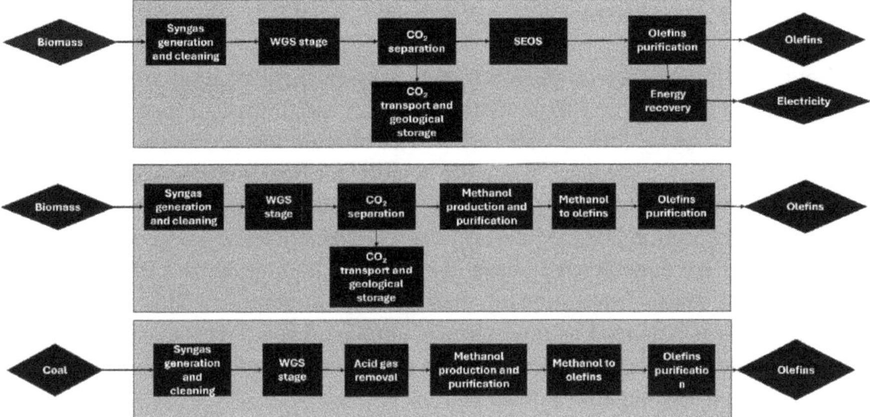

Fig. 2 System boundaries block diagram for olefin production from different pathways with a "cradle to gate" approach. Plant construction and wastewater treatment processes are excluded from all pathways

Multifunctionality is addressed via system expansion with substitution, where the environmental impacts of the conventional waste wood treatment are credited to the system.

2.3.2 Inventory Data

The life cycle inventory relies on a combination of detailed mass and energy balances obtained from process simulations, literature sources and data from the ecoinvent 3.10 database. Mass and energy balances derived from process simulations conducted as part of this work (described in Sect. 2.2 and reported in Sect. 3.1) are used to model the WtO processes, with and without SEOS. Data for methanol synthesis from syngas and MTO processes are sourced from ecoinvent 3.10. The background system activities, including the supply of materials and energy, are modeled using the ecoinvent 3.10 cut-off database to ensure consistency with the other system. This includes the production and delivery of chemicals used as gasification agents (e.g., oxygen), gas cleaning chemicals, and the carbon capture solvent (monoethanolamine, MEA); the net consumption and generation of thermal energy and electricity. Ecoinvent datasets were also used to model CO_2 transportation by lorry and sea tankers. CO_2 transport via pipeline and its injection into deep saline aquifers were modelled using inventory data from (Antonini et al. 2020). The United Kingdom as the geographical, and temporal context for all related cases was chosen.

2.3.3 Impact Assessment

The Environmental Footprint (EF) 3.1 impact assessment method was used due to its alignment with European Commission recommendations and its comprehensive coverage of mid-point impact categories. The climate change impact of biogenic carbon emissions is considered using the widely adopted approach "biogenic CO_2 neutral" framework endorsed by the IPCC (2019) and European PEF methodology (European Commission 2013; Materazzi et al. 2024). Under this approach, biogenic CO_2 emissions are considered to have zero climate change impact whilst biogenic CH_4 emissions cause lower impact than fossil counterparts. This assumes that released carbon is balanced by atmospheric uptake during the biomass growth. The permanent sequestration of biogenic carbon creates net atmospheric carbon removal, effectively achieving negative emissions.

3 Results and Discussion

To investigate the effect of various process parameters on the performance of the SEOS process, some key parameters have to be indicated. One of the key parameters is operating pressure, which is particularly important for the methanol synthesis step. Industrial processes for methanol production from syngas operate at pressure spanning from 80 to 150 bar, depending on different plant configurations and number of recycle loops (Bisotti et al. 2021). By removing water, one of the key intermediate products from equilibrium-controlled reactions, SEOS would be able to overcome thermodynamic barriers, significantly lowering the pressure requirements of the system.

The results presented in Fig. 3 show the conversion percentages of CO and CO_2 under different pressure regimes and process methodologies. The comparative analysis highlights significant differences between the conventional two-step process (syngas-to-methanol and MTO) and the combined process (DSO) with and without adsorption technology. In the conventional two-step process, there is a clear correlation between increased pressure and higher conversion rates. At 10 bar (STM-10 bar, MTO-1 bar), the conversion is minimal (~7%). When pressure is increased to 50 bar, the conversion reaches 63%, while at 80 bar, the maximum conventional conversion of 73% is achieved. In all these cases a recirculation of unreacted syngas is needed, increasing costs and complexity of the plant. The combined process, on the other hand, presents particularly interesting characteristics. At 10 bar, DSO shows improvement compared to the conventional low-pressure process, reaching approximately 19% conversion. The most significant breakthrough is observed in the combined process with the adsorption system (SEOS), which achieves an extraordinary conversion of 95% at same pressure. This remarkable increase can be attributed to the adsorbent's ability to selectively remove water from the reaction system, allowing almost complete utilization of the CO and CO_2 in syngas. Sorbent regeneration is made via temperature swing, where thermal energy is fully recovered by energy

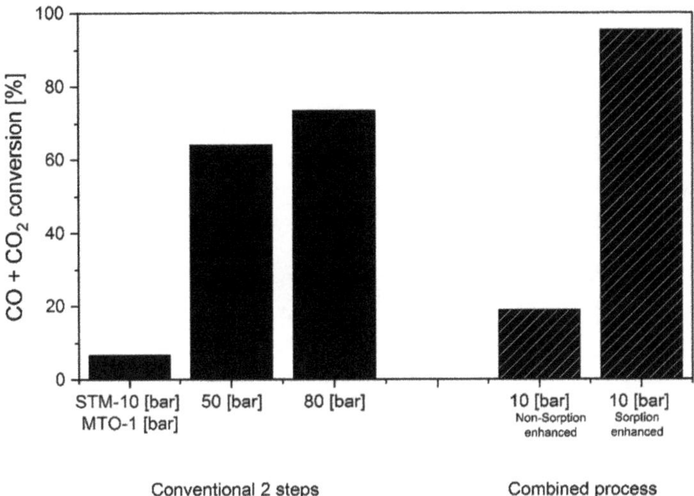

Fig. 3 CO and CO$_2$ conversion efficiency comparison between conventional two-step processes and combined processes at different operating pressures

integration within the plant. These results demonstrate how the integration of adsorption technologies in methanol synthesis processes, particularly when implemented in a fluidized bed reactor configuration, can represent a significant technological advancement.

The fully integrated WtO performance with SEOS is presented in Fig. 4: Sankey diagram showing energy flows for the waste to olefins pathway via Sorption-Enhanced Olefin Synthesis (SEOS). This shows that from a plant treating approximately 100,000 tpa of waste wood is possible to produce approximately 17,117 tpa of refined olefins and 91,000 tpa of liquified CO$_2$ for storage. This is a significant advancement against state-of-art of current renewable MTO technologies (+30% product yield), offering important opportunities for future bio-refinery concepts and small-medium scale applications with carbon capture and storage technologies (CCS). Figure 5 presents a comparative analysis of the three olefin production routes, across a set of 9 selected environmental categories. These categories were chosen to provide a comprehensive, multi-dimensional understanding of the systems, capturing key aspects such as climate change, resource consumption, ecosystem pressure and human health risk. The results indicate a clear trend: WtO with SEOS demonstrates the lowest environmental impacts in six out of nine categories, owing to its higher process efficiency and reduced energy demand, as it operates at lower pressure than the conventional MTO route with higher olefin production. Although WtO with SEOS exhibits a lower environmental impact than its counterpart without SEOS in all categories, both routes show significantly higher impacts than CtO (Coal-to-Olefins) in the ionization radiation category, and higher impacts in the categories land use and ozone depletion.

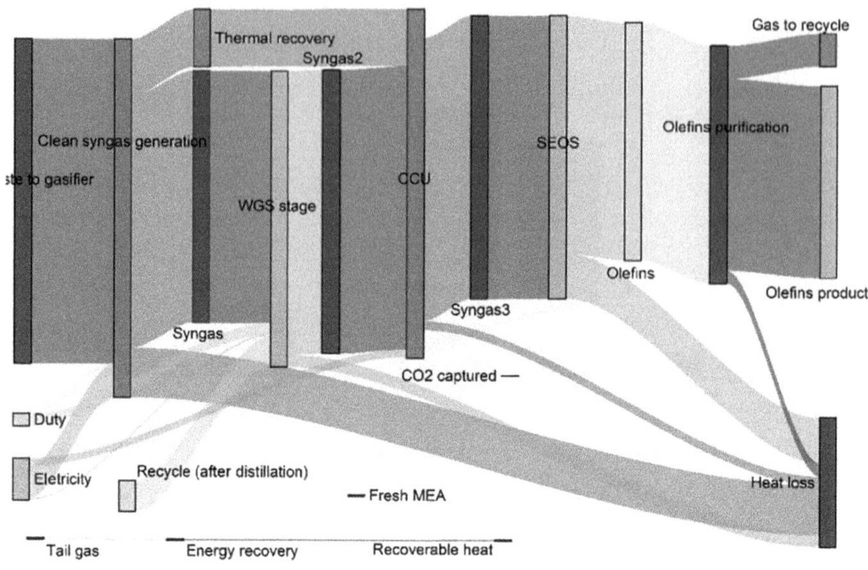

Fig. 4 Sankey diagram showing energy flows for the waste to olefins pathway via Sorption-Enhanced Olefin Synthesis (SEOS)

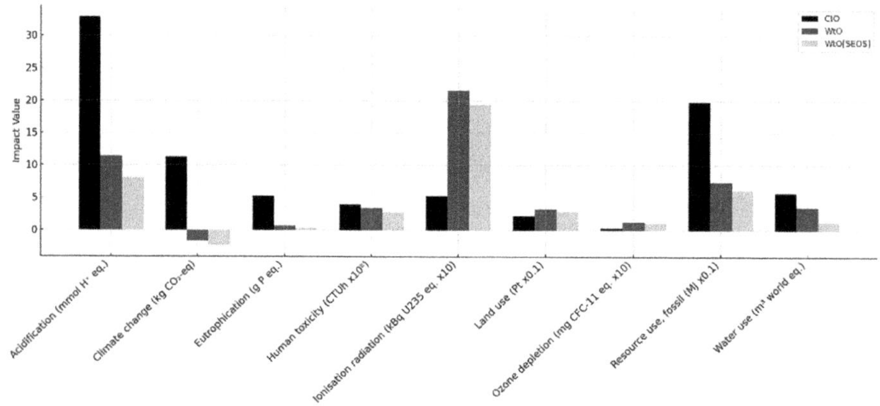

Fig. 5 Environmental impacts of one kg olefin via three alternative production routes. To read the chart, y-axis values should be multiplied by the values reported in the x-axis. For example, the environmental impact in the ionising radiation category for WtO with SEOS equals to 1.94 kBq U235 eq

As expected, the CtO pathway yields the highest global warming potential (GWP) (11.3 kg CO_2-eq/kg olefin), primarily due to its carbon-intensive coal feedstock and the lack of CCS. Notably, 95% of these emissions originate from the coal gasification process. In contrast, both waste WtO routes demonstrate net-negative climate change impacts, attributable to biogenic carbon uptake and CCS implementation. Among these, WtO with SEOS achieves the lowest emissions (-2.4 kg CO_2-eq/kg olefin) benefiting from a single-step DSO conversion, which reduces auxiliary fuel demand by operating at lower pressure and higher conversion. Both WtO with and without SEOS share identical life cycle inventory data for wood gasification and syngas generation with CCS (-2.36 kg CO_2-eq/kg olefin). However, the conventional WtO pathway (without SEOS) incurs higher emissions from two downstream processes: syngas-to-methanol conversion (0.17 kg CO_2-eq) and MTO synthesis (0.61 kg CO_2-eq). CtO consistently scores highest in acidification (32.9 mmol H^+ eq.), freshwater eutrophication (5.3 g P eq.), and human toxicity (4×10^8 CTUh), driven by emissions from coal gasification and large-scale material throughput due to high utility consumption. WtO with SEOS delivers the lowest values for acidification (8.1 mmol of H + eq.), freshwater eutrophication (0.4 g P eq.) and human toxicity (2.8 $\times$ 10^8 CTUh). Compared to the other environmental impact, ionisation radiation is the only category where the WtO pathway (without SEOS 2.2 kBq U235 eq.; and with SEOS 1.9 kBq U235 eq.) show significantly higher impacts than CtO (0.5 kBq U235 eq.). This is attributed to the greater reliance on electricity in wood gasification processes (lower energy density), which is in part sourced from nuclear power in the underlying LCA assumptions (i.e. Great Britain grid mix). The higher electricity demand of biomass gasification compared to coal gasification is also supported by (Liu et al. 2020). On the other hand, coal gasification is more heat-driven, with energy often sourced from coal itself. All three routes have moderate land use impacts, with WtO without SEOS displaying the highest impact, followed by WtO with SEOS and CtO. The slightly higher land use impact observed in WtO can be attributed to upstream supply chain of wood waste, which often includes shared burdens from forestry operations, and land occupation associated with wood processing activities. In contrast, coal mining, while intensive in other environmental impacts, typically registers lower direct land occupation per unit of output. Regarding ozone depletion, WtO with and without SEOS outperform CtO because wood gasification emits N_2O and halogenated compounds (from biomass chlorine content) during combustion/gasification, which contribute to stratospheric ozone depletion (Huijbregts et al. 2017). Coal gasification, however, produces minimal ozone-depleting substances owing to its inherently low chlorine content.

CtO has the highest fossil resource consumption (198 MJ), directly related to coal feedstock. WtO with and without SEOS (62 and 74 MJ, respectively) consume significantly fewer fossil resources, with the SEOS pathway contributing further reductions owing to its lower temperature and pressure requirements.

Additionally, CtO has the highest water demand (5.73 m^3), likely driven by coal washing, steam generation, and cooling requirements. On the other hand, WtO with SEOS shows the lowest water consumption (1.28 m^3), pointing to potential process integration or water recycling benefits enabled by SEOS. WtO with SEOS outperforms both CtO and conventional WtO across most EF impact categories, particulate matter emissions, photochemical ozone formation, ecotoxicity, and mineral resource use is comparable or slightly improved.

4   Conclusions

This study presents a comprehensive life cycle assessment of a novel SEOS technology for converting waste wood into olefins. By integrating in-situ water removal into the syngas-to-olefin process, the SEOS pathway significantly enhances carbon utilisation efficiency by at least 30% and allows high conversion at lower operating pressures, reducing energy demand.

The LCA results demonstrate that WtO with SEOS consistently outperforms both the conventional WtO and CtO pathways across most environmental impact categories. Notably, WtO with SEOS achieves net-negative climate change impacts and lowest values for acidification, eutrophication, human toxicity, fossil resource use, and water consumption. However, both WtO routes show higher impacts in ionising radiation, attributed to the electricity-intensive nature of biomass gasification and the use of nuclear power in the electricity mix. Additionally, land use and ozone depletion impacts are moderately higher due to upstream forestry-related activities and biomass composition.

References

Amaya-Santos G, Chari S, Sebastiani A, Grimaldi F, Lettieri P, Materazzi M (2021) Biohydrogen: a life cycle assessment and comparison with alternative low-carbon production routes in UK. J Clean Prod 319:128886. https://doi.org/10.1016/j.jclepro.2021.128886

Antonini C, Treyer K, Streb A, van der Spek M, Bauer C, Mazzotti M (2020) Hydrogen production from natural gas and biomethane with carbon capture and storage–a techno-environmental analysis. Sustain Energy Fuels 4(6):2967–2986. https://doi.org/10.1039/D0SE00222D

Bisotti F, Fedeli M, Prifti K, Galeazzi A, Dell'Angelo A, Barbieri M, Pirola C, Bozzano G, Manenti F (2021) Century of technology trends in methanol synthesis: any need for kinetics refitting? Ind Eng Chem Res 60(44):16032–16053. https://doi.org/10.1021/acs.iecr.1c02877

Cuevas-Castillo GA, Michailos S, Akram M, Hughes K, Ingham D, Pourkashanian M (2024) Techno economic and life cycle assessment of olefin production through CO_2 hydrogenation within the power-to-X concept. J Clean Prod 469:143143. https://doi.org/10.1016/j.jclepro.2024.143143

European Commission (2013) https://green-forum.ec.europa.eu/green-business/environmental-footprint-methods_en

Fsadni M, Royle M, Gibson E, O'Brien S (2025) Circular economy of olefins roadmap for the UK—Executive summary

Gabruś E, Nastaj J, Tabero P, Aleksandrzak T (2015) Experimental studies on 3A and 4A zeolite molecular sieves regeneration in TSA process: aliphatic alcohols dewatering–water desorption. Chem Eng J 259:232–242. https://doi.org/10.1016/j.cej.2014.07.108

Huijbregts MAJ, Steinmann ZJN, Elshout PMF, Stam G, Verones F, Vieira M, Zijp M, Hollander A, van Zelm R (2017) ReCiPe2016: a harmonised life cycle impact assessment method at midpoint and endpoint level. Int J Life Cycle Assess 22(2):138–147. https://doi.org/10.1007/s11367-016-1246-y

ISO, ISO 14040:2006—Environmental management—Life cycle assessment—Principles and framework, (2006). https://www.iso.org/standard/37456.html

Lange J-P (2001) Methanol synthesis: a short review of technology improvements. Catal Today 64(1–2):3–8. https://doi.org/10.1016/S0920-5861(00)00503-4

Liu Y, Li G, Chen Z, Shen Y, Zhang H, Wang S, Qi J, Zhu Z, Wang Y, Gao J (2020) Comprehensive analysis of environmental impacts and energy consumption of biomass-to-methanol and coal-to-methanol via life cycle assessment. Energy 204:117961. https://doi.org/10.1016/j.energy.2020.117961

Mahmoudi E, Sayyah A, Farhoudi S, Bahranifard Z, Behmenyar G, Turan AZ, Delibas N, Niaei A (2024) Advances in catalysts for direct syngas conversion to light olefins: a review of mechanistic and performance insights, J CO_2 Util, 86. 102893, https://doi.org/10.1016/j.jcou.2024.102893

Materazzi M, Chari S, Sebastiani A, Lettieri P, Paulillo A (2024) Waste-to-energy and waste-to-hydrogen with CCS: methodological assessment of pathways to carbon-negative waste treatment from an LCA perspective. Waste Manag 173:184–199. https://doi.org/10.1016/j.wasman.2023.11.020

Sphera (2025) https://sphera.com/

Xiang D, Yang S, Li X, Qian Y (2015) Life cycle assessment of energy consumption and GHG emissions of olefins production from alternative resources in China. Energy Convers Manag 90:12–20. https://doi.org/10.1016/j.enconman.2014.11.007

Yadav S, Choudhary AK, Yadav P, Pal DB (2024) Lignocellulosic biomass gasification for bio-circular economy sustainability: a multiple criteria analysis framework. Biomass Convers Bior. https://doi.org/10.1007/s13399-024-06387-3

Zhu H, Babkoor M, Coppens M-O, Materazzi M (2025) Thermochemical technologies for conversion of biomass and waste into light olefins (C_2–C_4). Fuel Process Technol 267:108174. https://doi.org/10.1016/j.fuproc.2024.108174

Life Cycle Assessment of PVC Recycling: Expanding Recyclable Fractions Through the CIRC-PVC Project

Jean Metzmacher and Angélique Léonard

Abstract This study evaluates the environmental impacts of a novel physical recycling technology for polyvinyl chloride (PVC), developed within the CIRC-PVC consortium. This technology is an upgraded VinyLoop™ dissolution process (VinyLoop™-D), that enables the removal of legacy additives (DEHP plasticisers and lead and cadmium stabilisers) while recovering high-quality recycled PVC comparable to virgin PVC. A life cycle assessment is conducted to compare the environmental impacts of this recycling method with those of conventional virgin PVC production via suspension polymerisation. Steam and solvent consumptions are identified as the primary contributors to environmental impacts of the recycling process. Considering data uncertainties, no clear environmental advantage or disadvantage of recycling compared with conventional production can be identified.

1 Introduction

Polyvinyl chloride (PVC) is one of the most widely used synthetic polymers globally, ranking third in production volume among all plastics due to its versatility, durability, and cost-effectiveness (Goodfish Group 2025; EuP Egypt 2025). Its widespread adoption across diverse sectors—including construction, healthcare, packaging, and consumer goods—underscores the importance of understanding its environmental implications throughout its life cycle (Bird et al. 2024). As environmental impacts become a central concern in material selection and waste management, life cycle assessment (LCA) emerges as a critical methodological framework for evaluating the environmental burdens associated with PVC production, use, and end-of-life scenarios (ECVM 2025).

PVC is a thermoplastic polymer composed of repeating vinyl chloride monomer units. The presence of chlorine atoms imparts unique properties such as chemical resistance and mechanical rigidity (Cruz 2025). However, the base resin is rarely used

J. Metzmacher (✉) · A. Léonard
Chemical Engineering Research Unit, PEPs-Product, Environment, and Processes Group, University of Liège, Liege, Belgium
e-mail: jmetzmacher@uliege.be

M. Traverso et al. (eds.), *Life Cycle Management from Global to Local*,
https://doi.org/10.1007/978-3-032-17987-6_22

in its pure form. To tailor its properties for specific applications, PVC is compounded with a wide range of additives (Bird et al. 2024): plasticizers, stabilizers, fillers, lubricants, pigments. These additives significantly influence the material's performance, processing behavior, and environmental footprint.

Plasticisers, particularly phthalate esters such as bis(2-ethylhexyl) phthalate (DEHP) and diisononyl phthalate (DINP) are commonly used to impart flexibility to soft PVC formulations (Elgharbawy 2022; Ledniowska et al. 2022). Thermal stabilisers, historically based on lead or cadmium, are now increasingly replaced by calcium-zinc or organotin compounds due to health and environmental concerns (VinylPlus 2023). While additives are essential for performance, they also introduce complexity in recycling and raise concerns about potential leaching and toxicity during use and disposal (Thornton 2002).

PVC recycling presents both opportunities and challenges. Mechanical recycling, which involves shredding, cleaning and remelting, is the most established method and is particularly effective for rigid PVC waste (VinylPlus 2025). However, it is limited by contamination and the presence of additives, often resulting in downcycling to lower-value applications (Ait-Touchente et al. 2024). Physical recycling, including solvent-based processes such as VinyLoop™ technology (Vandenhende et al. 2006), offers the potential to address PVC waste that cannot be effectively treated by mechanical recycling. By selectively dissolving the PVC waste, it allows to recover high-quality PVC from flexible waste streams (Wagner et al. 2024). Despite its promise, physical recycling is energy-intensive and economically constrained by solvent recovery and process complexity (Ait-Touchente et al. 2024; Ügdüler et al. 2020). Moreover, separating legacy additives from old PVC products remains a significant technical challenge (Ügdüler et al. 2020).

Within the CIRC-PVC consortium, INEOS Inovyn aims to upgrade VinyLoop™ dissolution technology to VinyLoop™-D in order to separate legacy additives, such as DEHP plasticisers and lead and cadmium stabilisers, from PVC waste. This study is conducted as part of this initiative and seeks to compare the environmental impacts of physically recycled PVC with those of conventionally produced virgin PVC. The Walloon Region funds the consortium through the Greenwin cluster.

In this context, the present work provides a scientifically rigorous and application-specific evaluation of the environmental impacts of PVC physical recycling, in the frame of the CIRC-PVC consortium, using life cycle assessment. By integrating industrial data, commercial database and literature sources, the study identifies key impact categories, evaluates trade-offs between virgin and recycled PVC, and supports the development of more environmentally-friendly material management strategies. The results are intended to inform decision-maker within the consortium and contribute to the broader discourse on circularity and environmental responsibility in polymer use.

2 Materials and Methods

The life cycle assessment (LCA) is conducted in accordance with ISO 14040 and 14044 standards. This study compares an upgraded VinyLoop™-D dissolution method based on the VinyLoop™ dissolution technology, which includes the extraction of legacy additives, with the conventional production of PVC through suspension polymerisation using a virgin material.

2.1 *Goal and Scope*

The objective of this study is to assess the environmental impacts associated with the physical recycling of PVC. This analysis is intended to serve as a decision-support tool for the partners of the CIRC-PVC project, helping them to reduce the environmental impact of the new dissolution technology.

In this study, a product-oriented perspective is adopted. The functional unit is defined as the production of 1 kg of PVC powder in Belgium.

The study scenario involves the physical recycling of old post-consumer PVC waste into recycled PVC powder (R-PVC), incorporating the removal of legacy additives. The system begins with the collection of waste and ends with the production of 1 kg of recycled PVC powder, as illustrated in Fig. 1. The waste is first transported to a collection facility where it is shredded. It is then sent to a sorting unit. Subsequently, it is transported to the recycling site, where it undergoes a dissolution process. During this stage, impurities are removed, and most of the lead and cadmium stabilisers are separated along with a large portion of the fillers as insoluble residues. These residues are landfilled, while the impurities are incinerated with energy recovery. The wastewater (mother liquor) generated during dissolution is treated via a dedicated process.

The output of this stage is a PVC powder that is free of heavy metals but still contains plasticisers. This intermediate product is then processed in an extraction unit, where plasticisers are removed. The extracted plasticisers are subsequently incinerated with energy recovery.

In order to evaluate the environmental performance of the study scenario, it is compared to a reference scenario representing conventional PVC production from virgin raw materials via suspension polymerisation (S-PVC). The multifunctionality of the study system is addressed through system expansion by addition (basket of products) (Arena et al. 2003; Garcia et al. 2025). The co-function of the recycling process, PVC waste management, is added to the reference system. Waste incineration with energy recovery is considered as the conventional waste treatment. The system boundaries of the reference scenario are illustrated in Fig. 2.

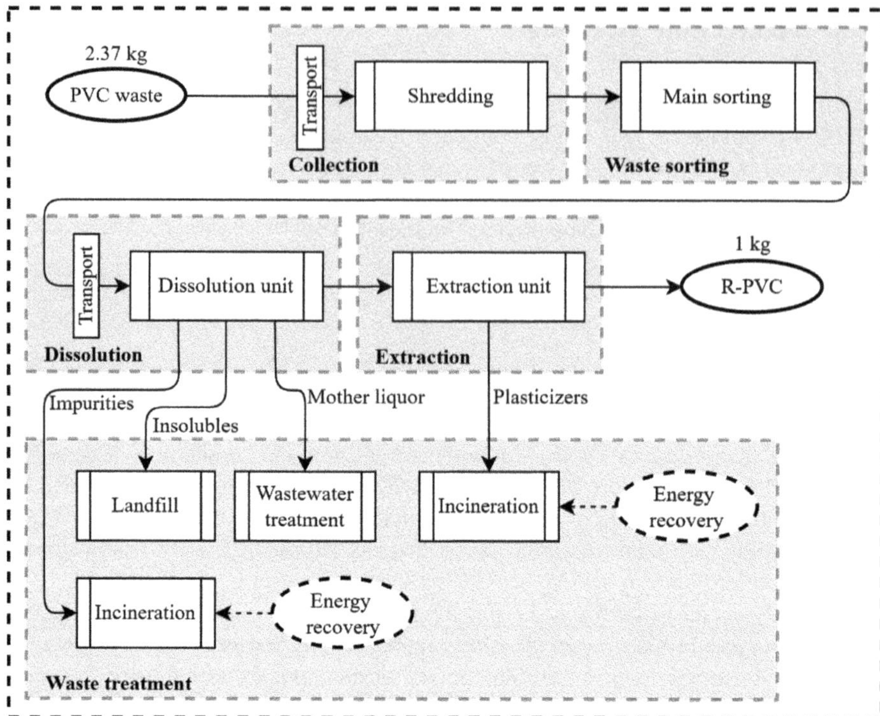

Fig. 1 System boundaries of the study scenario

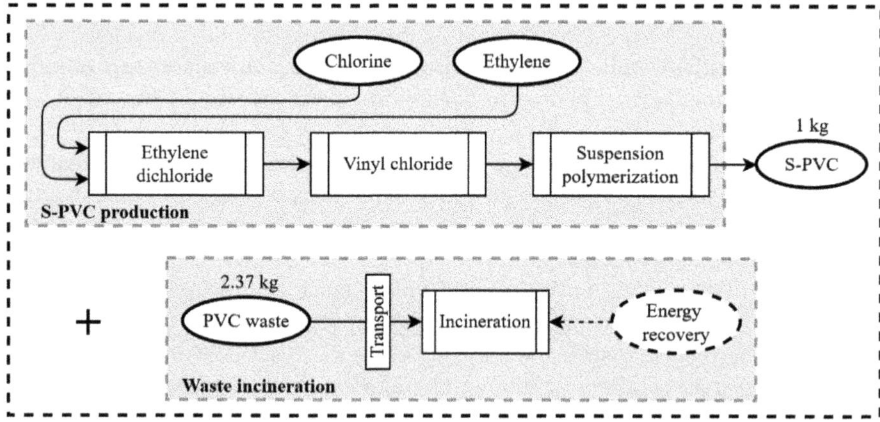

Fig. 2 System boundaries of the reference scenario

2.1.1 Assumptions and Neglected Processes

The infrastructures associated with the foreground processes are excluded from the system boundaries. Given the high production volumes typically associated with industrial-scale operations, the relative contribution of infrastructures to the overall environmental impacts is considered negligible.

The cut-off approach is used to address the end-of-life multifunctionality of the PVC waste. They therefore enter burden-free in the considered systems.

Based on data provided by INEOS Inovyn, the composition of the PVC waste stream is assumed to be 10% impurities and 90% PVC compounds. The latter consists of 46.9% suspension-grade PVC, 32.8% plasticisers (DEHP), 18.8% calcium carbonate ($CaCO_3$) as filler, and 1.5% stabilisers containing lead and cadmium.

For the waste sorting step, lab-scale estimates are upscaled to industrial scale using an exponential scaling approach. The auxiliary consumption for each input i is scaled using the following equation

$$\frac{C_{2,i}}{C_{1,i}} = \left(\frac{O_2}{O_1}\right)^R$$

where O_1 and O_2 are the capacities (in kg/h) of the prototype and the industrial-scale sorting unit, respectively, $C_{1,i}$ and $C_{2,i}$ are the auxiliary consumptions (e.g., electricity in kWh) for the prototype and the scaled-up unit, and R is the scaling exponent. Based on commercial sorter specifications, the industrial capacity O_2 is assumed to be 1 tonne per hour. In the absence of empirical data for this specific case, a scaling exponent $R = 0.65$ is adopted. This value corresponds to the average value for cost-based scale-up of the conveyor (Remer et al. 1993) and it is chosen because the impacts of the sorting device are mainly related to its conveyor.

A solvent recovery rate of 97% is applied to the dissolution unit, based on operational data from the former VinyLoop™ facility. For the plasticiser extraction process, a recovery rate of 95% is assumed for the substance that is used to perform the extraction, in line with supplier claims indicating minimal losses. Due to the lack of specific data, the incineration of extracted plasticisers is modelled using polyethylene terephthalate incineration as a proxy. Impurities present in the PVC waste are modelled as a generic mixture of non-PVC plastics. The insoluble fraction, primarily composed of fillers and most of the stabilisers, is assumed to be landfilled. In the absence of more detailed data, this is modelled using the Ecoinvent 3.10 inventory for the landfilling of lead smelter slag, which has a heavy metal content comparable to the expected concentrations of lead and cadmium in the insolubles. Wastewater generated during the recycling process is modelled using average municipal wastewater treatment, due to the lack of specific data on the treatment required.

2.2 Life Cycle Inventory and Impact Assessment

For both scenarios, background processes are modelled using the Ecoinvent 3.10 database. To ensure geographical representativeness, the electricity mixes associated with relevant processes are adapted to reflect the regional energy profiles where the activities occur. Recovered energy from municipal waste incineration is modelled as avoided electricity and heat, based on Ecoinvent 3.10 data and adjusted to the location of the incineration facilities.

In the study scenario, foreground processes are modelled using primary data provided by industrial partners of the CIRC-PVC consortium: INEOS Inovyn, Vanheede Environmental Logistics and Rovi-Tech. The collection step is based on operational data from Vanheede Environment's facilities.

The waste sorting step is modelled using estimates for a lab-scale prototype developed by Rovi-Tech. These estimates are upscaled to industrial scale as explained in Sect. 2.1.1.

The dissolution stage is modelled using data from the former industrial Viny-Loop™ facility in Ferrara (Italy), which operated from 2002 to 2018, and from pilot-scale experiments extrapolated to industrial scale using ASPEN simulation software. The extraction stage is modelled based on operational data provided by the supplier of the industrial-scale unit implemented in the project.

In the reference scenario, the production of suspension-grade PVC (S-PVC) is modelled using the inventory "polyvinylchloride production, suspension polymerisation" from Ecoinvent 3.10. The production of chlorine is modified to reflect exclusive use of membrane cell chlor-alkali electrolysis, in line with the technology employed by INEOS Inovyn. Waste incineration is modelled using Ecoinvent 3.10 inventories for municipal solid waste incineration with fly ash treatment, corresponding to the composition of PVC waste.

The life cycle analysis is performed using Brightway 2.5 in combination with Activity Browser. Environmental impacts are assessed using the Environmental Footprint 3.1 method. Emissions released more than 100 years after the activity (*i.e.*, "long-term" emissions) are excluded from the analysis.

2.3 Data Uncertainties

For background processes modelled using the Ecoinvent 3.10 database, the default uncertainty values provided within the database are retained. These include both statistical and methodological uncertainties inherent to the dataset.

For foreground processes, uncertainties are assumed to follow log-normal distributions. They are quantified using the pedigree matrix approach, as implemented in the Activity Browser tool. This method accounts for data quality indicators such as reliability, completeness, temporal and geographical correlation, and technological representativeness.

The composition of the PVC waste stream is considered fixed in this analysis. No uncertainty is applied to its formulation, in order to isolate and assess the variability associated with the consumption of auxiliary inputs such as energy and materials.

3 Results and Discussion

3.1 Study Scenario

The characterisation results for the environmental impacts of the study scenario are presented in Fig. 3. The dissolution stage of PVC waste clearly emerges as the most environmentally impactful across nearly all damage categories, with the exception of ionising radiation. This high contribution is primarily driven by the consumption of steam and the use of solvent, despite the latter being recovered at a rate of 97%. In the ionising radiation category, the extraction stage is the dominant contributor due to its relatively high electricity consumption.

The waste treatment stage contributes to impact reduction in most categories through energy recovery. However, in several categories—such as climate change, freshwater ecotoxicity, eutrophication, and human toxicity, non-cancer—it still accounts for a significant share of the impacts (ranging from 9 to 41%, depending on the category), mainly due to emissions from incineration.

This overview already highlights two key processes that should be prioritised to reduce the environmental footprint of physical recycling: steam consumption and solvent use. The steam consumption is modelled using the Ecoinvent 3.10 European inventory for "market for steam, in chemical industry" which is based on Eurostat data from 2011–2015. The fuel mix for steam production includes a substantial share of fossil fuels—36% natural gas and 15% oil and petroleum products (Eurostat 2025). Reducing the fossil fuel share or switching to electric boilers powered by low-carbon electricity could significantly lower the associated impacts. Concerning the solvent use, further improvements may be technically challenging as a high recovery rate (97%) is already assumed.

3.2 Comparison with the Reference Scenario

To focus the analysis on the most relevant environmental issues, only the damage categories with the highest normalised impacts were retained for comparison. As shown in Fig. 4, four categories clearly stand out: climate change, freshwater ecotoxicity, human toxicity, cancer, and resource use, fossils. The impacts for these categories in both scenarios are summarised in Table 1.

In the climate change category, impacts in both scenarios are dominated by CO_2 emissions to air, which account for approximately 95% of the total impact. Shared

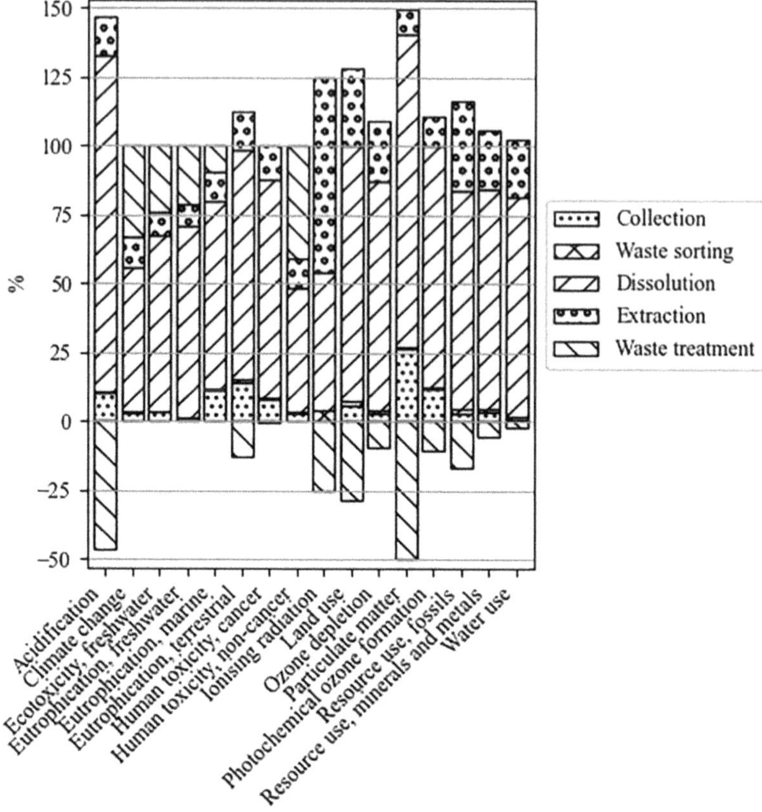

Fig. 3 Characterisation of the environmental impacts of the study scenario. The impacts are assessed using the Environmental Footprint 3.1 method

contributions include emissions from the incineration of impurities and plasticisers, representing around 30% of the total impact in each case. In the study scenario, the remaining emissions are primarily associated with steam consumption (28%), solvent production to compensate for losses (12%), and electricity use in the extraction unit (11%). In the reference scenario, the main contributors are the incineration of virgin PVC (31.2%) and the production of ethylene (18.5%).

A 10% reduction in climate change is observed in favour of the recycling. However, Monte Carlo analysis reveals a significant overlap in the distributions of results between both scenarios, indicating that this difference is not statistically significant.

In the freshwater ecotoxicity category, the study scenario achieves a 90% reduction in impacts compared to the reference scenario. This improvement is largely due to the PVC incineration in the reference scenario that leads to substantial chloride emissions to water, accounting for 90% of the total impact. The difference between both scenarios approaches one order of magnitude, and Monte Carlo analysis shows

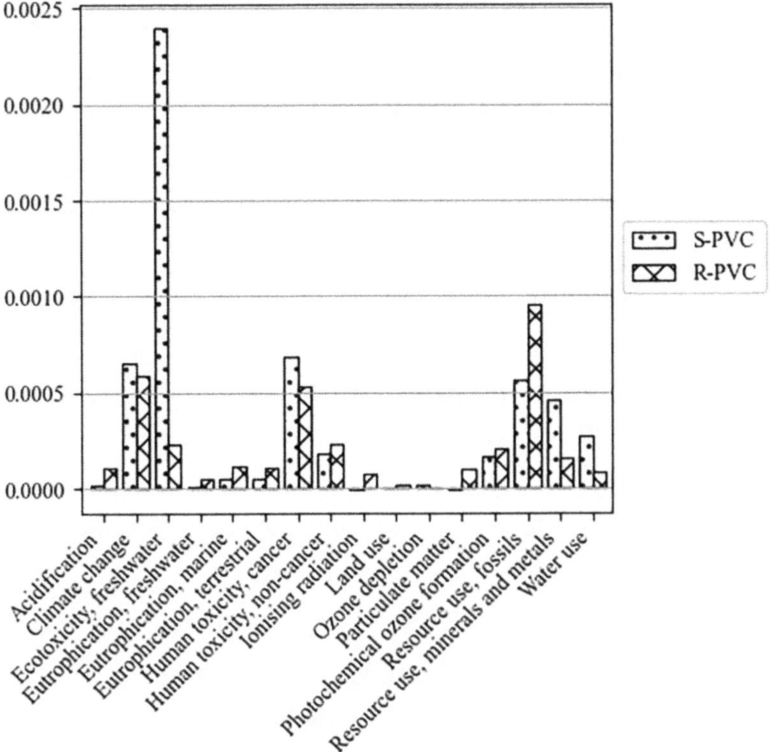

Fig. 4 Comparison between the normalised environmental impacts of the study scenario (R-PVC) and the reference scenario (S-PVC)

Table 1 Environmental impacts in the four most relevant damage categories for the study (recycled PVC) and reference (virgin PVC) scenarios

Damage category	Unit	Reference scenario	Study scenario
Climate change	kg CO_2 eq	4.95	4.43
Ecotoxicity, freshwater	CTUe	136	13.1
Human toxicity, cancer	CTUh	1.18E-08	9.2E-09
Resource use, fossils	MJ	36.4	62.1

minimal overlap between the distributions. A slight reduction in data uncertainty would be sufficient to confirm the statistical significance of this improvement.

In the human toxicity, cancer category, a 22% reduction is observed in the study scenario. In both scenarios, the dominant emission is anthracene to air, primarily originating from coke production used in upstream processes (74% of the impact in the study scenario and 57% of the reference scenario). However, due to the high uncertainty typically associated with toxicity-related categories, such differences are not considered significant unless they span at least one order of magnitude (Jolliet et al. 2015). This is confirmed by the Monte Carlo simulation, which shows a substantial overlap in the distributions of both scenarios.

In the resource use, fossils category, the reference scenario outperforms the study scenario, with a 70% lower impact. In the reference scenario, the impact is largely driven by ethylene production via steam cracking, which accounts for 87.6% of the total. In contrast, the high impact of the study scenario comes from its significant energy use, electricity consumption in the extraction unit (32.7%) and steam consumption in the dissolution unit (29.1%) being the two highest contributions. The solvent production to compensate for losses ranks third (25.8%). Although the mean impact is significantly higher in the study scenario, the Monte Carlo analysis reveals partial overlap between the distributions. This suggests that, while the difference is notable, it is not conclusively significant under current uncertainty levels. Reducing uncertainty would be necessary to confirm the extent to which physical recycling is more resource-intensive.

4   Conclusion

This study provides a first life cycle assessment of the VinyLoop™-D technology, an upgraded VinyLoop™ dissolution technology incorporating the removal of legacy additives. The results identify the dissolution stage as the primary contributor to environmental impacts, largely due to steam consumption and solvent losses.

Among the four most relevant damage categories, the recycled PVC scenario shows lower impacts than conventional PVC production in three. However, these differences are not statistically significant due to current data uncertainties. In the freshwater ecotoxicity category, a modest reduction in uncertainty would likely confirm a clear environmental benefit. Conversely, the study scenario shows higher fossil resource use, though this difference is also not statistically significant.

Overall, the study does not demonstrate a clear environmental advantage or disadvantage of recycling compared to conventional production. Reducing uncertainties and optimising steam use or production would improve the recycling scenario's performance.

This assessment is conducted as part of an ongoing research project. As such, further technical optimisations could be expected. Improvements could significantly reduce the environmental impacts of the recycling technology and strengthen its potential as a sustainable alternative to virgin PVC production.

Acknowledgements The authors gratefully acknowledge the financial support provided by the Walloon Region through the Greenwin cluster. We also extend our sincere thanks to the project partners—INEOS Inovyn, Vanheede Environnement, Rovi-Tech, Entreprises générales Louis Duchene, Eco-Dec, Avient Belgium, Centexbel and ULiège—QuantOM—for their valuable collaboration and for providing the data essential to this study.

References

Ait-Touchente Z, Khellaf M, Raffin G et al. (2024) Recent advances in polyvinyl chloride (PVC) recycling. Polym Adv Technol 35(1):e6228

Arena U, Mastellone ML, Perugini F (2003) Life cycle assessment of a plastic packaging recycling system. Int J Life Cycle Assess 8(2):92

Bird E, Griffiths M, Holden E et al. (2024) Polyvinyl chloride (PVC) additives: a scoping review. https://assets.publishing.service.gov.uk/media/65dc9dfcb8da63001dc861f8/Polyvinyl_Chloride_additives_scoping_review_-_report.pdf (Accessed 22.05.2025)

Cruz J (2025) PVC Part 1: It's All About Composition [White paper], The Madison Group. https://madisongroup.com/pvc-part-1-its-all-about-composition/ (Accessed 22.05.2025).

ECVM https://pvc.org/sustainability/eco-profiles-and-lca/ (Accessed 22.05.2025).

EuP Egypt (2025). https://eupegypt.com/blog/polyvinyl-chloride-pvc-plastic/ (Accessed 22.05.2025)

Elgharbawy A (2022) Poly vinyl chloride additives and applications—a review. J Risk Anal Cris Response 12:3

Eurostat https://ec.europa.eu/eurostat/databrowser/product/page/NRG_BAL_S (Accessed 7.06.2025)

Garcia ES, Huysveld S, Nachtergaele P et al. (2025) How multifunctionality modelling in LCA affects decision-making: The case of chemical recycling of plastic waste. Resour Conserv Recycl 218:108262

Goodfish Group (2025). https://www.goodfishgroup.com/pvc-in-the-construction-and-housing-sector (Accessed 22.05.2025)

Jolliet O, Saadé-Sbeih M, Shaked S, et al. (2015) Environmental life cycle assessment, CRC Press

Ledniowska K, Nosal-Kovalenko H, Janik W et al. (2022) Effective, environmentally friendly PVC plasticizers based on succinic acid. Polymers 14(7):1295

Remer DS, Chai LH (1993) Process equipment, cost scale-up, in: encyclopedia of chemical processing and design, CRC Press, 43, 306

Thornton J (2002) Environmental impacts of polyvinyl chloride (PVC) building materials, healthy building network. https://eldonjames.com/wp-content/uploads/2017/08/Environmental-Impacts-of_ThorntonRevised.pdf (Accessed 05.06.2025).

Ügdüler S, Van Geem KM, Roosen M et al. (2020) Challenges and opportunities of solvent-based additive extraction methods for plastic recycling. Waste Manage 104:148–182

Vandenhende B, Dumont JP (2006) Method for recycling a plastic material, U.S. Patent No. 7,056,956

VinylPlus, Progress Report 2025, (2025) https://www.vinylplus.eu/wp-content/uploads/2025/05/Progress-Report-2025-05-26_web.pdf (Accessed 05.06.2025)

VinylPlus, VinylPlus' contribution to the ECHA Investigation Report on PVC and its Additives, (2023). https://www.vinylplus.eu/wp-content/uploads/2023/08/Summary-of-VinylPlus-contributions-to-the-ECHA-Investigation-on-PVC-and-its-Additives.pdf (Accessed 05.06.2025)

Wagner S, Schlummer M (2024) Application of solvent-based dissolution for the recycling of polyvinylchloride flooring waste containing restricted phthalate plasticizers. Resour Conserv Recycl 211:107889

Enhancing Circularity of Strategic Value Chains Through Life Cycle Thinking—The Example of JRC Scientific Support to the Forthcoming Vehicles EU Regulation

Martina Orefice, Nacef Tazi, Thibaut Maury, Umberto Eynard, Daniele Candelaresi, Silvia Bobba, and Fabrice Mathieux

Abstract Electric mobility (e-mobility) is a key driver for the green transition and climate neutrality. Nonetheless, value chain of e-drive motors, a key component of e-vehicles, still presents significant circularity failures. Circular management of products should go beyond end-of-life management embracing the whole life-cycle to be effective. This is the approach of recent EU policies for e-mobility (e.g. the vehicles regulation proposal or the batteries regulation) as well as of recent JRC science-for-policy supports on e-mobility circularity. The paper presents and discusses two policy measures for enhancing e-drive motors circularity in applying a life cycle thinking approach, namely: (i) labelling and information folder and (ii) recycled content of permanent magnets. This contribution aims at showing how moving from a punctual approach to life-cycle approach along the value chain effectively enables circularity, of both materials and information, with benefits as for waste management operators as for manufacturers.

1 Introduction

EU policy making related to product-based legislation went through a novel and more comprehensive approach integrating life-cycle thinking (LCT) when designing new policy measures for circular economy. The recent regulation on batteries and waste batteries, (or, shortly, the batteries regulation) (EU) 2023/1542 (European Commission 2023c), has been the first-of-a-kind, with articles embracing from the

---

The work has been carried out at and on behalf of JRC.

M. Orefice (✉) · N. Tazi · T. Maury · U. Eynard · S. Bobba · F. Mathieux
European Commission—Joint Research Centre (JRC), Ispra, Italy
e-mail: Martina.Orefice@ec.europa.eu

D. Candelaresi
Seidor Italy S.R.L, Milan, Italy

M. Traverso et al. (eds.), *Life Cycle Management from Global to Local*,
https://doi.org/10.1007/978-3-032-17987-6_23

277

responsible sourcing of raw materials to durability, sustainability and safety, to waste collection and management. This life-cycle closes with requirements on recycled content to place recovered materials back into new batteries and, hence, into the value chain. Following that, additional product-based legislations were drafted considering a life-cycle approach, such as the circularity requirements for vehicles regulation proposal (VRP), 2023/0284 EC (European Commission 2023b), or the Critical Raw Materials (CRM) Act, (EU) 2024/1252 (European Commission 2024). In particular, the VRP highlights well this paradigm shift, expanding from the previous "end of life vehicle" (ELV) directive to a broader—encompassing life stages—"vehicle" regulation. The VRP contains, in its article 6, links to mandatory recycled content targets that were built also based on recent Joint Research Centre (JRC) reports (Maury et al. 2022; Tazi et al. 2025).

Circularity aspects are also crucial when designing product-based legislations with a life cycle-thinking approach. In the years, the introduction of policies like the Ecodesign and energy labelling framework (European Commission 2009) or the more recent batteries regulation reinforced the concept that circularity is more than recycling, targeting waste management operators (WMOs), and potential measures at design and manufacturing stage, targeting manufacturers (OEMs) (Bobba et al. 2019; Mathieux et al. 2020). In fact, circularity of a (regulated) product also refers to the set of provisions that would aim at making the product more sustainable along its life-cycle by boosting resource efficiency (Bobba et al. 2021). Examples of circularity measures can be related to waste collection and recycling, as well as to design for disassembling and labelling.

Circularity measures in product-based legislations need to be assessed against potential circularity failures, i.e. failures observed when the circularity of a material in a component is not maximised (Tazi et al. 2023). At electric vehicle (EV) level, the end-of-life recycling input rate (EOL-RIR, (Blengini et al. 2017)) of neodymium (Nd) and dysprosium (Dy), essential for e-drive motors, was < 1% in 2022 in the EU (RMIS 2024). The forthcoming VRP would thus need to incorporate the necessary measures to mitigate such circularity failures.

This paper summarises and reflects on the recent JRC approach to assess and develop circularity measures for products and highlights what has developed so far to enhance circularity of e-drive motor, its components and materials in the context of the forthcoming VRP. This paper is structured as following: first, the methodology of the study is illustrated. Results are discussed about the methodology application to solve a circularity failure of information flow and of materials flow. Lessons learned from the two case-studies, as well as insights for the policy process are discussed too. Finally, conclusions of the work and possible outcomes close the paper.

2 Method

The methodology of this work is illustrated in Fig. 1 and described as follows.

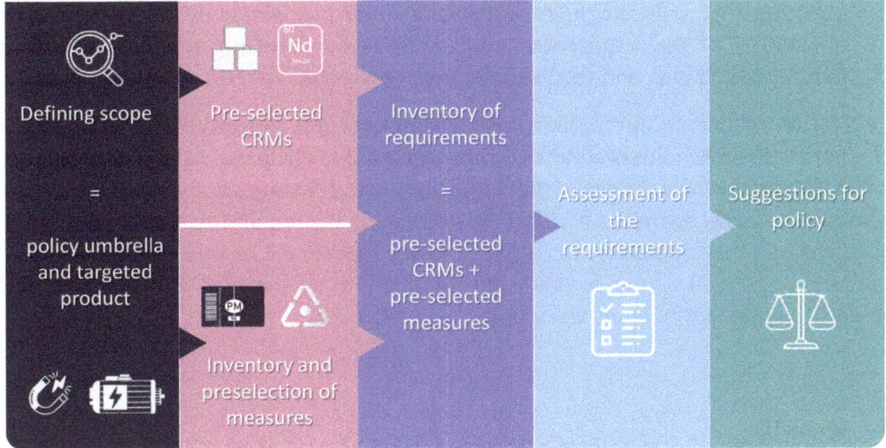

Fig. 1 Methodology adopted for this research, adapted from (Tazi et al. 2023, 2025)

Pre-selection of Materials and Measures. After having defined the product level (e-vehicle) and the component level (e-drive motor), a pre-selection of CRMs was made and circularity failures were analysed at the pre-selected material level. The criteria of preselection included (Tazi et al. 2025): (1) the classification of materials as 'critical' or 'strategic' for the EU, (2) material relative abundance at product and component level, (3) the presence of a circularity failure, such as a missed collection of EOL products or a lost information from designers to dismantlers. An inventory of potential circularity measures was developed based on the identified failures. Requirements, defined as the coupling of a preselected CRM and a corresponding circularity measure, were then identified.

Design and Assessment of the Requirements. Each requirement to enhance circularity of the materials targeted players either at the upstream or at the downstream of the value chain, or both. Feasibility assessments of the possible requirements included a qualitative analysis of the material flows and technology readiness; environmental and socio-economic effects were qualitatively analysed too. Literature studies as well as dialogue with key stakeholders were conducted and analysed in order to feed the design of the requirements. Stakeholders were selected for their position in the value chain and their representativeness in the EU context. In particular, representativeness of stakeholders was ensured in terms of the type of organisation (industry, association, academia), position along the value chain (balance of upstream and downstream players), geographical location, etc. Adopting a LCT perspective, the interviews to stakeholders were of two types:

(i) operators in the segment of the value chain targeted by the requirement (e.g. OEMs for labelling requirement) were mainly asked to provide data and feedback on the related circularity failure and on what might be needed to design the requirement, to maximise circularity as well as to facilitate their implementation.

(ii) operators in another segment of the value chain and beneficiary from the implementation of the requirement (e.g. WMOs for labelling requirement) were mainly asked data and feedback on how they could support the requirement.

Next to enhancing circularity of e-drive motors, the objective of the (iterative) dialogue with stakeholders at the two ends of the value chain was to find an optimum, in the data to disclose, between reducing administrative burden and reducing confidentiality risks for operators who have to implement the requirement, and the benefits obtained by other operators. In fact, this paper reports the results of the feasibility assessment as well as the benefits and implications of the LCT for upstream and downstream operators along the whole value chain.

3   Results

In the preselection of CRMs for which a circularity failure exists, materials from e-drive motors were chosen as not covered by the previous ELV Directive of 2000 (European Commission 2000) (likewise, EV batteries were neither covered but the new batteries regulations largely made up for that circularity failure). Namely, the chosen materials were Nd, Dy and electric steel (Carrara et al. 2023). While Nd and Dy are CRMs that make up rare-earth permanent magnets (REPMs) of e-drive motors (European Commission 2024), the strategic importance of electric steel sheet (that contain the critical raw material silicon metal) for clean technologies and its potential supply dependencies make it relevant to consider. Copper is also a main (strategic) material of e-drive motors but was discarded since not critical with the EU facing currently low supply dependencies (Carrara et al. 2023). Identified preselected circularity measures were: (a) labelling and information folder, i.e. the set of information to associate to a label, a product passport, a database, (b) recycling efficiency and recovery of materials, (c) recycled content, (d) design provisions for ease of disassembly and removal. In coupling pre-selected CRMs and measures, electric steel was discarded since its functional recycling was considered not feasible yet, as learned from bilateral discussions with recyclers, and so neither measure to improve the recycling. Requirements on design and removal would be at component (e-drive motor), and not material level, and hence affecting all (pre-selected and not) component materials.

Among the analysed requirements, two of them have been identified as particularly significant as actions to be considered circular and not punctual from the perspective of LCT: (1) labelling and information folder of rare-earth permanent magnets (REPMs), and (2) recycled content of REPM, as they, among other requirements, connect the two ends (design and EoL management) of the value chain through the cooperative action of WMOs and OEMs.

3.1 Labelling and Information Folder: A Circularity Measure Developed Together with Manufacturers for the Benefit of WMOs

WMOs, as car dismantlers and recyclers, highlighted in the interviews that EOL REPMs and e-drive motors are not properly collected and sorted since it is not possible for WMOs to recognise if the EoL e-drive motors contain REPMs or to quantify costs and benefits of a possible recycling process.

A requirement on labelling and information folder can definitely enhance the circularity of e-drive motors and benefit WMOs. However, to ensure the effectiveness of the labelling in the following life-cycle stages, the REPM label and information should be added to the e-drive motor at the manufacturing stage, which means that this requirement targets OEMs.

The design of the requirement was an exercise of balance between the granularity of information desired by the WMOs and the need for confidentiality of the information reported by the OEMs. While the information is available at the level of the magnet manufacturer, the label shall be added by the e-drive motor producer: hence, the authors interviewed magnet manufacturers as well as Tier 1 operators who integrate magnets in the e-drive motors for the OEMs. Detailed results on the suggested labelling appearance and the details to add in the information folder can be found in (Tazi et al. 2025), together with more extensive results of the requirement assessment. Table 1 summarises the impacts on the environment, circular economy and innovation at both levels of OEMs and WMOs.

As shown in Table 1, the assessed requirement involves potential impacts on innovation at both OEMs and WMOs level and it proves how thinking punctually, i.e. focusing only on specific stages of the value chain represents a limitation in fostering circular economy. A LCT does not only make the requirement more effective by optimising its design, but it can also open to more circular options and business opportunities. In fact, among the different circularity strategies and possible business models, it has also been considered that the labelling and information folder would improve the collection and sorting at WMO level in a way that products (e-drive motors) and components (REPMs) might be put back to the market through e.g. remanufacturing or repurposing.

Notably, it is also the first time that a labelling is conceived to provide information to WMOs for the circularity of metals and in particular of CRMs. While examples of mandatory declaration of quantities of CRMs in products (e.g. in (Talens Peiró et al. 2020)) and examples of labelling to improve the EOL of packaging (non-metal) materials (e.g. in (European Commission 1997)) or to indicate the presence of hazardous metals are known, at the best of the authors' knowledge it is the first time that a label on a product is suggested to improve the recovery of CRMs with relevant consequences also on the creation of a competitive and resilient market of secondary materials in the EU.

Table 1 Qualitative assessment of impacts of the measure 'Labelling and information folder' on environment, creation of secondary market and innovation at both levels of OEMs and WMOs. Information folder is intended as the set of information to associate to a label, a product passport, a database

Impact category	Type of impact	Description of the impact
Environment	Indirect effects	Limited considering new labelling processes, yet energy and materials could be improved by using sustainable options
	Direct effects	Contributes to avoid downcycling, loss of materials (incl. to landfill) and resource depletion. Promotion of circular economy by improving WMOs' knowledge at product/material/process level
Circular economy	EU policies	Synergies with CRM Act, Green Claim Directive, ESPR and eco-design legislation files
	Secondary market creation	Contribution to creation of well-defined e-motors flows (with and w/o REPM) and of business models based on repurposing, remanufacturing, or (closed loop) recycling
Innovation	OEM level	Use of sustainable label materials might foster innovation and/or trigger possible, related standardisation activities
	WMO level	Contributes to foster innovation towards new, more circular business models as well as to the development of manual and (semi-)automated methods to effectively sort/remove REPM within e-drive motors

3.2 Recycled Content: A Circularity Measure Explored Together with Waste Management Operators for the Benefit of Manufacturers

The other requirement example also links the operators at the two ends of the value chain. The recycled content closes the loop between upstream and downstream operators through a circular flow of materials. Following the example of the EU batteries regulation, the calculation and declaration of recycled content should be the responsibility of the OEM who places the product on the market. Nonetheless, the WMOs are the ones who recover the materials from the waste to make them available for a second life cycle. Such an innovative requirement was introduced for battery materials under article 8 of the batteries regulation (European Commission 2023c). This requirement was also proven to be in principle effective to enhance circularity of plastics in vehicles. The measure seeks to ensure a minimum demand volume for recycled plastics (entering the manufacturing of new vehicles), while increasing long-term certainty for recyclers thereby triggering investments to produce greater quantities and higher-quality recycled plastics. The latter was considered under article 6 of the VRP and paved the way towards a greater integration of recycled content for other materials in vehicles as well, such as steel, aluminium and CRMs.

For the latter, no requirement was formulated yet, but the feasibility of a requirement for recycled content targeting rare-earths and other materials in REPMs of

Table 2 Qualitative assessment of impacts of the measure 'Recycled content' on environment, creation of secondary market and innovation at both levels of OEMs and WMOs

Impact category	Requirement: recycled content	Requirement: recycled content coupled with recycling efficiency and recovery of materials
Environment	Limited environmental impact corresponding to avoided uptake of primary materials	Much lower environmental impact made with a combination of recycling (avoiding losses) and uptake of secondary materials (less incorporation of primary materials)
Circular economy (EU policies)	Limited effectiveness and competitiveness boost if applied alone	Best effectiveness in combination with information and recycling requirements to support both EU competitiveness and resilience
Circular economy (creation of a secondary market)	Limited performance without promotion of recycling, with a risk of supply–demand misbalance	Higher performance with the combination of three measures to create conditions for secondary materials market and their uptake
Innovation	Increased innovation on design, EoL and recycling processes	Increased innovation on design, EoL and recycling processes

e-drive motors was expected. The exercise in this case consisted in assessing the availability of secondary materials as well as of recycling technologies. Interestingly, the availability of secondary materials would be increased thanks to the availability of an effective labelling system.

The impact assessment (Table 2) proved that a recycled content requirement for REPM in e-drive motors will have, for most categories, high benefits only if coupled with other circularity requirements such as recycling efficiency and/or recovery of materials. At the same time, it improves the performance of the recycling efficiency and recovery of materials requirements: as such, recycled content is an incentive to promote high-quality recycling with mandatory reintegration of recycled materials into strategic technological applications.

3.3　Generalising the Approach: How Science-for-Policy Interventions Can Support Better Circularity of the EU Industrial Value Chains

The development of the two requirements for e-drive motors, labelling and information folder and recycled content, circularise the value chain connecting upstream and downstream players, through flows of information (from OEMs to WMOs) and materials (from WMOs to OEMs) based on LCT. This new systemic approach can enable a harmonised and standardised implementation of circular economy requirements for different, strategic technologies (Fig. 2).

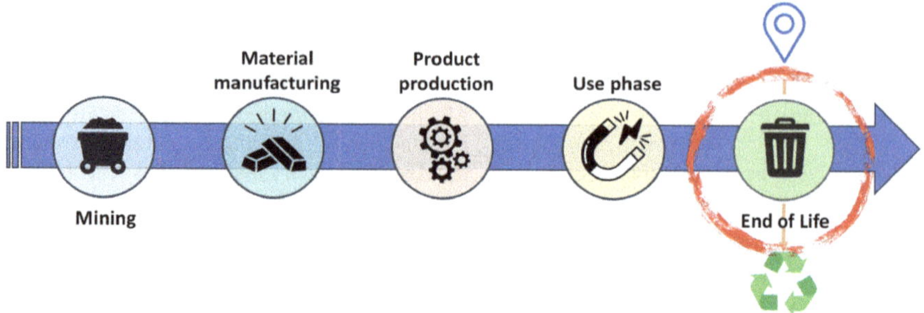

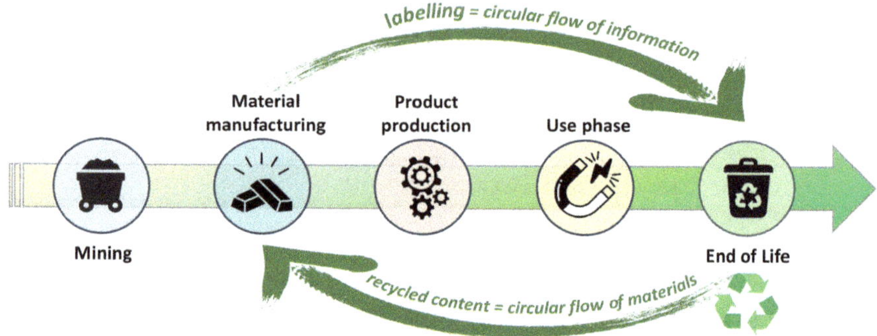

Fig. 2 Life-cycle thinking enables to implement circular flows of information and materials, effectively circularising the value chain compared to punctual interventions on the value chain

Policy Objectives. Going beyond the two cases of this paper, enhancing circularity in EU supply chains requires a well-calibrated policy mix to ensure that both ends of the value chains are covered, and to maximise benefits while keeping low administrative burden. The requirements are highly dependent on the policy objectives to be addressed, which might be waste management practices; reducing environmental footprint; mitigating supply risks and import dependencies.

Scope of the Intervention. The choice of the appropriate circularity measures should be built on a solid understanding of flow dynamics at multiple levels:

(1) Material level, by including appropriate material quality and sectors demand (share of Nd/Dy going in REPM production);
(2) Product level, by considering product-specific characteristics such as composition, lifespans, and market trends as forecasted demand as well as forecasted import/export;
(3) Waste stream level, by considering the composition of various waste streams and the management, from collection to recycling technologies and practices;

(4) Information level, by also figuring the associated administrative burden and looking for an optimum between data and information granularity, confidentiality and simplification.

Ambition of the Policy Intervention. Once a measure (recycled content, information disclosure, recycling efficiency) is selected, the proportionality of the potential policy intervention should also be assessed to set the appropriate requirement. A criterion to consider is the ambition of the policy intervention. Three levels can be defined (European Commission, 2023a):

(1) The starting point are usually soft policy tools such as implementing voluntary pledges or voluntary declarations from manufacturers;
(2) Secondly, and before requiring interventions at technology and practices level, mandatory information requirements can be defined, ensuring also verifiability. These might concern product performance or information to improve waste management. A requirement as the suggested REPM labelling is often useful prior to implementing performance requirements based on quality or quantity thresholds;
(3) Thirdly, when circularity failures are not likely to be solved by soft tools or mandatory information, performance requirements can be introduced in the form of targets, such as recycling efficiency or minimum recycled content targets. The target level corresponds in this case to the policy ambition.

Timeline. An additional key criterion is the timeline of the policy intervention. Ideally, the adaptation time, for the value chain, should be adequately proportional to the ambition of the policy intervention; moreover, review clause might also be introduced to review both the ambition and the timeline. Alignment of timelines of policy interventions interfering with each other, as the two cases of this paper, should be ensured too: reasonably, the circular flow of information should be implemented prior to the circular material flow.

Cross-sectorial and regional influences. Cascading effects on other sectors is important too. For instance, increasing the demand for recycled materials in one value chain (e.g. automotive) may reduce their availability for other sectors.

Enhancing policy coherence. Finally, when investigating the feasibility of a policy intervention dealing with circularity, it is important to ensure coherence between different legislative acts, such as the VRP, the CRM act, and potentially the ESPR for permanent magnets. This includes coherence in the scope, but also complementarity/synergies of the intended goals when focusing on intermediate products (REPM) versus final products (e-drive motors).

4   Conclusions

Policies to enhance the circularity of (critical) raw materials for strategic technologies have been focused for years mainly on waste management and recycling. However, this refers only to a single stage of the life-cycle and can be defined as a punctual rather than a circular requirement when not effectively coupled with requirements as design-for-recycling, material recovery or quality of recycling. In this paper, two circularity requirements for e-drive motors were assessed, together with their benefits (positive effects) and implications (negative effects), considering a life-cycle thinking approach. Namely, the labelling and information folder is a requirement designed together with manufacturers and producers for the benefit of waste management operators. The objective was to initiate at the manufacturing stage a circular flow of information to enhance the recovery of materials at the end of life. Dialogue with both upstream and downstream players optimized the balance between the needed granularity of information for recovery and the confidentiality of manufacturers.

An effective circular flow of information also benefits the circular flow of materials, increasing the availability of secondary materials that can be reintroduced at the product manufacturing. The requirement on recycled content had the objective of supporting such circular flow of materials. In a similar and opposite way to the previous case, the effective requirement design would rely on the dialogue with the players who make the flow (of materials) available (the WMOs) to optimise the benefits for players at the other end of the value chain (the material producers and the manufacturers).

This work highlights the importance of a systematic approach to harmonise and standardise the process of setting feasible requirements to foster circular economy of e-drive motors. As such approach is, in principle, applicable across various technologies, also considering their technological readiness levels, the JRC is currently developing a new methodological approach that can be applied to different products and strategic technologies for the EU. It will serve as a tool for the horizontal enabling of circularity policies, to ensure consistency in defining and drafting requirements at material level for different technologies. The next steps will involve defining and formalising the methodology and applying it to two or three different products or components (e.g. WEEE, REPM in applications other than e-drive motors, and/or others) to validate the methodology itself.

References

Blengini GA et al (2017) Assessment of the methodology for establishing the EU List of critical raw materials—background report. Jt Res Cent. https://doi.org/10.2760/73303

Bobba S et al (2021). Sustainable Use of Materials through Automotive Remanufacturing to Boost Resource Efficiency in the Road Transport System (SMART). https://doi.org/10.2760/84767

Bobba S, Mathieux F, Blengini GA (2019) How will second-use of batteries affect stocks and flows in the EU? A model for traction Li-ion batteries. Resour Conserv Recycl 145:279–291. https://doi.org/10.1016/j.resconrec.2019.02.022

Carrara S et al (2023) Supply chain analysis and material demand forecast in strategic technologies and sectors in the EU—A foresight study. https://doi.org/10.2760/386650

European Commission. Decision—97/129—EN—EUR-Lex. 1997. https://eur-lex.europa.eu/eli/dec/1997/129/oj/eng (Accessed 10.06.2025)

European Commission. End of Life Vehicle Directive—2000/53. (2000) https://eur-lex.europa.eu/legal-content/EN/TXT/?uri=celex%3A32000L0053 (Accessed 16.07.2025)

European Commission. Ecodesign and Energy Label (2009) https://energy-efficient-products.ec.europa.eu/ecodesign-and-energy-label_en (Accessed 14.06.2025)

European Commission. Better regulation: guidelines and toolbox (2023a) https://commission.europa.eu/law/law-making-process/better-regulation/better-regulation-guidelines-and-toolbox_en (Accessed 18.06.2025)

European Commission. Proposal for a Regulation of the European Parliament and of the Council on circularity requirements for vehicle design and on management of end-of-life vehicles, amending Regulations (EU) 2018/858 and 2019/1020 and repealing Directives 2000/53/EC and 2005. 2023b

European Commission. Regulation (EU) 2023/1542 of the European Parliament and of the Council of 12 July 2023 concerning batteries and waste batteries, amending Directive 2008/98/EC and Regulation (EU) 2019/1020 and repealing Directive 2006/66/EC (Text with EEA relevance). 2023c. https://eur-lex.europa.eu/eli/reg/2023/1542/oj (Accessed 16.10.2024)

European Commission. Regulation (EU) 2024/1252 establishing a framework for ensuring a secure and sustainable supply of critical raw materials and amending Regulations (EU) No 168/2013, (EU) 2018/858, (EU) 2018/1724 and (EU) 2019/102. 2024.

Mathieux F, Ardente F, Bobba S (2020) Ten years of scientific support for integrating circular economy requirements in the EU ecodesign directive: Overview and lessons learnt. Procedia CIRP. 90:137–142. https://doi.org/10.1016/j.procir.2020.02.121

Maury T et al (2022) Towards recycled plastic content targets in new passenger cars. Tech Propos Anal Impacts Context Rev ELV Dir. https://doi.org/10.2838/834615

RMIS. Raw Materials Information System (2024) https://rmis.jrc.ec.europa.eu/

Talens Peiró L et al (2020) Advances towards circular economy policies in the EU: The new Ecodesign regulation of enterprise servers. Resour, Conserv Recycl. 154:104426. https://doi.org/10.1016/j.resconrec.2019.104426

Tazi N et al (2023). Initial Analysis of Selected Measures to Improve the Circularity of Critical Raw Materials and Other Materials in Passenger Cars. https://doi.org/10.2760/207541

Tazi N et al (2025). Circularity Measures on Critical Raw Materials and e-Drive Motors in Vehicles. https://doi.org/10.2760/6053264

Assessing and Communicating Potential Environmental Benefits of the Circular Economy: A Case Study on Ski Core Circularity

Ulrike Kirschnick, Sara Casale, Herbert Gruber, Roland Pomberger, and Ewald Fauster

Abstract In the light of the Circular Economy, this study assesses the environmental impacts of a ski core circularity measure to enhance resource efficiency in the wintersport industry via Life Cycle Assessment. The circular ski is modelled with four distinctions regarding the closed-loop recycling of the core materials (two user specific scenarios, the 100/0 approach and the Circular Footprint Formula), and compared to the non-circular virgin ski to determine the change in environmental impacts of the circularity measure using EF 3.1 as the impact assessment method. The results highlight the influence of methodological choices with changes in impacts varying from e.g. a reduction of 11.8% in global warming when comparing the circular to the non-circular ski following one-time circularity at the level of private consumers, to a reduction of 23.6% when modelling the circular ski according to the 100/0 approach. Furthermore, trade-offs between impact categories can be observed, as eutrophication, freshwater and ionizing radiation increase in the circular ski compared to the non-circular one. The conformity of methodological choices (derived from differences in understanding the case study) with terminology of the R-strategies under the Circular Economy framework is analysed. Consequences of modelling choices on communication of the different LCA results to the internal and external stakeholders are discussed focusing on legal compliance, the LCA models' comprehensiveness, inference on product competitiveness, and activation of the desired consumer behaviour to partake in the return system. Future harmonization efforts should focus on aligning legal frameworks concerning terminology and providing guidelines for the communication of results.

U. Kirschnick (✉) · E. Fauster
Processing of Composites and Design for Recycling, Technical University Leoben, Leoben, Austria
e-mail: ulrike.kirschnick@unileoben.ac.at

S. Casale · H. Gruber
Head Sport GmbH, Schwechat, Austria

R. Pomberger
Waste Processing Technology and Waste Management, Technical University Leoben, Leoben, Austria

M. Traverso et al. (eds.), *Life Cycle Management from Global to Local*,
https://doi.org/10.1007/978-3-032-17987-6_24

1 Introduction

The Circular Economy framework aims to improve resource usage by extending the life cycle of products through different measures with the goal to reduce wastes, decrease environmental impacts and create further (economic) value. The R-strategies of the Circular Economy framework provide a collection of measures to extend the lifespan of a product and its parts through reuse, repair, refurbishment and remanufacturing, if reducing the consumption of resources is not possible in the first place (Potting et al. 2017). The implementation of successful circularity measures requires the active participation of customers as well as suppliers in multiple industrial sectors. The winter sport industry is representative for a variety of consumer good sectors, consisting of small and medium enterprises (SME) producing multi-material products in the European legal context as well as in a globalized market with complex value chains. Customers for winter sport equipment, such as ski, are both commercial (through rental stores) as well as private persons.

As an effort to increase the circularity of skis according to the Circular Economy framework, the design has been adapted using a reversible glue which allows for the re-utilization of the ski core at End-of-Life (EoL). The latter is reached when the outer materials of the ski can no longer provide the sufficient performance due to wear and abrasion. While in the first life the ski is manufactured with virgin materials, the customer (private or commercial) should return the ski at EoL to the manufacturer, who will dismantle and clean the ski with the aim to use the Glass Fiber-Reinforced Polymer (GFRP), wood core and acrylonitrile butadiene styrene (ABS) tip protector in manufacturing the next ski. The circular ski exhibits the same functionality and performance as the virgin ski in the first life. Overall, the ski core can be used a maximum of five times over the course of around ten years to manufacture a ski before a change in ski design and technology occurs.

One motivation of this circularity measure is the potential reduction of environmental impacts, which require verification using Life Cycle Assessment (LCA) (ISO 14040). Therefore, this study aims to quantify the environmental impacts of the circularity measure while discussing different methodological choices in LCA when comparing the non-circular ski as a benchmark to the circular one. The circularity of the material is considered in four models, which vary concerning functional unit and system boundary selection.

Next to the discussion of the applicability of these models to the case study, communication to management and to various consumers (retailers, rental and private customers) is crucial to enhance likeliness of the consumer returning the ski to the manufacturer and enabling circularity. The communication should respect scientific best practice and legal requirements implemented in the Proposal for a Directive on Green Claims to address greenwashing in the New Circular Economy Action Plan and the New Consumer Agenda (European Commission 2023). This proposal links to political efforts in empowering consumers for the green transition through better protection against unfair practices and better information. Different ways to communicate the results of the four LCA models are discussed in regards to legal compliance, comprehensiveness, inference on product competitiveness, and activation of the desired consumer behaviour (rental and private owners) to partake in the return system.

2   Material and Methods

The goal and scope of the LCA is the assessment of the environmental impacts of the ski core circularity measure by comparing the non-circular ski to a circular one, which can be modelled using four different EoL approaches. The interpretation investigates the compatibility of the LCA modelling approaches for the circular ski with terminology of the R-strategies, before proposing and discussing three options of communication of LCA results to customers. The flow chart of the virgin ski and circularity measure for an exemplary two lives of the ski is visible in Fig. 1. Technically, it is also possible to circulate the ski core more than one time at non-destructive consumer behaviour. Also, the ski in the second life can be bought by the same or different type of private or commercial customer. The virgin ski represents the first life and serves at the end of the first use phase as a benchmark for the comparison to the circular ski. Its functional unit is one pair of skis (either non-circular or circular) with a weight of 3.17 kg that are designed for alpine skiing and last according to the consumer behaviour an average of two seasons in ski rental and 10 seasons at private customers before being discarded for incineration and potential open-loop recycling. In order to use the core in the circular ski after the first life, a complex collection system is needed to return the ski to the manufacturing site, and dismantling and cleaning of the ski (closed-loop circularity) as well as additional materials to assure re-processability (circular support materials) and another set of virgin outer ski materials. Next to material composition, the virgin and circular ski also differ regarding manufacturing. The lamination of the ski core is necessary only in the virgin ski manufacturing, leading to a reduction in manufacturing time and inputs in the circular ski manufacturing stage.

Four models were defined to assess the environmental impacts of one pair of circular ski in comparison to a virgin counterpart:

(1) **One time circular**: Equal repartition of total impacts of one time circularity among the two lives of the ski. This scenario is likely for the ski being bought and returned by private customers.

(2) **Rental sector**: Equal repartition of total impacts for the potential five lives possible in the rental sector. Considering an expected consumer behavior in the rental sector of two seasons, the ski can be returned four times for circularity until the circular core materials can no longer be used in the closed-loop cycle.

(3) **100/0 approach**: Differentiation between first and second life. The circular ski core materials enter the second life burden-free and only incorporate the impacts associated with the circularity measure (collection and dismantling).

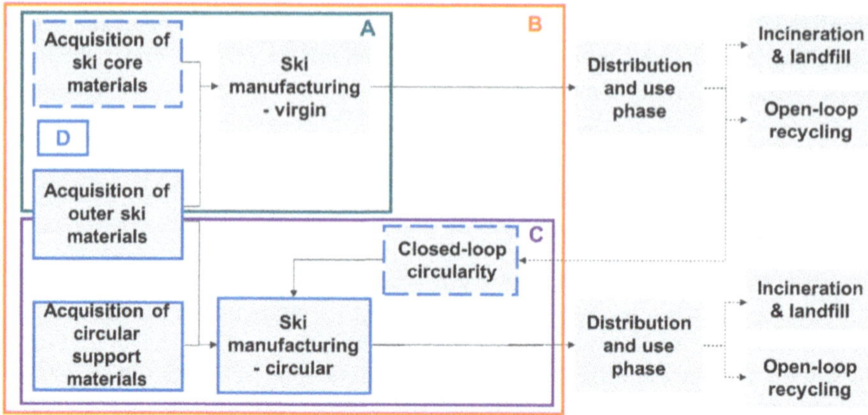

Fig. 1 Flow chart of two exemplary ski lives from virgin ski via ski core circularity to alternative treatment at EoL (represented by dashed lines) with the differing system boundaries for **a** the non-circular reference ski, **b** the recycled ski in the one-time circular and rental sector scenarios, **c** the recycled ski modelled with the 100/0 approach, and **d** the recycled ski modelled using parameters of the CFF. Dashed lines for the processes included in the CFF model indicate a repartition of environmental impacts between first and second ski life according to A factors

(4) **Circular Footprint Formula (CFF)**: Differentiation between first and second life through repartition of impacts using the CFF (Zampori und Pant 2019). The A factors derived from Annex C are 0.5 for GFRP and polymers, and 0.8 for wood part while there is no loss in ski core quality. The environmental burden of the recycling process (circularity) is allocated according to mass of the respective material.

The reference virgin ski and the four models for the circular ski differentiate regarding lifespan as well as system boundaries and consequently, inventory data as visible in Fig. 1. The system boundaries for the virgin ski (A) and the recycled ski following 100/0 approach (C) cross through the "acquisition of outer ski materials" process implying that each model requires this process one time, contrary to the system boundary of the recycled ski in the one-time circular and rental sector scenarios (B), where the outer ski materials are required two and five times, respectively. As the circularity measure has no influence on distribution and use phase, these are omitted from the system boundary in all models. Similarly, the ski will always reach an ultimate EoL with incineration and (potential) open-loop recycling, therefore all models including the non-circular benchmark follow cradle-to-gate system boundaries.

The Life Cycle Inventory (LCI) comprises primary data on material (GFRP, different polymers, wood, steel) and energy types and amounts, which were modelled selecting market providers from the ecoinvent 3.11 cut-off database as proxies. Specific transport was modelled additionally from material suppliers to the factory, and for collection of the skis at EoL within the circularity phase. While the ski core production took place in Austria with electricity supplied from hydropower, the finishing of the ski and the entire circularity measure were carried out in Czech Republic using natural gas and grid electricity. The selected Life Cycle Impact Assessment method is EF 3.1 with 16 midpoint impact categories due to its relevance for the European context and to discuss potential trade-offs between impact categories (Andreasi Bassi et al. 2023). No normalization and weighting was applied to receive a single score, as normalization factors neglect geographical specificities and do not consider natural resource constraints (Matuštík et al. 2024).

3   Results and Discussion of the LCA Models

The results for all 16 impact categories of the four models to evaluate the environmental impacts of a pair of circular ski when compared to the benchmark non-circular ski are depicted in Table 1. Negative numbers indicate a reduction in environmental impact of the circular ski compared to the benchmark while positive numbers mark an increase, making the circularity environmentally disadvantageous rather than beneficial. Across 14 of the 16 impact categories, ski core circularity leads to a reduction in impacts regardless of the LCA modelling approach. Only freshwater eutrophication and ionizing radiation exhibit higher impacts in the circular ski models mainly due to the composition of the Czech electricity market. While the treatment of spoil from lignite mining, which is used in coal-fired power plants, is the main cause for the increase freshwater eutrophication, ionizing radiation is related to nuclear power in the Czech electricity grid.

The influence of the methodological choices is visible when comparing the approaches to each other per impact category. The highest difference exhibits material resources, metals and minerals with -90% in the 100/0 approach and -45% in the CFF, whereas modelling-related differences are least pronounced in ionizing radiation, where one time circular model exhibits comparative lowest differences to the benchmark with + 4% and the 100/0 approach exhibiting + 8%. Overall, the 100/0 approach model leads to the highest reduction in 14 impact categories compared to the other three modelling approaches. The most conservative impact reductions of the ski core circularity provide one-time circular and CFF LCA models in eight and six impact categories respectively. The existing small differences between the two approaches are due to the A factor for wood being 0.8. Generally, the appropriateness of the A factors used for the circular ski core materials is debatable as the acquisition of recycled materials in sufficient quality proves difficult, thus indicating a supply-constraint from the ski manufacturer point of view.

The consideration of the R-strategies of the Circular Economy framework can potentially help to reflect on the applicability of the methodological choices in LCA but the identification of the correct R-strategy to describe the ski core circularity measure is difficult. The term "repair" can be used referring to the extended life of the ski core, which applies to both outside and within the waste characteristic (intention to discard the product). If the circularity measure is perceived as repair, the lifetime extension approaches (one-time circular and rental sector) seem most

Table 1 Change in environmental impacts of the four models with different EoL approaches to evaluate the circular in comparison to the benchmark non-circular ski

Impact category	One-time circular (%)	Rental sector (%)	100/0 approach (%)	CFF (%)
Climate change	−11.8	−18.9	−23.6	−12.4
Eutrophication, freshwater	+ 4.7	+ 7.4	+ 9.3	+ 6.7
Eutrophication, terrestrial	−20.1	−32.2	−40.3	−20.4
Eutrophication, marine	−15.4	−24.7	−30.8	−15.4
Ionizing radiation	+ 3.9	+ 6.3	+ 7.9	+ 6.0
Ozone depletion	−28.5	−45.6	−57.0	−30.6
Energy resources, non-renewable	−9.7	−15.4	−19.3	−9.8
Material resources, metals & minerals	−44.9	−71.9	−89.8	−44.6
Land use	−41.0	−65.6	−82.0	−63.9
Water use	−7.2	−11.5	−14.3	−6.2
Human toxicity, carcinogenic	−25.1	−40.1	−50.1	−24.8
Human toxicity, non-carcinogenic	−35.0	−56.0	−70.1	−34.6
Acidification	−18.0	−28.8	−36.0	−17.4
Particulate matter formation	−21.9	−35.0	−43.8	−23.5
Photochemical oxidant formation	−18.0	−28.8	−36.1	−18.1
Ecotoxicity, freshwater	−13.2	−21.1	−26.3	−11.3

appropriate. On the other hand, the ski core circularity can be described by the term "remanufacture" as a new ski is manufactured using a discarded part of a waste ski. For this notion, allocation approaches such as the 100/0 approach and the CFF seem more suitable emphasizing on the clear cut between the two (or multiple) lives. Regardless of the selected term and model, consistent modelling, transparency of assumptions, and critical interpretation of results and trade-offs are pivotal to producing meaningful LCA results of Circular Economy measures as outlined by (Peña et al. 2021).

4 Communication of the LCA Model Results

Communication of LCA results is required both company-internal as well as to the various types of customers to enable informed decision-making but there exist different framework conditions for communication to these groups and sub-groups.

Company-internal decision-makers usually have a certain amount of time dedicated to understanding LCA models and results as the communication usually takes place in a format of a meeting with presentations. Therefore, it is possible to inform the management about the four LCA models and explain differences among them as well as discuss about trade-offs between impact categories—especially by highlighting optimization potentials to reduce impacts in the respective categories. However, it is also important to highlight the potential of the measure as the potential lever for change and increase investment in circularity. Here, the 100/0 approach model results in highest differences especially in the climate change impact category, which is still the focus in company-wide sustainability targets and strategies.

At the same time, the most realistic model should be selected and management should (and can) be sensitized for LCA modelling choices and their influence on results. The rental sector model offers the advantage of differentiating between rental and private customers, which reflects the company's sale structure and allows for nuanced reflection of the full potential of the ski core circularity measure over the due-course of the ski models entire lifetime. On the other hand, the CFF is promoted by the Product Environmental Footprint (PEF) framework representing a solid and standardized base for external comparisons (Zampori and Pant 2019). While the CFF equation might look confusing at the first glance when presented to managers, it also provides clear indications on calculations, despite some methodological challenges of A factor applicability to the use case and repartition of impacts of the ski core circularity measure. In external communication, it also ensures fair comparisons with competitors and for external benchmarking given that the modelling of circularity and EoL treatment is included and transparently standardized within the PEF Category Rules (PEFCR).

While the above-mentioned discussion points can be highlighted to company-internal stakeholders, the communication pathways with external customers buying the ski differ according to the target group: The established contact (and their purchase volume) with ski rental companies allows for more exhaustive communication, elaborating on potential reduction in impacts while highlighting the feasibility of the return system and logistics. Similarly, retailers as intermediary points in the value chain need to be aware of the circularity measure in a way that it can help them reach their internal sustainability goals and targets. Usually, both consumer groups do not require LCA results as presented in Table 1 but in absolute values per impact category and per pair of ski for their own reporting and consideration.

For the winter sport industry, this is different for individual private customers standing at the retailer with the aim to leave the shop in a reasonable timeframe with a new pair of skis. As they tend to have limited time to be informed about the LCA results and need communication within simple, meaningful and concise statements, e.g. printed on the product's packaging. According to the Green Claims Directive, this statement should be informative, clear (e.g. by using specific numbers on specific reduction or recycling rates), reliable, transparent and verified. Next to legal requirements, it should also be comprehensive regarding language and core message (keeping it simple), and should not have a negative impact on the competitiveness of the product when compared to similar products of other brands. Lastly, the consumer should be activated to partake in the return system without being confused by the return system's and LCA's complexity. Especially for customers not familiar with the LCA methodology who typically have few minutes to consider environmental impacts in their purchase decision, it is difficult to balance these four principles when communicating externally. The following three statements depict a range of communication options to discuss compliance with the multiple principles of private customer communication.

> Returning this product for closed-loop remanufacturing can lead to a change in environmental impacts between -90% to +9% depending on model choice and impact category

This statement is accurate on the verified LCA results and terminology, which underlines legal compliance und full transparency of results calculated in this study. There are both positive and negative effects on product's competitiveness from communicating trade-offs between impact categories. On the other hand, this phrase is little comprehensive because "model choice", "closed-loop" and "impact category" might be not self-explanatory terms. Similarly, the presentation of trade-offs can have adverse effects on the consumer's willingness of returning the product at EoL creating confusion to whether the ski core circularity is environmentally beneficial.

> CO_2 emissions linked to this product can be reduced by 20% when you return your ski

This statement fulfils some legal requirements as the impact reduction result is reliable and clear, yet there is a lack of information and transparency on trade-offs between multiple impact categories. Technically, the statement is not true in the first life of the ski but only in the second life. Similarly, consumer activation can remain limited due to communication of future effects, which might occur in a couple of years only. On the other hand, this statement exhibits high comprehensiveness and competitiveness because climate change is a well-known impact category and the reduction in impacts underlines the positive nature of the product design.

> Return this product and receive a 100 € voucher for your next ski while reducing environmental impacts by up to 90%

While this sentence activates consumers to take place in the return system by giving the highest environmental incentive as well as an additional financial one, the legal compliance is debatable. While the statement is informative, (somewhat) clear and transparent it lacks reliability through the selective presentation of the (arbitrary) highest number, which can be perceived as greenwashing. Concerning comprehensiveness, the term "environmental impacts" might be ambiguous and also presenting the result as "up to 90%" can lead to follow-up questions. At the same time, emphasizing the most optimistic potential of the measure can lead to advantages of product competitiveness.

5 Conclusion and Outlook

Reflecting back on the goal and scope of the study, the influence of methodological choices on LCA results and trade-offs between impact categories are visible when evaluating the environmental impacts of the ski core circularity measure. There are multiple limitations of the LCA models that should be addressed in further studies, such as verification of the collection system and other inventory data. The inclusion of all life stages helps obtaining a more holistic picture of skiing as an activity. Additionally, the return system for ski core circularity can lead to an increase also in higher open-loop recycling due to the source-separated collection and dismantling of the ski. Monetary valuation (ISO 14008) can increase the understanding and relevance of different impact categories and put trade-offs into perspective. Absolute Environmental Sustainability Assessment can provide meaningful insights by referencing planetary boundaries (Rodrigues et al. 2021).

Concerning communication of LCA results, the framework conditions differ among internal and external stakeholder groups. For external communication, it is important to keep the main messages visual, clear and competitive while also following the legal and fair competition code of conduct. It is possible to extend concise statements on products by providing additional information on LCA results and the product via supplementary material and QR codes, i.e. increased traceability by allowing the consumer to track the journey of the returned ski.

Finally, the existing legal frameworks applicable to the case study, such as Circular Economy framework (Potting et al 2017), Green Claims Directive (European Commission 2023) and PEF (Zampori and Pant 2019) should align on terminology and methodology concerning EoL treatment of consumer goods and include information on communication of results to customers, e.g. by providing examples and guidance. This does not only enable informed decision-making but furthermore ensures fair competition. Verification mechanisms have to be established to facilitate the transition between standardizing methodology, such as PEFCR, and application and communication by companies. Accreditation procedures from the energy and waste management sector can provide inspiration for the development of a similar processes concerning LCA conduction and communication.

Acknowledgements The authors acknowledge the financial support through project WINTRUST—"Wintersport Resource Efficiency and improved Circular Economy" provided by the Austrian Ministry for Climate Action, Environment, Energy, Mobility through the "Collective Research" program administered by the Austrian Research Promotion Agency (FFG).

References

Andreasi Bassi S, Biganzoli F, Ferrara N, Amadei A, Valente A, Sala S, Ardente F (2023) Updated characterisation and normalisation factors for the environmental footprint 3.1 method, Publications Office of the European Union

European Commission, Proposal for a Directive of the European Parliament and of the Council on substantiation and communication of explicit environmental claims (Green Claims Directive), 2023/0085 (COD), (2023)

Matuštík J, Paulu A, Kočí V (2024) Is normalization in Life Cycle Assessment sustainable? Alternative approach based on natural constraints. J Clean Prod 444:141234

Peña C, Civit B, Gallego-Schmid A, Druckman A, Pires AC, Weidema B, Mieras E, Wang F, Fava J (2021) Canals L M i, Cordella M, Arbuckle P, Valdivia S, Fallaha S, Motta W, Using life cycle assessment to achieve a circular economy. Int J Life Cycle Assess 26:215–220

Potting J, Hekkert E, Worrell E, Hanemaaijer A (2017) Circular economy: measuring innovation in the product Chain, PBL Publishers

Rodrigues J, Gondran N, Beziat A, Laforest V (2021) Application of the absolute environmental sustainability assessment framework to multifunctional systems—The case of municipal solid waste management. J Clean Prod 322:129034

Zampori L, Pant R (2019) Suggestions for updating the Organisation Environmental Footprint (OEF) method, Publications Office of the European Union

LCM Supporting Product and Organizational Reporting

Application of LCA for Pharmaceuticals: Lessons from a Ten-Year Programme in AstraZeneca

Aoife Calnan, Simon Aumônier, Andy Whiting, Chloë Smithers, Caireen Hargreaves, Michael Collins, Donald Reid, and Hugh Whetherly

Abstract AstraZeneca (AZ) has developed a comprehensive product life cycle assessment (LCA) programme to support its sustainability goals. The programme aims to generate robust, product-specific environmental data to inform reduction strategies and corporate reporting. Using ISO 14040/44-aligned methods, the LCAs cover > 90% of AZ's commercial portfolio by 2024 revenue. These studies integrate primary data from operations and suppliers and are updated regularly to reflect system changes, such as packaging format updates. Results from the programme reveal diverse environmental hotspots across product types and life cycle stages, with solvents, energy use, and distribution as key drivers. The Product Sustainability Index (PSI) uses LCA to enable internal benchmarking and prioritisation. The programme demonstrates how LCA can guide emissions reduction, improve scope 3 reporting, and support broader applications such as clinical trials assessments.

1   Introduction

The emissions, waste, and consumption of resources associated with supply, production, packaging, distribution, use stage activities and waste management all contribute to the environmental impacts of products and services. These impacts are becoming increasingly significant to companies and society within the context of planetary boundaries. AstraZeneca (AZ), a global provider of pharmaceutical products (www.astrazeneca.com), is no exception. AZ seeks to make a positive impact on the health of people, society and the planet through taking action on climate and nature, health equity and health systems resilience.

A. Calnan (✉) · M. Collins · D. Reid · H. Whetherly
ERM—Environmental Resources Management, London, UK
e-mail: aoife.calnan@erm.com

S. Aumônier
Aumônier Consulting Limited, Oxford, UK

A. Whiting · C. Smithers · C. Hargreaves
AstraZeneca, Macclesfield, UK

© The Author(s) 2026   301
M. Traverso et al. (eds.), *Life Cycle Management from Global to Local*,
https://doi.org/10.1007/978-3-032-17987-6_25

Life cycle assessment (LCA) is a method for appraising some of these impacts, standardised through ISO 14040/44, which set out guidance on principles and frameworks, and requirements and guidelines, respectively (ISO 2006a, b). Numerous other related methods, for example, product carbon footprinting, apply the same 'life cycle' logic in quantifying contributions to environmental impacts from 'cradle to grave'. These methods share a common ambition of reporting a holistic footprint. This approach helps to avoid potential counter-productive outcomes that might occur if decisions about interventions to reduce impacts are based on a more limited scope of analysis.

Environmental footprints indicate life cycle stages that contribute most to environmental impacts (hotspots) and help evaluate the effectiveness of measures aimed at reducing these impacts. In the case of greenhouse gas (GHG) emissions, they can also guide pathways to Science-Based Targets and Net Zero.

Healthcare accounts for almost 5% of global GHG emissions, with medicines estimated to represent a quarter of the sector's climate change impact (NHS 2024). Pharmaceutical products are diverse, but can broadly be categorised into small molecules and biologics (large molecules). Products can be manufactured via different methods and administered in different presentations (e.g. drug product format or device), and often have long and complex supply chains. Previous reviews (Chen et al. 2024) have identified significant hotspots for specific drugs. However, these focus mainly on small molecules and do not provide information on how to implement a wider programme.

AZ's comprehensive evidence-based product LCA programme, and the application of LCA across other business functions, has allowed the company to navigate complexity in developing effective reduction and communication strategies in light of the diversity in pharmaceutical products' impact hotspots.

2 Methods

2.1 Timeline

In 2011, the National Health Service (NHS) in the UK brought together a group of pharmaceutical and medical device companies to seek solutions to the significant component of the NHS's organisational carbon footprint residing in its purchases of their products. The group, now the Sustainable Healthcare Coalition (SHC), published its guidance on footprinting these products a year later (SHC 2012).

As a SHC member, AZ instituted a programme of product LCA studies, beginning with pilot projects on small molecule pharmaceuticals. The results of its first assessment of its lipid modulator drug were published in AZ's 2015 Sustainability Report, shown in Fig. 1. LCA studies of inhalers and biologics products were introduced later in the programme. 2015 was also the original baseline year for AZ's scope 1 and 2 GHG reduction targets, which are now to achieve 98% absolute reduction by

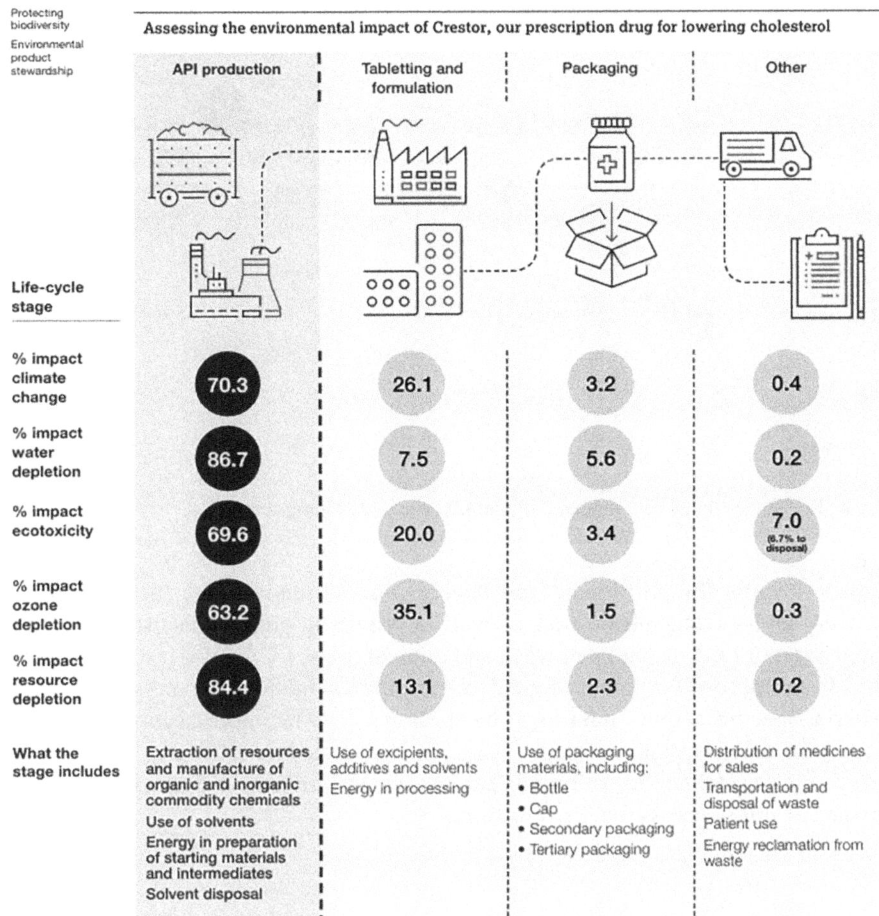

Fig. 1 Reporting the first LCA study in AZ's 2015 sustainability report (AstraZeneca 2015)

2026 in operations and to become carbon negative across the value chain by 2030. The baseline year has since been updated to 2019.

AZ's LCA studies have now delivered >400 product impact profiles, spanning a range of products, supply chains, and markets, with many studies updated on one or more occasions to account for new data, supply chain changes, operational developments and efficiencies. In 2025, coverage exceeded 90% of AZ's product portfolio by 2024 revenue. The insights gained from viewing AZ's whole portfolio illustrate the diversity of product impact profiles and the interventions required to meet corporate goals. The cadence of product assessment updates is roughly every three years. The evolution of the LCA programme is illustrated in Fig. 2.

The programme provided the foundation for development and deployment of an eco-design tool to inform AZ's product development functions. Product LCAs

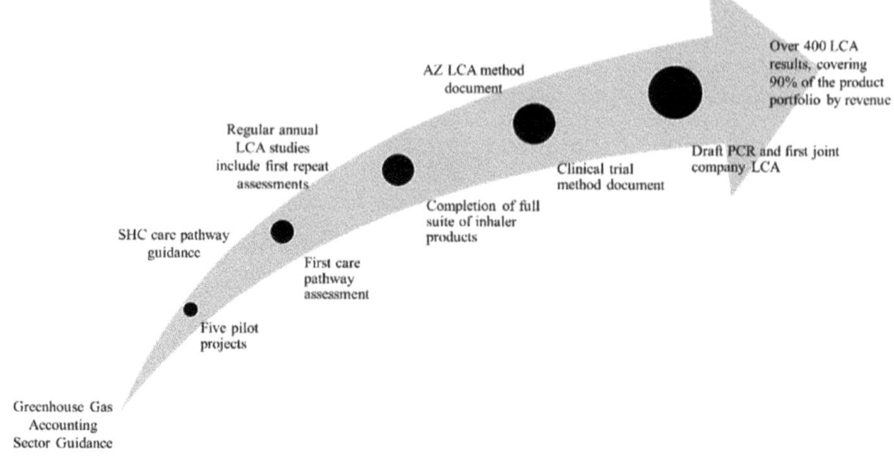

Fig. 2 Key stages in the development of AstraZeneca's LCA programme

provide a basis for prioritising and testing process innovations, such as solvent recovery and solvent substitution, as well as providing product data to support the application of LCA to care pathways and clinical trials. LCAs also provide bottom-up data to improve corporate scope 3 accounting estimations, where supplier- and material-specific activity data have been shown to offer more accurate emission factors for GHG accounting than spend-based approaches. LCA is a fundamental part of AZ's sustainability ambitions across all core functions and provides multiple streams of utility across the organisation.

2.2　Collaboration and Integration

AZ's first LCA studies were carried out for its Global Sustainability and Safety, Health and Environment (GS&SHE) function, aimed at internal sustainability professionals within the business, seeking to understand the key drivers of its environmental impact as its Sustainability Strategy grew in ambition. Assessments are increasingly requested across the business due to customer demands and internal GHG reduction targets for product environmental data such as carbon footprints.

In 2022, the NHS introduced a weighting on Net Zero and Social Value in its procurements. Medicines were originally excluded from this weighting until 2024, under the Evergreen assessments. In 2028, the NHS is expected to request individual product footprints from suppliers (NHS 2023). Healthcare organisations in France and the Nordic countries are also introducing similar measures.

AZ's earliest studies relied on secondary data to characterise its supply chain, for example drawing on patents and reported information on analogous chemistries. In some cases, it was difficult to obtain information on its own operations, where this was not routinely collected or where the data available were insufficiently granular.

Over time, an increasing proportion of AZ's contract manufacturers have engaged with the programme and offered information on their processes and potential improvements. At the same time, provision of data from AZ's operating sites to LCA practitioners has become more routine, in part because of AZ's commitment to reducing its direct GHG emissions and an increasing awareness in employees of the company's Net Zero ambitions. AZ has also engaged in collaborative research with device manufacturers and within alliance agreements with other pharmaceutical companies, where there is a mutual interest in product environmental profiles. With the growth of the programme, LCA data have been made available to the company as a whole, with growing interest in product teams outside the GS&SHE function. Engagement and support from senior leadership also helps to drive improvements in data collection, improving output quality.

2.3 Assessment Focus

The SHC footprinting guidance and the global LCA standards together provided the methods underpinning AZ's early studies. As the programme matured, LCAs increasingly focused on climate change and water use, and the company developed its own scoring system, the Product Sustainability Index (PSI), tailored to align with its corporate goals and streamlined to facilitate the engagement of technical teams. The PSI allows AZ to benchmark product footprints–guided by well-established standards–and report to the wider business through a simple score focusing on key metrics. The PSI enables product teams to prioritise more effectively efforts to minimise product environmental impact and to establish clear reduction targets.

Directed by NHS England, Office of Life Sciences, Sustainable Markets Initiative Health Systems Task Force members, and the Pharma LCA Consortium, the British Standards Institution (BSI) has developed PAS 2090 (currently in draft), a standardised method for conducting LCAs of pharmaceutical products. It sets common assumptions to ensure sector-wide consistency (BSI 2024).

AZ has applied draft PAS 2090 in recent product LCA updates, providing feedback on its practicability and utility as the method is finalised. In due course, it is expected that the NHS will advise suppliers in providing product footprints, potentially using the same, or a closely aligned, approach as PAS 2090. AZ's LCA experience helped to highlight critical decisions involved in conducting pharmaceutical LCA studies and key factors influencing outcomes, underscoring the need for standardisation. This insight contributed to the development of the PAS 2090 standard, in collaboration with other members of the PEG LCA consortium.

3 Results

The climate change hotspot variation across AZ's portfolio is shown in Fig. 3, aggregated by product type (small molecules, inhalers and biologics), with distribution and use including patient travel. The figure presents the instances where 20% or more of total impact arises in that life cycle stage. For example, API manufacture is a hotspot in approximately 50% of small molecule studies. There is often a correlation in hotspots among different life cycle impact categories, but it is important to study more than only climate change, as a reduction in one impact category may result in an increase in another as a result of burden-shifting.

A table for 15 biologic products, showing the climate change impact by life cycle stage, is presented in Table 1. Despite all results being within the same product category, the hotspots for each product vary, depending on a number of factors, including the amount of drug substance per dose, supply chain, and mode of transport to the market. The range of potential hotspots shows the importance of collecting primary data and conducting individual LCA studies, to paint a holistic view both of individual products and of product groups. LCAs help to create a focus for product teams to understand where real differences lie and how reductions in impact can reliably be made.

4 Discussion

4.1 Product Footprints

With a high level of coverage of its product portfolio, AZ's LCA programme has been able to shed light on variations in the drivers of environmental impacts across therapeutic areas, modalities, and supply chains. Pharmaceutical products are inherently complex and different starting chemicals, synthesis routes, manufacturing locations (including those of contract manufacturers), utilities consumptions, packing formats, and administration methods, and inter-site and downstream distribution routes mean that it is difficult to arrive at simple 'rules of thumb' for understanding the principal levers of the footprint. For example, the complexity and number of synthesis steps for an active pharmaceutical ingredient (API) is a poor analogue for the final drug product and its environmental profile.

Nonetheless, energy consumption and solvent use are typically major contributors to the footprint and lie behind AZ's drive for a rapid transition to renewable energy at its sites, including its innovative adoption of the supply of biomethane for heating (Future Biogas 2025). Periodically reviewing and updating LCAs shows the quantitative impact savings on the product that AZ is making, through changes such as moving to biomethane from natural gas or improving process efficiency. With respect to solvents, LCA investigations of their reuse/recycling and switching to bio-based solvents are all responses to their identification as a key hotspot in many

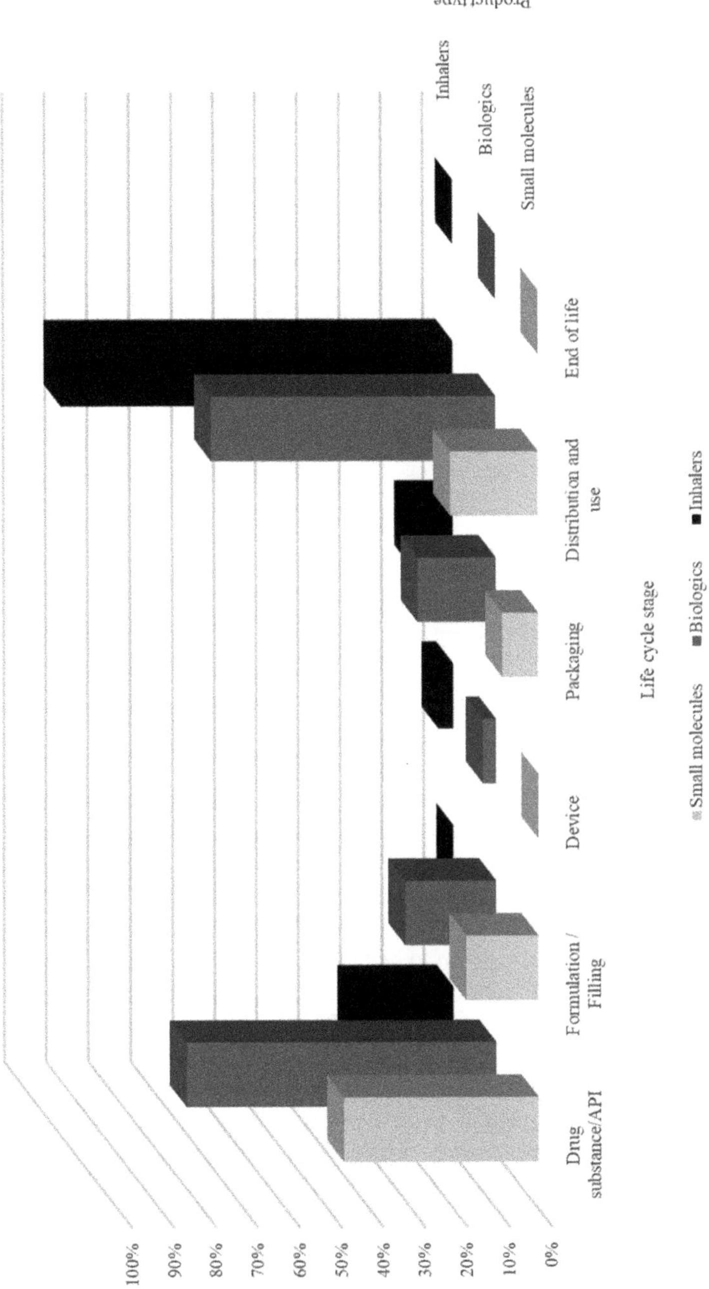

Fig. 3 Small molecules, biologics and inhaler climate change hotspots, taken from results over AZ's ten-year LCA programme

Table 1 Biologic climate change hotspot distribution for 15 example products (percentage contribution to total impact by each life cycle stage)

Product	Drug substance (%)	Formulation / filling (%)	Device (%)	Packaging (%)	Distribution/ use (%)	End of life (%)
1	57	3	< 1	2	37	1
2	33	2	< 1	1	62	2
3	15	44	1	2	38	< 1
4	13	10	6	32	38	2
5	3	7	1	21	68	< 1
6	37	7	< 1	3	53	< 1
7	31	< 1	< 1	5	63	< 1
8	15	44	1	1	38	< 1
9	31	44	< 1	10	13	1
10	46	8	< 1	16	24	5
11	24	4	< 1	1	68	3
12	69	6	< 1	24	1	< 1
13	35	16	38	5	1	6
14	21	48	< 1	< 1	18	13
15	74	5	< 1	< 1	21	< 1

LCA studies of AZ products. Inclusion of targeted sensitivity and scenario analyses in product studies also provides AZ with insights as to how changes to distribution mode or supply chain, for example, can affect product impact.

At a high level, trends in hotspots can be seen for similar pharmaceutical product types. However, each product has its own unique characteristics, and the precise picture varies considerably as a result. Broadly speaking, the climate change impact of synthetic API is significantly driven by solvent use and end of life management during API production. Nonetheless, the formulation, packaging and distribution stages of the life cycle are also shown to be hotspots in some product LCA studies.

The broad trend with biologic drug substances is for the climate change impact to be driven by high energy use as a result of HVAC (heating, ventilation, and air conditioning) requirements for clean room environments. Inter-site and downstream distribution can also make a significant contribution due to air freight. The use of active cooling containers is noted as particularly significant.

In the case of inhalers, hotspots vary considerably, depending on inhaler type (i.e. dry powder vs metered dose). The majority of metered dose inhaler climate change impact is attributable to the release of propellant gas during inhaler use and end of life. These findings drove AZ to commit to launching a next-generation metered dose inhaler with a novel near-zero global warming potential propellant.

Patient travel to collect prescriptions or to receive treatments at clinics and hospitals is found to make a significant contribution to the total impact in some LCA studies. Although its inclusion is debated in the LCA community, the contribution of patient travel is within the scope of AZ's LCA studies as it is an important and variable component of a product's life cycle impacts, and one which can be influenced by product design.

4.2 Care Pathways

Although AZ's programme focuses on its own products, where it has the greatest opportunities to intervene and to make reductions, it has also conducted wider 'care pathway' studies, examining products in the context of the entire patient care pathway. This broader life cycle allows a study to shed light on the implications of the use of a product beyond its direct footprint, for example as more efficacious drugs and devices reduce visits to general practitioner (GP) surgeries or to hospitals. In some cases, healthcare innovations might appear detrimental to the environment, for example, through the addition of sensor electronics and batteries to inhaler devices, but the overall impacts of the care pathway may fall as a result of improved adherence to prescriptions, more than counterbalancing the investment.

AZ was one of the first companies to adopt the care pathways guidance developed by the SHC and has contributed to the collection and publication of additional data on specific pathways relevant to its product portfolio to support the wider adoption of a care pathway lens to sustainability, better health outcomes, and lower environmental impact. The system boundary for AZ's first care pathway study of paediatric asthma is shown in Fig. 4.

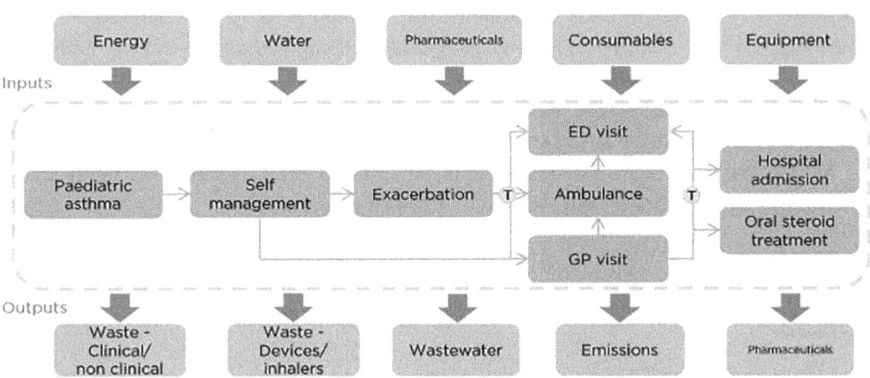

Fig. 4 Care pathway study of paediatric asthma (SHC 2017)

4.3 Clinical Trials

Clinical trials are an essential part of pharmaceutical product development and regulatory approval. Although believed to be a significant source of the sector's GHG emissions, until recently they had been only sparsely studied. Following a call in the Lancet journal (Adshead et al. 2021) for the measurement and reduction of the footprint of clinical trials, AZ conducted LCA studies of a sample of its own trials (Mackillop et al. 2023). The assessments provided clear indications of the hotspots in the trial footprint and the opportunities for intervention, and subsequently became the foundation for a cross-functional effort in developing a comprehensive roadmap for decarbonisation. The application of LCA identified significant opportunity for reduction in the GHG emissions of clinical trials and brought to the fore the consequences of decisions made in their design.

4.4 Eco-Design

Typically, LCA studies are lengthy, and require a degree of expertise in the technique itself. Recognising the value of more rapid assessments to inform its research & development functions, AZ with ERM has developed an online modular eco-design tool allowing non-experts to perform efficient 'LCA-lite' assessments, drawing on an extensive database of starting materials, consumables, auxiliaries and packaging materials developed through the product LCA programme.

Users can model complex, branching reaction pathways for synthetic API and detailed multi-step processing chains for biologics API production and for formulating the final drug product through to the product end of life, as shown in the schematic in Fig. 5. LCA-lite results are calculated rapidly and provided with sufficient detail to identify hotspots in the product life cycle and inform decisions about environmental footprint reduction which can allow early interventions in the product life cycle. These results are not as accurate as those of an ISO-conformant study, and are not appropriate for communication outside of AZ, but have been shown nonetheless to be a reasonable approximation of a full study.

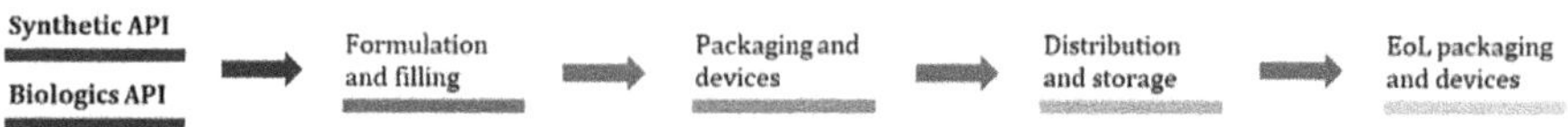

Fig. 5 Schematic structure of the eco-design tool

4.5 Scope 3 GHG Emissions

AZ was one of the first companies to have GHG reduction targets verified against the Science Based Targets initiative Net-Zero Corporate Standard. Its Ambition Zero Carbon programme aims to deliver a 50% reduction in emissions across its entire value chain by 2030 (AstraZeneca 2023, 2024). The LCA programme strongly complements Ambition Zero Carbon, providing robust and credible activity-based emission factors for sources outside AZ's direct control in scope 3 and a credible basis for strategic decisions regarding decarbonisation roadmaps. The LCA programme's high coverage of the portfolio provides a robust basis for developing proxy emission factors for products yet to be assessed. Specific, activity-based, emission factors developed from LCA studies offer greater confidence in making investment decisions about GHG reductions than generic and sector-average spend-based factors. An LCA approach with supplier engagement provides specific good quality data and a mechanism by which the scope 3 footprint can respond to changes in the supply chain and development of partnerships to reduce GHG emissions for all parties. This is an improvement on commonly applied spend-based and industry-average calculation methods.

5   Conclusion

AZ's ten-year LCA programme began with five pilot projects, since evolving to cover >400 products. The programme's expansion has led to greater understanding of product environmental impacts and their drivers throughout AZ, resulting in increased stakeholder engagement in reduction initiatives. The programme has made LCA data available across the business, enabling diverse applications including care pathway assessments, clinical trial appraisals, eco-design and corporate footprint reporting, and demonstrating the value of the method in many different parts of the company. The programme shows the potential and various benefits of expanding LCA from initial pilot studies to a company's entire product portfolio and in deploying the method as a versatile strategic management tool.

Acknowledgements The authors would like to dedicate this paper to the memory of Alex Mullen, whose invaluable contributions and dedication over the past decade were instrumental to the success of this programme. The authors acknowledge the contribution of many other current and former AstraZeneca colleagues in the Global Environmental Protection group to the work outlined in this publication: Nigel Budgen; Leonie Campbell; James Cunningham; Barry Dillon; David Ennis; Katerina Maslova; Bryan Mulchinock; Jason Snape; Miriam Turner; and Wesley White. The authors also acknowledge the contribution of current and former ERM colleagues: Rodrigo Alvarenga; Dáire Archbold; Tom Costelloe; Saori Galley; Lyndsey Lamb; Doyin Oladele; Peter Shonfield; Carolyn Thomas; and Zimo Wu.

References

Adshead F et al (2021) A strategy to reduce the carbon footprint of clinical trials. The Lancet 398(10297):281–282

AstraZeneca (2015) https://www.astrazeneca.com/content/dam/az/PDF/2015/Sustainability_update_2015.pdf, (Accessed 08.05.2025)

AstraZeneca (2023) https://www.astrazeneca.com/sustainability/environmental-protection/ambition-zero-carbon.html, (Accessed 08.05.2025)

AstraZeneca (2024) https://www.astrazeneca.com/content/dam/az/Sustainability/2025/pdf/Greenhouse-Gas-Methodologies-2024.pdf, (Accessed 23.05.2025)

BSI (2025) https://pharmaenvironment.bsigroup.com/, (Accessed 27.05.2025)

Chen Z et al (2024) Application of life cycle assessment in the pharmaceutical industry: A critical review. J Clean Prod 459(142550):0959–6526

Futurebiogas.com (2025) https://www.futurebiogas.com/about/content-hub/news/future-biogas-and-astrazeneca-bring-the-uk-s-first-unsubsidised-biomethane-plant-online/, (Accessed 09.05.2025)

ISO (2006a) 14040, Environmental management—life cycle assessment—principles and framework (ISO 14040:2006 + ISO 14040:2006/Amd 1:2020). International Organization for Standardization

ISO (2006b) 14044, Environmental management—life cycle assessment—requirements and guidelines (ISO 14044:2006 + ISO 14044:2006/Amd 1:2017 + ISO 14044:2006/Amd 2:2020). International Organization for Standardization

Mackillop N et al (2023) Carbon footprint of industry-sponsored late-stage clinical trials. BMJ Open 13:e072491

NHS (2023) https://www.sps.nhs.uk/articles/net-zero-and-social-value-requirements-in-nhse-medicines-tenders/, (Accessed 14.05.2025)

NHS (2024) https://www.england.nhs.uk/greenernhs/a-net-zero-nhs/areas-of-focus/, (Accessed 12.05.2025)

SHC (2012) https://ghgprotocol.org/sites/default/files/tools/Guidance-Document_Pharmaceutical-Product-and-Medical-Device-GHG-Accounting_November-2012.pdf, (Accessed 27.05.2025)

SHC (2017) https://shcoalition.org/wp-content/uploads/2019/12/Digital-Adherence-Monitoring-in-Poorly-Controlled-Paediatric-Asthma-Final.pdf, (Accessed 27.05.2025)

Unlocking Synergies Between PCF and EPD Reporting: Industry's Path to a Sustainable Future or Mere Fantasy

Manel Sansa, Florian Ansgar Jaeger, Benjamin Canaguier, Johannes Wunderlich, and Matt Seiler

Abstract Life cycle assessment (LCA) manifests in two parallel streams in industries; Design focused Environmental Product Declarations (EPDs) and procurement driven Product Carbon Footprints (PCFs), each with different objectives, data and verification methods. This paper investigates how to align and integrate supplier specific PCFs into EPDs efficiently, aiming to reduce redundancy, improve data quality and enhance reporting clarity. A structured analysis is conducted in this paper investigating three integration scenarios, namely: Hybrid, scaling, and separate approach. A case study on electrical switchgear components showcased that the hybrid approach yielded a 7.5% reduction in GWP with higher supplier data utilization. Hybrid integration offers a viable pragmatic path forward balancing accuracy and simplicity. Broader adoption will require standardization of data formats and verifier acceptance.

1 Introduction

Product Life Cycle Assessment (LCA) has become central to evaluating environmental performance, driven by standards such as ISO 14040 (ISO 2006a) and 14044 (ISO 2006b), product category rules, and initiatives like the Science Based Targets initiative (SBTi) (Science Based Targets initiative 2024). In industrial practices, two main LCA application contexts have emerged, design and procurement. Design-oriented LCAs have matured over time, supporting performance improvements, and enabling Environmental Product Declarations (EPDs) that provide multi-indicator disclosures. Meanwhile, procurement practices are increasingly influenced by Scope 3 emissions targets, Indeed, scope 3 categories such as "purchased goods

M. Sansa (✉) · B. Canaguier · M. Seiler
Schneider Electric, Rueil-Malmaison, France
e-mail: manel.sansa@se.com

F. A. Jaeger · J. Wunderlich
Siemens AG, Berlin, Germany

© The Author(s) 2026

M. Traverso et al. (eds.), *Life Cycle Management from Global to Local*,
https://doi.org/10.1007/978-3-032-17987-6_26

and services" are often accounting for the largest share of a company's carbon footprint. Hence, procurement departments are now central to achieving decarbonization goals. As a result, suppliers are being asked to provide Product Carbon Footprints (PCFs) (Goldstein et al. 2020). PCFs are mainly supported by initiatives, namely:

(1) Catena-X (Catena-X Automotive Network 2025), a German-led automotive focused data ecosystem aiming to create PCF exchange across the supply chain.
(2) Together for Sustainability (Together for sustainability 2025), a global initiative driven by chemical companies to standardize supplier sustainability assessments, including carbon reporting.
(3) And PACT (WBCSD 2023), a cross-industry initiative that provides the pathfinder framework, a standardized approach for exchanging PCF data across value chains.

Both applications differ in purpose, data needs, and outcomes. Design LCAs guide environmental decisions and feed into EPDs, while procurement LCAs focus on PCFs, emphasizing primary data to assess supplier-specific emissions. As companies adopt both EPD and PCF reporting, there is growing interest in integrating these frameworks to reduce duplication and enhance transparency (Würz and Finkbeiner 2022). However, aligning them presents methodological, procedural, and operational challenges.

The growing emphasis on environmental accountability across supply chains, spurred by regulations like the EU Green Deal and voluntary efforts such as the SBTi, has made emissions disclosure a strategic priority. EPDs and PCFs have emerged as key tools for quantifying and communicating product-level emissions. Despite their shared goals of transparency and comparability, they have evolved separately, leading to methodological inconsistencies that complicate integration.

Despite the parallel development of EPDs and PCFs as tools for environmental disclosure, there is currently no widely accepted methodology for integrating PCFs into certified EPDs without compromising the multi-indicator consistency required by Type III declarations. Hence, the main research question: How can supplier specific PCFs be integrated into EPDs in a way that improves data representativeness for scope 3 emissions while maintaining data accuracy and verification compatibility?

To answer this question, this paper explores how synergies between EPDs and PCFs can be realized, examining the benefits, barriers, and conditions for convergence. It aims to provide guidance for industry professionals, LCA practitioners, and policymakers navigating this evolving landscape. We assess the hybrid approach that incorporates supplier specific PCFs into EPDs and argue that, despite existing challenges, alignment is achievable through coordinated efforts and shared standards.

2 State of the Art and Conceptual Framing

Several studies have explored the data alignment of PCF and EPD frameworks. Curran (2013) provides a foundational overview of LCA methodology, outlining the balance between detailed and simplified approaches. Zampori and Pant (2019) examine the Product Environmental Footprint (PEF) method's potential to unify diverse environmental metrics. Finkbeiner (2014) critiques policy-driven footprinting, stressing the need for scientific rigor and stakeholder engagement. Wernet et al. (2016) present the Ecoinvent 3 database, emphasizing the role of data quality and primary data in improving footprint precision.

Figure 1 presents a comparative bar chart illustrating an example of the share of primary data used in the PACT framework and EPD. The PACT framework demonstrates a higher primary data share at around 60% compared to EPD. Figure 1 demonstrates the variation in data quality and specificity. This variation is because EPD product category rules rely more on secondary data based on environmental databases generic datasets while PCFs are mainly based on supplier specific data (WBCSD, 2023).

Moreover, Table 1 compares key aspects of (PCFs) and (EPDs), highlighting their methodological differences. PCFs typically adopt a cradle-to-gate system boundary and focus solely on Global Warming Potential (GWP) as the impact indicator, using market-based electricity modelling and prioritizing primary, supplier-specific data. Verification is corporate-led or self-declared (WBCSD 2023). Nevertheless, Catena-X and TFS are initiating PCF third party certification program that is estimated to be published by September 2025 (Catena-X Automotive Network 2025). In contrast, EPDs generally follow a cradle-to-grave approach, include multiple environmental indicators, rely on location-based electricity modelling, and use secondary data

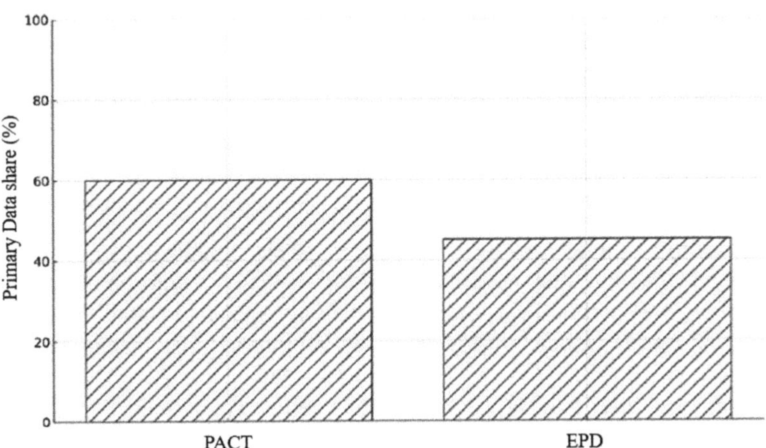

Fig. 1 Data completeness PACT versus EPD programs

Table 1 Methodological differences between PCF and EPD programs

Aspect	PCF	EPD
System boundaries	Often cradle-to-gate	Typically cradle-to-grave
Impact indicators	GWP only	Multiple indicators
Electricity modelling	Market-based	Location-based
Data source preference	Primary (supplier-specific)	Secondary (databases)

from databases. EPDs are also subject to third-party verification, ensuring greater standardization and credibility.

The above analysis highlighted the complementary nature of both frameworks and the challenges in aligning them.

A key obstacle in harmonizing PCF and EPD reporting lies in their differing methodological foundations. PCFs typically use supplier-specific, market-based data, whereas EPDs adhere to standardized, location-based LCA models as mandated by Type III declarations (e.g., EN 15804 (European committee for standardization 2019)). These contrasting approaches can result in divergent global warming potential outcomes. The variation between GWP values across life cycle stages is mainly caused by the system boundaries, data resources and modelling assumptions.

Despite these insights, existing literature lacks concrete examples of integrating supplier specific PCFs into EPDs without compromising methodological integrity. This study addresses that gap by presenting practical integration scenarios informed by real-world industrial practices.

3 Methodology

Our approach is grounded in a qualitative assessment of current industry practices and technical documentation related to EPDs and PCFs. Drawing on insights from collaborative initiatives such as PACT, Together for Sustainability (TfS), and Catena-X through publicly available guidance documents, we identify key challenges and propose integration strategies that reflect both methodological rigor and operational feasibility. These strategies are evaluated across three distinct scenarios:

(1) Integrating supplier-specific PCFs into EPDs frameworks: This scenario explores the potential to enrich existing EPDs by incorporating granular, supplier-level carbon data, thereby enhancing specificity and alignment with Scope 3 reporting needs. This approach is supported by Catena X, and Pact framework.

(2) Maintaining separate PCFs and EPDs reporting streams: This approach preserves the independence of each framework, allowing organizations to meet different stakeholder and regulatory requirements without forcing methodological convergence. This approach is mainly supported by the standard (i.e. EN15804).

Table 2 Advantages and drawbacks of integration scenarios

Scenario	Advantages	Drawbacks
Hybrid	– High representativeness – Transparency – Corporate alignment	– Indicator asymmetry – Verifier acceptance uncertain
Separate	– Methodologically independent – Avoids data merging	– Customer confusion – Resource intensive
Scaling	– Simple to model	– Distorts results – Lacks scientific rigor

(3)　Scaling additional impact categories using PCF-to-average GWP ratios. This hybrid strategy proposes using supplier-specific GWP data as a scaling factor to estimate other environmental impacts, offering a pragmatic compromise between data availability and reporting completeness. However, this approach is criticized in the literature for introducing inconsistency between impacts categories (Laurent et al. 2010).

Table 2 highlights the advantages and drawbacks of each scenario.

Among these, Scenario 1 is examined in greater depth due to its promising balance between methodological coherence and practical applicability. It offers a pathway to bridge the gap between design-phase LCA and procurement-driven carbon accounting. Effective integration of supplier specific PCFs into EPD reporting requires consensus across a broad range of stakeholders, including manufacturers, suppliers, certification bodies, and LCA software providers. Figure 2 presents a conceptual map of stakeholder roles in bridging methodological and operational gaps. In fact, it begins with suppliers, who provide essential raw material data, followed by manufacturers, who process received data into product-level assessments. Certifiers then verify the accuracy and reliability of the data, ensuring compliance with established standards. EPD program operators manage the publication and registration of verified declarations, while standardization bodies define the overreaching methodologies and frameworks that guide the entire process.

In the following section, we will introduce a practical example of how integrating supplier-specific Product Carbon Footprints (PCFs) into Environmental Product Declarations (EPDs) can enhance the accuracy of carbon reporting in a manufacturing context using the first scenario.

4　Illustrative Example: Integrating Supplier PCFs into EPDs

To demonstrate the practical implications of integrating supplier specific PCFs into EPDs, we consider a hypothetical case involving a mid-sized manufacturer of electrical switchgear.

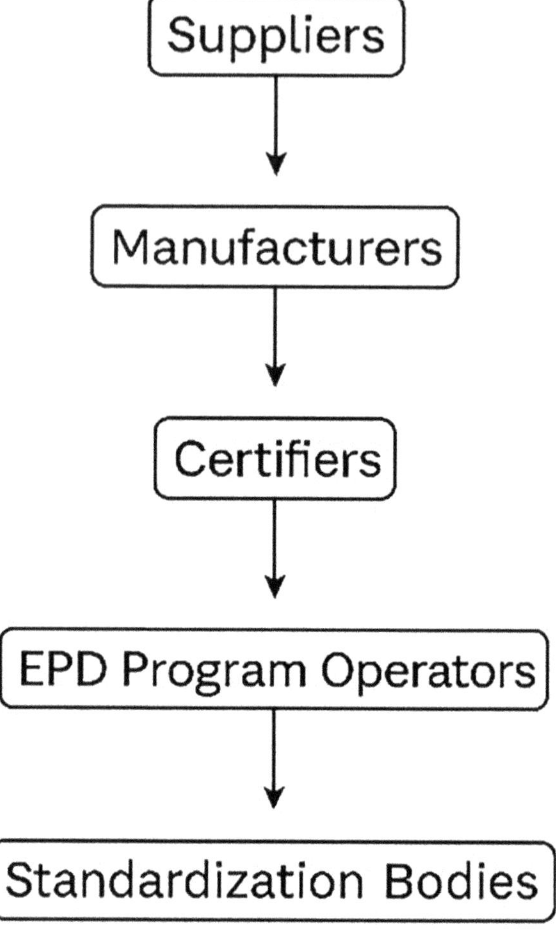

Fig. 2 Stakeholder roles in enabling PCF-EPD integration

The company sources steel components from two suppliers. Supplier A provides average emission data from an industry database, while Supplier B reports a verified PCF based on Catena-X standards. The goal is to integrate the PCF from Supplier B into the product's EPD following the processes described in Fig. 3.

Using a hybrid approach, the manufacturer replaces the steel module's GWP value in the EPD by the supplier's PCF, maintaining secondary data for the other indicators (e.g., water use, ozone depletion). The calculations are displayed in Table 3.

The GWP characterization method used in the example is the IPCC 2013 GWP 100a (100-year time horizon) (IPCC, 2013). Secondary data is extracted from the ecoinvent 3.9 (Wernet et al. 2016) (Europe average) and supplier B data includes actual energy mix and logistics. The resulting EPD shows a 12% reduction in GWP compared to the baseline scenario for the steel frame with a 7.5% reduction in the total GWP. The declaration also includes a disclaimer regarding the asymmetry of primary

Fig. 3 Hybrid data integration process

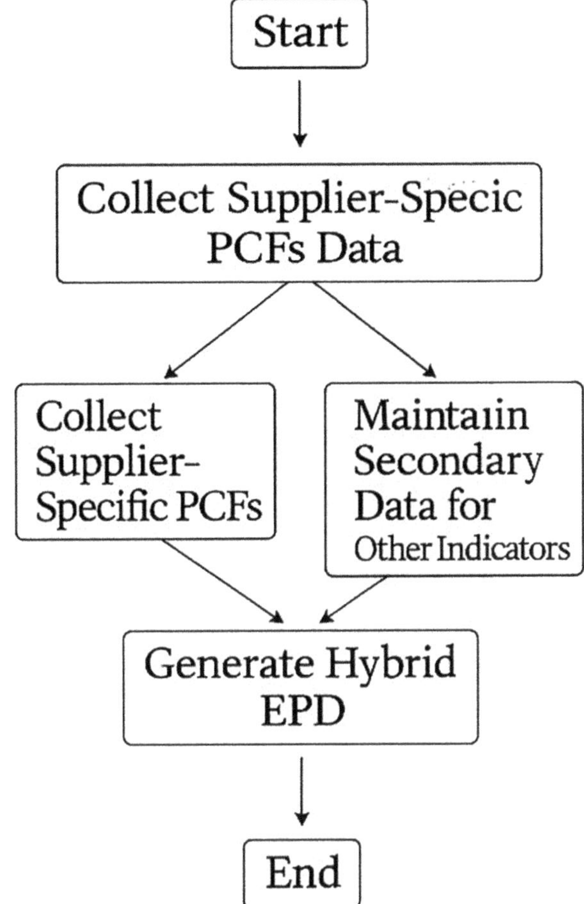

Table 3 Component-level GWP comparison: baseline versus hybrid approach

Component	Source	GWP (Baseline)	GWP (Hybrid)
Steel frame (supplier A)	Industry average (database)	120 kg CO_2 eq	–
Steel frame (supplier B)	Supplier-specific PCF	–	105 kg CO_2 eq
Plastic housing	Database	35 kg CO_2 eq	35 kg CO_2 eq
Wiring	Database	20 kg CO_2 eq	20 kg CO_2 eq
Assembly energy	Database	25 kg CO_2 eq	25 kg CO_2 eq
Total GWP	–	200 kg CO_2 eq	185 kg CO_2 eq

data coverage across indicators. This example illustrates the benefits and limitation of the hybrid approach: Indeed, the hybrid approach improved the representativeness for carbon emissions, but this improvement is subject to a methodological inconsistency across the full impact profile. If these results are used to inform downstream design decisions, where different impact categories must be weighed against each other, this asymmetry could lead to inconsistent outcomes. However, in procurement contexts, this issue is largely irrelevant in practice, as there are typically no Scope 3 targets or similar objectives set for categories such as abiotic resource depletion.

5   Limitations and Future Work

While our analysis identifies potential pathways for integrating PCFs into EPDs, several limitations persist. First, there is no established methodological consensus among PCF initiatives and EPD program operators, making it difficult to generalize integration strategies across sectors or product types. Second, the proposed scenarios have not yet been validated through large-scale industrial applications involving third-party verification. Their feasibility may vary depending on the maturity of supplier data systems and the availability of high-quality primary data. Future research should prioritize the formation of cross-sector consortia to pilot hybrid integration models, engage with standardization bodies to promote methodological alignment, and enhance modelling tools to manage multi-source datasets effectively. Verification remains a cornerstone of credible environmental declarations. EPDs, particularly under Type III programs that require third-party validation to ensure methodological consistency and data reliability. In contrast, PCFs are published at much higher volumes and therefore frequently self-declared or internally verified, leading to potentially inhomogeneous data transparency and quality. It is important to note that the upcoming PCF program certification will have the potential to solve this issue in the long term, but at this moment, this discrepancy complicates the integration of supplier PCFs into certified EPDs, especially for public disclosure. To address this, we propose developing cross-recognition mechanisms whereby PCF programs can be pre-approved by EPD operators. This would facilitate integration, reduce verification burdens, and uphold data integrity. Also, the remaining scenarios, the scaling and the separate approaches will be further examined. The separate approach reflects current industry practice; producing both PCFs and EPDs can be time-consuming and resource-intensive, with redundancy in the cradle-to-gate phase. Additionally, differences in life cycle impact assessment methods may lead to significant variations in GWP results, potentially confusing customers. Lastly, scaling other impact categories from secondary datasets based on CO_2 emissions from supplier PCFs is attractive from a management point of view. However, methodologically, it is the most complex issue. Even if the methodology used for the supplier PCF and the secondary dataset were identical, variations in process efficiency or entirely different production routes could cause higher or lower PCFs. These different routes can also lead to significantly different ratios between impact categories (Laurent et al. 2010).

Trade-offs, such as those seen when replacing fossil-based with bio-based materials, could be misjudged. While scaling based on CO_2 eq may be valid for some materials, products, and processes, it is entirely unsuitable for many others. Therefore, the main challenge in the scaling approach is defining criteria for a suitable categorization method to ensure consistency.

6 Conclusion

The findings of this study indicate that hybrid integration, combining supplier specific PCFs for GWP with secondary data for other indicators, offers a feasible route to enhance the precision and relevance of EPDs. However, successful implementation hinges on broader stakeholder alignment, updates to standardization frameworks, and the capabilities of digital tools. Regulatory support could help close methodological gaps, especially in sectors facing stringent disclosure mandates. We recommend that policymakers and program operators develop adaptable templates such as compliant digital tools that enable partial PCF integration while preserving LCA consistency. Additionally, industry-led pilot programs are essential to validate these models under real-world verification and operational conditions.

References

Catena-X, Catena-X Automotive Network (2025). https://catena-x.net/. Accessed 20 Jun 2025

Curran MA (2013) Life cycle assessment: a review of the methodology and its application to sustainability. Curr Opin Chem Eng 2(3):273–277

European Committee for Standardization (CEN) (2019). EN 15804:2012+A2:2019) Sustainability of construction works—environmental product declarations—Core rules for the product category of construction products, Brussels

Finkbeiner M (2014) Product environmental footprint—breakthrough or breakdown for policy implementation of life cycle assessment? Int J Life Cycle Assess 19(7):266–271

Goldstein B, Gounaridis D, Newell JP (2020) The carbon footprint of household energy use in the United States. Proc Natl Acad Sci USA 117(32):19122–19130

IPCC (2013) Climate change 2013—physical science basis, Geneva

ISO (2006a) ISO 14040:2006—Environmental management—Life cycle assessment—Principles and framework, Geneva

ISO (2006b) ISO 14044:2006—Environmental management—Life cycle assessment—Requirements and guidelines, Geneva

Laurent A, Olsen SI, Hauschild MZ (2010) Carbon footprint as environmental performance indicator for the manufacturing industry. CIRP Ann Manuf Technol 59(1):37–40

SBTi, Science Based Targets initiative (2024) Near-term and net-zero criteria, science based targets initiative. https://sciencebasedtargets.org. Accessed 20 Jun 2025

Together for sustainability (TfS) (2025). https://tfs-initiative.com/. Accessed 20 Jun 2025

WBCSD (2023) Pathfinder framework v1.0: guidance for product-level GHG emissions accounting and exchange, world business Council for Sustainable Development., https://www.wbcsd.org/Programs/Climate-and-Energy/Climate/Reducing-Emissions/Pact. Accessed 20 Jun 2025

Wernet G, Bauer C, Steubing B, Reinhard J, Moreno-Ruiz E, Weidema B (2016) The ecoinvent database version 3 (part I): overview and methodology. Int J Life Cycle Assess 21(7):1218–1230

Würz A, Finkbeiner M (2022) From product carbon footprint to environmental product declaration: exploring the potential of methodological alignment. Int J Life Cycle Assess 27:1302–1316

Zampori L, Pant R (2019) Suggestions for updating the product environmental footprint (PEF) method, EUR 29682 EN. Publications Office of the European Union, Luxembourg

Sustainable Innovation or Merely Reporting? State of the Art of Assessing Scope 3 Emissions of Organizational Procurement

Widiene Essouid, Johannes Zobel, Philippe Loubet, Stéphane Trébucq, Sandra Köhler, Andrea Thorenz, Axel Tuma, and Guido Sonnemann

Abstract Amid growing regulatory pressure and sustainability goals, organizations face challenges in accurately assessing procurement-related carbon emissions. Scope 3 emissions from purchased goods and services are often calculated using either Life Cycle Assessment (LCA) or Environmentally Extended Input–Output (EEIO) analysis with spend-based data. While EEIO offers ease of use, it lacks the granularity needed for effective decarbonization. LCA provides precision but remains resource-intensive and limited by data availability. Through a systematic literature review, this study analyzes 54 case studies to assess the suitability of these methods and their alignment with decarbonization strategies. Results reveal a disconnection between stated goals and chosen methodologies, with many organizations relying on imprecise data despite ambitious targets. Hybrid approaches emerge as a pragmatic compromise, combining scope with feasibility. This contribution invites a shift beyond basic reporting toward harmonized, reliable methods that enable actionable sustainability strategies.

W. Essouid (✉) · P. Loubet · G. Sonnemann (✉)
University of Bordeaux, CNRS, INP, ISM, Talence, France
e-mail: Widiene.essouid@u-bordeaux.fr

G. Sonnemann
e-mail: Guido.sonnemann@u-bordeaux.fr

W. Essouid · S. Trébucq
University of Bordeaux, IAE School of Management - Bordeaux, IRGO, Bordeaux, France

J. Zobel · S. Köhler · A. Thorenz
Resource Lab, Institute of Materials Resource Management, University of Augsburg, Augsburg, Germany

A. Tuma
Chair for Production and Supply Chain Management, University of Augsburg, Augsburg, Germany

© The Author(s) 2026
M. Traverso et al. (eds.), *Life Cycle Management from Global to Local*,
https://doi.org/10.1007/978-3-032-17987-6_27

1 Introduction

Organizations face growing pressure to reduce their environmental footprint, with an increasing focus on indirect emissions from activities in the up- or downstream value chain classified as Scope 3 under the Greenhouse Gas Protocol (WRI and WBCSD 2011). These emissions, particularly from purchased goods and services, often represent around 75% of a company's total carbon footprint across sectors but occur outside of the organization's control. The share of Scope 3 emissions differs depending on the sector, with manufacturing organizations having higher Scope 3 emissions than service-oriented organizations (Alvarez et al. 2014; Blanco et al. 2016). Accurately quantifying procurement-related emissions is thus essential for effective climate strategies.

Regulatory initiatives, such as the European Union's Corporate Sustainability Reporting Directive (CSRD), now require Scope 3 emissions disclosure as a part of broader sustainability reporting (European Commission 2023). However, wide variations in quantification methods complicate comparability and decision-making (Baehr et al. 2024). Common approaches include Life Cycle Assessment (LCA), which provides product-level insights but is resource-intensive (Sonnemann and Margni 2016), and Environmentally Extended Input–Output (EEIO) analysis, which offers broader system coverage but often lacks granularity (Suh 2009).

To address these trade-offs, hybrid methods combining LCA and EEIO or integrating monetary and physical data have emerged as practical solutions (Cimprich and Young 2023; Finogenova et al. 2019). Yet, terminology inconsistencies, misaligned system boundaries, and varying inventory approaches and impact assessment methods remain widespread. As a result, many assessments fall short of supporting meaningful decarbonization, instead of serving primarily as compliance or reporting functions (SBTi 2024).

This paper applies the PRISMA (Preferred Reporting Items for Systematic Reviews and Meta-Analyses) framework (Page et al. 2021) to examine 54 case studies on procurement-related carbon accounting. It analyzes methodological choices, data sources, and goals of each study, evaluating their alignment with effective sustainability strategies. To guide this review, the following research questions are posed:

RQ1: What carbon accounting methodologies are most applied to assess procurement emissions in the scientific literature, and what are their main strengths and limitations?

RQ2: What inconsistencies can be observed across standards and methodologies, and how do these affect the reliability of procurement emission assessments?

RQ3: How well are the employed methods aligned with the stated sustainability goals of the studies, and how suitable are they for supporting actionable decarbonization strategies?

By addressing these questions, this study supports efforts to harmonize procurement emissions accounting as there is, to the best of our knowledge, no comprehensive review regarding procurement emission calculations. This work seeks to support a shift from emission reporting to impact-driven actions.

2 Methodology

This review adheres to the PRISMA framework (Page et al. 2021) through a seven-step protocol to locate, screen, and analyze peer-reviewed studies on procurement-related Scope 3 carbon accounting.

Step 1: Web of Science and Scopus were queried in May 2024 using keywords and synonyms related to "organizational LCA," "EEIO," "carbon accounting," "Scope 3," "procurement," and "supply chain" while no regional limitations were defined, yielding 384 articles published between the years 2009 and 2024; Step 2: Duplicates, non-English or non-peer-reviewed articles were removed; Step 3: Papers not addressing carbon accounting or procurement and reviews were removed; Steps 4—5 entailed a full-text screening. Only articles with explicit methodological detail on procurement-focused carbon accounting were included; Step 6: A backward search in Google Scholar added further relevant case studies; Step 7: The final pool consisted of 44 articles covering 54 case studies.

The methodological analysis entails the categorization of carbon accounting methods as well as data and inventory approaches. In the application analysis, goal alignment, standards, system boundaries, and data sources are analyzed in accordance with the LCA standard. The structured review matrix is provided as supporting material.

3 Results

This section presents the key findings from the 54 selected studies on methods, data, inventories, and the standards used to assess procurement-related carbon emissions.

3.1 Inventory Approaches and Emission Quantification Methods

Procurement emission assessments in the reviewed case studies rely on three main inventory approaches: top-down, bottom-up, and hybrid. In top-down inventories, activity data are collected at the organizational level from aggregated sources such as procurement or accounting systems, utility bills, or fuel records, without allocation to specific products or activities. This approach is well suited to EEIO analysis, which matches monetary spend data with sector-based emission factors to provide economy-wide coverage. While EEIO is a common top-down method, not all top-down approaches rely on monetary data; physical data sources are also widely used. In fact, 34% of top-down studies in our review employed physical activity data. In contrast, bottom-up inventories use physical data, such as weight, volume, or quantity of products purchased, to assess emissions at a more granular level. This approach is usually adopted in LCA-based models and allows for more detailed hotspot analysis and product-level interventions. However, it often requires intensive data collection efforts and is limited by data availability across diverse procurement portfolios (Rimano et al. 2021; Toniolo et al. 2023).

Although, the term hybrid is used variably across studies and can refer to: (1) combining LCA and EEIO models in a single assessment (Finogenova et al. 2019); (2) merging supplier-specific Scope 1 and 2 data with generic upstream emission factors (Aigner 2024; WRI & WBCSD 2011); (3) integrating top-down and bottom-up inventory structures (UNEP 2015); or (4) blending monetary and physical activity data to improve completeness (de Camargo et al. 2019). This diversity highlights the need to clarify what hybridization entails in each context.

Hybrid inventory approaches aim to overcome the limitations of top-down and bottom-up methods by combining their strengths. For example, organizations may apply a top-down model for broad procurement and a bottom-up model for high-impact purchases. Nine case studies adopted such hybrid structures, sometimes integrating different data types to improve accuracy (Finogenova et al. 2019).

Regarding emission quantification, LCA was used in 46.3% of studies, offering robust, product-level insights suitable for supplier engagement and eco-design (Cucchi et al. 2023). However, its application was often limited by data availability and complexity. EEIO, used in 22.2% of cases, enabled broad system coverage and ease of use through national or global input–output tables (Kjaer et al. 2015; Suh 2009), but lacked granularity. Hybrid methods, adopted in 20.4% of the studies, sought to balance detail and scalability by integrating LCA and EEIO methods (Finogenova et al. 2019). Figure 1 illustrates how different inventory structures, data types, and accounting methods interact within an organizational context.

These variations reveal the fragmentation in hybrid modeling practices. Although hybridization helps address the limitations of LCA and EEIO, the absence of clear definitions and guidelines hinders comparability and replication. This highlights the need for standardized frameworks and harmonized reporting to improve transparency and support decision-making in carbon accounting.

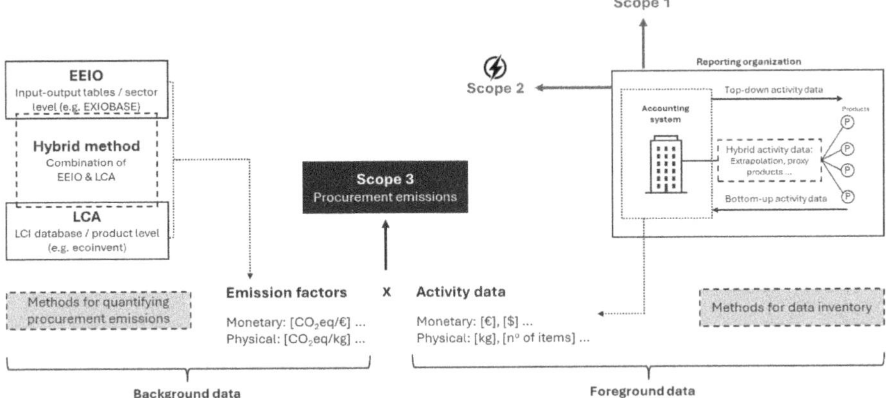

Fig. 1 Sketch of methods for data inventory and procurement emissions quantification

3.2　Carbon Accounting Practices in Case Studies

3.2.1　Classification and Analysis of Case Study Goals

Every study should be guided by a clear goal and scope definition and employ appropriate methods to meet its goals (UNEP 2015). To compare these goals, we used deductive analysis to group individual goals into 11 generic categories defined by the OLCA standard. Most studies pursued multiple goals, though prioritization was unfeasible due to inconsistent descriptions. While 57% focused solely on carbon emissions, the rest applied multi-criteria approaches across impact categories. OLCA categorizes goals as analytical (e.g., quantifying hotspots and reduction opportunities), managerial (e.g., informing procedures, reporting, and marketing), or societal (influencing external stakeholders). The most common goals, hotspot identification, strategic decision support, and impact reduction, spanned all three categories. Related goals include risk assessment, procedure improvement, and performance tracking. Less frequently cited were cost reduction, marketing, supply chain management, and stakeholder motivation (UNEP 2015).

3.2.2　Carbon Accounting Standards and System Boundary Definitions

The use of internationally recognized standards varied considerably across the reviewed case studies. Out of the 54 analyzed cases, 16 explicitly adopted the Organizational Life Cycle Assessment (OLCA) framework (UNEP 2015), 8 referred to the GHG Protocol Standard (WRI and WBCSD 2011), and 3 applied the European Commission's Organizational Environmental Footprint method. These frameworks provide structured guidance on system boundaries, data quality, allocation procedures, and reporting formats.

However, over one-third of the studies did not mention any standard explicitly, and several others only cited them without demonstrating operational integration into their accounting practices. This lack of formal adherence raises concerns about the transparency, especially for studies aimed at informing organizational decisions or supporting regulatory compliance (Baehr et al. 2024).

Differences were also observed in the interpretation of system boundaries. Some studies limited their scope to Tier 1 upstream suppliers, while others applied cradle-to-gate or cradle-to-grave perspectives, sometimes without clearly specifying the cut-off criteria. These inconsistencies further highlight the need for clearer methodological documentation and standardized reporting.

3.2.3 Foreground Data and the Role of Accounting Systems

The assessment of procurement-related emissions begins with collecting foreground data, flows directly controlled or recorded by the organization, such as purchases, quantities, or specifications, and are essential for linking procurement actions to environmental impacts. Their availability, quality, and structure strongly influence method selection and results accuracy (Essouid et al. 2026).

Across the reviewed case studies, limited data availability was a major constraint for applying the bottom-up approach. As a result, most organizations use multiple internal and external data sources, often varying in granularity and format, which created integration challenges and affected assessment robustness (Cucchi et al. 2023).

Accounting and procurement records (e.g., invoices, purchase orders, and stock inventories) were the most commonly used foreground sources, observed in approximately one-third of the cases. These records supported both monetary and physical data types (Fig. 1). Other sources included internal documents (e.g., technical sheets), supplier data, surveys, literature proxies, and on-site measurements. Enterprise systems such as ERP and MRP often helped consolidate procurement data and support automated classification across multiple sites. Yet, data silos between environmental and procurement systems limited Scope 3 resolution and detailed modeling.

In summary, foreground data availability and data types, especially from accounting systems, shaped methodological choices. These findings highlight the strategic role of digital procurement infrastructures and the integration of environmental data into purchasing workflows to enable scalable and reliable carbon accounting.

3.2.4 Emission Factors and Background Data Sources

Background data represent upstream processes and emissions outside the organization's direct control and are essential for quantifying embedded impacts across the supply chain. Typically sourced from standardized environmental databases, they are used to assign emission factors especially when supplier-specific data are unavailable. Two main types of emission factors (Fig. 1) were identified in the reviewed studies: monetary emission factors (e.g., kg CO_2eq/€) or physical emission factors (e.g., kg CO_2eq/quantity (i.e., kg, kWh, L…)).

Monetary emission factors, commonly used in top-down and EEIO-based approaches, were derived from databases such as EXIOBASE, USEEIO, and CEDA. These provide wide coverage by linking economic flows with environmental extensions, but rely on sector averages that may not reflect the product- or supplier-specific emissions (Suh 2009). Their accuracy depends on database granularity (e.g., EXIOBASE provides EU country-specific data, while USEEIO is U.S.-focused), affecting interpretability and reliability of results (Keil 2023; Kjaer et al. 2015).

Physical emission factors, mostly used in bottom-up models, were sourced from LCI databases like ecoinvent or GaBi. They enable detailed modeling linked to specific materials and processes (Cucchi et al. 2023), and are better suited for high-impact purchases. However, they require precise foreground data and involve greater modeling effort, limiting their use in broad organizational assessments. Several hybrid studies combined supplier-specific or process-level data (e.g., energy use, transport distances) with background emission factors from standardized databases to enhance specificity. However, many lacked documentation on database versioning, system boundaries, allocation rules, and regional matching (Finogenova et al. 2019).

Overall, while background data are essential, the studies revealed inconsistencies in selection, reporting, and integration. Assumptions about data age, region, and representativeness were often missing, hindering comparability and interpretation.

These findings underscore the need for transparent reporting, alignment between foreground and background data, and standardized criteria for database selection. Developing region- and industry-specific datasets, for both monetary and physical emission factors, is the key to improving accuracy and relevance in procurement carbon accounting.

3.2.5 Impact Assessment Methods

The selection of impact assessment methods is critical for interpreting environmental results; however, reporting varied widely across the reviewed studies. Only about half of the cases using physical data (typically in LCA or hybrid models) clearly specified the impact assessment method applied; others omitted or partially reported this information. In contrast, studies based solely on monetary data (e.g., EEIO) often inherited predefined methods from the databases used, limiting transparency and methodological control. The IPCC Global Warming Potential was the most frequently used method (32%), particularly for mono-criteria assessments, together with ReCiPe. Other common methods included CML (15%), TRACI (5%), EF (5%), and IMPACT 2002 + (5%), though these were cited less frequently. Although method choice was sometimes category-specific, methodological justification was rarely discussed, underscoring the need for more consistent and transparent impact method selection.

4 Discussion

This review synthesizes methods for quantifying Scope 3 procurement emissions, highlighting key inconsistencies and trade-offs across 54 studies and proposing a framework for methodological harmonization.

4.1 *Inconsistencies in Procurement Emission Calculations*

Significant inconsistencies were observed across the reviewed case studies, affecting data quality and transparency. Many did not clearly document system boundaries, inventory structure, impact assessment methods, or emission factor sources. Some procurement categories were excluded without justification, particularly in GHG Protocol-based approaches. Variability also arose from differing treatments of monetary data (e.g., basic prices versus purchase) and inconsistent classification of purchased electricity (e.g., Scope 2 versus Scope 3).

Foreground data gaps were common. Limited access to supplier-specific data led to reliance on secondary sources or assumptions. Emission factors ranged from broad EEIO averages to product-specific LCA values, contributing to inconsistent results. Methodologies varied significantly: EEIO offered coverage but lacked granularity; LCA was precise but data-intensive; hybrid methods aimed to balance the two but often lacked integration clarity (Finogenova et al. 2019; Kjaer et al. 2015; Suh 2009). Use of outdated factors and missing uncertainty assessments further weakened robustness.

Differences in standard interpretation also played a role. "Hybrid" was defined inconsistently, ranging from mixed data types to combined inventory structures. OLCA emphasized cradle-to-gate or cradle-to-grave flows, while GHG Protocol studies often used activity-based groupings with limited product-level detail. These divergences complicated benchmarking, particularly in the service sector where downstream impacts are frequently excluded.

4.2 *Methodological Trade-Offs and Goal Alignment*

A key insight from the reviewed case studies is the frequent mismatch between the goals stated by organizations (e.g., hotspot identification, supplier engagement, or emissions reduction) and the data or methods actually used. Although many studies pursued ambitious decarbonization objectives, they often relied on top-down monetary data, which lacks the resolution necessary for actionable insights. Conversely, bottom-up physical data, better suited for process-level interventions, was underused due to demanding data requirements. Hybrid inventories that combined data types offered a practical balance but were inconsistently applied and often under-documented. As shown in Fig. 2, studies frequently used top-down monetary data even for detailed goals, while more appropriate bottom-up or hybrid approaches were applied less systematically. Overall, methodological choices appeared more influenced by data availability than by alignment with stated sustainability objectives.

4.3 Pathways for Harmonization

To enhance consistency and strategic relevance in procurement carbon accounting, a harmonized framework is needed. Figure 3 proposes a structured integration of organizational and product-level perspectives, adapted from Aigner (2024). It distinguishes between three categories of emissions: core activities (direct product-related operations), supporting activities (indirect but attributable emissions such as packaging or logistics), and overhead activities (non-product-specific emissions like employee commuting or investments). Core and supporting activities are typically product-related and offer direct mitigation potential through emission assessments with physical data or bottom-up inventories. Overhead activities are less controllable and primarily relevant for transparency and screening purposes, where less accurate top-down and monetary emission calculations suffice. This structured differentiation helps organizations identify active emissions that are manageable and reducible, as opposed to those that are passively disclosed. It also facilitates better integration of environmental and procurement systems, supports clearer data governance, and aligns reporting structures with operational realities. By promoting a shared logic between Scope 3 activity categories and product-oriented emission flows, this approach advances the strategic use of carbon data, shifting procurement accounting from a compliance tool to a platform for decarbonization action.

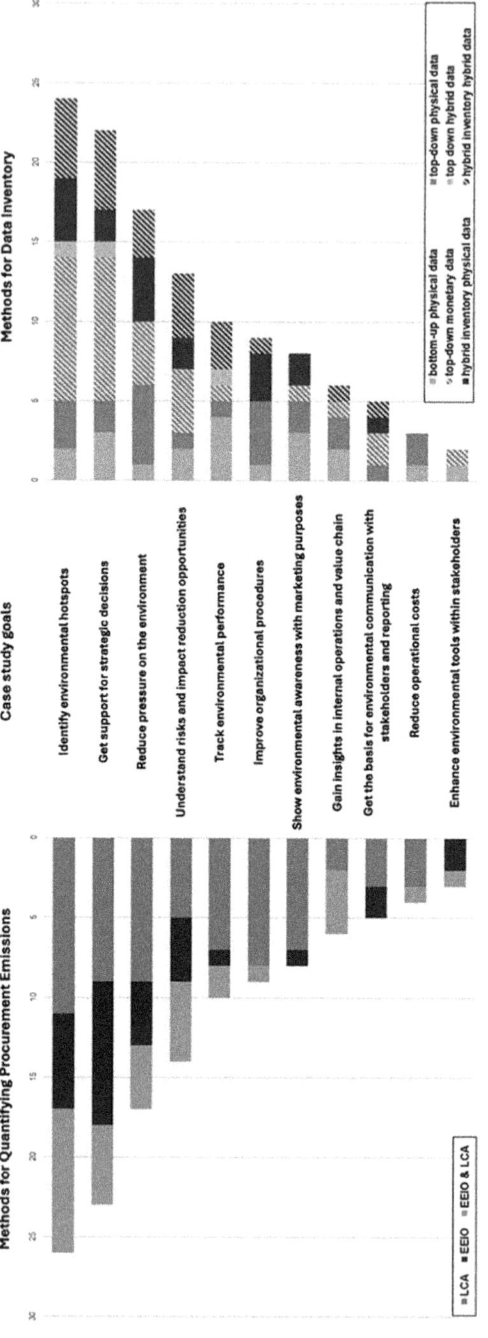

Fig. 2 Alignment of goals, inventories, and emission quantification methods

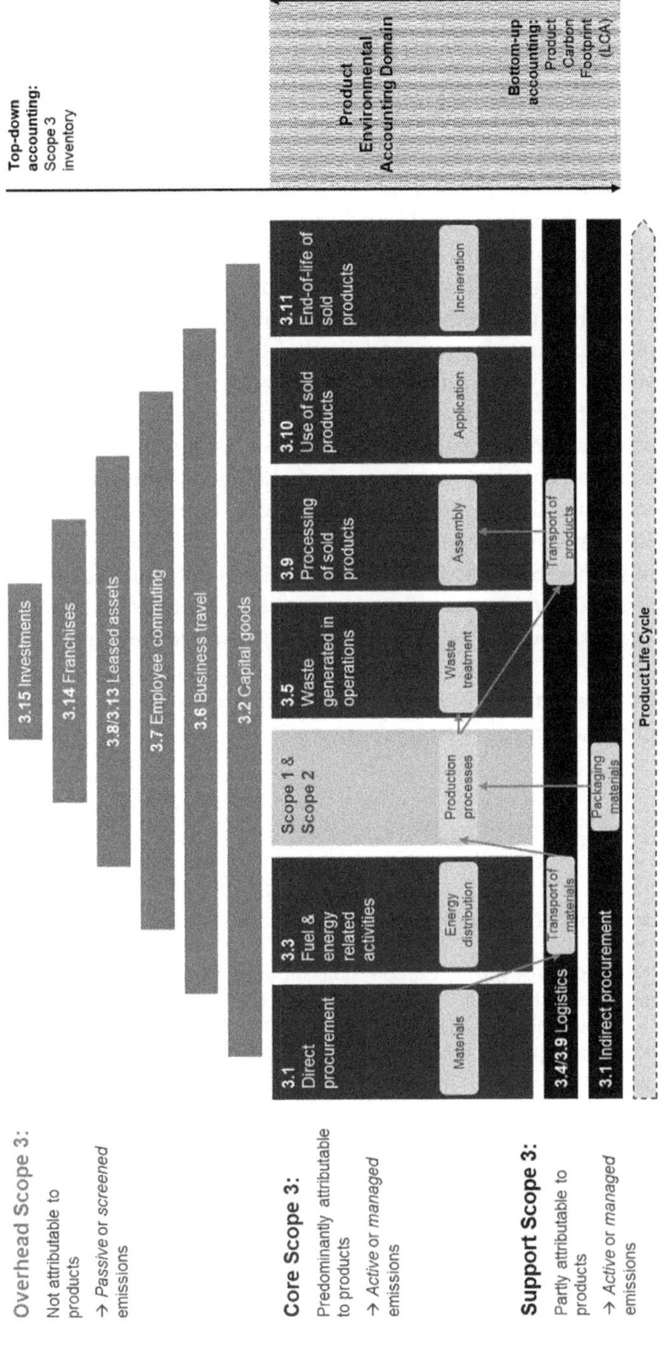

Fig. 3 Alignment of organizational and product-based accounting (Adapted from Aigner 2024)

5 Conclusion

This review examined current methodologies for accounting Scope 3 procurement emissions, with a focus on LCA, EEIO, and hybrid approaches. LCA provides detailed, product-level insights but is often constrained by data availability and modeling complexity. EEIO methods offer broader system-level coverage and scalability but lack the granularity needed for supplier- or product-specific decision-making. Hybrid approaches attempt to balance these trade-offs by integrating monetary and physical data or combining top-down and bottom-up inventories. However, their application remains inconsistent, and their effectiveness in supporting decarbonization efforts is not yet well demonstrated.

The findings highlight persistent challenges related to supplier data availability, reliance on generic emission factors, and limited transparency in methodological choices. Many studies showed a misalignment between their goals and the methods applied, often driven more by data availability than by strategic intent. Integrating economic performance metrics, such as Life Cycle Costing (LCC), could enhance the decision-making potential of procurement carbon accounting. Geographic and sectoral representation in the literature remains uneven, limiting the generalizability of current insights. Existing standards, including the GHG Protocol, should be re-evaluated in light of methodological developments and contextual needs.

To improve accuracy and impact, future efforts should focus on enhancing data quality, promoting open-access databases, and leveraging technologies such as AI and real-time monitoring. Harmonizing standards and encouraging organizational engagement beyond reporting will be key to unlocking the full potential of Scope 3 carbon accounting as a strategic sustainability tool.

References

Aigner J (2024) Workshop 03: Navigating methodological inconsistencies in Scope 3 accounting & LCA, FSLCI e.V

Alvarez S, Blanquer M, Rubio A (2014) Carbon footprint using the compound method based on financial accounts. Case Sch for Eng, Tech Univ Madr, J Clean Prod 66:224–232

Baehr J, Zenglein F, Sonnemann G, Lederer M, Schebek L (2024) Back in the driver's seat: how new EU greenhouse-gas reporting schemes challenge corporate accounting. Sustain 16:9

Blanco C, Caro F, Corbett CJ (2016) The state of supply chain carbon footprinting: Analysis of CDP disclosures by US firms. J Clean Prod 135:1189–1197

Cimprich A, Young SB (2023) Environmental footprinting of hospitals: Organizational life cycle assessment of a Canadian hospital. J Ind Ecol 27:5

Cucchi M, Volpi L, Ferrari AM, García-Muiña FE, Settembre-Blundo D (2023) Industry 4.0 real-world testing of dynamic organizational life cycle assessment (O-LCA) of a ceramic tile manufacturer, Environmental Science and Pollution Research, 30, 60, 124546–124565

de Camargo AM, Forin S, Macedo K, Finkbeiner M, Martínez-Blanco J (2019) The implementation of organizational LCA to internally manage the environmental impacts of a broad product portfolio: An example for a cosmetics, fragrances, and toiletry provider. Int J Life Cycle Assess 24:1

Essouid W, Salem H, Trébucq S, Loubet P, Sonnemann G (2026) Improving scope 3 procurement emissions accounting: key challenges and a strategic approach using MICMAC and the sustainability balanced scorecard. Corp Soc Responsib Environ Manage. https://doi.org/10.1002/csr.70380

European Commission, The European Green Deal (2019)

European Commission, Corporate sustainability reporting, (2023)

Finogenova N, Bach V, Berger M, Finkbeiner M (2019) Hybrid approach for the evaluation of organizational indirect impacts (AVOID): Combining product-related, process-based, and monetary-based methods. Int J Life Cycle Assess 24(6):1058–1074

Forin S, Martínez-Blanco J, Finkbeiner M (2019) Facts and figures from road testing the guidance on organizational life cycle assessment. Int J Life Cycle Assess 24(5):866–880

Herth A, Blok K (2023) Quantifying universities' direct and indirect carbon emissions – the case of Delft University of Technology. Int J Sustain High Educ 24(9):21–52

ISO, 2006, ISO 14040, Environmental Management—Life Cycle Assessment—Principles and Framework, International Organization for Standardization.

Keil M (2023) The greenhouse gas emissions of a German hospital—A case study of an easy-to-use approach based on financial data. Cleaner Environmental Systems 11:100140

Kjaer LL, Høst-Madsen NK, Schmidt JH, McAloone TC (2015) Application of environmental input-output analysis for corporate and product environmental footprints—learnings from three cases. Sustain 7:9

Nordman M, Lassen AD, Christensen LM, Trolle E (2024) Tracking progress toward a climate-friendly public food service strategy: Assessing nutritional quality and carbon footprint changes in childcare centers. Nutr J 23(1):13

Page MJ, McKenzie JE, Bossuyt PM, Boutron I, Hoffmann, Moher D (2021) The PRISMA 2020 statement: An updated guideline for reporting systematic reviews, BMJ, n71

Rimano M, Simboli A, Taddeo R, Del Grosso M, Raggi A (2021) The environmental impact of organizations: a pilot test from the packaging industry based on organizational life cycle assessment. Sustain 13:20

SBTi, Aligning Corporate Value Chains to Global Climate Goals, (2024)

Sonnemann G, Margni M (2016) Life Cycle Management, Springer

Suh S (2009) Handbook of Input-Output Economics in Industrial Ecology, 23, Springer

Toniolo S, Marson A, Fedele A (2023) Combining organizational and product life cycle perspective to explore the environmental benefits of steel slag recovery practices. Sci Total Environ 867:161440

UNEP (2015) Guidance on organizational Life Cycle Assessment

WRI and WBCSD, Corporate Value Chain (Scope 3) Accounting and Reporting Standard (2011)

User Studies on Visualisation and Reporting of Life Cycle Sustainability Performance of Products

Olubukola Olumuyiwa Tokede, Anastasia Globa, and Winter Jesse

Abstract Visualisation of Life Cycle Sustainability Assessment (LCSA) results is crucial in communicating sustainability preferences to various stakeholders and policymakers. Augmented reality (AR) and virtual reality (VR) offer a more intuitive means to interpret the life cycle sustainability performance of products and projects and constitute an invaluable approach to engaging users with no prior scientific knowledge. User studies are vital in appraising and assessing technological interventions to improve usability, engagement and functionality of novel tools. Currently, there have been limited user studies that have been deployed in developing the vast array of tools used in life cycle sustainability reporting. This paper undertakes comprehensive user studies of the LCSA Tracker—a novel AR tool used in visualising life cycle sustainability outcomes. The user studies comprise questionnaire surveys with different user groups (students and professionals). It was found that the LCSA Tracker performs well in usability, engagement and interactiveness. However, the user studies identified opportunities for development with regard to data format, compatibility with existing platforms and data availability. Future work will seek to enhance user experience by addressing the issues identified through the user studies and will seek to build a community of practice for the LCSA Tracker.

O. O. Tokede (✉)
School of Architecture and Built Environment, Geelong Waterfront Campus, Deakin University, Geelong, Australia
e-mail: olubukola.tokede@deakin.edu.au

A. Globa
Sydney School of Architecture, Design and Planning, The University of Sydney, Sydney, Australia

W. Jesse
Institute of Social Neuroscience, Ivanhoe, Australia

© The Author(s) 2026

M. Traverso et al. (eds.), *Life Cycle Management from Global to Local*,
https://doi.org/10.1007/978-3-032-17987-6_28

1 Introduction

Advances in reporting and visualisation of Life Cycle Sustainability Assessment (LCSA) results have enhanced the effectiveness in the communication of LCSA outcomes (Traverso et al. 2012a; Backes et al. 2023; Ashby 2024) and subsequently supported the opportunities for more sustainable decision-making (Apellániz et al. 2024; Tokede and Globa 2025). Current visualisation methods using Microsoft Excel® or commercial LCA software, however, tend to be limited to simplistic graphs and sometimes fail to provide infographics that accurately depict LCSA results (Tokede and Globa 2025). The LCSA triangle developed by Finkbeiner et al. (2010) aims to define the weights of three aspects of sustainability while the LCSA dashboard developed by Traverso et al. (2012b) provides visualised results with different colour scales to ensure comparability across the different dimensions of sustainability. More recently, the development of the LCSA Wheel by Backes et al. (2023) has compared the performances of existing LCSA tools but still fails to provide a more dynamic platform for interacting with LCSA outcomes and data. Hollberg et al. (2021) concluded that LCSA visualisation through interactive dashboards and immersive technologies have immense potential in facilitating the interpretation of LCSA results and achieving collaborative outcomes.

Despite the best intentions of LCSA practitioners, many LCSA reporting and visualisation templates use static platforms (Hollberg et al. 2021; Tokede and Globa 2025), such that users can only assimilate the results without recognising opportunities for improvements. Furthermore, there is often a lack of interaction and effective communication between different stakeholders involved in the complex calculations of LCSA, thereby resulting in a 'black-box' syndrome and, therefore, experiencing limited user engagement (Backes et al. 2023; Tokede and Globa 2025). Even when results are understandable, it can be hard to achieve robust interpretations without benchmarking with alternative options, as users are often overwhelmed just by the abundance of graphs, indicators and outcomes (Apellániz et al. 2024; Ashby 2024). In addition, given the different levels of development in LCSA dimensions and different data formats, there is often a lack of scalable functionality in many LCSA reporting and visualisation templates.

The absence of appropriate software has been a frontline issue across many LCSA studies. Software issues are multifaceted and could include the level of technology development (Kühnen and Hahn 2018), availability of cutting-edge tools and technologies and costs of maintenance and implementation (Pesonen and Horn 2013), to mention a few. Technology development has often been considered in many sustainability settings, based on the level of participation (Ramirez et al. 2016); potential for technology transfer (Ren et al. 2015) and lack of skilled resources (Lehmann et al. 2013). Nevertheless, societal conditions impact the novel technologies (van Haaster et al. 2017). Sani et al. (2023), therefore, concluded that there is a missing link between technological enhancements and economic barriers.

This paper reports on the development of the LCSA Tracker application using an interactive digital technology and conducts user studies using questionnaire survey to ascertain its effectiveness in engaging practitioners and non-experts. Three key processes are adopted. Firstly, an appraisal of existing tools is conducted to critique their functionality and applicability for LCSA reporting and visualisation. Secondly, an augmented reality (AR) tool is developed to support the engagement of relevant stakeholders, and thirdly, an appraisal of stakeholder opinions on the LCSA Tracker is conducted using a questionnaire survey. Finally, the discussion emphasises the learnings from the multiple layers of research data to support future development of the LCSA Tracker and in developing a community of practice to adopt its usage.

2   Methodology

An interactive digital data visualisation tool was developed as a medium to facilitate easy-to-use communication of LCSA metrics and results for a wide range of possible stakeholders (professionals and students). The initial iteration of the LCSA Tracker tool, as reported by Tokede and Globa (2025), was developed into a mobile application using AR technology to enable intuitive and engaging user-interactions, following the process depicted in Fig. 1. The application has three main functionalities—tracking, mapping and geolocation. This implies that users can interpret life cycle sustainability performance by (i) assessing an overall impact for each dimension of sustainability (economic, environmental and social); (ii) selecting and reporting on individual metrics for each dimension of sustainability and (iii) tracking the performance of individual sustainability metrics based on the location. The application was developed for different hardware outputs including desktop, web browsers and mobile devices. The mobile version of the application uses AR to allow users to overlay real, physical/printed world maps and objects with a layer of augmentation—digital projections, enabling dynamic and engaging data exploration and visualisation. The functionality of the application allows users to upload and visualise custom data files in 'csv.' format and to manipulate parameters dynamically in real-time.

The functionality, interactions and design of the App was informed by several existing tools that have been developed to communicate LCSA results. Figure 1 provides a detailed overview of the procedural methodology for the study showcasing the problem definition, App development and user-testing. The initial phase consisted in an appraisal of the existing tools and techniques and the work of Backes et al. (2023) provided an informative account of the development of the LCSA visualisation tools. At the second phase, we decided to identify relevant and streamlined data, and to benchmark data to achieve comparative quantitative numbers across the dimensions of sustainability. To this end, systematic data from Dong et al. (2021) was utilised for the Life Cycle Assessment (LCA) of buildings, while similar literature sources were used for Life Cycle Costing (LCC) and Social Life Cycle Assessment (S-LCA) in step 1, following procedures described in Tokede and Globa (2025). Step 2 of the second phase required normalising the results based on the best and worst scenarios, across the 9-colour scale, similar to the work of Traverso et al. (2012b). Step 3 involved the

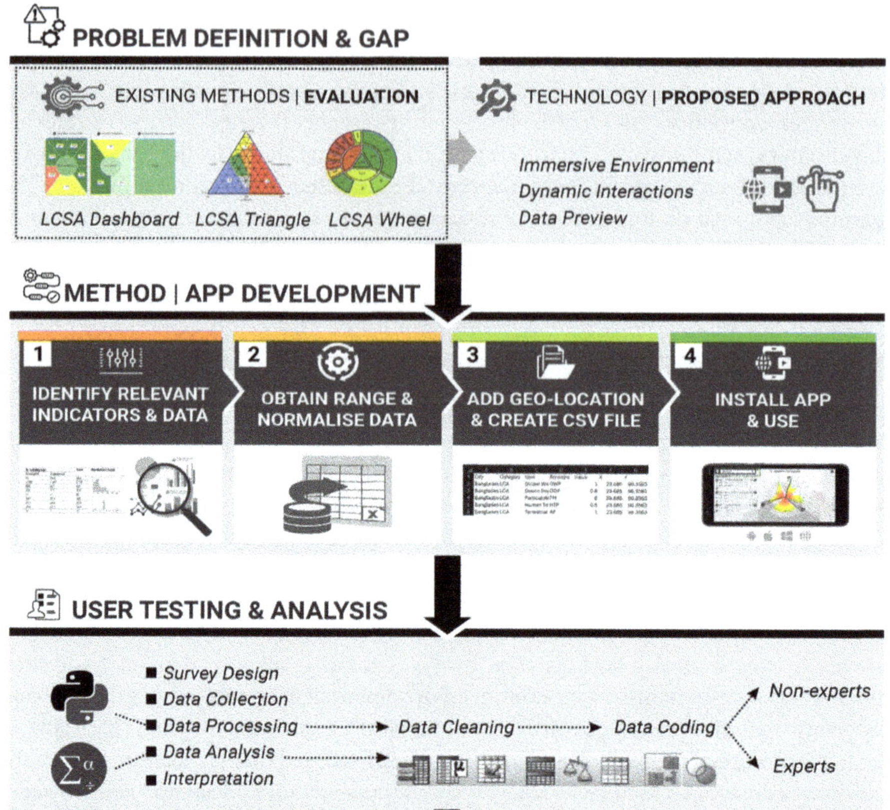

Fig. 1 Tool development and LCSA tracker

assembling of relevant geolocation data (i.e. GPS coordinates) to ensure that the tool was able to overlay the values on the LCSA data on a physical location. This was intended to enhance the relevance and relatability of the results. Finally, we developed a mobile application using a game engine (Unity) to facilitate the opportunities for users to interact with the LCSA Tracker, on their personal devices. Step 4 involves the process to install the app and to execute its usage.

This interactive digital technology is a new approach involving the use of augmented reality techniques. Mobile applications have the advantage of being updated to incorporate new features, hence future iterations of the application can be made easily available. The application was tested and holistically evaluated through a series of experimental user studies. During the experiment, users learned the basics of user-interface of the application and were provided with the sample LCSA datasets that they were asked to explore and complete several step-by-step tasks. After their experience the users were provided with the link to the online survey that consisted of the following components:

(1) User background information (experience, age, gender, occupation and cultural background).
(2) Usefulness and functionality.
(3) Application content and data metrics evaluation.
(4) User interactions.
(5) System usability scale (Baumgartner et al. 2021).
(6) Immersion, engagement and presence.
(7) Opportunities and limitations.

Currently, there is no validated measure to assess user engagement with AR tools within the LCSA field. The I-group Presence Questionnaire (IPQ) is a 14-item Likert scale that has been used in previous studies and translated across five languages; however, it often focuses on virtual environments rather than AR. As a result, several questions pertain specifically to fully immersive realities, such as virtual reality (VR) and are not relevant to AR. Several items pertaining to the presence were used from the IPQ and were adjusted to reference AR over VR (Bareišytė et al. 2024).

3 Results

This questionnaire was designed as an indicative cross-sectional study. An in-person questionnaire survey was conducted between August and December 2024, at locations in Delhi, Chennai and Mumbai (India) as well as Ho Chi Minh City (Vietnam). All respondents used either a mobile phone or a smart device for testing and responding to the questionnaire survey. The questionnaire consisted of two parts. The first part described items related to the respondent's demographic and individual profile. The second part included four items for the engagement construct. Overall, 246 responses were received. Data was retrieved from Qualtrics and then imported and analysed in jamovi 2.6. In total, 216 complete and valid responses were obtained for both countries comprising 175 for India and 41 for Vietnam. Table 1 provides the demographic profiles of the samples including whether they have had experience with a sustainability tool and the duration over which the survey was completed.

It was found that the respondents remained engaged whether or not they had previous experience with this style of tool. Prior to the analyses, participants were screened to ensure sufficient and comparable information had been provided. The focus was restricted to Vietnam and India, given that other countries were not significantly represented in the data collected so far. The data was checked for accuracy to ensure errors were not made in data entry or coding. Participants who had incomplete survey responses were excluded, either due to completing demographics but no other measures, or due to ambiguous responses received. After data cleaning and assumption testing using standard post hoc procedures, there were 176 analysed responses that were employed for the t-tests. The hypothesis was tested using independent samples t-tests along with a confirmatory factor analysis. It was found that gender, ethnicity and occupation were not limiting factors to user engagement. Significant

Table 1 Demographic characteristics of respondents

Characteristics	India	Vietnam
Occupation		
Student	165 (94.3)	34 (82.9)
Professional	10 (5.7)	7 (17.1)
Gender		
Male	139 (79.4)	23 (56.1)
Female	33 (18.9)	18 (43.9)
Prefer not to say/non-binary	3 (1.7)	0
Experience using sustainability visualisation tool		
Yes	22 (12.6)	9 (22.0)
No	153 (87.4)	32 (78.0)
Engagement (Duration)		
< 2 min	8 (4.6)	0 (0.0)
2 min–5 min	33 (18.9)	2 (4.9)
> 5 min	98 (56.0)	37 (90.2)
Incomplete	36 (20.6)	2 (4.9)
Total	175	41

[*] Numbers in brackets represent the percentages of the total group

effects are supported by the absence of zero within the confidence intervals (Hayes and Scharkow 2013), or with a p-value of less than 0.05.

The LCSA Tracker user's engagement was assessed through four items on a Likert scale of 1 to 7, where 1 was 'not at all' and 7 was 'very much'. The four items pertained to engaging visualisation (The LCSA visualisation tool interactions involved/engaged me.), dynamic responsiveness (The sense of dynamic responsive interactions of the visualisation was compelling), involving interactions (I was involved in the visualisation interactions), and enjoyment experience (I enjoyed using this tool). Table 2 shows the high level of self-reported engagement with the LCSA Tracker from the respondents.

These engagement items were adapted from the unified UX in IVE questionnaire (IVE) where the engagement subscale of 3 items has a Cronbach's alpha: $\alpha = 0.759$ (Tcha-Tokey et al. 2016). This questionnaire again was VR-focused, further

Table 2 Summary of responses on engagement questions

Constructs	Mean (out of 7)	Std dev	Variance
Engaging visualisation	5.48	1.31	1.73
Dynamic responsiveness	5.49	1.42	2.01
Involving interactions	5.63	1.50	2.26
Enjoyable experience	5.65	1.52	2.30

highlighting a gap in the field. However, the authors also concluded that it was 'a non-definitive tool and still needs adjustments' (Tcha-Tokey et al. 2016). Thus, several items were adopted with wording adjusted from VR terminology to enhance face validity for the AR audience. Moreover, the authors suggested using this questionnaire as a guide when assessing prototypes within the technological field (Tcha-Tokey et al. 2016).

4 Discussion

Previous work by Backes et al. (2023) has assessed existing LCSA visualisation tools on the basis of expressivity, effectiveness and appropriateness. The LCSA Wheel succeeds in improved expressivity and appropriateness, but it is still doubtful in its effectiveness in reaching a broad range of stakeholders, as no published user studies have been conducted on any of the existing LCSA tools. The findings from the survey in this work revealed that technology can be an equalising mechanism for understanding and interacting with data on LCSA in India and Vietnam, where there is currently little awareness of LCSA. Besides, the opportunity to achieve shared understanding of LCSA outcomes (Apellániz et al. 2024) the LCSA Tracker provides an intuitive approach to assimilating complex information. The LCSA Tracker was found to be advantageous in providing real-time visualisation and offering an immersive and participatory experience for users. A recent study by Apellániz et al. (2024) found that results from a 3D intelligence dashboard accounted for between 5–35% of preference over traditional approach. The LCSA Tracker was perceived positively across multiple User experience indicators—engagement, responsiveness, inclusivity, and enjoyment. Another recent survey conducted by Cappelletti et al. (2023) involving 102 respondents in Europe on LCA revealed that there was still a limited level of awareness on LCA, with 28% recognising that economic barriers are significant in implementing relevant tools. Another survey of 71 respondents conducted by Backes et al. (2021) on LCSA application revealed that 14% believed that the lack of suitable software was a challenge in engaging with LCSA. Consistent with the studies by Toyin et al. (2025), the use of AR has also been found to foster the development of spatial awareness, problem-solving and attention to details. Hence, the LCSA Tracker consolidated information awareness and decision support facilitated by LCSA data and digital representation. It is also remarkable that previous user studies on life cycle visualisation tools have had limited engagement represented by the number of surveys (Backes and Traverso 2023; Ashby 2024), and hence, user-feedback has been notably absent in developing a robust LCSA tool.

One dominant challenge with the LCSA Tracker lies in automating data access to support LCSA (Fnais et al. 2022). Artificial intelligence (AI) and machine learning techniques could offer valuable potential for processing and collecting data (Hollberg et al. 2021) needed for the LCSA Tracker. AI can be used to simulate data from multiple sources that can help LCSA professionals to compare different decision matrices (Chofreh et al., 2021). The process of collecting these LCSA impact data is capital-intensive, and a large portion of the data is confidential to the manufacturer due to intellectual property and competition issues. Tools like EC3 (Embodied Carbon in Construction Calculator) and One Click LCA EPD Generator have also been widely used to enhance EPD and PCR development as they aid in the seamless data collection and analysis process and significantly reduce time and effort. Other technologies, such as enterprise resource systems (ERP), the Internet of Things (IoT) and blockchain (Chofreh et al. 2020; Ferrari et al. 2021) have also been proposed as valuable technologies but there have been little or no practical implementations in published case studies. Particularly, the area of AI integration has seldom been researched (Fnais et al. 2022), and this can be easily integrated with the LCSA Tracker to facilitate interactive and collaborative decision-making in several contexts.

5 Conclusion and Recommendation

This research reports on user studies using questionnaire surveys on an interactive digital technology (LCSA Tracker) based on augmented reality (AR). The LCSA Tracker helps to report and visualise LCSA results to support decision-making and improve stakeholder engagement. The LCSA Tracker has three main functionalities—tracking, mapping and geolocation intended to respectively (i) appraise an overall impact for each dimension of sustainability (economic, environmental and social); (ii) select and report on individual metrics for each dimension of sustainability and (iii) track the performance of individual sustainability metrics based on the location. Data from 176 valid questionnaires were analysed using descriptive statistics, independent samples t-tests and confirmatory factor analyses. Results from the user studies reveal that the LCSA Tracker is an engaging and useful tool for LCSA audiences. The results also indicate that demographic factors such as occupation, gender and ethnicity did not constitute limiting factors to user engagement. The engagement constructs of engaging visualisation, dynamic responsiveness, involving interactions and enjoyable experience were valid indicators that contribute meaningfully to the overall engagement measure. Hence future studies will also involve qualitative information and well-validated questionnaires to test the LCSA Tracker in multiple contexts.

References

Apellániz D, Alkewitz T, Gengnagel C (2024) Visualisation of building life cycle assessment results using 3D business intelligence dashboards. Int J Life Cycle Assess 29(7):1303–1314

Ashby M (2024) LCSA: how to carry it out and integrate and communicate its results – illustration with four case studies, Handbook on Life Cycle Sustainability Assessment, 1st ed, Edward Elgar Publishing

Backes J, D'Amico A, Pauliks N, Guarino S, Traverso M, Brano VL (2021) Life cycle sustainability assessment of a dish-stirling concentrating solar power plant in the Mediterranean area. Sustain Energy Technol Assess 47:101444

Backes JG, Steinberg LS, Weniger A, Traverso M (2023) Visualization and interpretation of life cycle sustainability assessment—existing tools and future development. Sustain 15(13):10658

Bareišytė L, Slatman S, Austin J, Rosema M, van Sintemaartensdijk I, Watson S, Bode C (2024) Questionnaires for evaluating virtual reality: A systematic scoping review. Comput Hum Behav Rep 16:100505. https://doi.org/10.1016/j.chbr.2024.100505, (Accessed03.02.2025)

Baumgartner J, Ruettgers N, Hasler A, Sonderegger A, Sauer, J (2021) Questionnaire experience and the hybrid system usability scale: using a novel concept to evaluate a new instrument. Int J Hum Comput Stud 147:102575. https://doi.org/10.1016/j.ijhcs.2020.102575

Cappelletti F, Roberto M, Marta R, Michele G (2023) Comparison between LCA results and consumers-perceived environmental sustainability of three swimming products. Int J Interact des Manuf (IJIDeM) 17(4):1905–1932

Chofreh A, Goni F, Klemeš J, Malik M, Khan H (2020) Development of guidelines for the implementation of sustainable enterprise resource planning systems. J Clean Prod 244:118655

Dong Y, Ng ST, Liu P (2021) A comprehensive analysis towards benchmarking of life cycle assessment of buildings based on systematic review. Build Environ 204:108162. https://doi.org/10.1016/j.buildenv.2021.108162, (Accessed04.02.2025)

Ferrari A, Volpi L, Settembre-Blundo D, García-Muiña F (2021) Dynamic life cycle assessment (LCA) integrating life cycle inventory (LCI) and enterprise resource planning (ERP) in an industry 4.0 environment, J Clean Prod, 286, 125314

Finkbeiner M, Schau E, Lehmann A, Traverso M (2010) Towards life cycle sustainability assessment. Sustain 2(10):3309–3322

Fnais A, Rezgui Y, Petri I, Beach T, Yeung J, Ghoroghi A, Kubicki S (2022) The application of life cycle assessment in buildings: Challenges, and directions for future research. Int J Life Cycle Assess 27(5):627–654

Hayes A, Scharkow M (2013) The relative trustworthiness of inferential tests of the indirect effect in statistical mediation analysis: Does method really matter? Psychol Sci 24:1918–1927. https://doi.org/10.1177/0956797613480187, (Accessed03.01.2025)

Hollberg A, Kiss B, Röck M, Soust-Verdaguer B, Wiberg AH, Lasvaux S, Galimshina A, Habert G (2021) Review of visualising LCA results in the design process of buildings. Build Environ 190:107530

Kühnen M, Hahn R (2018) From SLCA to positive sustainability performance measurement: A two-tier Delphi study. J Ind Ecol 23(19):615–634

Lehmann A, Zschieschang E, Traverso M, Finkbeiner M, Schebek L (2013) Social aspects for sustainability assessment of technologies—challenges for social life cycle assessment (SLCA). Int J Life Cycle Assess 18(8):1581–1592

Pesonen H, Horn S (2013) Evaluating the sustainability SWOT as a streamlined tool for life cycle sustainability assessment. Int J Life Cycle Assess 18(9):1780–1792

Ramirez P, Petti L, Brones F, Ugaya C (2016) Subcategory assessment method for social life cycle assessment. Part 2: Application in Natura's cocoa soap, The International Journal of Life Cycle Assessment, 21(1), 106–117

Ren J, Manzardo A, Mazzi A, Zuliani F, Scipioni A (2015) Prioritization of bioethanol production pathways in China based on life cycle sustainability assessment and multicriteria decision-making. Int J Life Cycle Assess 20(6):842–853

Sani M, Amin M, Siddique A, Nasif S, Ghaley B, Ge L, Wang F, Yong J (2023) Waste-derived nanobiochar: A new avenue towards sustainable agriculture, environment, and circular bioeconomy. Sci Total Environ 905:166881

Tcha-Tokey K, Christmann O, Loup-Escande E, Richir S (2016) Proposition and validation of a questionnaire to measure the user experience in immersive virtual environments. Int J Virtual Rity 16(1):33–48

Tokede O, Globa A (2025) Life cycle sustainability tracker: A dynamic approach. Eng Constr Archit Manag 32(5):2971–3006

Tokede O, Roetzel A, Ruge G (2021) A holistic life cycle sustainability evaluation of a building project. Sustain Cities Soc 73:103107

Toyin J, Sattineni A, Wetzel E, Fasoyinu A, Kim J (2025) Augmented reality in US construction: Trends and future directions. Autom Constr 170:105895

Traverso M, Asdrubali F, Francia A, Finkbeiner M (2012a) Towards life cycle sustainability assessment: An implementation to photovoltaic modules. Int J Life Cycle Assess 17:1068–1079

Traverso M, Finkbeiner M, Jørgensen A, Schneider L (2012b) Life cycle sustainability dashboard. J Ind Ecol 16(5):680–688

van Haaster B, Ciroth A, Fontes J, Wood R, Ramirez A (2017) Development of a methodological framework for social life-cycle assessment of novel technologies. Int J Life Cycle Assess 22(3):423–440

Social Sustainability Performance in Industry

Advancing Social Life Cycle Assessment (S-LCA) for Public Services: Evaluating Water Supply Systems in Venice and Treviso

Bernadette Sidonie Libom, Alessandro Manzardo, and Stefania Presta

Abstract This study explores applying social life cycle assessment (S-LCA) to Piave Servizi S.p.A, a public utility managing water supply in Treviso and Venice. The research adapts the UNEP methodology to evaluate social impacts on six stakeholder groups. A custom framework with 13 stakeholder categories, 45 impacts subcategories, and 203 indicators was developed. Data was collected through surveys and documents. Findings show positive performance regarding water quality, employment stability, and environmental awareness. However, gaps in equal opportunity, communication, and stakeholder engagement were identified. Results support S-LCA's relevance in public services despite data access and stakeholder diversity challenges. The results support the relevance of S-LCA in public services, despite the challenges associated with access to data and the diversity of stakeholders. The scores for each stakeholder were calculated, and the results generally indicate good social performance across all stakeholder groups, except for the local community, which shows high risks related to service transparency and communication channels that require urgent attention. The study then offers methodological innovation aligned with SDGs 1, 10, and 11. It highlights the need for better communication, technological integration, and stakeholder inclusion. Future integration of ISO 14075:2024 is recommended for comparability, robustness, and harmonization of terminology.

1   Introduction

In the context of increasing attention to the sustainability of public services, social life cycle assessment (S-LCA) has proven to be a valuable tool for assessing the social and socio-economic impacts. While S-LCA is widely applied in industry, its adoption in public services, such as water supply, remains limited (Benoît et al. 2009; Martínez-Blanco et al. 2015; Norris et al. 2020). This study investigates the application of

B. S. Libom · A. Manzardo (✉) · S. Presta
Department of Civil, Environmental, and Architectural Engineering, DICEA, Padova University, Padova, Italy
e-mail: alessandro.manzardo@unipd.it

© The Author(s) 2026
M. Traverso et al. (eds.), *Life Cycle Management from Global to Local*,
https://doi.org/10.1007/978-3-032-17987-6_29

S-LCA to Piave Servizi S.p.A, a public company that manages the entire urban water cycle in 39 municipalities in Treviso and Venice (Piave Servizi SpA 2022). The choice of this case is particularly relevant given the strategic importance of water as a common good and the complex socio-institutional environment in which public utilities operate. Public water utilities have broader social responsibilities than private companies, including equitable access, transparency, and long-term sustainability, which influence their social performance and expose them to specific risks. Water supply services also raise important social issues such as affordability, service quality, and the inclusion of vulnerable groups. This study applies the S-LCA framework to Piave Servizi to assess its social impacts and contribute to improving methods for evaluating the social sustainability of public services.

2 Methodology

It is important to point out that this study was carried out before the ISO 14075 standard was published. This study applies the four-phase structure of the UNEP guidelines for social life cycle assessment (S-LCA): (1) goal and scope definition, (2) life cycle inventory, (3) impact assessment, and (4) interpretation (UNEP 2020). To adapt this structure to public services, a specific framework was developed based on findings from the literature and empirical considerations.

2.1 Goal and Scope Definition

In this framework, the public service "water supply" is treated as a "product." The stakeholder categories, impact subcategories, and performance indicators (understood here by components) are selected based on their relevance to the service and following UNEP guidelines. New components have been developed to better align them with the water supply. This adapted approach maintains methodological consistency while considering the unique social and institutional characteristics of public utilities. Figure 1 illustrates the main steps to be considered in the S-LCA of public services.

This study aims to assess the potential social impact of the water supply services managed by Piave Servizi S.p.A. The functional unit is defined as $1\ m^3$ of good-quality drinking water provided by Piave Servizi for drinking and general consumption.

The system boundaries include the phases of water production, treatment, distribution, and consumption. Wastewater and sanitation services are excluded from the analysis.

The study considers six key stakeholder categories: workers, users, local community, children, society, and public regulators. These groups represent a comprehensive range of social actors affected by or involved in the water service.

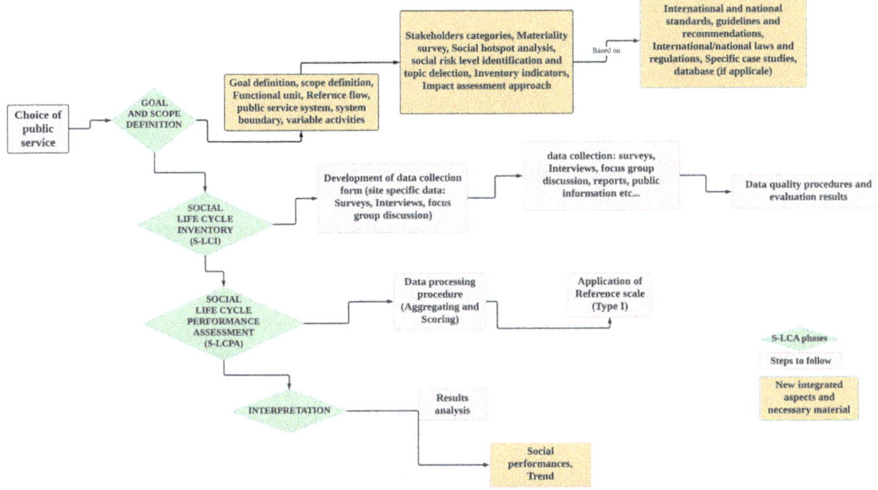

Fig. 1 S-LCA methodological framework with a further breakdown of the four phases for implementation in the S-LCA of the public services (inspired by UNEP 2020)

For the assessment of social life cycle performance, the study applies the Reference Scale Type I (RS S-LCIA) method. This approach enables the evaluation of social impacts using qualitative and semi-quantitative indicators.

Social subcategories and indicators were identified through a triangulation of sources. These included international standards, national legislation, literature reviews, and previous S-LCA case studies. The selection focused on issues such as equitable access, water quality, health and safety, and stakeholder engagement.

However, the study faced limitations in data collection. Access to site-specific information and to downstream stakeholders was restricted, which constrained the depth of the social assessment.

There is currently no literature that applies the social life cycle assessment (S-LCA) framework specifically to the water sector in Venice and Treviso. Nevertheless, several relevant studies (Cekodhima et al. 2023; Costa and Salandin 2023; Buffa et al. 2023; D'Andrea 2017; D'Amore et al. 2021; García-Sánchez and Güereca 2019; Sánchez et al. 2023; Libom et al. 2024) addressing social issues of the water sector in Italy and elsewhere were identified and analysed (although not designed for the S-LCA of public services). As there were no studies at the local level, the investigation was extended to the national level, maintaining the sectoral focus on water supply systems. Although these studies do not explicitly apply the S-LCA methodology, they provide useful insights into the social challenges of the sector. Two main selection criteria guided the literature review: geographical scope (national or sectoral relevance) and thematic focus (water supply). The identified social problems were categorized into three risk levels: high, medium, and low. Problems that were mentioned frequently (more than three times) were classified as high risk, those that were mentioned twice were classified as medium risk, and those that were mentioned only once were classified as low risk. In addition, compliance with national legislation was also considered: Issues reflecting full non-compliance with legislation were classified as high risk, while partial implementation was classified as medium risk. This classification allowed for a structured and context-specific identification of social hotspots in the Italian water sector, as shown in Table 1. The identification of social hotspots was also carried out with the help of the PSILCA database. However, it is important to clarify that this database was not designed for the S-LCA of public services. In this database, public services are grouped under the broader category of "Public administration and defence; compulsory social security." While the data does not distinguish between specific types of public services, it nonetheless offers a general overview of the sector. Regarding the water sector, very high risks are observed in the following areas: The existence of sufficient security measures, the living wage per month (AV), certified environmental management systems, total embodied value added, and corruption in the public sector.

Following the analysis, the water supply systems of Treviso and Venice considered 13 stakeholder categories, 4 derived from the UNEP guidelines and 9 newly

Table 1 Social hotspot analysis from the literature review

Social aspects	Risk level
Protection of biodiversity and the environment	High risk
Efficient management of water resources (reduce water loss and misuse)	High risk
Corruption	High risk
Quality of drinking water	High risk
Stakeholder engagement	High risk
Social initiatives	High risk
Economic contribution	Medium risk
Customer satisfaction	Medium risk
Supplier assessment for impacts on society	Medium risk
Service continuity	Medium risk
Tariff equity	Medium risk
Ethics and integrity	Medium risk
Road and traffic disruption	Medium risk
Human rights (regulatory compliance)	Low risk
Protection and enhancement of heritage sites	Low risk
The health of local inhabitants	Low risk
Impact on landscape and the community	Low risk
Protection of drinking water resource	Low risk
Diversity, equal opportunities, and employee well-being	Low risk
Health and safety at work	Low risk

Stakeholder categories	Subcategories	Indicators	Total	Stakeholder categories	Subcategories	Indicators	Total
Worker	Freedom of association and collective bargaining	6	69	User	Health and safety	7	43
	Child labor	3			Feedback mechanism	9	
	Fair salary	7			Customer Satisfaction	4	
	Working hours	5			Quality and continuity of service	12	
	Forced labor	3			Quality of water	2	
	Equal opportunities/ discrimination	4			Equity and support for vulnerable users	7	
	Health and safety	13			Affordability	2	
	Social benefits/social security	5		Society	Public commitments to sustainability issues	4	22
	Employment relationship	2			Contribution to economic development	3	
	Sexual harassment	4			Technology development	3	
	Training and professional development	4			Corruption	3	
	Employee well-being	4			Poverty alleviation	1	
	Dialogue with stakeholder	3			Corporate ethics and integrity	8	
	Employment stability	2		Government	Regulatory compliance and risk management	1	4
	Knowledge of social and environmental OBJ	2			Knowledge of social and environmental OBJ	1	
	Contribution to social and environmental OBJ	2			Collaboration with authorities	1	
Local community	Access to material resources	17	54		Participation in Public Policies	1	
	Access to immaterial resources	6		Children	Education provided in the local community	5	11
	Delocalization and migration	2			Health issues for children as consumers	3	
	Cultural heritage	4			Children concerns regarding marketing practices	3	
	Safe and healthy living conditions	5					
	Secure living conditions	3		Legend	From UNEP Guidelines (27)		
	Respect of indigenous rights	3			New subcategories (18)		
	Knowledge of social and environmental OBJ	1					
	Local community satisfaction	1					

Fig. 2 Impact subcategories and number of performance indicators

identified, along with 45 impact subcategories, including 27 from the UNEP framework and 18 additional ones. In total, 203 performance indicators were considered, as shown in Fig. 2.

2.1.1 Stakeholder Categories

For complete analysis, 13 categories of potential stakeholders were considered, selected based on their relevance to the service. Data collection is a limitation, especially for public services, as the number of stakeholders to be reached can be very high, and most are not easy to access. Feedback was received from six stakeholders: Workers, the local community, users, children, society and government, public regulators, and local authorities. These six stakeholders were ultimately considered in the analysis.

2.1.2 Impact Subcategories and Performance Indicators

Figure 2 shows the stakeholder groups, the corresponding impact subcategories, and the number of indicators that were developed after an initial screening and were selected for analysis. The selection of these elements was based on their relevance to the specific characteristics of the public water supply. The impact sub-categories and indicators were selected through a combination of literature review, analysis of social risks identified at the national level, and alignment with the UNEP S-LCA guidelines. Priority was given to indicators reflecting the specific social dynamics and risks in the Italian water sector, particularly those relevant to the provision of public services. In the final selection, care was taken to ensure balanced consideration of the 6 stakeholders and social impacts. In Fig. 2, impact subcategories from the UNEP guidelines are shown in white, while the newly developed impact subcategories are highlighted in grey.

2.2 Social Life Cycle Inventory

The data collection combined interviews with stakeholders and internal documentation to assess Piave Servizi's social performance. A structured questionnaire served as a guide to ensure coherent and comparable data collection for all stakeholder groups. The questions targeted qualitative, quantitative, and semi-quantitative indicators derived from previously identified sub-categories of impact. Six questionnaires were developed, tailored to the different stakeholder categories. After expert validation and stakeholder consultation, the questionnaires were distributed online via the LimeSurvey platform from July 31 to September 1, 2024. In parallel, internal documents were reviewed from January to July 2024, including sustainability reports, internal guidelines, audit results, and governance documents.

The results of the inventory are presented by stakeholders and come mainly from questionnaires, the company's sustainability report, the organization, management and control model document, and the code of ethics. Given the complexity of the data from the individual stakeholders, they are not presented here. Apart from the analysis of the survey response rate, presented in this section. The rest of the data will be discussed directly in the social life cycle impact assessment session (Sect. 2.3).

Figure 3 shows the rates of completed and uncompleted responses from the participants. The completion rate (62.18%) shows that most respondents participated in the entire process. This figure indicates an overall positive level of engagement and reflects the perceived interest or relevance of the questions asked. However, the significant dropout rate (37.82%) raises questions about the factors that may have hindered the completion of the survey.

This imbalance could indicate the presence of obstacles, such as the length of the questionnaire, the complexity of the questions, or perhaps a lack of motivation or time on the part of certain participants. Such a distribution suggests that methodological adjustments should be considered to optimize the completion rate in future surveys. Only questions that were answered by all the respondents of the respective stakeholders were considered in the analysis.

2.3 Social Life Cycle Impact Assessment

The data collected was processed using a standardized 1–5 scoring system adapted to each question type. Responses on a 10-point scale were converted to 5-point scores to ensure consistency. Scoring was based on social performance criteria aligned with relevant national and international standards, and weighted averages were used to calculate final scores. For example, for the "Health and safety" indicator, data

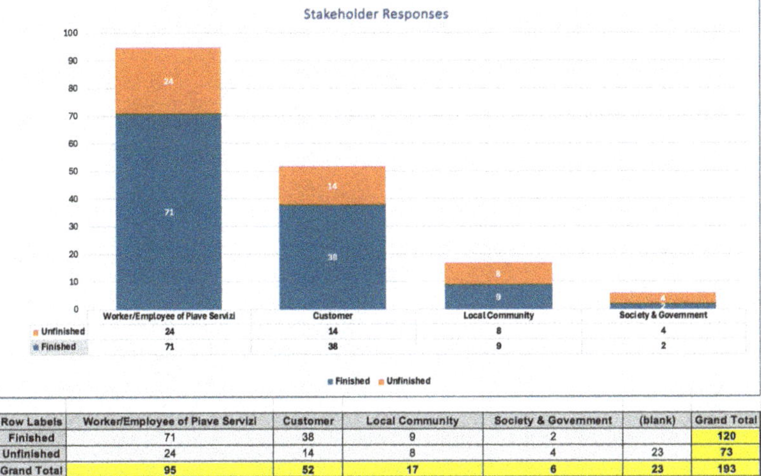

Row Labels	Worker/Employee of Piave Servizi	Customer	Local Community	Society & Government	(blank)	Grand Total
Finished	71	38	9	2		120
Unfinished	24	14	8	4	23	73
Grand Total	95	52	17	6	23	193

Fig. 3 Survey response rate analysis

from the Piave Servizi report on accidents at work and hours of safety training per employee were collected. These data were assessed in relation to national labor legislation and ILO standards. By applying the weighting criteria (% of accidents, % of training, and % of certification), the final score was calculated using a weighted average, resulting in an aggregated performance score for the indicator. These scores were then classified using a colour scale (from red to green: poor [red] (1–1.5), poor [orange] (1.5–2.5), fair [yellow] (2.5–3.5), good [light green] (3.5–4.5), very good [green] (4.5–5)) and converted to a reference scale of $+2$ to -2 to allow a synthetic, comparable interpretation of stakeholder responses (Fig. 4). This approach to data processing was inspired by the methodology developed in the pilot projects on the social life cycle guidelines by the Life Cycle Initiative and the Social Life Cycle Alliance (2022).

Given the complexity of the data, the results will not be presented in a table summarizing the various indicators and corresponding scores. Instead, they will be directly analysed and discussed according to each stakeholder group.

1-1.5	1.5-2.5	2.5-3.5	3.5-4.5	4.5-5
-2	-1	0	+1	+2

Fig. 4 Conversion of scores to a reference scale capitalized

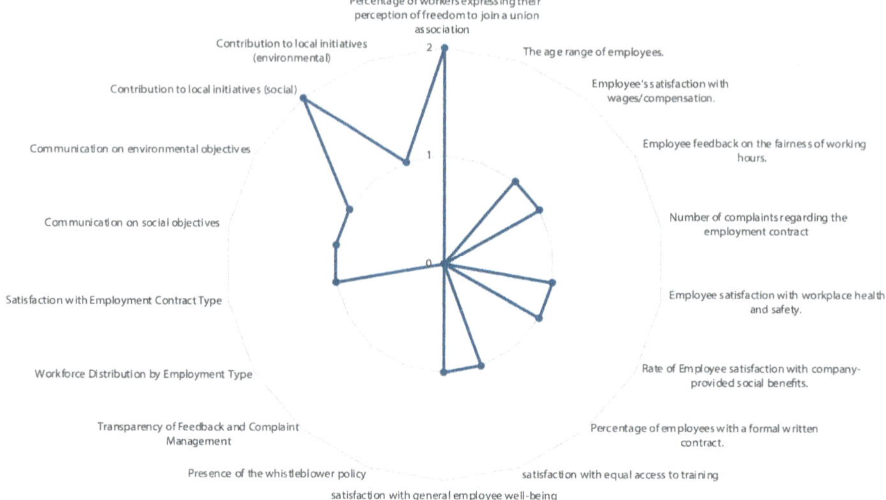

Fig. 5 Workers' results

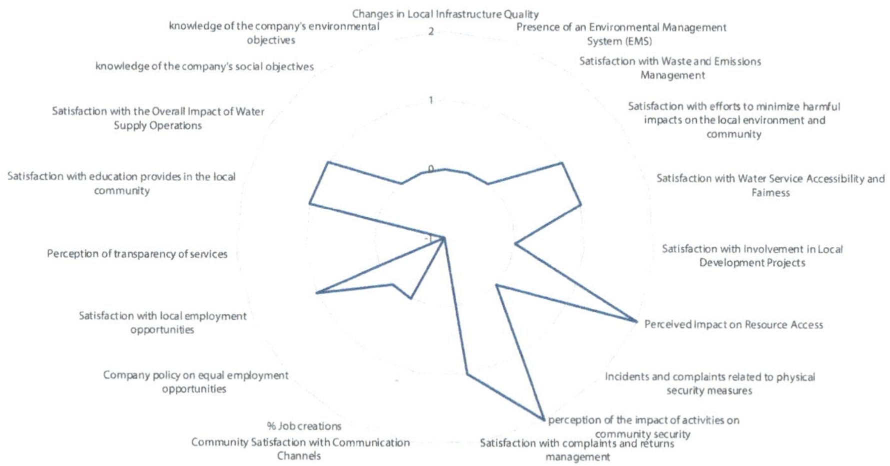

Fig. 6 Results for the local community

2.3.1 Workers' Results and Discussion

The social performance of Piave Servizi regarding workers' social impacts is generally positive (Fig. 5). Regarding freedom of association, the lack of official trade union representation limits the effective implementation of these rights and poses a medium risk (0). The absence of child labor demonstrates very good social performance (2). In terms of fair pay, overall satisfaction is positive, but expectations regarding salary progression are not met, which is a low risk (1). Working hours are generally considered fair, but a lack of flexibility and low policy transparency may lead to future dissatisfaction, which is a medium risk (0). No forced labor was reported, reflecting very good social performance (2). In terms of equal opportunities, perceived inequality of treatment of employees is a high risk (-1). Health and safety management is effective, although there is a low risk (1) due to the desire not to cause accidents. Benefits are generally well perceived, but unfulfilled expectations among a minority of employees pose a low risk (1). Sexual harassment prevention systems are in place, but the perception of inadequate complaint management is a medium risk (0). Training and development inequalities in access result in a medium risk (0). Employee well-being initiatives are valued, but improvements in equal access and communication are needed, which corresponds to a low risk (1). Stakeholder dialog transparency is limited by a lack of visibility in reporting, a medium risk (0). Employment stability with most permanent full-time contracts reflects very good performance (2). Knowledge of social and environmental objectives communication is generally good, with a low risk (1). The contribution to objectives is very positive, with high employee participation, indicating very good performance (2). Overall, improvements are needed in certain areas, particularly equal opportunities and grievance management.

2.3.2 Results and Discussion of the Local Community

Piave Servizi's social performance in terms of impact on the local community is generally positive (Fig. 6). Access to material resources shows good results (+2) thanks to sustainability efforts and equitable access to water (+1). However, issues such as water imbalance and average waste management signal a medium risk (0). Cultural heritage receives a neutral rating (0). Safe and healthy living conditions are rated positively (+1), while safe living conditions show very good performance (+2), as there are no complaints. However, community engagement shows weaknesses: Complaint handling and communication are rated negatively (-1), although overall engagement is seen as positive (+1). Local employment is well established but perceived as average overall (0), with some appreciation of employment opportunities (+1). Satisfaction with the impact of services is good (+1), but transparency of services is a weak point (-1). Awareness of social and environmental objectives is average (0), indicating a need for clearer communication.

2.3.3 Customers' Results and Discussion

Overall, user performance is satisfactory, with a high appreciation for water quality (+2) despite some preferences for bottled water (+1) (Fig. 7). Contact channels, accessibility of services, and complaint handling are in line with expectations (0), although communication could be improved (+1). Users are generally satisfied with the quality and continuity of services (+1), and the response to emergencies is also rated positively (+1). Water quality is well managed (+1), and there are measures

Fig. 7 Customers' results

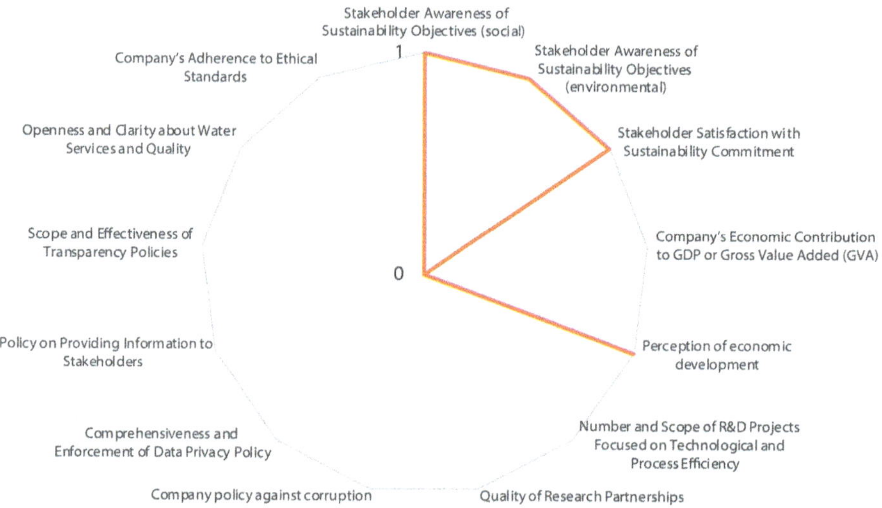

Fig. 8 Results of society

in place to promote equity, such as water houses and support programs, although awareness could be improved (0); satisfaction with water houses is particularly high (+2). Tariffs are fair and below the national average, with users perceiving pricing as positive (+1). Awareness of social and environmental objectives is moderate (0), indicating a need for better communication. Cooperation with educational institutions is perceived positively (+1), and the company's efforts to educate children about the environment are in line with policy (0).

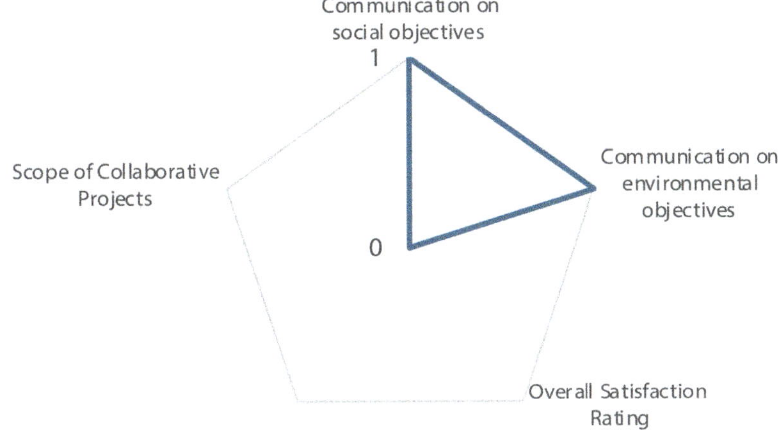

Fig. 9 Results of government, public regulators, and local authorities

2.3.4 Results and Discussion of Society

Overall, Piave Servizi shows a satisfactory social performance (Fig. 8). Stakeholders show a good awareness of the company's social and environmental objectives (+1), which reflects its alignment with sustainability goals. The company's economic contribution is in line with expectations in terms of jobs created (0), while R&D efforts and partnerships are present but limited in scope (0). Anti-corruption, data protection, and transparency measures meet legal standards (0), but communication with stakeholders could be clearer. Ethical commitments are met, but openness and commitment could be improved (0).

2.3.5 Results and Discussion of Government, Public Regulators, and Local Authorities

Piave Servizi has satisfactory performance with government agencies, regulators, and local authorities. Compliance and risk management are rated as moderate (0), indicating that while legal standards are met, some aspects of cooperation could be improved (Fig. 9). Cooperation projects are in place but remain average (0) in scope. Communication of social and environmental objectives is perceived positively (+1 for both), indicating good awareness among public actors, although further efforts are needed to maintain and improve this understanding.

3 Interpretation, Discussion, and Conclusion

Piave Servizi's overall performance toward stakeholders is generally satisfactory, with notable strengths in areas such as water quality, user satisfaction, and environmental commitment. However, some aspects point to areas of risk, in particular, the lack of responsiveness and clarity in communication with certain user groups, and aspects of technological innovation and partnerships with local authorities falling short of expectations. While transparency and compliance with ethical and regulatory standards are generally respected, continued efforts are needed to deepen the company's social and environmental impact and strengthen engagement and communication with key stakeholders. The scores were converted into weights using proportional weighting, with all stakeholders assigned the same weighting (1) in the overall assessment, ensuring no stakeholder is considered more important than another.

This study introduces a methodological innovation aligned with the UNEP guidelines (2020) by proposing new stakeholders, subcategories, and indicators tailored to public services. The results of the analysis show that the proposed model contributes to reducing social inequalities in access to services (SDG 10), while improving living conditions for vulnerable populations (SDG 1). Furthermore, strengthening the quality and availability of water supply services through participatory approaches aligns with the targets of SDG 11, particularly those related to equitable access to basic services. Therefore, the model supports a socially sustainable and territorially grounded approach. Despite limitations related to data privacy, the complexity of public services, and limited access to some stakeholder groups, the research provides a solid basis for improving social life cycle assessments in the public sector. Future work should enhance data collection with detailed socio-economic information, integrate digital tools such as GIS mapping and mobile surveys, establish feedback mechanisms via dedicated platforms, and systematically engage stakeholders through workshops and regular consultations.

References

Benoît C, Norris GA, Valdivia S, Ciroth A, Moberg Å, Bos U et al (2009) The guidelines for social life cycle assessment of products: a social LCA framework. UNEP/SETAC Life Cycle Initiative

Buffa G, Fantinato E, Preo SM (2023) Monitoraggio e valorizzazione del patrimonio naturalistico del Bosco Belvedere e dei prospicenti laghetti di Marteggia nel Comune di Meolo (VE). In: www.corila.it. Consorzio per il coordinamento delle ricerche inerenti al sistema lagunare di Venezia. https://www.corila.it/wpcontent/uploads/2023/11/Relazione_intermedia_Vegetazione_ottobre2023.pdf. Accessed 06 Sept 2024

Cekodhima A, Marciano R, Gross A (2023) Zero-impact rehabilitation of a water main in Conegliano Veneto. In: www.eventiiatt.it. The second edition of the European NO DIG conference, Milan, Lombardy, Italy. https://www.eventiiatt.it/download/papers/P15%2011_00.pdf

Costa L, Salandin P (2023) Tridimensional vulnerability assessment of wells supplying drinking water in agricultural areas. In: EGU general assembly 2023, Vienna, Austria, 24–28 April 2023, EGU23-15708. https://doi.org/10.5194/egusphere-egu23-15708

D'Amore G, Landriani L, Lepore L (2021) Ownership and sustainability of Italian water utilities: the stakeholder role. Util Policy J 71:101228. https://doi.org/10.1016/j.jup.2021.101228

D'Andrea A (2017) Applying GRI sustainability reporting in the water sector: evidence from an Italian company. Int J Bus Adm 8(3):10. https://doi.org/10.5430/ijba.v8n3p10

García-Sánchez M, Güereca LP (2019) Environmental and social life cycle assessment of urban water systems: the case of Mexico city. Sci Total Environ 693:133464. https://doi.org/10.1016/j.scitotenv.2019.07.270

Libom BS, Traverso M, Mankaa RN, Manzardo A (2024) Development and design perspective of a model for analyzing the social life cycle of public organizations: examination of existing models. Sustainability 16(16):6925. https://doi.org/10.3390/su16166925

Martínez-Blanco J, Lehmann A, Chang YJ et al (2015) Social organizational LCA (SOLCA)—a new approach for implementing social LCA. Int J Life Cycle Assess 20:1586–1599. https://doi.org/10.1007/s11367-015-0960-1

Norris CB, Traverso M, Neugebauer S, Ekener E, Schaubroeck T, Garrido SR, Berger M, Valdivia S, Lehmann A, Finkbeiner M et al (eds) (2020) Guidelines for social life cycle assessment of products and organizations 2020. UNEP, Nairobi, Kenya

Piave Servizi SpA (2022) Bilancio di Sostenibilità

Sánchez MG, Padilla-Rivera A, Güereca LP (2023) Social life cycle assessment of mexico city's water cycle. Adv Sustain Syst 7(6). https://doi.org/10.1002/adsu.202300024

Traverso M, Mankaa MN, Valdivia S, Roche L, Luthin A, Garrido SR, Neugebauer S et al (eds) (2022) Pilot projects on guidelines for social life cycle assessment of products and organizations. In: Life cycle initiative and social life cycle alliance

Towards Sustainable Steelmaking: Selecting Social Impact Categories and Developing Reference Scales for the German Steel Industry

Gijs Krekel, Julian Suer, Nils Jäger, and Marzia Traverso

Abstract Social Life Cycle Assessment (S-LCA) enables organizations to analyse the social performance and impacts of products and services in a holistic manner. Such an assessment is especially important for steelmaking, due to due diligence requirements and independent reporting standards. This paper outlines the implementation of the initial two stages of the reference scale approach, as specified in ISO 14075 for Social Life Cycle Performance Assessment (S-LCPA), within the product system of thyssenkrupp Steel Europe, a leading German steel producer. Relevant impact categories for the assessment are defined based on the results of a social hotspot assessment, a double materiality assessment and expert knowledge. Furthermore, the determination of industry-specific reference scales and indicators is presented based on an example for the impact subcategory worker health and safety.

1 Introduction

Steel is defined as an alloy of iron and carbon, containing a carbon content less than 2 wt%. It is a valuable raw material used in numerous applications ranging from the construction industry to food packaging. In 2023, 1.9 billion tonnes of steel were produced to meet the growing global demand (WSA 2024). Roughly 70% of this amount was produced via the Blast Furnace—Basic Oxygen Furnace (BF-BOF) route (WSA 2024). The remaining 30% stem from the Electric Arc Furnace (EAF) route (WSA 2024).

The BF-BOF route, also known as primary steelmaking, produces steel from iron ore. Iron is most commonly found within the earth's crust in an oxidized form such as haematite (Fe_2O_3) or magnetite (Fe_3O_4). These iron oxides are reduced and melted

G. Krekel (✉) · M. Traverso
Institute of Sustainability in Civil Engineering, RWTH Aachen University, Aachen, Germany
e-mail: Gijs.krekel@inab.rwth-aachen.de

G. Krekel · J. Suer · N. Jäger
Innovation and Quality Center—Sustainable Steel Production, Thyssenkrupp Steel Europe AG, Duisburg, Germany

M. Traverso et al. (eds.), *Life Cycle Management from Global to Local*,
https://doi.org/10.1007/978-3-032-17987-6_30

in the blast furnace process using metallurgical coal and coke. The produced hot metal is carbon rich (ca. 4 wt%) and is refined into crude steel in the basic oxygen furnace. Steel scrap is added as a cooling material to counteract the exothermic reaction between carbon and oxygen.

In contrast, the EAF route produces steel from steel scrap. Since the scrap has already undergone reduction in its previous lifecycle, the iron is present in its metallic state. Therefore, it requires only to be remelted by means of an electric arc. Carbon and oxygen are injected into the bath to promote slag foaming, which shields the refractory material of the furnace and acts as insulation against radiative heat losses.

The global steelmaking industry relies on intricate supply chains, which are essential for sustaining the sector by ensuring a steady provision of energy and resources. Given the complexity of these global supply chains, the steelmaking industry must adhere to the principles of due diligence, ensuring that their operations are conducted responsibly (OECD 2018; Proposal for a Directive of the European Parliament and of the Council on Corporate Sustainability Due Diligence and Amending Directive (EU) 2019/1937 2022). This obligation extends beyond their own production processes to encompass the mitigation of risks that occur upstream in the supply chain, thereby promoting sustainable and ethical business practices across the industry (OECD 2018; Proposal for a Directive of the European Parliament and of the Council on Corporate Sustainability Due Diligence and Amending Directive (EU) 2019/1937 2022). Additionally, independent parties such as the World Steel Association (WSA) or ResponsibleSteel initiative require the assessment of social aspects such as work-related injuries or responsible sourcing (ResponsibleSteel 2022; WSA 2025). It is becoming more and more relevant for the steelmaking industry to assess the social impacts of steel products in a holistic manner.

An effective method for addressing this issue is Social Life Cycle Assessment (S-LCA). The S-LCA methodology, defined in the ISO 14075, provides a systematic assessment framework for analysing the social impacts and performance of products and services along their entire life cycle (ISO 2024). By applying S-LCA, valuable insights can be generated, allowing organizations to improve their impact on relevant stakeholders.

While the investigation of environmental impacts through life cycle assessment has become standard practice within the steel industry (Suer et al. 2022), a limited application of S-LCA is highlighted by previous research (Krekel et al. 2024). Most publications in this field focus on the assessment of social performance of steelmaking, by applying Social Life Cycle Performance Assessment (S-LCPA) (Chandola et al. 2022; Hadler et al. 2023; Singh and Gupta 2018; Smith 2020; Vinci et al. 2023). A key step in this type of analysis is the definition of reference scales and performance reference points (PRP), which allow practitioners to analyse the social performance of a product or service in relation to a defined benchmark (ISO 2024; UNEP 2020). However, few authors define reference scales themselves, instead of opting for database assisted forms of impact assessment (Chandola et al. 2022; Hadler et al. 2023; Vinci et al. 2023). This type of analysis is conducted at a country and sector level, and is often paired with lower assessment granularity, which raises concerns if the performance of a specific organization is accurately measured (Hadler et al.

2023). To address the lack of industry-specific reference scales, the present research outlines a methodological approach for developing specific scales tailored to BF-BOF steelmaking, based on industry data and best practices of a major German steel producer: thyssenkrupp Steel Europe (tkSE).

2 Methodology

This research follows the methodological framework laid out by the ISO 14075 for S-LCPA (ISO 2024). More specifically, it documents the first two steps of the reference scale approach and includes the definition of impact (sub-)categories, in addition to the development of indicators and reference scales. This paper focuses on applying these steps for the processes of an integrated steel site in Germany, but the method can also be applied to processes situated in the value chain. For the definition of reference scales, a preliminary example is provided for the stakeholder category 'workers', and impact subcategory 'health and safety'. Other reference scales will be published at a later date.

2.1 Definition of Impact Categories

The definition of impact categories is based on the assessment framework defined in the guidelines for Social Life Cycle Assessment of Products and Organizations 2020 (UNEP 2020). This framework defines six stakeholder categories, each encompassing multiple social topics, referred to as impact subcategories. Definitions of each impact subcategory are determined in the methodological sheets for S-LCA (UNEP 2021). A composite approach was implemented to identify impact subcategories relevant to steelmaking. In total, three different methods were combined to make the final selection. These include a social hotspot assessment conducted for steelmaking, a double materiality assessment conducted by the company in 2023, as well as based on expert knowledge.

The social hotspot assessment was conducted for the production of hot rolled coil, and combined import data from the German steel industry with social data at country and sector level from the social hotspot database (Benoit-Norris et al. 2022). The framework and results of the social hotspot assessment are described in greater detail in a previously published research paper (Krekel et al. 2024). The assessment provides valuable insights into social issues relevant to steelmaking and its value chain. However, it focuses exclusively on social risks. A distinctive aspect of S-LCA is its ability to capture both positive and negative impacts (UNEP 2020). Since the social hotspot assessment does not account for positive performance, additional sources are necessary to achieve a more comprehensive selection of impact subcategories.

The second process for identifying relevant impact subcategories was through detailed analysis of the double materiality assessment conducted for environmental, social, and governance (ESG) reporting by tkSE. The content of this assessment is restricted to internal stakeholders and has yet to be published externally. A total of 24 ESG topics were assessed and ranked on a scale from zero (no impact) to five (high impact), considering both inside-out and outside-in perspectives. The evaluation was conducted through an online questionnaire and supplemented by personal interviews. Participants in the materiality assessment included company internal and external experts, in addition to the tkSE supervisory and executive boards. Employees assessed the accuracy of the inside-out evaluation using an online questionnaire. Overall, more than 1,000 participants were involved. High impact social topics identified by the double materiality assessment were used in selection of impact subcategories. First, non-social-related subjects, e.g., air pollution, were filtered out of the analysis results. The remaining material subjects were then matched to the definitions of impact subcategories based on the methodological sheets for S-LCA (UNEP 2021).

Finally, to address the selection of impact categories and data collection strategies, a meeting was arranged with steel industry experts and S-LCA practitioners. A preliminary list of relevant social topics was compiled based on findings from the social hotspot and double materiality assessments. Feedback was provided from both sides and discussed. Ultimately, this led to the inclusion of additional impact subcategories based on expert knowledge.

2.2 Definition of Indicators and Reference Scales

Following the selection of impact subcategories, a list of indicators was compiled to measure them with. In this assessment, indicators could be either quantitative, semi-quantitative or qualitative, with all three being deemed acceptable. No explicit requirements were set regarding the number or proportion of each within individual impact subcategories. Indicator selection started with a brainstorming session based on the guidelines and methodological sheets for S-LCA (UNEP 2020, 2021), the handbook for product social impact assessment (Goedkoop et al. 2020), previously published S-LCA studies for steelmaking (Singh and Gupta 2018; Smith 2020) and other applications (Bonilla-Alicea and Fu 2022; Gompf et al. 2020; Siebert et al. 2018b, 2018a), as well as industry experience. An extensive list was compiled, detailing each proposed indicator along with its description, desired direction, the required data for implementation, potential data sources and designated company contacts for data collection. This list was then refined during a number of meetings with industry experts and S-LCA practitioners. Wherever appropriate, multiple indicators have been established for a single impact subcategory. This facilitates a comprehensive assessment and enables the description of more diverse social topics. For example, when assessing work place discrimination, indicators are defined to reflect different aspects of discrimination based on race, gender or political belief.

The process of defining indicators and reference scales is currently ongoing. Therefore, a first example is presented for the impact subcategory of worker health and safety, with the lost time injury frequency rate (LTIFR). The selection of LTIFR as an indicator is based on previous S-LCA studies in steelmaking and its common use in corporate sustainability communications. The WSA defines lost time injuries (LTI) as all injuries resulting in the worker being unable to return to their next work shift (WSA 2025). Accordingly, the LTIFR is defined as the ratio of LTI (including fatalities) per million hours worked (WSA 2025).

The use of the LTIFR in S-LCA studies has been highlighted previously for steel in the works of Singh and Gupta, as well as Smith (Singh and Gupta 2018; Smith 2020). These authors define a reference scale for this indicator; however, no information is provided for the definition of the performance reference points (PRP). Therefore, this step is reiterated in this paper. The LTIFR indicator is evaluated using a 5-point Likert scale, spanning from -2, representing worst performance, to $+2$, reflecting ideal performance/industry best practice (UNEP 2020). PRP are defined based on quartiles of the LTIFR reported in a 5 year time span (2020–2024) by major European BF-BOF steelmakers.

3 Results and Discussion

This section outlines and examines the established impact subcategories, and presents a preliminary reference scale for the LTIFR indicator. A complete presentation of the selected impact subcategories is provided. The definition of indicators and reference scales is restricted to a singular example, as this process is currently ongoing.

3.1 Definition of Impact Categories

The final selection of impact subcategories according to the guidelines for S-LCA is presented in Table 1

In total 14 of the 40 predefined impact subcategories have been selected for further study. A majority of these subcategories relate to the stakeholder categories 'workers' and 'local community', with six and four impact subcategories, respectively. However, certain topics relating to 'value chain actors', 'society' and 'children' as stakeholders were considered as well.

The results of the social hotspot assessment have been published in a previous research paper (Krekel et al. 2024). For the steelmaking process, social risks are identified for the subcategories 'worker health and safety', and 'delocalization and migration'. The majority of selected impact subcategories stem from the double materiality assessment. Here, more worker related topics such as 'fair salary', 'social benefits' and 'discrimination' have been identified. Additionally, the relevancy of tkSE close relationship to the local community is highlighted with subcategories

Table 1 Final selection of relevant impact subcategories for German steelmaking

Stakeholder category	Impact subcategory	Social hotspot assessment	Double materiality assessment	Expert knowledge
Workers	Child labour	–	–	X
Workers	Fair salary	–	X	–
Workers	Working hours	–	–	X
Workers	Equal opportunities/ discrimination	–	X	–
Workers	Health and safety	X	X	X
Workers	Social benefits/ social security	–	X	–
Local community	Access to material resources	–	X	–
Local community	Delocalization and migration	X	–	–
Local community	Safe and healthy living conditions	–	X	–
Local community	Community engagement	–	X	–
Value chain actors (not incl. consumers)	Supplier relationships	–	X	–
Society	Technology development	–	X	–
Society	Corruption	–	X	–
Children	Education provided in the local community	–	X	–

such as 'safe and healthy living conditions', 'community engagement' and 'education provided in the local community'. Finally, as a result of consultations with industry experts, the existing range of impact subcategories has been expanded. This now includes additional topics related to worker health and safety, as well as the inclusion of subcategories addressing child labour and working hours, further enhancing the original selection.

3.2 Definition of Indicators and Reference Scales

Each of the methods used to select impact categories, identified worker health and safety as a key topic for the assessment. This impact subcategory was therefore chosen to discuss the development of reference scales, using the LTIFR indicator as an example.

In total, the annual reports of nine different European primary steelmakers were investigated. These include ArcelorMittal (ArcelorMittal 2022, 2023, 2024, 2025), Dillinger Hütte (AG der Dillinger Hüttenwerke 2024), Liberty Steel (Liberty Steel Group Holdings UK Ltd 2024), Saarstahl (Saarstahl AG 2022, 2023, 2024), Salzgitter (Salzgitter AG 2025), SSAB (SSAB AB 2025), Tata Steel NL (Tata Steel Nederland B.V. 2024), Tata Steel UK (Tata Steel UK 2024), thyssenkrupp (thyssenkrupp AG 2024) and Voestalpine (Voestalpine AG 2024). Currently thyssenkrupp reports at group level, for the final S-LCPA study, tkSE specific values will be used. While ArcelorMittal includes subcontractors in its LTIFR reporting, the majority of companies limit their data to workers. Therefore, the development of the reference scales has been restricted to workers as well. Individual LTIFR values have been summarized in Table 2. ArcelorMittal's LTIFR have been included in the table due to their relevancy to the European market. Nevertheless, their values are excluded from the initial reference scale development due to concerns regarding comparability.

As the documented LTIFR were heavily skewed towards two companies, PRP were determined using quartiles from the data spanning the defined 5 year period. The PRP for the LTIFR indicator are shown in Table 3.

PRP were chosen conservatively, so that only the best in class can achieve a performance level of $+2$. This corresponds to the minimum value, or lowest LTIFR

Table 2 Lost time injury frequency rates of major primary European steelmakers

Lost time injury frequency rate (LTIFR)—workers						
Organization	2020	2021	2022	2023	2024	Source
ArcelorMittal[a]	1.07	1.19	1.11	1.46	1.34	ArcelorMittal S.A. (2022, 2023, 2024, 2025)
Dillinger Hütte	–	–	1.5	2.5	–	AG der Dillinger Hüttenwerke (2024)
Liberty Steel	1.9	2.0	1.9	1.7	–	Liberty Steel Group Holdings UK Ltd (2024)
Saarstahl	1.8	2.3	1.8	1.9	–	Saarstahl AG(2022, 2023, 2024)
Salzgitter	8.42	8.57	6.78	7.6	7.4	Salzgitter AG (2025)
SSAB	3.2	1.6	0.69	0.72	0.65	SSAB AB (2025)
Tata Steel NL	0.99	0.93	1.01	0.72	0.81	Tata Steel Nederland B.V. (2024)
Tata Steel UK	–	1.93	2.1	3.29	–	Tata Steel UK (2024)
Thyssenkrupp[b]	–	–	–	2.4	2.4	thyssenkrupp AG (2024)
Voestalpine[b]	9.4	8.2	9.2	8.8	7.4	Voestalpine AG (2024)

[a]LTIFR reported for employees and subcontractors; excluded from initial reference scale development

[b]LTIFR reported at group level

Table 3 Defined reference scale for the lost time injury frequency rate indicator

Indicator: lost time injury frequency rate (LTIFR)					
Performance level	-2	-1	0	$+1$	$+2$
Performance reference point	7.09–9.40	2.0–7.09	1.55–2.0	0.65–1.55	≤ 0.65

across all companies and years. The remaining performance levels are divided into four distinct intervals: a performance level of $+1$ is achieved from the minimum value to the first quartile (Q1), 0 from Q1 to the median (Q2), -1 from Q2 to the third quartile (Q3), and -2 from Q3 to the maximum value. Singh and Gupta previously defined ideal performance based on an LTIFR less than 0.1, however based on the data collected here, and LTIFR reported by the WSA, this level currently appears unrealistically low (Singh and Gupta 2018; WSA 2025).

It should also be mentioned, that the scale defined in Table 3 is preliminary, and could be subject to change in the future. At the time of writing, some aspects regarding this indicator remain unclear. While the WSA provides a precise definition of the LTIFR, the investigated companies appear to lack a uniform approach when reporting this metric. For example, it is currently uncertain if fatal injuries are included in the LTIFR definitions of ArcelorMittal, Dillinger Hütte, Liberty Steel, Saarstahl, Tata Steel NL and Tata Steel UK. Further investigation is required to provide a more precise distinction between reported values and to allow for a robust comparison across the industry.

4 Conclusions and Outlook

This paper applied the first two steps of the reference scale approach according to the ISO 14075 (ISO 2024). Following the framework laid out in the guidelines for S-LCA, a selection of impact categories relevant to primary steelmaking has been identified (UNEP 2020). Fourteen impact subcategories have been determined for further study based on the results of a social hotspots assessment, a double materiality assessment and expert knowledge. These relate mainly to the stakeholder categories 'workers', and 'local communities', with selected topics relating to 'value chain actors', 'society' and 'children' being identified as well. Additionally, the definition of industry-specific indicators and reference scales is presented with the LTIFR indicator used as an illustrative example. A deeper investigation is needed to distinguish reported LTIFR figures more accurately and strengthen industry-wide comparisons. Future research aims to investigate this, in addition to continue the definition of other indicators and reference scales. Finally, it is foreseen to apply these in a case study for steel production at tkSE.

References

AG der Dillinger Hüttenwerke (2024) Geschäftsbericht Dillinger 2023 [annual report]. AG der Dillinger Hüttenwerke. https://www.dillinger.de/app/uploads/sites/3/2024/07/GB_2023_de_r ev1.pdf

ArcelorMittal S.A. (2022) ArcelorMittal annual report 2021 [annual report]. ArcelorMittal S.A. https://corporate.arcelormittal.com/media/xm4blr5z/annual-report-combined-2021.pdf

ArcelorMittal S.A. (2023) ArcelorMittal annual report 2022 [annual report]. ArcelorMittal S.A. https://corporate.arcelormittal.com/media/obsd1lud/annual-report-2022.pdf

ArcelorMittal S.A. (2024) ArcelorMittal annual report 2023 [annual report]. ArcelorMittal S.A. https://corporate.arcelormittal.com/media/upipeqnl/annual-report-2023.pdf

ArcelorMittal S.A. (2025) ArcelorMittal sustainability report 2024 [sustainability report]. Arcelor-Mittal S.A. https://corporate.arcelormittal.com/media/3fwar2wu/2024-sustainability-report.pdf

Benoit-Norris C, Bennema M, Norris G (2022) The social hotspot database supporting documentation update 2022 (V5)

Bonilla-Alicea RJ, Fu K (2022) Social life-cycle assessment (S-LCA) of residential rooftop solar panels using challenge-derived framework. Energy, Sustain Soc 12(1):7. https://doi.org/10.1186/s13705-022-00332-w

Chandola V, Bennema M, Kumaraguru R (2022) Social organizational life cycle assessment of Jamshedpur Steel plant of Tata Steel. In: Traverso M, Mankaa MN, Valdivia S, Roche L, Luthin A, Garrido SR, Neugebauer S (eds) Life cycle initiative, pilot projects on guidelines for social life cycle assessment of products and organisations 2022, 1st edn. UNEP, pp 160–179. https://www.lifecycleinitiative.org/wp-content/uploads/2022/05/Pilot-projects-on-UNEP-SLCA-Guidelines-12.5.pdf

Goedkoop MJ, de Beer IM, Harmens R, Saling P, Morris D, Florea A, Hettinger AL, Indrane D, Visser D, Morao A, Musoke-Flores E, Alvarado C, Rawat I, Schenker U, Head M, Collota M, Andro T, Viot JF, Whatelet A (2020) Product social impact assessment handbook. https://www.social-value-initiative.org/wp-content/uploads/2021/04/20-01-Handbook2020.pdf

Gompf K, Traverso M, Hetterich J (2020) Towards social life cycle assessment of mobility services: systematic literature review and the way forward. Int J Life Cycle Assess 25(10):1883–1909. https://doi.org/10.1007/s11367-020-01788-8

Hadler M, Brenner-Fliesser M, Kaltenegger I (2023) The social impact of the steel industry in Belgium, China, and the United States: a social lifecycle assessment (s-LCA)-based assessment of the replacement of fossil coal with waste wood. J Sustain Metall 9(4):1499–1511. https://doi.org/10.1007/s40831-023-00742-w

ISO (2024) ISO 14075:2024—environmental management—principles and framework for social life cycle assessment. ISO. https://www.iso.org/standard/61118.html

Krekel G, Suer J, Traverso M (2024) Identifying the social hotspots of German steelmaking and its value chain. Sustain Prod Consum 51:222–235. https://doi.org/10.1016/j.spc.2024.09.013

Liberty Steel Group Holdings UK Ltd. (2024) Liberty steel sustainability report 2023 [sustainability report]. Liberty Steel Group Holdings UK Ltd. https://libertysteelgroup.com/wp-content/upl oads/2023/11/Liberty-Steel-Sustainability-Report-2023.pdf

OECD (2018) OECD due diligence guidance for responsible business conduct. https://mneguidel ines.oecd.org/OECD-Due-Diligence-Guidance-for-Responsible-Business-Conduct.pdf

Proposal for a directive of the European Parliament and of the Council on corporate sustainability due diligence and amending directive (EU) 2019/1937, 2022/0051(COD) (2022) https://eur-lex.europa.eu/legal-content/EN/TXT/?uri=CELEX%3A52022PC0071

ResponsibleSteel (2022) ResponsibleSteel international standard version 2.0. https://assets-global.website-files.com/6538e481169ed7220c330f0a/66034300556ac7c2610cd8d0_ResponsibleS teel-Standard-2.0.pdf

Saarstahl AG (2022) Executive summary of the financial statement Saarstahl 2021 [annual report]. Saarstahl AG. https://en.saarstahl.com/app/uploads/2024/03/20221026031313-Executive-Sum mary-of-the-Financial-Statement-Saarstahl-2021.pdf

Saarstahl AG (2023) Executive summary of the financial statement Saarstahl 2022 [annual report]. Saarstahl AG. https://en.saarstahl.com/app/uploads/2023/07/Geschaeftsbericht_SAG_Englisch_2022_Einzeln.pdf

Saarstahl AG (2024) Executive summary of the financial statement Saarstahl 2023 [annual report]. Saarstahl AG. https://en.saarstahl.com/app/uploads/2024/10/Executive-Summary-of-the-Financial-Statement-Saarstahl-2023.pdf

Salzgitter AG (2025, May 22) Ordentliche Hauptversammlung 2025. https://www.salzgitter-ag.com/fileadmin/footage/MEDIA/SZAG/investor_relations/hauptversammlung/2025/de/20250522-SalzgitterAG-Hauptversammlung-2025-de.pdf

Siebert A, Bezama A, O'Keeffe S, Thrän D (2018a) Social life cycle assessment: in pursuit of a framework for assessing wood-based products from bioeconomy regions in Germany. Int J Life Cycle Assess 23(3):651–662. https://doi.org/10.1007/s11367-016-1066-0

Siebert A, Bezama A, O'Keeffe S, Thrän D (2018b) Social life cycle assessment indices and indicators to monitor the social implications of wood-based products. J Clean Prod 172:4074–4084. https://doi.org/10.1016/j.jclepro.2017.02.146

Singh RK, Gupta U (2018) Social life cycle assessment in Indian steel sector: a case study. Int J Life Cycle Assess 23(4):921–939. https://doi.org/10.1007/s11367-017-1427-3

Smith L (2020) The sustainability of steel manufacturing—liberty speciality steels. The University of Sheffield. http://libertysteelgroup.com/uk/wp-content/uploads/sites/2/2021/11/The-Sustainability-of-Steel-Manufacturing-Liberty-Speciality-Steels-2020.pdf

SSAB AB (2025) SSAB annual report 2024 [annual report]. SSAB AB. https://www.ssab.com/en/news/2025/03/ssabs-annual-report-2024-is-published

Suer J, Traverso M, Jäger N (2022) Review of life cycle assessments for steel and environmental analysis of future steel production scenarios. Sustainability 14(21), Article 21. https://doi.org/10.3390/su142114131

Tata Steel Nederland B.V. (2024) Tata Steel Nederland annual report and accounts 2023–2024 [annual report]. Tata Steel Nederland B.V. https://www.tatasteelnederland.com/sites/default/files/tata-steel-jaarverslag-2023-2024-digitaal.pdf

Tata Steel UK (2024) Tata Steel's UK business sustainability report 2021–2023 [sustainability report]. Tata Steel UK. https://www.tatasteeluk.com/sites/default/files/tata-steel-uk-sustainability-report-2021-2023.pdf

thyssenkrupp AG (2024) Thyssenkrupp Geschäftsbericht 2023/2024 [annual report]. Thyssenkrupp AG. https://www.thyssenkrupp.com/de/investoren/berichterstattung-und-publikationen

UNEP (2020) Guidelines for social life cycle assessment of products and organizations 2020. https://www.lifecycleinitiative.org/wp-content/uploads/2021/01/Guidelines-for-Social-Life-Cycle-Assessment-of-Products-and-Organizations-2020-22.1.21sml.pdf

UNEP (2021) Methodological sheets for subcategories in social life cycle assessment (S-LCA) 2021. https://www.lifecycleinitiative.org/wp-content/uploads/2021/12/Methodological-Sheets_2021_final.pdf

Vinci G, Ruggeri M, Barricella C (2023) Assessing the life cycle performance in the metallurgical sector: a case study in Italy. Int J Innov Technol Explor Eng 12(9):22–27. https://doi.org/10.35940/ijitee.I9702.0812923

Voestalpine AG (2024) Voestalpine corporate responsibility report 2023/24 [corporate responsibility report]. Voestalpine AG. https://reports.voestalpine.com/2024/cr-report/_assets/downloads/entire-va-crr24.pdf

WSA (2024) World steel in figures 2024. WSA. https://worldsteel.org/data/world-steel-in-figures/world-steel-in-figures-2024/

WSA (2025) Safety and health in the steel industry data report 2025 [industry report]. WSA. https://worldsteel.org/wp-content/uploads/Safety-and-Health-Data-Report-2025.pdf

Integrating Social Aspects into Safe and Sustainable by Design (SSbD): A Mapping of European Project Contributions

Federica Silveri, Manuela D'Eusanio, and Luigia Petti

Abstract The Chemicals Strategy for Sustainability (CSS) marks a turning point in the EU's approach to chemical safety and sustainability, with the Safe and Sustainable by Design (SSbD) framework as a key innovation tool. This paper aims to explore how EU-funded projects (Horizon Europe 2025; Horizon 2020) are integrating the social dimension, via Social Life Cycle Assessment (S-LCA), into the SSbD framework, with a focus on underrepresented sectors such as plastics and specifically PVC (PolyVinyl Chloride). A mapping and screening of 66 EU projects was conducted, based on publicly available data and expert input. Of these, 18 projects demonstrated some methodological or applicative contribution to social assessment. Results show that whilst awareness of social sustainability is growing, its integration remains fragmented and preliminary. Concrete approaches, such as stakeholder engagement and social hotspot analysis, were identified only in a few cases. The lack of sector-specific guidance and data availability, particularly in early-stage technologies, hinders broader uptake of S-LCA. Nonetheless, initiatives such as ISO 14075:2024 and Social Product Declarations (SPDs) offer promising pathways for developing Product Category Rules (PCRs) and harmonizing practices. Overall, the study highlights the need for structured, cross-sectoral guidance to strengthen the role of social sustainability in SSbD.

F. Silveri · M. D'Eusanio · L. Petti
Department of Economic Studies (DEC), University "G. d'Annunzio", Pescara, Italy
e-mail: luigia.petti@unich.it

M. D'Eusanio (✉)
Department for the Promotion of Human Sciences and Quality of Life, San Raffaele Roma University, Rome, Italy
e-mail: manuela.deusanio@unich.it

© The Author(s) 2026
M. Traverso et al. (eds.), *Life Cycle Management from Global to Local*,
https://doi.org/10.1007/978-3-032-17987-6_31

1   Introduction

The European Union (EU) Chemicals Strategy for Sustainability (CSS), adopted in October 2020 as part of the European Green Deal, represents a paradigm shift in European chemicals policy. It aims to redefine the way chemicals are produced, used and regulated to better protect human health and the environment, whilst stimulating innovation for safe and sustainable alternatives (European Commission 2020). The CSS identifies pollution, alongside climate change and biodiversity loss, as one of the triple planetary crises, advocating for a 'toxic-free environment' as a long-term vision for the European Union.

The CSS builds upon existing regulations, such as REACH (Registration, Evaluation, Authorisation and Restriction of Chemicals) and CLP (Classification, Labelling and Packaging), but goes further by introducing preventive and precautionary measures. These include a ban on the most harmful chemicals in consumer products (e.g. endocrine disruptors, persistent substances and substances affecting the immune or neurological systems), particularly when vulnerable groups, such as children, are concerned (ECHA 2021). The strategy also promotes grouping approaches in chemical assessments, streamlining regulatory processes whilst preventing regrettable substitutions.

1.1   *The Safe and Sustainable by Design and the ANALYST Project*

One of the most innovative aspects of the CSS is its promotion of Safe and Sustainable by Design (SSbD) principles, which are designed to guide new technologies in their early stages towards chemicals and materials that are intrinsically safer and more sustainable throughout their life cycle. This supports the EU's ambition to become a global leader in chemical safety and circularity, enabling green industrial transformation in sectors such as textiles, construction, electronics and packaging (Caldeira et al. 2022). The SSbD framework was developed by the Joint Research Centre (JRC) in 2022: it consists in a step-by-step approach ensuring that safety, sustainability and circularity are considered from the earliest stages of research and innovation. The first three steps refer to safety assessments, whilst the fourth focuses on environmental sustainability. The fifth step, which is optional, addresses socioeconomic aspects.

In the context of SSbD, the ANALYST project was launched to accelerate the transition to a safer and more sustainable industrial value chain by advancing the SSbD framework. The project will develop and validate an integrated impact assessment approach, covering health, environmental, social and economic dimensions, through three use cases focused on the PolyVinyl Chloride (PVC) value chain. This project pursues three key innovations: (1) ANALYST-Box: Development of robust methodologies to assess health, environmental, social and economic impacts across the plastic value chain; (2) ANALYST-Digi: Creation of an open-access platform featuring a Digital Multi-criteria Decision-Making Support tool (DMDMS) and a FAIR (findable, accessible, interoperable and reusable) database to facilitate impact-informed decision-making and support replicability pathways; (3) ANALYST-T-Pack: a validation programme grounded in practical case studies from the plastics value chain, specifically targeting the automotive and construction sectors.

1.2　S-LCA in SSbD and Plastic Sector

One of the first attempts to integrate the four dimensions of SSbD was carried out by Caldeira et al. (2024). Their review provides valuable insights into the direction of SSbD and its application. However, it becomes clear that there is no standardized approach to sustainability aspects, indicators and corresponding calculation methods, especially for the socioeconomic dimension.

The ANALYST-BOX proposes a comprehensive framework for assessing chemicals, materials, products and processes through an integrated lens that includes safety, environmental, economic and social dimensions, aligning with the SSbD approach. Whilst safety and environmental dimensions are relatively mature, the social dimension remains underdeveloped. This is largely due to methodological gaps, particularly in the application of S-LCA during early stages of innovation, when data is scarce and uncertainty is high (Hannouf et al. 2024).

Some studies, such as Stoycheva et al. (2022), have attempted to address this by combining S-LCA Guidelines (UNEP 2020) with Multi-Criteria Decision Analysis techniques, notably the Multi-Attribute Value Theory (MAVT), offering a self-assessment tool for early-stage decisions. However, these efforts remain limited in scope, and their implementation across diverse sectors is still rare.

A notable gap in the literature concerns the plastics sector, especially materials like PVC, which have not been addressed by S-LCA studies. Existing applications focus mainly on agriculture (e.g. tomatoes, honey, pork) (Rezaei Kalvani et al. 2021). The rarely research in the plastic sector has predominantly focused on plastic packaging (HDPE non-beverage bottles, flexible plastics, medical gowns, etc.) (i.e. Albrecht et al. 2013; Papo and Corona 2022; Zhao et al. 2022; Haslingher et al. 2024), bioplastics (e.g. bio-based fuels and chemicals) (Spierling et al. 2018), aeronautical recycling sector (e.g. carbon fibres) (Pillain et al. 2019) and electronics (e.g. laptop production) (Ekener-Petersen 2013). In conclusion, the fifth step of the SSbD framework, which should address socioeconomic sustainability, is currently optional and lacks specific implementation guidelines. Critical barriers remain, including the absence of Product Category Rules (PCRs), lack of methodological standardization in defining system boundaries and selecting indicators, and the inadequacy of S-LCA methods for low Technology Readiness Levels (TRLs). Additionally, social indicators are often undervalued in decision-support tools dominated by safety and environmental metrics. Data accessibility is also a concern, with most social databases being either paywalled, outdated or incomplete. The ANALYST-DIGY initiative aims to address this by developing an open-access tool. Furthermore, stakeholder engagement, essential to robust S-LCA, is inconsistently applied in SSbD efforts, and there is a lack of clear procedures for incorporating public perception, social acceptance and risk awareness. Lastly, the absence of demonstrator projects in sectors such as PVC or complex chemicals limits the empirical validation of S-LCA within SSbD. Addressing these challenges is crucial for ensuring that the social dimension is treated with the same rigour as the environmental and safety aspects, and for advancing truly sustainable innovation by design.

The aim of this study is to map how current European projects (i.e. Horizon Europe 2025; Horizon 2020) contribute to the methodological development of SSbD framework. Special attention is devoted to exploring how these projects incorporate the social dimension into their methodologies and assessment strategies.

2 Methodology

Since the development of the SSbD framework, the EU has funded several projects aimed at developing safer, more sustainable and circular materials and substances that protect human health and the environment. In this context, the SSbD approach is promoted to support the implementation of these projects, encouraging the adoption of a life cycle perspective that considers environmental, social and economic impacts throughout all stages of a product's life cycle. To understand how social aspects can be integrated within the SSbD framework, especially given the lack of S-LCA studies in SSbD, a mapping of various European projects focused on SSbD was conducted. EU Projects were identified via the CORDIS database by using keywords 'SSbD' and filtered by 'projects'. A total of 59 projects emerged from the review.

From these results, projects initiated and completed prior to 2022, the year in which the framework was published, were excluded (i.e. NANORIGO and SSBD SMALL). The ANALYST project was also excluded from the results. Following a preliminary screening, the WoodTreat was likewise excluded from the results, as it was not aligned with the SSbD framework.

In addition, some projects were added to the list at the suggestion of the ANALYST project partners, who also work on other SSbD-related projects (i.e. MARCHES, VALESOR, SABYDOMA, SbD4Nano, SUNSHINE, ASINA, SAByNA, PROMICON and REDONDO). Finally, even if the ORIENTING project isn't focused directly on SSbD, it was added to the list. The reason lies in the fact that the ORIENTING project developed an integration of the three dimensions of sustainability, and it included circularity in their assessment. The final review of EU projects related to the SSbD topic identified 66 projects relevant to our objective. Amongst the 66 projects, a screening process was conducted to assess their relevance to the application of social sustainability.

For each project, the description, objectives, the website and, whenever available, preliminary results (i.e. deliverables and peer-reviewed articles) were analysed.

Table 1 Extract of table analysing EU projects

EU project	Full title	Main aim of the project	Consideration on S-LCA	Sector
DESIDERATA Grant agreement ID: 101178011 01/25–12/28	Integrated pathways: Advancing safe and sustainable by design material innovation through collaborative wisdom	The project will test and refine the EU'S SSbD framework focusing on alternatives to substances of concern, the project aims to advance safer value chains and provide policy recommendations to improve the framework	– Social acceptance – Assessing stakeholder needs and concerns	Industry: Plasticisers, flame retardants and surfactants

Projects were listed in a table with the name and the Grant Agreement, duration, objectives, application and social considerations.

An extract of the grid is illustrated in Table 1. The complete list can be provided by the authors upon request.

Ultimately, 18 projects were identified as relevant for both methodological and applied developments within the social dimension: AGRILOOP, ALCHE-MISSTs, BLUECOAT, DESIDERATA, ENKORE, GUESS-WHY, INTEGRANO LIGNOFUN, NOUVEAU, ORIENTING, PARC, PLANETS, PROPLANET, RADAR, SURPASS, SURFs UP, TORNADO and ZEROF.

Between these projects, one was closed in 2024 (ORIENTING), two have yet to start (i.e. BLUECOAT and LIGNOFUN) and the rest are still open.

3 Results

3.1 Overview of SSbD-Related EU Projects and Their Fields of Application

The review of EU projects related to the SSbD topic identified 66 projects relevant to the research objective. Amongst these, 18 were selected as relevant for methodological developments in the social dimension. These projects span a highly diverse range of fields, including bio-based products, plasticisers, flame retardants, coatings, nano-materials and others. The AGRILOOP project promotes collaboration between the EU and China by converting agricultural residues into high-value bio-based products through SSbD processes. Its goals include enhancing resource efficiency, creating new markets and reducing waste by employing innovative biorefinery approaches. The ALCHEMISSTs project applies the SSbD framework to develop safer alternatives to substances considered high-risk, such as surfactants, plasticisers and flame retardants. Similarly, the BLUECOAT project aims to develop twelve safe, sustainable and bio-based coating formulations that emit low carbon and serve demanding sectors like marine, textile and construction, addressing concerns associated with fossil-based coatings. The DESIDERATA project focuses on testing and refining the SSbD framework through industrial case studies, emphasizing alternatives to substances of concern and striving to advance safer value chains whilst providing policy recommendations to improve the framework. The ENKORE project tackles the development of eco-design frameworks for environmentally compliant devices and eco-responsible packaging that minimize environmental impact, reduce carbon footprints and maximize resource preservation.

3.2 Specific Innovations and Technological Targets in SSbD Projects

The GUESS-WHy project works to improve sustainability and safety in Fuel Cell and Hydrogen (FCH) systems by promoting SSbD guidelines across seven different FCH systems and various technology readiness levels along the hydrogen value chain. INTEGRANO seeks to develop a standardized quantitative approach for evaluating specific nanomaterials in alignment with current SSbD guidelines. LIGNOFUN aims to replace fossil-based aromatic compounds with sustainable alternatives derived from lignin to address environmental challenges. The NOUVEAU project intends to develop advanced solid oxide cells with reduced reliance on rare earth elements by applying SSbD principles, advanced coatings and predictive modelling to optimize material performance, support industrial deployment and comply with regulations such as REACH. The ORIENTING project developed an operational methodology for Life Cycle Sustainability Assessment (LCSA), integrating environmental, social and economic impact analyses. The Partnership for the Assessment of Risks from Chemicals (PARC) fosters a large European collaboration aimed at evaluating risks chemicals pose to human health and the environment, involving nearly 200 institutions. The PLANETS project demonstrates the SSbD framework by developing safer, sustainable and economically viable alternatives to plasticisers, flame retardants and surfactants. PROPLANET is committed to the development of safe and sustainable coatings. The RADAR project develops renewable and eco-friendly chemicals to replace harmful substances such as Bisphenol A (BPA) and phenol, utilizing biomass side streams and cutting-edge technologies. The SURPASS project drives the shift towards Safe, Sustainable and Recyclable by Design (SSRbD) polymer materials targeting the building, transport and packaging sectors. SURFs UP focuses on the development and scale-up of cost-effective second-generation bio-surfactants from wood and food or industrial waste. The TORNADO project addresses the creation of non-toxic coatings for the food, agriculture and biotechnology sectors. Lastly, the ZeroF project develops SSbD coating alternatives to replace PFAS compounds in food packaging and upholstery textiles.

3.3 Current Status and Challenges in Addressing the Social Dimension

Regarding the social dimension within these projects, limited detailed information and applied developments are currently available. This is primarily because most projects have only recently commenced and have yet to publish deliverables or preliminary results. Social dimension considerations are mostly inferred from project descriptions, with only a few projects, specifically ORIENTING, SURPASS, PROPLANET and ZeroF, providing more concrete insights.

The ZeroF project centres its social sustainability assessment on evaluating social acceptance, focusing particularly on consumer and stakeholder perspectives. To capture both attitudes and acceptance levels towards the new coating technology, the project employs two complementary frameworks. The Technology Acceptance Model (TAM) is used to examine consumer perceptions, focusing on perceived ease of use and acceptance, but since TAM mainly addresses behavioural intention rather than actual behaviour change, the COM-B model is applied to bridge this gap. The COM-B framework identifies Capability, Opportunity and Motivation as the primary factors influencing Behaviour, assessing consumer awareness, understanding of sustainability aspects of coated materials and the conditions enabling action. The methodology involves a quantitative survey of 1500 participants from Finland, Luxembourg (including the Greater Region) and Spain to segment consumers based on acceptance and sustainability-related behaviours, as well as to compare attitudes across Northern, Central and Southern Europe. Additionally, up to 30 semi-structured interviews with value chain stakeholders will support the project's human-centered approach by identifying further adoption drivers and barriers, including economic and policy-related factors.

The ORIENTING project developed an operational LCSA methodology that integrates environmental, economic and social assessments throughout the entire product life cycle. Its S-LCA approach is grounded in the UNEP Guidelines (2020) and the Handbook for Product Social Impact Assessment (PSIA). The methodology uses a Reference Scale Approach (RSA), combining qualitative and quantitative evaluations to analyse social performance and risks along the product value chain. ORIENTING implements three assessment levels—entry, intermediate and advanced—to accommodate different complexities depending on data availability and objectives. The methodology was tested in five industrial case studies, which highlighted the benefits of S-LCA in providing transparent insights into product value chains, identifying data gaps and supporting mitigation actions necessary for compliance with corporate sustainability regulations. In summary, ORIENTING significantly contributed to integrated product sustainability assessment, with a particular focus on social dimensions.

In the SURPASS project, a temporal perspective for assessing social sustainability indicators was proposed, reflecting the dynamic nature of innovation processes. Although UNEP's guidelines for S-LCA do not explicitly integrate innovation dynamics or temporal dimensions, SURPASS introduced a phased approach across early, mid and late stages of development. During the early stage, the assessment focused on a preliminary set of social parameters such as health and safety, child labour, working conditions, fair competition, supplier relationships, access to material resources, responsible communication, transparency, end-of-life responsibility, delocalization and migration, impacts on affected communities and local community education. These parameters were selected based on their perceived relevance by project experts and initial identification of potential social issues relevant to the sectors addressed. The early evaluation was valuable for guiding the design process but acknowledged as not exhaustive. As innovation progresses, the project recommends expanding social indicators and engaging stakeholders to support selection and validation of relevant parameters to ensure a robust and context-sensitive social sustainability assessment.

PROPLANET combined a literature review and insights from a chemical ethanol case study to better understand commonly addressed impact categories and social themes in S-LCA studies. A preliminary set of social subcategories was identified based on expert judgement, focusing on relevance and data availability. Within the 'Worker' stakeholder category, subcategories such as child labour, fair salary, fatal accidents and forced labour were prioritized. For the Local Community category, unemployment was highlighted due to its potential significance in assessing social impacts of emerging technologies and supply chains. To support prioritization, an initial social hotspot analysis was conducted using the Product Social Impact Life Cycle Assessment (PSILCA) database, applying the Social Impacts Weighting Method. This method quantifies social risks across countries and sectors and expresses results in medium-risk hours, enabling high-level mapping of social hotspots along the value chain. These results serve as a foundation for more detailed, context-specific assessments and help integrate social considerations early in innovation processes.

4 Frontiers, Conclusions and Future Perspectives

The SSbD framework aims to integrate safety and sustainability—covering safety, environmental, economic and social aspects—into the development of chemicals and materials from the outset. Whilst progress has been made in environmental and safety assessments, the social dimension remains underdeveloped, both methodologically and in practice.

A review of 66 EU-funded SSbD-related projects shows that only 18 meaningfully address social sustainability, with current efforts often fragmented or preliminary.

Key challenges include lack of sector- and product-specific guidance for applying S-LCA, especially in complex sectors like plastics and chemicals; early stages of project implementation limiting data availability; limited transparency and access to methodological deliverables; variability in scope, sectors and TRLs across projects.

The recent ISO 14075 (2024) standard represents a major advancement. Built on S-LCA principles and aligned with PCRs, it offers harmonized and comparable reporting of social performance at the product level. For complex sectors, Social Product Declarations (SPDs) can significantly improve the integration of social aspects in SSbD.

To move forward, the following actions are recommended. First of all, the development of standardized social indicators tailored to the SSbD context and different TRL stages. Furthermore, the incorporation of temporal dynamics to enable iterative social impact assessments during innovation. In addition, to improve data collection and context-sensitive hotspot mapping. Finally, supporting capacity building through training, platforms and collaboration networks.

In conclusion, whilst the integration of social sustainability into SSbD is still in its infancy, foundational work and emerging standards like ISO 14075 offer a promising path forward towards more holistic, dynamic and actionable sustainability assessments.

Acknowledgments and Disclaimers This study was financially supported by the European Union's Horizon Europe research and innovation programme under Grant Agreement No 101138548 (ANALYST). Views and opinions expressed are however those of the authors only and do not necessarily reflect those of the European Union and of the Consortium Partners of ANALYST Project. Neither the European Union nor the granting authority can be held responsible for any use that may be made of the information it contains.

References

Albrecht S, Brandstetter P, Beck T, Fullana-i-Palmer P, Grönman K, Baitz M et al (2013) An extended life cycle analysis of packaging systems for fruit and vegetable transport in Europe. Int J Life Cycle Assess 18:1549–1567

Caldeira C, Abbate E, Moretti C, Mancini L, Sala S (2024) Safe and sustainable chemicals and materials: a review of sustainability assessment frameworks. Green Chem 26(13):7456–7477

Caldeira C, Farcal LR, Garmendia Aguirre I, Mancini L, Tosches D, Amelio A et al (2022) Safe and sustainable by design chemicals and materials. Framework for the definition of criteria and evaluation procedure for chemicals and materials. Publications Office of the European Union

Ekener-Petersen E, Finnveden G (2013) Potential hotspots identified by social LCA—part 1: a case study of a laptop computer. Int J Life Cycle Assess 18:127–143

European Commission (2020) Chemicals strategy for sustainability—towards a toxic-free environment , COM (2020) 667 final, 14 Aug 2020

European Commission: Directorate-General for Research and Innovation (2025) Europe's budget—Horizon Europe (2028–2034), Publications Office of the European Union, https://data.europa.eu/doi/10.2777/8979807

European Commission (2020) Horizon 2020—the framework programme for research and innovation, COM (2011) 808 final, 30 Nov 2011

Hannouf MB, Padilla-Rivera A, Assefa G, Gates I (2024) Social life cycle assessment (S-LCA) of technology systems at different stages of development. Int J Life Cycle Assess 1–16

Haslinger AS, Huysveld S, Cadena E, Dewulf J (2024) Guidelines on the selection and inventory of social life cycle assessment indicators: a case study on flexible plastic packaging in the European circular economy. Int J Life Cycle Assess 1–18

International Organization for Standardization (2024) ISO 14075: environmental management—principles and requirements for social responsibility in LCA. ISO, Geneva

Papo M, Corona B (2022) Life cycle sustainability assessment of non-beverage bottles made of recycled high density polyethylene. J Clean Prod 378:134442

Pillain B, Viana LR, Lefeuvre A, Jacquemin L, Sonnemann G (2019) Social life cycle assessment framework for evaluation of potential job creation with an application in the French carbon fiber aeronautical recycling sector. Int J Life Cycle Assess 24:1729–1742

Regulation (EC) No 1907/2006 of the European Parliament and of the Council of 18 December 2006 concerning the Registration, Evaluation, Authorisation and Restriction of Chemicals (REACH), OJL 396, 30 Dec 2006, pp 1–849

Rezaei Kalvani S, Sharaai AH, Abdullahi IK (2021) Social consideration in product life cycle for product social sustainability. Sustainability 3(20):11292

Spierling S, Knüpffer E, Behnsen H, Mudersbach M, Krieg H, Springer S et al (2018) Bio-based plastics—a review of environmental, social and economic impact assessments. J Clean Prod 185:476–491

Stoycheva S, Zabeo A, Pizzol L, Hristozov D (2022) Socio-economic life cycle-based framework for safe and sustainable design of engineered nanomaterials and nano-enabled products. Sustainability 14(9):5734

UNEP (2020) In: Benoît Norris C, Traverso M, Neugebauer S, Ekener E, Schaubroeck T, Russo Garrido S et al (eds) Guidelines for social life cycle assessment of products and organizations 2020. United Nations Environment Programme

Zhao X, Klemeš JJ, Saxon M, You F (2022) How sustainable are the biodegradable medical gowns via environmental and social life cycle assessment? J Clean Prod 380:135–153

Socio-economic and Climate Impacts of Increased Legume Production for Human Consumption in Norway

Rannvá Danielsen, Clara Valente, Anna Birgitte Milford, and Simen Wilsher-Lohre

Abstract This paper examines the effect on employment and Greenhouse Gas (GHG) emissions if legume production and consumption in Norway increase and replace animal-based proteins. Three scenarios—maximising employment and cultural landscapes, minimising GHG emissions and evenly reducing each type of livestock—were analysed. Results show that increasing legume production and consumption can significantly reduce GHG emissions but also reduce employment in the livestock sector. This demonstrates a trade-off between reducing GHG emissions versus maintaining cultural landscapes and employment opportunities in rural areas where other income opportunities may be scarce.

1 Introduction

A dietary shift towards less red and processed meat and more plant-based food, including legumes, is called for by numerous international and national environmental and health organisations and public authorities (UNEP 2023, WHO 2023). Meat from ruminants is an important source of global methane emissions which contributes to climate change (Willett et al. 2019), and excess consumption of red and processed meat is associated with an increased risk of non-communicable diseases (WHO 2023). Legumes, on the other hand, are associated with beneficial health effects and have a low environmental impact compared to red and processed meat (Singh et al. 2017, Stagnari et al. 2017).

Official dietary guidelines in numerous countries recommend a limited intake of red meat (European Commission 2025), and some also recommend an increased intake of legumes (Hughes et al. 2022). Despite these recommendations, diets in many developed countries, such as Norway, are dominated by animal-based food,

R. Danielsen (✉) · C. Valente
NORSUS—Norwegian Institute of Sustainability Research, Fredrikstad, Norway
e-mail: rad@norsus.no

A. B. Milford · S. Wilsher-Lohre
NIBIO—Norwegian Institute of Bioeconomy Research, Division of Food Production and Society, Bergen, Norway

© The Author(s) 2026

M. Traverso et al. (eds.), *Life Cycle Management from Global to Local*,
https://doi.org/10.1007/978-3-032-17987-6_32

with around 70% of protein consumed by Norwegians derived from meat and dairy (Svanes et al. 2022). However, the consumption of plant-based protein is increasing across Europe, with plant-based food consumption growing by 49% between 2018 and 2020 (Smart Protein 2021). In Norway, the market for plant-based products has expanded, driven by environmental, health and animal welfare considerations (Gonera and Millford 2018, Gonera et al. 2021) but many legumes-based products in Norway rely on imported ingredients, such as soy and pea concentrate (Stagnari et al. 2017).

Due to Norway's unfavourable climate and topography, peas and faba beans are the only legumes commercially grown in Norway. Cultivation is limited to a few regions and almost exclusively for animal feed. Farmers growing legumes face challenges such as adverse weather events and imported legumes such as soybeans are generally lower priced than domestic alternatives. Expanding domestic legume production could mitigate supply chain disruptions and foster regional economic development. Furthermore, moving from animal to plant-based protein could significantly reduce Norway's Greenhouse Gas (GHG) emissions, which would be in line with its commitment to the Paris agreement. Studies indicate that the carbon footprint of dried peas and faba beans is substantially lower than that of meat-based proteins, ranging from 0.57 to 0.62 kg CO_2-eq/kg for peas and faba beans compared to 4.03 kg CO_2-eq/kg for chicken, 3.76 kg CO_2-eq/kg for pork, 30.3 kg CO_2-eq/kg for beef and 34.9 kg CO_2-eq/kg for lamb and mutton (Svanes et al. 2022). Increasing the production of legumes could also foster more sustainable agricultural practices in cereal production, due to legumes' ability to fix nitrogen which reduces the need for fertilisers (Abrahamsen et al. 2019).

The Norwegian livestock sector is highly dependent on imported plant proteins used as feed concentrate, including for cattle and sheep (Norwegian Agriculture Agency 2025). However, cattle and sheep also eat grass from pastures during the summer months and from grass harvesting. This ancient practice has led to the creation of cultural landscapes which are important habitats for several different species and contribute to cultural heritage and beauty (Beyene et al. 2024). Although abandonment of grazing practices and cultural landscapes over the last decades has happened whilst meat consumption has increased (Beyene et al. 2024), one could expect that a transition away from animal to legume-based proteins could accelerate this development. Furthermore, a reduction in meat consumption could impact employment. Ruminant livestock production is labour-intensive and takes place in all areas of Norway where there are available grazing resources, including in rural districts where other income opportunities often are scarce. There is concern that with a transition away from animal-based food these sources of employment will be lost (Farstad et al. 2021). On the other hand, increased production of food products from domestically grown legumes could also create new job opportunities, both in primary production and in processing (Stagnari et al. 2017).

The aim of this paper is to look at the employment effects of increasing national legume production for human consumption. Using estimates of the potential production capacity for legumes, we create different scenarios depending on which animal protein products the legumes will replace. We also look at the impact on GHG emissions of different production scenarios.

2 Materials and Methods

To develop employment estimates, we used data from NIBIO's Operations Planning Handbook (Hovland 2022) as the basis for calculating the labour input per kilogram of legumes, livestock and dairy. The results are presented in Full-Time Equivalent (FTE) employment. The number of actual persons employed will not be the same as FTE as some persons are employed on a seasonal basis or working reduced hours. One FTE position was assumed to be 1845 h a year (Hjukse 2022). Our estimates for labour intensity are presented as time spent per kilogram per product, from which we estimated FTE employment per tonne of produced legumes, livestock or dairy.

To assess the effects on employment of increasing legume production and reducing livestock and dairy production, we created three different scenarios for primary production of legumes, livestock and dairy in Norway, with three varying volumes of legume production. For each scenario the calorie intake for the Norwegian population was held constant, i.e. when the production (and consumption) of legumes increases, we assume an equivalent reduction in the consumption (and therefore production) of livestock and dairy so that calorie intake is the same in every scenario. For these calculations, we used raw data on food intake and per capita calorie intake in Norway (Norwegian Directorate of Health 2024). The population number for Norway we used is 5,521,000, kept constant.

The basis for our scenarios is an increase in the production and consumption of legumes in Norway. According to Abrahamsen et al. (2019), an average of 37,800 acres were used for cultivating peas and faba beans in Norway in 2016–2018, which produced on average 6400 tonnes of peas and 6600 tonnes of faba beans a year. They estimate that the potential yields for peas and faba beans if all suitable areas of land in Norway were utilised is 54,500 and 39,700 tonnes, respectively. In total, the production of legumes could increase from 13,000 tonnes a year to 94,200 tonnes, of which 40–60% is estimated to have edible quality. We used the higher end of that range. Assuming that the volume of imported legumes consumed (19,200 tonnes) stays constant, the total volume of legumes consumed in Norway will be approx. 75,600 tonnes.

We estimated the employment effects in Norwegian primary production under three different variations: a) business as usual where production of legumes and livestock remains unchanged (baseline), which means that consumption is 19,200 tonnes (all assumed to be imported), and the 13,000 tonnes of domestically produced peas and faba beans all goes to animal feed; b) production of legumes is doubled to 26,000 tonnes, of which 60% is edible (15,600 tonnes), meaning the total domestic consumption increases to 34,000 tonnes, which includes the 19,200 tonnes imported. The production of livestock is reduced so that calorie intake per capita remains constant; c) production of legumes is at the maximum that Norwegian farmland can support (94,000 tonnes), and the consumption increases to 75,600 tonnes (including imported legumes) whilst the production of livestock is reduced to keep calorie intake constant. We refer to these three variations as baseline, double, and maximum, respectively (see Table 1).

We developed three scenarios, each comprising the three before-mentioned variations of legume production (see Table 1). In each scenario with increased legume consumption, different combinations of livestock and dairy consumption are reduced so that the average daily calorie intake is constant (see Table 2). Livestock and dairy consumption are adjusted with estimates for net imports (Rustad 2024b) and food import numbers (Statistics Norway 2024), so that only employment effects in Norway are calculated. The three scenarios are as follows: The first scenario is focused on maximising employment in the agricultural sector and maintenance of cultural landscapes. Ruminants maintain cultural landscapes through grazing and are also more labour-intensive than pork and chicken; therefore, this scenario implies that the legumes replace pork and chicken whilst the production of sheep and cattle remains unaltered. The second scenario is focused on minimising GHG emissions. Ruminants, who emit methane gases, are associated with far higher GHG emissions than monogastric animals (Svanes et al. 2022). This scenario thus implies reducing sheep and cattle to compensate for increased legumes consumption whilst pork and chicken production remain unaltered. In the maximum legume production variation of this scenario, dairy is also reduced since the number of dairy cows is reduced. In the third scenario, each type of livestock is reduced with the same percentage to compensate for the increased legumes consumption. Dairy remains unaltered (see Table 2).

Figure 1 shows the labour requirements (time) for legumes, livestock and dairy production. Production of feed is not considered in livestock production. Figure 2

Table 1 Three scenarios with three variations each were developed for estimating employment effects in primary production

Scenario 1: Maximising employment and cultural landscapes
1a. Baseline production of legumes, livestock and dairy
1b. Double production of legumes and reduction of chicken and pork
1c. Maximum production of legumes and reduction of chicken and pork
Scenario 2: Minimising GHG emissions from production
2a. Baseline production of legumes, livestock and dairy
2b. Double production of legumes and reduction of emissions-intensive livestock
2c. Max. Production of legumes and reduction of emissions-intensive livestock
Scenario 3: Even percentage reduction of all livestock
3a. Baseline production of legumes, livestock and dairy
3b. Double production of legumes and even reduction of all livestock production
3c. Max. Production of legumes and even reduction of all livestock production

Table 2 The production of legumes, livestock and dairy in the three scenarios and their variations

	Legumes (%)	Pork(%)	Chicken (%)	Sheep (%)	Cattle (%)	Dairy (%)
Scenario 1: Maximising employment and cultural landscapes						
1a. Baseline	100	100	100	100	100	100
1b. Double	200	87	87	100	100	100
1c. Maximum	725	53	53	100	100	100
Scenario 2: Minimising GHG emissions from production						
2a. Baseline	100	100	100	100	100	100
2b. Double	200	100	100	74	74	100
2c. Maximum	725	100	100	68	62	92
Scenario 3: Even percentage reduction of all livestock						
3a. Baseline	100	100	100	100	100	100
3b. Double	200	92	92	92	92	100
3c. Maximum	725	70	70	70	70	100

Fig. 1 Production time for legumes, livestock and dairy at farm gate. *Data source* Hovland (2022)

shows the estimated emissions per kg of product for legumes, livestock and dairy. These were used to calculate net GHG emissions in scenarios 1–3.

3 Results

Figure 3 shows the estimated employment in our three scenarios. In scenario 1, we see that maximising employment and cultural landscapes in Norway had little impact on overall estimated employment in all three variations. Total employment was reduced by 142 FTE positions (−0.6%) when legume production doubled (scenario 1b) and by 350 FTE positions (−1.4%) when legume production was at maximum (1c). The reason for the relatively small effect of this scenario is that the employment reduction in the pork sector—where employment was reduced by 618 FTE positions (−47%)—is to a large extent compensated by an increase in legume production where employment increased by 391 FTE positions (+625%).

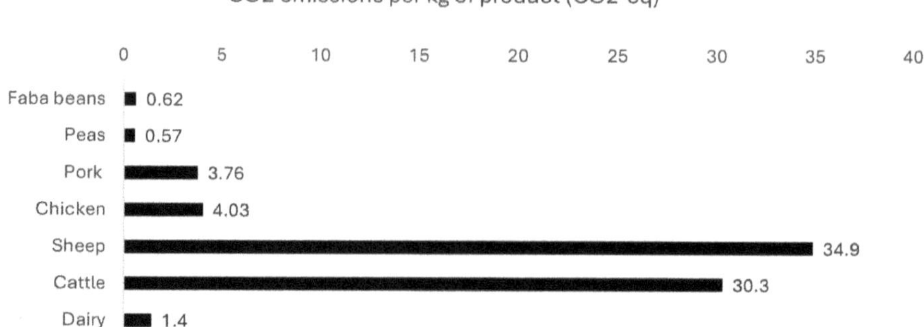

Fig. 2 CO_2-emissions per kg of legumes, livestock and dairy. *Data sources* Møller and Samson-stuen (2024), Svanes et al. (2022)

In scenario 2, we see that minimising GHG emissions from production had the greatest impact on overall employment. Total FTE employment is reduced by 4720 FTE positions (-18%) when legume production doubled (scenario 2b) and by 7062 FTE positions (-27%) when legume production was at maximum (2c). The explanation is that sheep, cattle and dairy are both the most emissions-intensive and the most labour-intensive types of production (see Figs. 1 and 2). The biggest single employment effect in the emissions-reductions approach was found in cattle production where employment was reduced by 3227 FTE positions (-26%) in scenario 2b and by 4716 FTE positions (-38%) in 2c.

In scenario 3, we see that an even percentage reduction of all livestock also has a significant impact on employment, but the overall effect is weaker than in scenario 2. In scenario 3b, total FTE employment is reduced by 1535 FTE positions (-6%) and in 3c by 5600 FTE positions (-22%). The biggest single reduction was again found in cattle production where estimated employment was reduced by 993 FTE positions (-8%) in scenario 3b and by 3723 FTE positions (-30%) in 3c.

Effects on GHG emissions are shown in Fig. 4. In scenario 1, we see that maximising employment and cultural landscapes in Norway had little impact on GHG emissions in all variations. The biggest reduction in this scenario was in 1c, which had a reduction of 406 CO_2-eq. tonnes (-7%) from baseline. GHG emissions from pork and chicken production are nearly halved (-47% for both), but overall, reducing pork and chicken consumption will not reduce GHG emissions substantially because they are quite low in GHG emissions compared to other meat products.

In scenario 2, we see that minimising emissions from production and doubling legume production reduced GHG emissions substantially (-17%) but the biggest reduction of all scenarios was found in scenario 2c, which had a reduction of 1494 CO_2-eq. tonnes (-25.4%) from baseline. The primary reason is reduced emissions from cattle production, which had a reduction of 1134 CO_2-eq. tonnes, followed by sheep, which had a reduction of 322 CO_2-eq. tonnes (-38% for both), and finally dairy (-463 CO_2-eq. tonnes, -8%).

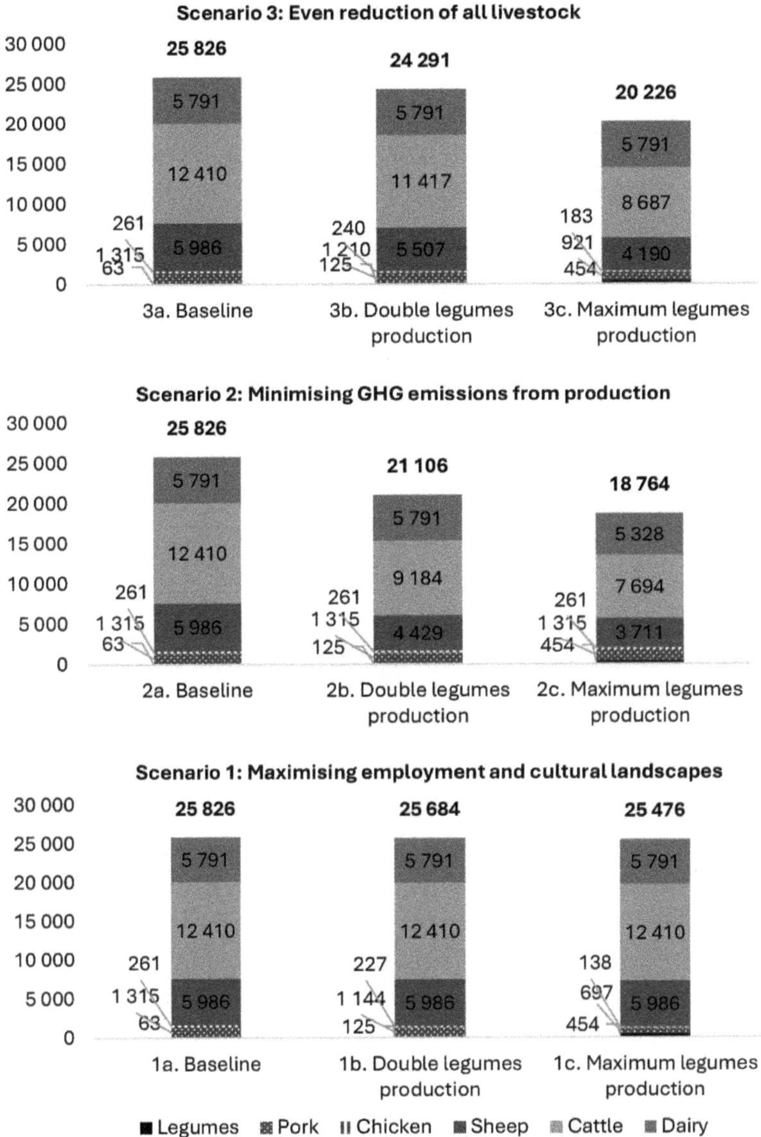

Fig. 3 Estimated employment effects in scenario 1–3. Total employment estimates are displayed at the top of each column

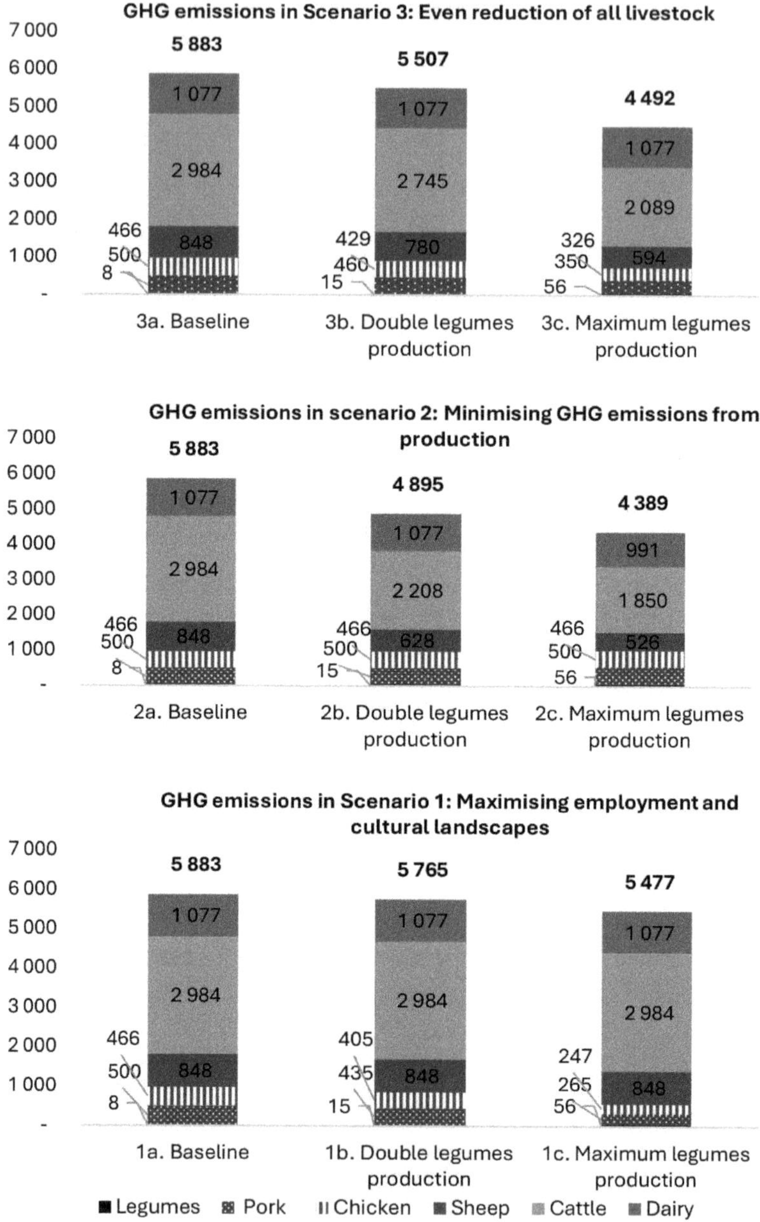

Fig. 4 Estimated GHG emissions in scenarios 1–3. Total emissions are displayed at the top of each column

In scenario 3, we see that an even reduction of all livestock only has a modest reduction of GHG emissions when legume production is doubled (−6% from baseline). However, when legume production is at maximum (3c), the reduction in GHG emissions is much greater—a reduction of 1392 CO_2-eq. tonnes (−23.6% from baseline). This reduction is 1.75 percentage points (103 tonnes) less than the reduction in scenario 2c where the reduction was mainly in cattle and sheep production as opposed to evenly across all livestock. Thus, you can achieve almost the same GHG emissions-reductions in 3c and 2c, but in 3c the impact from reduced employment is spread evenly (in percentage terms) between sectors instead of being concentrated in cattle and sheep production.

4 Discussion and Conclusions

There is a potential for increasing the production of legumes in Norway to replace meat in the Norwegian diet, which would reduce GHG emissions and improve public health stemming from a healthier diet. A substantial reduction of meat consumption in favour of legumes as a source of protein can in some of our scenarios reduce GHG emissions from Norwegian food production dramatically (scenarios 2c and 3c), but even a modest reduction of cattle and sheep can reduce GHG emissions from food production substantially (scenario 2b). Due to the nitrogen fixing effect of legumes, such a transition would also contribute to increasing the sustainability of cereal production in Norway.

However, as we have shown, an increase in the consumption of legumes produced in Norway, paired with calorie-equivalent decreases in the consumption of meat and dairy, will lead to a net reduction in employment in the agricultural sector at the primary level. This is in line with a similar study (Mittenzwei et al. 2020). When the increased consumption of legumes is met with a reduction in sheep, cattle and dairy consumption (scenario 2c), the impact on employment is far stronger than if only pork and chicken production is reduced. Chicken production, when not counting the production of the feed, is less labour-intensive than legume production. Thus, our results show that there is an inherent trade-off between aims of maximising employment and cultural landscapes (scenario 1) and minimising GHG emissions from production (scenario 2). A reduction in GHG emissions cannot be achieved without impacting employment, but scenario 3c shows that one can achieve a significant emissions reduction whilst spreading the employment impact more evenly between sectors, on the one hand, and rural and urban areas, on the other.

To what extent a reduction in employment has a negative societal impact depends on whether the country has a situation of surplus labour and need for employment opportunities, or, the contrary, a lack of available labour force. If the aim is to maintain employment and income opportunities in rural areas, the societal impact also depends on to what extent a reduction in meat consumption is met with a reduction in livestock production in rural or more centralised areas with more employment opportunities. This will depend on the type of agricultural policies and subsidies that are implemented, and to what extent they are favouring livestock production in the rural areas. It should also be considered that cattle and sheep production is highly subsidised in Norway (sheep farmers get 67% of their income from subsidies) (Rustad 2024a), which means that reduced production would release funds that could be used for instance to finance more extensive agricultural practices that generate more employment and better maintain cultural landscapes.

Producing legume-based products in Norway can generate new employment opportunities at the processing level if they are used in the production of, for instance, vegetarian burgers or bread spreads. However, employment will also disappear at the processing level for livestock and dairy products, and we do not know if new legume-based employment will compensate for these. Moreover, whilst factories processing legumes are generally located in urban areas, cattle and sheep production generate employment in rural areas where there are fewer employment opportunities. And, in addition to their role as employment generators, cattle and sheep are grazing animals that can maintain cultural landscapes, which can contribute to biodiversity and have important sentimental values for many people. On the other hand, all meat produced in Norway is highly reliant on the import of concentrate feed, which is putting environmental pressure on land areas in other countries.

To conclude, there is a trade-off between improving public health and reducing GHG emissions in food production, on the one hand, and maintaining employment in food production and cultural landscapes in the districts of Norway, on the other. If pork and chicken production is reduced, there is less reduction in employment, but also less reduction in GHG emissions. If sheep and cattle is reduced, the impact on GHG emissions is far greater, but there is also a far greater effect on employment, which often will be in the rural districts of Norway. An even reduction of all livestock production presents a potential middle ground where the emissions reduction is still significant, but the employment impact is spread more evenly between sectors and rural and urban areas.

References

Abrahamsen U, Uhlen AK, Waalen W, Stabbetorp H (2019) Muligheter for økt proteinproduksjon på kornarealene, Jord-og Plantekultur 2019. Forsøk i korn, olje-og proteinvekster, engfrøavl og potet, vol 2018

Beyene SM, Naumov V, Angelstam P (2024) Long-term dynamics of grasslands and livestock in Norwegian cultural landscapes: implications for a sustainable transition of rural livelihoods. Landsc Ecol 39:171

European Commission (2025) Food-based dietary guidelines recommendations for meat, health promotion and disease prevention knowledge gateway

Farstad M, Vinge H, Straete EP (2021) Locked-in or ready for climate change mitigation? Agri-food networks as structures for dairy-beef farming. Agric Hum Values 38:29–41

Gonera A, Millford AB (2018) The plant protein trend in Norway: market overview and future perspectives. NOFIMA

Gonera A, Svanes E, Bugge AB, Hatlebakk MM, Prexl K-M, Ueland Ø (2021) Moving consumers along the innovation adoption curve: a new approach to accelerate the shift toward a more sustainable diet. Sustainability 13

Hjukse O (2022) Totalkalkylen for jordbruket. Jordbrukets totalregnskap 2020 og 2021. Budsjett 2022. Totalkalkylen for jordbruket, Ås: NIBIO

Hovland I (2022) Handbok for driftsplanlegging 2022/2023, Ås: NIBIO

Hughes J, Pearson E, Grafenauer S (2022) Legumes—a comprehensive exploration of global food-based dietary guidelines and consumption. Nutrients 14:3080

Mittenzwei K, Walland F, Milford AB, Grønlund A (2020) Klimakur 2030. Overgang fra rødt kjøtt til vegetabilsk og fisk

Møller H, Samsonstuen S (2024) Life cycle assessment of meat and egg—Nortura, Fredrikstad: NORSUS

Norwegian Agriculture Agency (2025) Kraftfôrstatistikk, Oslo

Norwegian Directorate of Health (2024) Utviklingen i Norsk Kosthold 2024, Oslo

Rustad LJ (2024a) Ti fakta om bønder og tilskudd, Ås: NIBIO

Rustad LJ (2024b) Totalkalkylen for jordbruket: Jordbrukets totalregnskap 2022 og 2023. Budsjett 2024. Totalkalkylen for jordbruket, Ås: NIBIO

Singh B, Singh JP, Shevkani K, Singh N, Kaur A (2017) Bioactive constituents in pulses and their health benefits. J Food Sci Technol 54:858–870

Smart Protein (2021) Plant-based foods in Europe: how big is the market, smart protein plant-based food sector report by smart protein project, European Union's horizon 2020 research and innovation programme

Stagnari F, Maggio A, Galieni A, Pisante M (2017) Multiple benefits of legumes for agriculture sustainability: an overview. Chem Biol Technol Agricu 4

Statistics Norway (2024) Table 08806: external trade in goods, Oslo

Svanes E, Waalen W, Uhlen AK (2022) Environmental impacts of field peas and faba beans grown in Norway and derived products, compared to other food protein sources. Sustainable Prod Consumption 33:756–766

UNEP (2023) What's cooking? An assessment of the potential impacts of selected novel alternatives to conventional animal products. Frontiers

WHO (2023) Red and processed meat in the context of health and the environment: many shades of red and green, information brief. World Health Organization

Willett W, Rockstrom J, Loken B (2019) Food in the anthropocene: the EAT-lancet commission on healthy diets from sustainable food systems. Lancet 393:2590

Policy, Regulatory Frameworks and Standards

Green Public Procurement:
A Decision-Support Tool for Quantifying Environmental Impacts of Asphalt Mixture Projects

Shaffie Juma, Ben Moins, Wim Van den bergh, and David Hernando

Abstract This paper presents a tool developed to quantify and analyse the environmental impact of asphalt mixture projects. It is intended to support green public procurement practices by aiding road owners in making informed decisions about the environmental impact of the asphalt mixtures they procure. The tool can assess 19 environmental impact categories, which can be combined into a single score using the European Union's recommended Product Environmental Footprint (PEF) methodology. It covers raw material acquisition, asphalt mix production, material transportation, construction, deconstruction and end-of-life phases. The tool allows users to select materials for the asphalt mix, energy sources, transportation modes and operational machinery configurations variables. Lastly, a sensitivity analysis provides insights into how the different variables affect the global warming potential and single score.

1 Introduction

Green Public Procurement (GPP) tools are vital for road stakeholders to quantify the environmental impacts of asphalt pavements. In the road infrastructure industry, such tools help assess whether the asphalt works procured are in line with environmental policy goals, such as the European Green Deal, EU Climate Laws or the Circular Economy Action Plan (European Commission 2019, 2020; European Parliament 2021; Moins et al. 2025). Tools such as DuboCalc Rijkswaterstaat (2000), AsPECT (Ortiz et al. 2009) and the regional Flemish agency for roads and traffic GPP tool (Van and Wegenbouwkunde 2016) fail to incorporate the complex trade-offs and choices

S. Juma (✉) · W. Van den bergh · D. Hernando
Sustainable Pavements and Asphalt Research (SUPAR), Faculty of Applied Engineering,
University of Antwerp, Antwerp, Belgium
e-mail: Shaffie.Juma@uantwerpen.be

B. Moins
Energy and Materials in Infrastructure and Buildings (EMIB), Faculty of Applied Engineering,
University of Antwerp, Antwerp, Belgium

© The Author(s) 2026

M. Traverso et al. (eds.), *Life Cycle Management from Global to Local*,
https://doi.org/10.1007/978-3-032-17987-6_33

available to contractors, such as alternative binder types, varying energy sources in the asphalt plant, recycled materials options, stockpile moisture and different classes of construction and deconstruction machines from higher to lower power ratings. Studies from the Federal Highway Administration (FHWA) have highlighted the importance of comprehensive LCA frameworks to address these limitations (Harvey et al. 2016). This paper introduces a decision-support tool for quantifying environmental impacts of asphalt mixture projects to address such challenges. The GPP tool is intended to support green public procurement practices in the European Union (EU) by aiding road owners in making informed decisions about the environmental impact of the asphalt works they procure. In contrast to conventional LCA tools typically used by LCA experts, the proposed GPP tool is tailored to road authorities and contractors. For example, it enables users to input key variables such as energy mixes and paving machinery types based on power ratings. It further allows the user to input multiple transportation modes such as trucks, barge, ships, trains and their respective fuel types. The tool is aligned with ISO 14040/44 and uses EF 3.1 methodology to assess 19 midpoint impact categories. A single score is also calculated through normalisation and weighting. Whilst the tool follows the Product Environmental Footprint (PEF) principles, no asphalt-specific PEF Product Category Rule (PEFPCR) exists yet (Nickel 2024). Nonetheless, this paper presents a GPP tool which can be adapted to follow the PEF methodology once the final PEFPCR is ready.

2 Tool Concept and Design

This section presents the tool's system boundaries, assessment method, user input and results output.

2.1 *System Boundaries*

The tool covers both cradle-to-gate (declared unit) and cradle-to-grave (functional unit). The cradle-to-gate analysis is expressed as environmental impacts per tonne of mixture and covers modules A1, C1, C2', C3', A2, A3 and D as shown in Fig. 1. In addition, the cradle-to-grave analysis is expressed in environmental impacts per square metre for a given layer thickness and covers modules A1, A2, A3, C1, C2', C3', A4, A5, C1 and D as shown in Fig. 2. Modules B1, B2 and C4 are excluded. The system boundaries ensure the environmental impacts of the most relevant stages of a pavement's life cycle which are evaluated considering the service life of the asphalt mix and pavement thickness.

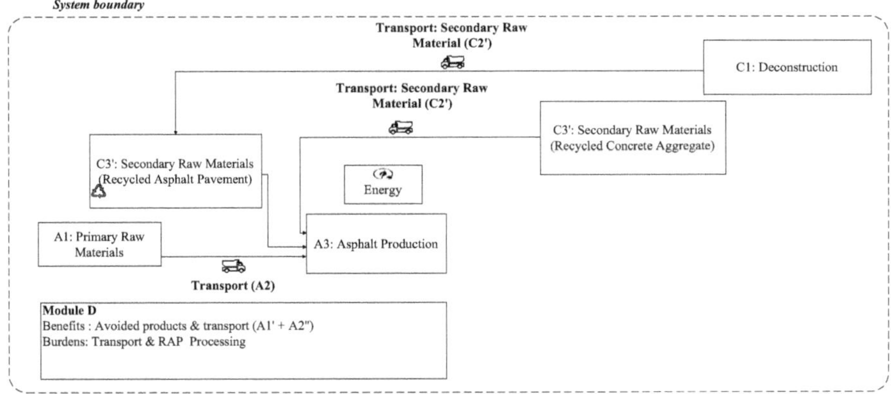

Fig. 1 System boundaries A1–A3, D (Declared unit: impact/ton)

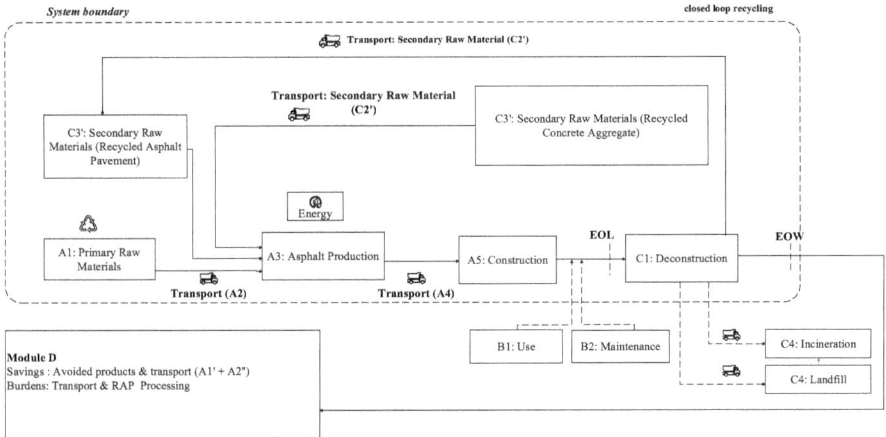

Fig. 2 System boundaries: A1–D (Functional unit: impact/m^2)

2.2 Environmental Impact Assessment Method

The LCA tool employs the EF 3.1 method to quantify 19 environmental impact midpoint categories. These categories range from climate change to water use. Whilst the characterisation models are based on EF 3.1, the midpoint impact categories are aligned with those defined in EN15804+A2. The life cycle inventory data is retrieved from eco-invent version 3.11. The standard attributional allocation cutoff by classification was used. The midpoint impact categories were subsequently normalised and aggregated into a single score which is expressed in milli point (mPt). This single score enables the comparison of products and systems based on the weighted aggregation of all impact categories.

2.3 User Input

The GPP tool's input section was organised to reflect real-world road construction activities. This means that a non LCA experts like contractors and road agencies can easily input information into the tool. The structure of the GPP tool is described below.

2.3.1 General Information

This section gathers project-specific information, including the mixture identification, mixture type, binder content and binder replacement rate (for Recycled Asphalt Pavement (RAP) mixtures).

2.3.2 Mixture Production

The tool gathers detailed information about the production of asphalt mixtures to ensure accurate environmental impact calculations. Users input the location of the asphalt plant and the type of plant, along with the average hourly production rate (tonne/h). To assess energy consumption, the tool requires the user to provide the following information:

- Primary and secondary energy sources for heating and drying processes, including diesel, natural gas and electricity.
- Electrical component utilisation (auxiliary power consumption at the plant).

The operational input parameters for the asphalt plant include the following:

- Mixture production temperature range. It depends on the type of mix; hot mix: 155–185 °C; warm mix: 110–145 °C; or cold mix: 60–95 °C).
- Bulk density of the mix, expressed in kg/m3 (i.e. used in scaling impacts per declared tonne).
- Average annual daily production temperature.

The thermodynamic energy principle quantifies the energy needed to heat and dry the raw materials by considering the specific heat capacities, moisture content and the production temperatures of the asphalt mix composition materials.

2.3.3 Layer Specifications

The user inputs the layer thickness, which is used to calculate the mass per functional unit by considering the bulk density. This allows the tool to model the cradle-to-grave impacts.

2.3.4 Material Composition and Transportation

The user provides the following information on asphalt mix components:

- Type and percentage weight of the asphalt mix components.
- Stockpile coverage (covered stockpiles have lower moisture content than uncovered ones).
- Biogenic origin (for asphalt binder).
- Origin of the material, distance to the asphalt plant and transportation of the asphalt mix to the construction site.
- Detailed transport routes (up to three legs).

2.3.5 Construction and Deconstruction

In construction stage, users specify the key operational parameters related to the machines. In this stage, the user selects the type of machine (i.e. paver, roller, excavator) used in the project according to its power rating. For the selected machines, the user must categorise the project as small-, medium- or large-scale, corresponding to the respective road construction project. Small-scale projects can be related to bicycle paths, medium-scale projects represent motorways, whilst large-scale projects can be airport runaways or highways construction. The project classification determines the energy the machine consumes because each machine within a project class will have a different power rating. Finally, the user inputs the fuel source (i.e. diesel, biodiesel or electric) for the respective machine. During the deconstruction stage, the user selects the type of cold in-place milling machines also according to project scale (small, medium or large). The user also has the option to choose a sweeper and specify the fuel source of the deconstruction machinery.

3 GPP Output

The results tab presents quantitative tables and visual dashboards across various life cycle stages and impact categories. This offers detailed environmental performance, allowing the user to interpret the results effectively. The results are presented in declared and functional units. For each system boundary, the tool computes and shows

(1) Single score (in mPt/tonne for declared unit and mPt/m^2 for functional unit).
(2) 19 environmental impacts (impacts/tonne and impacts/m^2).
(3) Contribution analysis in which GWP and single scores are broken down by life cycle stage (e.g. A1, A2, A3, D).
(4) GWP and single score results are presented according to activity (e.g. asphalt production, construction, deconstruction).
(5) Single score contributions by impact category.

3.1 GPP in Decision-Making

The results computed in the GPP tool are significant because

(1) Procurement authorities can assess asphalt mixtures based on the total PEF and GWP scores.
(2) Contractors can identify which activity within each life cycle stage contributes most significantly to the overall environmental impacts.
(3) The tool allows the evaluation of the environmental impact trade-offs in terms of benefits and burdens of using different asphalt mixture components (e.g. aggregates, types of binders, additives).

4 Sensitivity Analysis of Asphalt Mix Scenarios

A one-at-a-time sensitivity analysis was conducted for a reference mix to demonstrate how GPP tool users can make environmentally sustainable asphalt work choices whilst keeping another parameters constant.

4.1 Reference Mix Design and Transport Summary

The reference mix used for the analysis was a standard AC20 base layer with a thickness of 80 mm and a mix service life of 25 years. An overview of the asphalt mix components and their transportation summary are illustrated in Table 1.

Other parameters used include

- Production temperature: 155–185 °C.
- Asphalt plant energy source: Natural gas.
- Distance from plant to site: 43 km.

Table 1 Reference mix design per tonne asphalt mix and transportation summary

Component	Mode of transport	Source distance (km)	% Weight (%)
Virgin binder	Euro 5 truck	61	1.75
Virgin crushed coarse aggregate 1 (>2 mm)	Euro 5 truck	128	32.91
Virgin crushed coarse aggregate 2 (>2 mm)	Euro 5 truck	128	10.10
Virgin natural fine aggregate 1 (≤2 mm)	Euro 5 truck	95	5.30
Filler limestone residue	Euro 5 truck	124	0.09
Recycled asphalt pavement	Euro 5 truck	43	49.85

- Construction site machinery: Conventional diesel (one paver and two rollers).
- Deconstruction site machinery: Conventional diesel (one cold milling machine and one truck sweeper).

4.2   Sensitivity Scenarios

The one-at-a-time sensitivity analysis was conducted by changing five parameters of the GPP tool to assess the environmental impacts (GWP and Single score). The scenarios retained the same AC20 mix design, transportation distance and project layout to isolate the effects of the operational variables. Table 2 outlines the scenarios.

Table 2 Sensitivity analysis scenarios

Scenario	Variable adjusted	Alternative option
S1	Stockpile coverage	All stockpiles are covered before production
S2	Heating and drying energy source	Asphalt plant energy source changed from natural gas to heavy fuel oil
S3	Heating and drying energy source	Asphalt plant energy source changed from natural gas to hydrogen grey
S4	Heating and drying energy source	Asphalt plant energy source changed from natural gas to propane
S5	Heating and drying energy source	Asphalt plant energy source changed from natural gas to electricity (Belgian grid)
S6	Heating and drying energy source	Energy source changed from natural gas to electricity (green renewable)
S7	Transportation fuel	Truck fuel source changed from diesel to 100% biodiesel
S8	Transportation fuel	Truck fuel source changed from diesel to 100% electricity (Belgian grid)
S9	Transportation fuel	Truck fuel source changed from diesel to 100% electricity (green renewable)
S10	Machinery equipment	Machinery fuel source changed from diesel to 100% biodiesel
S11	Machinery equipment	Machinery fuel source changed from diesel to fully electric machinery (Belgian grid)
S12	Machinery equipment	Machinery fuel source changed from diesel to fully electric machinery (green grid)

4.3 Sensitivity Analysis Results Summary

The GPP tool calculated each scenario's GWP (in $kgCO_2$ eq/m^2) and the aggregated single score. Figures 3 and 4 illustrate the percentage change in impact (ΔGWP [%] and ΔSingle score [%]) resulting from one-at-a-time changes of key operational variables such as transport fuels, plant energy sources and stockpile management practices. All the scenarios results are expressed in functional unit of 1 m^2 of an 80 mm thick layer. It is important to note that the environmental impacts associated with the manufacturing of the paving and compaction machines and trucks (including conventional and electric) were not included in this assessment. Only the fuel and electrical consumption was considered.

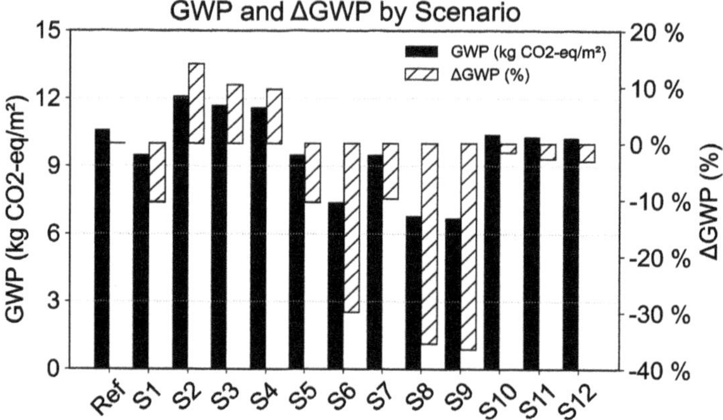

Fig. 3 Global warming potential scores and sensitivity analysis results

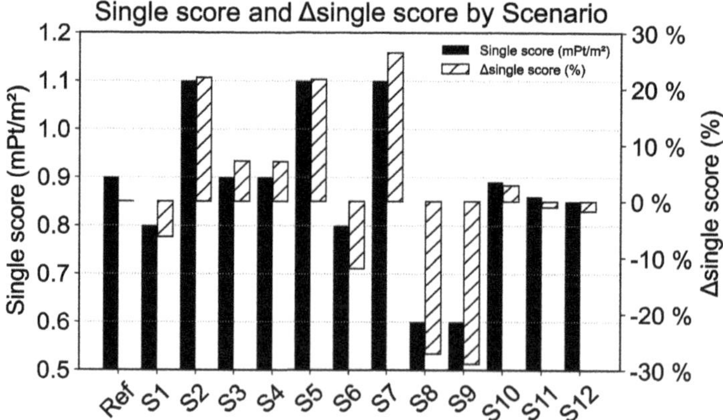

Fig. 4 Single scores and sensitivity changes results

4.3.1 Impact on Global Warming Potential (GWP)

As shown in Fig. 3, the electrification of transport (especially to green electricity in S9) shows significant GWP reductions of approximately 37%, followed by electrification using the Belgian electricity grid mix, which results in approximately 35% reduction in GWP as seen in S8. Considering operations in the asphalt plant, in S2, an asphalt plant powered by heavy fuel oil showed a GWP increase of approximately 11%, whilst replacing the plant energy source from natural gas to electric (S6) (powered by green renewables) results in a 29.8% reduction in GWP. The sensitivity results showed that covering aggregate stockpiles results in a 10% reduction in GWP, as seen in S1. This indicates that the energy required for drying and heating the aggregates is reduced by covering the stockpiles. Regarding the machinery equipment, using biodiesel (S10) results in a slight GWP reduction of −1.6%, whilst switching to Belgian grid electricity (S11) and green electricity (S12) resulted in slightly greater reductions of −2.7% and −3.1%, respectively.

4.3.2 Impact on Product Environmental Footprint (PEF) Score

A similar trend can be observed in the single score results (Fig. 4). Once again, the electrification of transport (especially to green electricity) demonstrates reductions of approximately 29%, with trucks powered by the Belgian grid following closely at a 27.1% reduction. Interestingly, trucks powered by 100% biodiesel exhibit a 26.4% increase in single score. This suggests that although biodiesels may reduce CO_2 emissions, the higher impacts are likely due to acidification, eutrophication, particulate matter and land use, due to agricultural and processing burdens in its lifecycle as shown in Fig. 5. This highlights the need of reporting all indicators alongside the aggregated single score to ensure all trade-offs are interpreted. Considering asphalt plant operations, switching the energy source from natural gas to heavy fuel oil (S2) increases the single score by 22%. This is likely due to high carbon content and sulphur impurities leading to higher impacts in climate change, acidification, particulate matter and ecotoxicity. Conversely, switching to electric power (from green renewables) in S6 results in an 11.9% reduction in single score although fresh water eutrophication, human toxicity (carcinogenic), particulate matter formation and water use greatly influenced the single score (see Fig. 6). The results show the importance of considering multiple impact categories beyond the climate change impact category. The electrification of construction machinery using the Belgian grid to green electricity leads to approximately a −1.0% and 1.7% reduction in single score. Lastly, the impact of covering stockpiles (S1) shows a 6.3% reduction in single score.

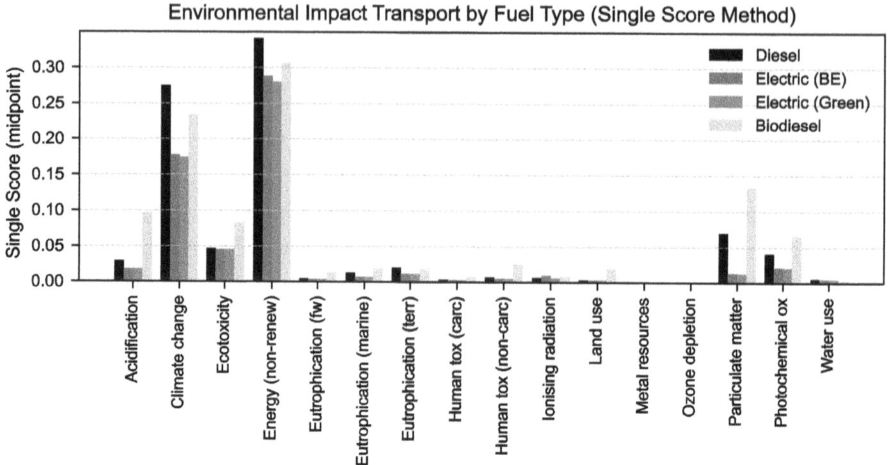

Fig. 5 Single score for each impact category for transportation fuels

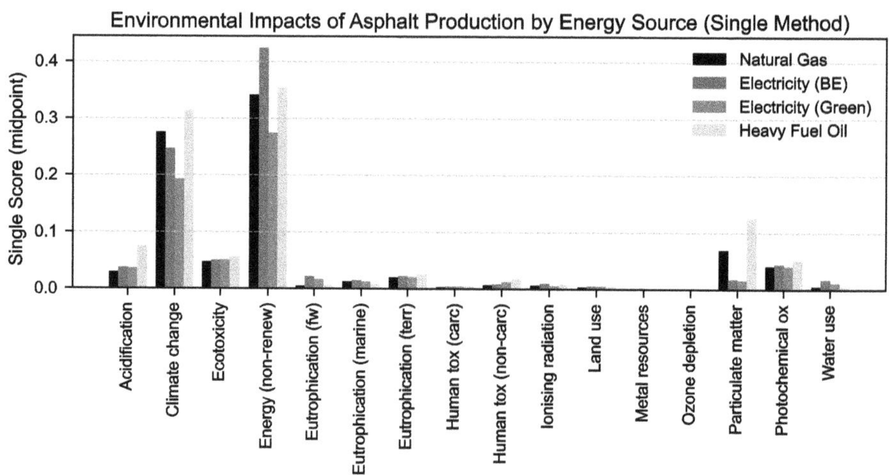

Fig. 6 Single score for each impact category for asphalt plant energy sources

5 Conclusion

This paper has presented a GPP tool which has been designed to quantify the environmental impacts of asphalt mixture projects. The tool's use of the EU's EF 3.1 model and ISO 14040/44 standards enables stakeholders to make well-informed decisions regarding environmental sustainability on the asphalt works they procure. The tool is simply designed for the use of contractors and non LCA experts and models real-world variables of road construction activities such as energy sources, transportation modes, transportation distances and machineries. The sensitivity analysis shows that implementing renewable electricity in asphalt production and transportation lowers the GWP, whilst the use of biodiesels may increase the aggregated single score. Whilst the tool applies the EU PEF 3.1 method, it does not fully comply with an official PEFCR specific to asphalt, as such PEFCR is not yet available. In addition, the tool was developed from Belgian contexts, using region-specific energy mixes and construction practices which affects the comparability of results. The accuracy of the results depends on the quality of input data (e.g. emission factors, transportation distances, energy consumption). In some cases, general information from the eco-invent database was used. The tool assumes static conditions over time and does not consider degradation rates of the machines, maintenance cycles or service life predictions of the asphalt mixture. Future research aims to incorporate such temporal dynamics.

Funding and Acknowledgements This research was funded by the European Union's Horizon 2020 research and innovation programme under Grant Agreement no 101037564 (PORTable Innovation Open Network for Efficiency and Emissions Reduction Solutions—PIONEERS). The authors would like to further acknowledge all our 47 partners in close coordination with the Port of Antwerp-Bruges for providing valuable insights for the tool. The GPP tool is currently being demonstrated within the Digital Information Management in the Infrastructure (DIMinfr@) COOCK+ project, which aims to support digital innovation and sustainability in the Flemish infrastructure.

References

European Commission (2019) The European green deal, Brussels

European Commission (2020) A new circular economy action plan: for a cleaner and more competitive Europe, Brussels

European Parliament (2021) Establishing the framework for achieving climate neutrality and amending regulations (EC) No 401/2009 and (EU) 2018/1999 ('European Climate Law'). Off J Eur Union, L 243:1–17

Harvey J, Meijer J, Ozer H et al (2016) Pavement life cycle assessment framework

Moins G, Dupont A, Janssen M (2025) Green procurement and LCA integration in asphalt projects. Int J Sustain Infrastruct 15:115–129

Nickel L (2024) Product environmental footprint (PEF) overview. Ecochain. https://ecochain.com/blog/product-environmental-footprint/. Accessed 12 July 2025

Ortiz O, Castells F, Sonnemann G (2009) Sustainability in the construction industry: a review of recent developments based on LCA. Constr Build Mater 23:28–39. https://doi.org/10.1016/j.conbuildmat.2007.11.012

Rijkswaterstaat (2000) What is DuboCalc. DuboCalc Portal. https://www.dubocalc.nl/en/. Accessed 13 July 2025

Van D, Wegenbouwkunde T (2016) Green public procurement. Accessed 13 July 2025

Bridging Standards: Harmonising Sustainability Assessment Methods for Biobased Products in Construction

Julia Weißert, Matthias Müller, and Simon Panik

Abstract The integration of sustainability principles in the construction sector, especially through biogenic products, is increasingly essential due to environmental concerns. However, there are various Life Cycle Assessment (LCA) methodologies and standards, hindering a clear and unified assessment and comparability between similar products. The European Union has established the Product Environmental Footprint (PEF) methodology, yet the specific PEF Category Rules (PEFCRs) for biogenic construction products are still pending. Meanwhile, the existing Product Category Rule (PCR) for EN 15804 offers established guidelines for Environmental Product Declarations (EPDs). This study aims to link PEF and EN 15804 for biogenic construction products, providing a methodology to assess these products according to PEF principles, by closing the gap and providing a clear methodology that enables the environmental assessment of biobased construction materials.

1 Introduction

As the construction industry grapples with the dual challenges of environmental degradation and resource depletion, integrating environmental sustainability principles has become paramount (UNEP 2024). Biogenic building products derived from renewable biological resources are essential in this transition (Dubois et al. 2016) by offering opportunities for reduced carbon emissions, even for temporary carbon storage, and enhanced resource efficiency. Furthermore, they contribute to circular economy through the biogenic nature of the materials (Chen et al. 2024; Fischer and Losacker 2025). However, the adoption of biogenic products is influenced by several factors, such as market awareness, consumer preferences and regulatory frameworks, but also the availability of standardised assessment methodologies for decision-making support is important to consider (Dong et al. 2023; Friedrich 2022; Bahramian and Yetilmezsoy 2020). The way biogenic products are assessed has a huge influence on the statements and recommendations derived from the assessment.

J. Weißert · M. Müller (✉) · S. Panik
Institute for Acoustics and Building Physics, University of Stuttgart, Stuttgart, Germany
e-mail: matthias.mueller@iabp.uni-stuttgart.de

M. Traverso et al. (eds.), *Life Cycle Management from Global to Local*,
https://doi.org/10.1007/978-3-032-17987-6_34

411

The current landscape of Life Cycle Assessment (LCA) methodologies is standardised by general frameworks such as ISO 14040 (DIN e.V. 2021a) and ISO 14044 (DIN e.V. 2021b), which provide essential guidelines for conducting LCA and ensure a robust assessment process. Yet LCAs carried out over the last 2 decades are hardly comparable due to divergent scopes (Bahramian and Yetilmezsoy 2020). Addressing this issue, the European Union (EU) has established the Product Environmental Footprint (PEF) as a comprehensive approach for evaluating the environmental performance of products, aimed at fostering consistency across the EU market (European Commission 2021). However, the specific PEF Category Rules (PEFCRs) for construction products are not yet developed, creating uncertainty in their assessment (European Commission, Green Forum—Product Environmental Footprint method). Conversely, the EN 15804 standard (DIN e.V. 2022) offers established guidelines for Environmental Product Declarations (EPDs) in the construction sector, providing a detailed framework for assessing environmental impacts of construction products and even providing a Product Category Rule (PCR) for timber products (The Norwegian EPD Foundation 2019). Nonetheless, significant discrepancies exist between the PEF and EPD methodologies, particularly in aspects such as end-of-life modelling and biogenic carbon management, exacerbated by the lack of uniform databases and requirements (Pscherer and Krommes 2025).

The necessity for a standardised assessment framework becomes evident when considering the implications of these differing methodologies. A coherent approach would not only facilitate accurate environmental reporting, but also enhance transparency and trust amongst stakeholders in the construction sector (Haverkamp et al. 2025). This study aims to bridge the gap between EPD and PEF standards, focusing specifically on biogenic construction products, with particular emphasis on the performance of a timber slab, in accordance with PEF principles. The overarching goal is to foster a user-friendly application of the PEF, ultimately supporting the circular economy from the earliest design stages, promoting comparable and transparent LCA results, whilst minimising the ecological footprint associated with biobased construction materials. The structure of this paper is as follows:

- Section 2 describes the state of the art by evaluating the existing frameworks, pointing out their similarities and differences.
- Section 3 deals with the research approach for the proposed methodological framework, harmonising EN 15804 and PEF.
- Section 4 presents the results of a case study applying the proposed methodology.
- Section 5 discusses the strengths and weaknesses of the proposed methodology.
- Section 6 terminates the study by presenting essential findings and outlining prospective directions for further research potentials.

2 State of the Art

In Europe's construction sector, environmental assessments are predominantly conducted using the EN 15804 standard (DIN e.V. 2022). This framework underpins Environmental Product Declarations (EPDs) (DIN e.V. 2011) and is widely adopted for its sector-specific relevance and structured modular reporting. In contrast, the Product Environmental Footprint (PEF) (European Commission 2021) was designed as a harmonised LCA-based methodology applicable across all product categories. Although methodologically robust, PEF remains relatively underutilised in the construction industry, partly due to the absence of finalised PEF Category Rules (PEFCRs) for building products, which intend to provide specific rules for a study's product, and the lack of programme operators to support its consistent application (Haverkamp et al. 2025; Salehi 2020). PEF emphasises comparability through normalisation and weighting, ultimately producing a single-score environmental impact metric. This design supports consumer-oriented communication and policy alignment across sectors (Salehi 2020). Conversely, EN 15804 focuses on traceability and modularity, catering to professionals in construction by providing stage-specific environmental data without aggregation (DIN e.V. 2022). Having two partly contradicting approaches decreases the comparability and reproducibility of studies (Haverkamp et al. 2025).

A core methodological distinction lies in the treatment of End-of-Life (EoL) scenarios. EN 15804 applies a 'cut-off' approach, allocating all environmental burdens to the product system and reporting recycling or recovery benefits externally in a separate life cycle module (Module D). This structure separates the core LCA results from future-oriented credits and burdens, which is particularly advantageous, trustworthy and transparent given the long-time horizons associated with building lifespans (Durão et al. 2020). PEF, on the other hand, integrates the Circular Footprint Formula (CFF) directly into the LCA results. The CFF allocates burdens and credits between life cycles based on an allocation factor 'A', which reflects market conditions for recyclability and recycled content. This difference can lead to significant variations in results (Salehi 2020). Whilst EN 15804 provides no standardised procedure for defining recycling rates, and PEF's default values are incomplete, both frameworks rely heavily on practitioner assumptions.

EN 15804 explicitly and transparently identifies the temporary storage of biogenic carbon that is embedded in organic matter after sequestration from e.g. atmosphere via photosynthesis. Biogenic carbon stored in the product is taken into account in the balance and is released again in the end-of-life phase (DIN e.V. 2022). In the PEF approach, both storage and release of the biogenic carbon is treated as neutral to avoid misleading double-counting and facilitate reporting (European Commission 2021).

In summary, whilst PEF offers a harmonised and theoretically consistent framework, EN 15804 remains more relevant and practical for the construction sector due to its tailored modular structure and widespread institutional support. Bridging these methodological gaps will be crucial for enhancing transparency, comparability and effectiveness in the environmental assessment of construction products and brings the development of a PEFCR closer.

3   Research Approach and Methodological Steps

In many aspects, PEF is more detailed and comprehensive than EN 15804, as seen in the definition of Functional Unit (FU) or benefits and loads integration into the product system (Manfredi et al. 2015). In other points, such as the cut-off criteria, EN 15804 is more generously allowing for 5% cut-offs instead of 3% (European Commission 2021; DIN e.V. 2022). As EN 15804 is the more commonly used approach to calculate LCAs in the building and construction context, the proposed methodology starts from the EN 15804 standards and points out linkages and bridges to PEF. Aiming to provide the reader with an understanding of what to consider when changing from EPD LCAs to PEF LCAs.

3.1   Goal and Scope Definition

The goal and scope of the LCA delineate its boundaries and objectives. According to EN 15804, the legitimacy of several system boundaries is contingent upon the consideration of different life cycle stages and comprises production (A1–A3), production and EoL (A1–A3 + C1–D) or the entire life cycle (A–D), amongst others. However, for a PEF study, the life cycle in its entirety must be taken into account. To align the approaches, the PEF scope is adopted. Moreover, this provides a more granular description of FU, which is in accordance with the definition based on EN 15804. It is incumbent upon the practitioner to respond to the following questions: who are the intended applicants, what is the study purpose and decision context, who is the target audience? To define FU, it is first necessary to define the element that is to be assessed (What?). The product must be quantified (How much?) and the duration of service life must be defined (How long?). Also, to accelerate the development of PEFCRs and increase harmonisation potential, a representative product must be declared, which is classified under CPA/NACE code (Eurostat n.d.).

3.2   Life Cycle Inventory

In the context of life cycle inventory, it is imperative to collect pertinent product-related data. This encompasses material quantities, energy consumption during the fabrication process and end-of-life routes. The collected data should be maintained in a modular format as foreseen in EN 15804, categorised according to materials and processes. In addition, the selection of an environmental database is to be made.

The authors propose—as long as there are major differences in data sets—a dual selection of an EN 15804 database and an EF database (e.g. LCA for Experts).

3.3　*Life Cycle Impact Assessment and Interpretation*

The authors suggest that the modular approach from EN 15804 (Modules A-D) be used to calculate impacts. This involves separating the different life cycle phases (Raw material, manufacturing, use phase, end-of-life) and distinguishing between material (A1) and fabrication (A3) impacts. This facilitates both a hotspot analysis and the use of the CFF, which is needed to calculate the PEF. The utilisation of the CFF facilitates the distribution of loads across multiple product systems. This is a fundamental difference between EPDs and PEFs. In the context of EPDs, the loads or benefits that extend beyond the boundaries of the system are delineated in module D. This is a feature that is superfluous when utilising the CFF. The CFF can be subdivided into 3 equations concerning the use of virgin and secondary material as well as material recycling (1), energy recovery (2) and waste disposal (3).

Material recycling

$$(1 - R_1) \cdot Ev + R_1 \cdot \left(A \cdot E_{recycled} + (1 - A)E_v \cdot \frac{Q_{Sin}}{Q_P} \right)$$

$$+ (1 - A) \cdot R_2 \cdot \left(E_{recyclingEoL} - E_v^* \cdot \frac{Q_{Sout}}{R_P} \right) \qquad (1)$$

Energy recovery

$$(1 - B) \cdot R_3 \cdot \left(E_{ER} - LHV \cdot X_{ER,heat} \cdot E_{SE,heat} - LHV \cdot X_{ER,elec} \cdot E_{SE,elec} \right) \qquad (2)$$

Waste disposal

$$(1 - R_2 - R_3) \cdot E_D \qquad (3)$$

where

A	allocation factor of burdens and credits
B	allocation factor of energy recovery processes
$Q_{S,in}$	quality of the ingoing secondary material
$Q_{S,out}$	quality of the outgoing secondary material
Q_P	quality of the primary (virgin) material
R_1	proportion of material recycled from a previous system
R_2	proportion of material recycled in a subsequent system
R_3	proportion of material undergoing energy recovery at EoL

$X_{ER,heat}$ efficiency of the energy recovery process for heat
$X_{ER,elec}$ efficiency of the energy recovery process for electricity
LHV   lower heating value of the material used in energy recovery
E_V emissions and resources from the extraction of virgin material
E_{*V} emissions and resources from the extraction of virgin materials assumed
               to be substituted by secondary materials
$E_{recycled}$ emissions and resources from the processing of recycled materials
$E_{recycling,\,Eol}$ emissions and resources from the recycling of materials at EoL
E_{ER} emissions and resources from energy recovery processes
$E_{SE,heat/elec}$ emissions and resources from the specific substituted energy source,
               heat and electricity respectively
ED   emissions and resources from the waste disposal.

After running the CFF (1), (2), (3), the impacts must be summed up with the transportation and processing impacts, pursuant to modules A2 and A3 in EPDs. Whereas EPDs distinguish different life cycle phases (modules) with respective environmental impacts, PEF discloses a single score requiring normalisation and weighting, aiming for a methodology that aligns with PEF, normalisation and weighing are inevitable to the proposed method. The interpretation follows with no further specification.

4   Materials and Methods

The project underlying this study aims to develop innovative point-supported timber slab structures in multistorey buildings, offering a more environmentally friendly alternative to reinforced concrete slabs, see Fig. 1. It focuses on creating a universally applicable and affordable building system using complex arrangements of wood lamellas for enhanced structural performance. As an initial step, the introduced methodology is going to be exemplarily showcased on a timber slab.

4.1   *Goal and Scope Definition—Timber Slab*

The goal of this study is to identify hotspots of a timber slab (6×6 m^2) under development, to recommend potential optimisation measures during the early design and development stage. Furthermore, the study aims to apply the proposed methodology. The audience of the study are developers of the timber slab as well as the scientific LCA researchers. The hereby generated results must be seen as preliminary results during the research process. According to the proposed method, the FU must be defined for the point-supported timber slab as seen in Table 1.

System boundaries must be disclosed in both EN 15804 and PEF to define the product system under study, specifying all relevant inputs and outputs, see Fig. 2.

Fig. 1 Base design of a point-supported timber slab

Table 1 Definition of functional unit

Element	Description
What?	Point-supported timber slab used as floor/ceiling in a multistorey timber building
How much?	1 m^2 of timber slab
How well?	3 [kN / m^2] payload (DIN e.V., 2023)
How long?	60 years (The Norwegian EPD Foundation, 2019)
NACE	41.20 (eurostat)

4.2   Life Cycle Inventory—Timber Slab

Both the design phase and the succeeding prototype production of the considered timber slab include high uncertainty and variability regarding functionality, material selection and manufacturing processes, which results in variable inventory data. Many iteration steps are necessary to ultimately meet the given requirements; hence, data acquisition and validation remain challenging and at a large scale. Yet, the abundance of degrees of freedom in early design stages facilitates the implementation of environmental criteria and enables resilient decision-making. Nevertheless,

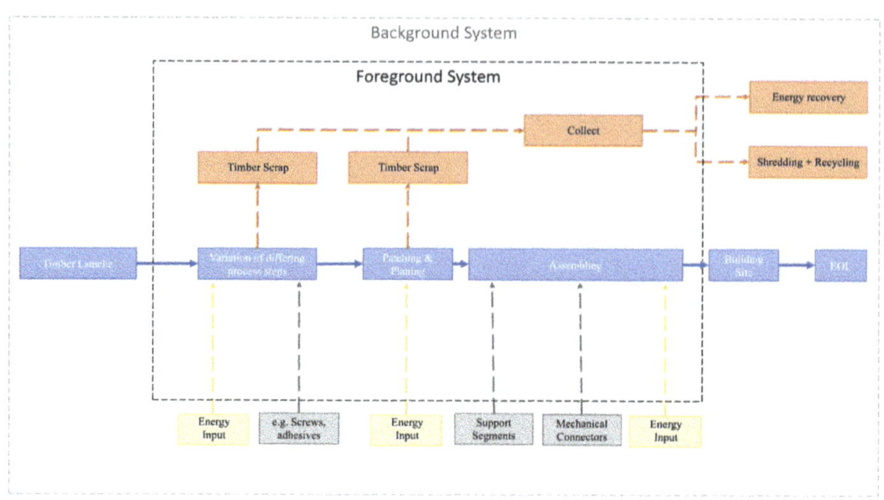

Fig. 2 Simplified system boundaries of the developed timber slab

according to the system boundaries, Table 2 presents the inventory of the latest technical development. Data is provided by project partners and comprises data collection from design simulations, enriched with generic background LCA information (e.g. transportation, grid mix, production processes) (Sphera Solutions GmbH, 2024).

The CFF parameters presented in Table 3 for timber, steel and adhesives applied in this study were defined based on the PEF method and accompanying guidance (European Commission 2019). Limitations arise for both timber and adhesive, where the suggested categories ('wood' and 'thermal insulation') and applications ('packaging—pallet' and 'PU glue—flat roof-fixing') do not reflect the precise use case in this study.

Table 2 Inventory of the timber slab

Material	Unit	Quantity	Dataset
Timber	kg	268.9	Glued laminated timber (Sphera Solutions GmbH 2024)
Steel	kg	10.6	Fixing material screw galvanised (Sphera Solutions GmbH 2024); steel sheet (Sphera Solutions GmbH 2024)
Adhesive	kg	1	Adhesive system polyurethane-prepolymer (Sphera Solutions GmbH 2024)
Energy	kWh	30	German electricity grid mix (Sphera Solutions GmbH 2024)

Table 3 CFF parameters

Parameter	Timber	Steel	Adhesive
A	0.8	0.2	0.5
B	0	0	0
$Q_{S,in}/Q_P$	N/A	1	N/A
$Q_{S,out}/Q_P$	N/A	1	N/A
R_1	0	0.107	0
R_2	0.3	0.95	0
R_3	0.7	0	1

5   Results

5.1   Life Cycle Impact Assessment—Timber Slab

16 environmental impacts are assessed following EF 3.1, using LCA for Experts datasets, Version 10.8.0.14—2024.2 (Sphera Solutions GmbH 2024). The characterised, normalised and weighted results are presented in Table 4.

Table 4 Characterised, normalised and weighted LCIA results

Impact category	Unit	Total	PEF
Acidification	Mole of H + eq	4.07E-01	0.027
Climate change—total	kg CO2 eq	1.10E+02	
Ecotoxicity, freshwater—total	CTUe	8.71E+02	
Eutrophication, freshwater	kg P eq	4.80E-04	
Eutrophication, marine	kg N eq	1.69E-01	
Eutrophication, terrestrial	Mole of N eq	1.85E+00	
Human toxicity, cancer—total	CTUh	1.03E-07	
Human toxicity, non-cancer—total		7.97E-07	
Ionising radiation, human health	kBq U235 eq	7.17E+00	
Land use	Pt	1.10E+05	
Ozone depletion	kg CFC-11 eq	1.52E-09	
Particulate matter	Disease incidences	5.62E-05	
Photochemical ozone formation, human health	kg NMVOC eq	5.90E-01	
Resource use, fossils	MJ	1.68E+03	
Resource use, mineral and metals	kg Sb eq	6.47E-05	
Water use	m^3 world equiv	3.07E+00	

5.2 Interpretation of the Results—Timber Slab

The most influential impact categories identified include Land use (40%), Particulate matter (31%) and Climate change—total (11%). The elevated impact in Land use is primarily driven by the resource-intensive nature of timber production, which requires significant land occupation for forestry activities. Even when sourced sustainably, these processes can lead to biodiversity loss and ecosystem disturbances. The high contribution to Particulate matter is largely attributable to emissions from energy-intensive manufacturing processes and transportation, which rely on fossil fuels and contribute to air pollution. For Climate change—total, the impact stems from greenhouse gas emissions during production and transport, despite the carbon sequestration potential of wood. These factors collectively highlight the environmental trade-offs associated with the use of timber as the primary material.

The identification of hotspots and the most relevant impact categories helps understanding and improving the product system at early design stages from an environmental perspective. The results suggest a stronger focus on sourcing timber from certified sustainable forestry to mitigate Land use impacts, transitioning to renewable energy sources and optimising logistics to reduce Particulate matter, and minimising fossil fuel use in production stage to address Climate change. The robustness and functionality of the PEF model was evaluated through completeness and consistency checks. This ensured that methodological choices such as system boundaries and data quality did not unduly bias the outcomes. Nonetheless, these findings underscore the importance of integrating environmental considerations into early design and supply chain decisions to enhance sustainability outcomes.

6 Discussion

The proposed method demonstrates a viable approach for the PEF assessment, utilising elements from EN 15804. This approach enables the systematic fulfilment of PEF requirements within the construction sector, leveraging a method (EN 15804) that has been thoroughly evaluated and validated. However, some obstacles make it challenging to establish a clear PEFCR.

Limitations lie within the traceability of CFF parameters. Both the lack of reliable default values for the utilised materials (esp. wood and adhesives) and scarce information on recycling and recovery rates lead to non-tangible, hence vague LCA results. Neither are number of reuse cycles accounted for in CFF nor are the allocation factors—mainly based on default values, if applicable—incorporated robustly. The single-score presentation of PEF is generally discussed and debated in literature (Finkbeiner 2014) and the industry is sticking to the well-established EPDs. These concerns are leading to the need for fundamental research in terms of data availability and reliability, as well as the need for further case studies to mainstream the applicability of CFF, followed by more comparable PEF studies.

Key principles of PEF studies, utilising modular EPD elements, are met with the proposed methodology and are highly relevant to achieve a higher comparability between LCA studies. Furthermore, the authors advocate the use of the PEF in the construction sector, as the focus has largely been on the climate change impact category, even when other impact categories have been identified (Dong et al. 2021). The PEF ensures that impacts other than climate change are considered.

7 Conclusion and Outlook

Harmonising standards and methodologies are key for the comparability and transparency of environmental concerns. This paper demonstrates the feasibility of integrating the standards EN 15804 and PEF. For the first time, PEF´s CFF is applied to timber slabs, suggesting a novel integrated methodological approach aiming to facilitate the development of PEFCR. The LCIA and identified hotspots provide deeper comprehension of the timber slabs under development, particularly beneficial for developer´s decisions in an early design stage. More granular data over time in combination with information about manufacturing processes leverage a single-score PEF to produce even more resilient LCA results. This study provides a foundation for the subsequent refinement of methodological procedures and the extension of their application to a range of biobased construction products.

Funding Funded by Horizon Europe and the European Innovation Council and SMEs Executive Agency (EISMEA) via the EIC Pathfinder Project 'UniversalTimberSlab' with GA Project Number 101161103. Views and opinions expressed are however those of the author(s) only and do not necessarily reflect those of the European Union. Neither the European Union nor the granting authority can be held responsible for them.

References

Bahramian M, Yetilmezsoy K (2020) Life cycle assessment of the building industry: an overview of two decades of research (1995–2018). Energy Build 219:109917

Chen X et al (2024) Biomaterials technology and policies in the building sector: a review. Environ Chem Lett 22:715–750

DIN e.V., DIN EN 15804:2022–03 – Nachhaltigkeit von Bauwerken – Umweltproduktdeklarationen – Grundregeln für die Produktkategorie Bauprodukte; Deutsche Fassung EN 15804:2012+A2:2019+AC:2021, Beuth Verlag GmbH, Berlin, 2022.

DIN e.V. (2023) DIN EN 1991-1-1:2023-04—Eurocode 1: Einwirkungen auf Tragwerke—Teil 1-1: Allgemeine Einwirkungen—Wichte von Baustoffen und Lagergütern, Eigengewicht von Bauwerken und Nutzlasten im Hochbau; Deutsche und Englische Fassung prEN 1991-1-1:2023, DIN Media GmbH, Berlin

DIN e.V. (2011) DIN EN ISO 14025:2011-10—Umweltkennzeichnungen und–deklarationen—Typ III Umweltdeklarationen—Grundsätze und Verfahren (ISO 14025:2006); Deutsche und Englische Fassung EN ISO 14025:2011, DIN Media GmbH, Berlin

DIN e.V. (2021a) Environmental management—life cycle assessment—principles and framework (ISO 14040:2006 + Amd 1:2020); German version EN ISO 14040:2006 + A1:2020, DIN Media GmbH

DIN e.V. (2021b) Environmental management—life cycle assessment—requirements and guidelines (ISO 14044:2006 + Amd 1:2017 + Amd 2:2020); German version EN ISO 14044:2006 + A1:2018 + A2:2020, DIN Media GmbH

Dong YH, Ng ST, Liu G (2021) A comprehensive analysis towards benchmarking of life cycle assessment of buildings based on systematic review. Build Environ 204:108162

Dong YH, Ng ST, Liu G (2023) Towards the principles of life cycle sustainability assessment: an integrative review for the construction and building industry. Sustain Cities Soc 95:104604

Dubois O, Gomez San Juan M (2016) How sustainability is addressed in official bioeconomy strategies at international, national, and regional levels—an overview. FAO, Rome

Durão V et al (2020) Assessment and communication of the environmental performance of construction products in Europe: comparison between PEF and EN 15804 compliant EPD schemes. Resour Conserv Recycl 156:104703

European Commission (2021) Commission recommendation (EU) 2021/2279 of 15 December 2021 on the use of the environmental footprint methods to measure and communicate the life cycle environmental performance of products and organisations, European Commission

European Commission (2019) Suggestions for updating the product environmental footprint (PEF) method, European Commission, Luxembourg

Eurostat, Statistische Systematik der Wirtschaftszweige in der Europäischen Gemeinschaft, eurostat, Luxembourg

Finkbeiner M (2014) Product environmental footprint—breakthrough or breakdown for policy implementation of life cycle assessment? Int J Life Cycle Assess 19:266–271

Fischer L-B, Losacker S (2025) How to build (in) the future? Legitimacy of socio-technical visions in a bio-based construction sector. Environ Innov Soc Transit 56:100996

Friedrich D (2022) Consumer and expert behaviour towards biobased wood-polymer building products: a comparative multi-factorial study according to theory of planned behaviour. Archit Eng des Manag 18:73–92

Haverkamp J et al (2025) Product environmental footprint in the construction sector—proposal and assessment of a representative product for carbon-reinforced concrete. J CO_2 Util 95:103078

LCA for Experts (2024) Sphera Solutions GmbH, CUP 2024.2, 10.8.0.14

Manfredi S et al (2015) Comparing the European commission product environmental footprint method with other environmental accounting methods. Int J Life Cycle Assess 20:389–404

Mutel C (2025) Environmental footprint methods—PEF method. https://green-forum.ec.europa.eu/environmental-footprint-methods/pef-method_en. Accessed 23 May 2025

Pscherer C, Krommes M (2025) LCA standards for environmental product assessments in the bioeconomy with a focus on biogenic carbon: a systematic review. Int J Life Cycle Assess 30:371–393

Publications Office, Supporting information to the characterisation factors of recommended EF Life Cycle Impact Assessment methods, 2018

Salehi M (2020) A comparative study of product environmental footprint (PEF) and EN 15804 in the construction sector concentrating on the end-of-life stage and reducing subjectivity in the formulas

The Norwegian EPD Foundation (2019) PCR Part B for wood and wood-based products for use in construction. The Norwegian EPD Foundation

United Nations Environment Programme (2023) Global status report for buildings and construction: beyond foundations—mainstreaming sustainable solutions to cut emissions from the buildings sector; United Nations Environment Programme (2024) Global status report for buildings and construction: beyond foundations—mainstreaming sustainable solutions to cut emissions from the buildings sector

Towards an Increasing Standardisation of LCA Data Reporting: The Experience of Aluminium's Industry

Sanaa Zinbi, Benedetta Nucci, Francesca Reale, Alessandra Zamagni, and Gioia Garavini

Abstract The increased demand for high-quality primary LCI/LCA data from downstream industries and regulatory bodies calls for consistent calculation approaches. The relevance of standardised methodologies and sector-specific guidance for the aim of consistency is clear, but practical implementation poses challenges, particularly for raw materials like aluminium, used across multiple markets. Several public and private initiatives exist (Catena-X, PACT, EN 15804, PEF, etc.) to ensure consistency and comparability of calculations. Since aluminium is one of the most widely used metals across various markets, these developments significantly impact the industry's efforts to standardise the reporting of LCI/LCA. This paper shares the aluminium industry's experience with these standardisation efforts, highlighting the primary challenges, lessons learned and future hurdles in achieving uniform LCA data reporting.

1 Introduction

The aluminium industry plays a significant role in the global economy as the world's most used non-ferrous metal. Valued for its unique combination of properties, notably its lightweight, durability and recyclability, aluminium is an important input for multiple sectors, including automotive, packaging, construction and clean energy technologies.

In recent decades, the growing demand for aluminium has been accompanied by an increasing need for transparent assessment and communication of environmental impacts associated with aluminium products. This need is primarily driven

S. Zinbi (✉) · B. Nucci
European Aluminium, Brussels, Belgium

B. Nucci
e-mail: nucci@european-aluminium.eu

F. Reale · A. Zamagni · G. Garavini
Ecoinnovazione, Bologna, Italy

© The Author(s) 2026

M. Traverso et al. (eds.), *Life Cycle Management from Global to Local*,
https://doi.org/10.1007/978-3-032-17987-6_35

by both regulatory requirements and growing market demand. Life Cycle Assessment (LCA) has appeared as an effective tool for evaluating and communicating environmental performance. However, inconsistent methodologies across policies and sectors hindered comparability, reducing the reliability of LCA results, and ultimately undermining accountability and fair competition when the environmental footprint is considered.

In this context, this paper examines the increasing need for harmonised LCA approaches, using the aluminium sector as a case study. It reviews key policy and market drivers, as well as the current offer and demand for Life Cycle Inventory (LCI) and LCA data in the European aluminium industry. The paper then presents an analysis of the current benchmarking efforts in the main aluminium end-use markets, namely automotive and construction. The analysis is intended to provide a concise overview on whether, and to what extent, key methodology aspects are defined in LCA standard and guidelines used for aluminium products. These aspects include, amongst others, data sources and quality, the handling of multi-functionality and energy modelling. The ultimate objective is to highlight areas where methodological divergences exist or may potentially emerge.

2 The Demand for Standardised LCA/LCI Approaches

The increasing use of LCA methodology across environmental policy and corporate sustainability has exposed critical methodological inconsistencies that compromise result comparability and decision-making utility. Variability in system boundaries, functional units, allocation procedures and impact assessment methods can produce substantially different outcomes for identical products or processes (Sala et al. 2020). Such divergence poses significant challenges for evidence-based environmental policy development and corporate sustainability reporting, where decision-makers require robust and comparable assessments. Harmonisation efforts and needs have been highlighted by many scholars.

Emerging approaches such as 'message-LCA', including the avoided emissions concepts and virtual chain-of-custody models, further complicate the landscape, raising concerns about the credibility of LCA studies in the absence of clearer harmonisation and guidance (Finkbeiner et al. 2025).

The European Commission's (EC) Environmental Footprint (EF) framework addresses some of these issues by offering standardised modelling rules, including Product and Organisation Environmental Footprint methods (PEF and OEF), tailored to product categories or sectors (EC Recommendation 2021/2279). However, the process of developing sector-specific rules is still ongoing for some applications. In this context, parallel private and/or public initiatives, such as PACT, Catena-X and EPDs, have initiated a process for harmonising the methodology of LCA within their sector or context. The multiple harmonisation efforts are resulting in the proliferation of methods, which sometimes exhibit different and contrasting requirements.

2.1 Overview of Policies and Market Demand for LCA Data Reporting

Providing information relating to environmental and/or carbon footprint (EF and CF) has shifted from voluntary efforts to legal requirements, particularly in the EU. Companies now face growing pressure from customers, consumers and public authorities to provide robust environmental data.

Relevant voluntary (i.e. non-regulatory) initiatives aiming to harmonise LCA reporting include PACT, Catena-X, the Global Battery Alliance and Environmental Product Declarations (EPDs) from various Programme Operators (POs). In parallel, global and EU policies have increasingly integrated LCA into sustainability governance, from the UN Sustainable Development Goals (SDGs) to the EU Green Deal and Circular Economy Action Plan, as illustrated by Sala et al. (2021).

In addition, over the past five years, the application of life cycle approaches and related methodologies has expanded from policy design to implementation, highlighting the role of LCA in making policies operational. A key example is the EU batteries Regulation, often called 'the blueprint for the use of LCA in EU policies' (2023/1542). Article 7 of the regulation sets CF requirements for Electric Vehicles (EVs) and industrial batteries, organised into three levels:

(1) Information requirement: mandatory reporting of the CF of batteries.
(2) Labelling requirement: mandatory classes of performance based on the value of the CF of the battery.
(3) Performance requirement: establishment of a maximum threshold for the CF of batteries put on the EU market.

This policy establishes that the CF of batteries must be calculated according to a delegated act based on the PEF recommendation (EC Recommendation 2021/9332).

Similar provisions appear in other EU laws, including the Critical Raw Materials Act (CRMA) (EU 2024/1252), Ecodesign for Sustainable Products Regulation (ESPR) (EU 2024/1781) and Construction Products Regulation (CPR) (EU 2024/3110). These EU laws generally complement national or sometimes regional, regulations and market-specific practices.

However, the growing number of initiatives, both private and public, each with varying methods, risks resulting in greater fragmentation. These requirements, whilst focused on final products, also impact upstream supply chains, creating inconsistencies in EF and/or CF data for intermediate and semi-finished products.

The need for harmonisation and consistency in the calculation is thus therefore evident, both when it comes to the calculation of impacts for final products and intermediate products. Harmonised methodologies, complemented by the availability of high-quality secondary data on the environmental profile of materials, processes and components, would contribute to increased comparability, accountability, reliability and finally fair competition when the EF or the CF are considered.

2.2 *Progress and Gaps in LCA Studies in the Aluminium Industry*

In the context of increasing regulatory requirements and market demand for environmental impact transparency, European Aluminium (EA) surveyed its members in 2024, which are aluminium industry stakeholders across Europe, to assess the state of LCI and LCA data demand, delivery and verification. Most respondents reported frequent customer requests, especially for CF data, although interest in broader environmental impacts is increasing. Challenges include inconsistent data formats and limited third-party verification. When acting as downstream clients, aluminium companies face similar issues when sourcing data, namely inconsistent formats, insufficient details and limited third-party verification.

These findings suggest a general industry need for harmonised reporting guidelines across the aluminium value chain to meet current demand and anticipate stricter future requirements.

The aluminium industry has a long history of promoting life cycle thinking. The IAI has published LCI datasets since 1998, with updates every five years, now covering over 200 sites worldwide (IAI 2022). In Europe, EA has released Environmental Profile Reports since the 1990s, with the latest LCI data representing over 160 sites (EA 2024). In North America, the Aluminum Association (AA) has led similar efforts, with datasets dating back to 1998 and the latest covering over 100 sites across the United States and Canada (AA 2022).

Beyond data provision, industry associations have also worked to harmonise LCA methodologies, offering guidance and training. Notably, the IAI has issued CF calculation guidelines (IAI 2021); EA has contributed to the development of the PEF method (EC Recommendation 2021/9332) and published its guidelines for evaluating the environmental performance of intermediate and semi-finished products (EA 2023).

In summary, the aluminium industry has made substantial progress in data quality and methodological alignment. However, gaps remain, requiring continuous coordination amongst industry stakeholders and with regulators. Broader harmonisation efforts are also underway across the metal sector as detailed by Santero and Hendry (2016), though achieving full alignment remains a challenge due to the diversity of existing initiatives as highlighted in Sect. 3.

3 Description of Benchmarking Efforts

Tables 1 and 2 provide a benchmarking of key initiatives relevant to the automotive and building sectors, two of the primary end-use markets for aluminium. Whilst the list is not exhaustive, it gives an idea of the scale of the differences and an overview of what are the aspects that should be analysed when considering whether two or more initiatives provide consistent results.

Table 1 Analysis of LCA-based standards and sectorial guidelines: key descriptive features

	PEF (EC 2021/9332)	PACT V.3 (2025)	Catena-X v.3 (2024)	EN15804:A2 (2019)	Draft DA —CFB-EV[a]
Context	EC initiative to set a harmonised LCA methodology to display and benchmark the environmental performance of products, services and companies	Harmonise product carbon footprint (PCF) exchange across industries, based on existing standards	Standardisation of product carbon footprint (PCF) data within the automotive sector and of their exchange	Define rules for EPDs for construction products	Define rules to calculate the carbon footprint of batteries for EVs to satisfy the requirements of article 7 of the batteries regulation
Sector	Cross-sectoral	Cross-sectoral	Automotive	Construction and building products	Automotive (EV battery)
Alignment with standards	ISO 14040–14044	ISO 14064, GHG protocol, PEF	ISO 14067	ISO 14040–14044–14025	PEF
Additional documents	PEFCR needed	Standalone	Standalone, unless a PCR is available	PCR needed	Standalone

[a] Draft carbon footprint methodology for EV batteries delegated act

In the following, a short description of each initiative is provided, with details about its use.

Product Environmental Footprint (PEF)—An EU Commission initiative to standardise LCA methodology for legislation since 2013 (EC 2013/169). The latest version is published as EC Recommendation 2021/9332 requires PEFCRs for comparability. A PEFCR Guidance has been published to support the development of specific PEFCR. Developed as a stand-alone Recommendation, it is now widely used as a reference in EU policies involving LCA.

PACT Methodology for Product Carbon Footprints (in short: PACT)—The PACT Methodology builds on existing frameworks and standards to guide accounting, verification and exchange of cradle-to-gate PCFs to create more accurate, granular and comparable emissions data (PACT 2025).

Catena-X—An open and collaborative data ecosystem for the automotive industry, enabling standardised, and secure data exchange across the value. It includes a 'Product Carbon Footprint Rulebook', for consistent carbon footprint calculations for automotive components (Catena-X 2024).

EN15804:A2—A standard, for developing EPDs for building products, initially published in 2012 and revised in 2019. EPDs require complementary Product Category Rules (c-PCRs) and are based on ISO 14025, along with specific PCR rules.

Table 2 Analysis of LCA-based standards and sectorial guidelines: key methodological features

	PEF (EC 2021/9332)	PACT V.3 (2025)	Catena-X v.3 (2024)	EN15804:A2 (2019)	Draft DA—CFB-EV[a]
Modelling of recycling	Circular footprint formula (CFF). A material-specific (from 0.2 to 0.8)	Cut-off	Cut-off	Co-product allocation in Mod. A and if a recycling scenario is considered at EoL, its burdens and benefits are reported in module D (net environmental burden and benefits)	CFF. A material-specific (from 0.2 to 0.8)
Database	Environmental footprint database, if datasets available	None recommended	None recommended. Wish to achieve the construction of a harmonised database free and accessible to all	None recommended	Environmental footprint database if datasets available
Requirements for primary data collection	For processes run by the company performing the study: company-specific data and creation of company-specific datasets with minimum DQR	Primary data always preferred. Activity data that is used to calculate a PCF shall be company-specific process-based data (i.e. primary data)	Primary data is always preferred to secondary datasets	Use of primary data at least for processes the manufacturer of the assessed product has influence over	Mandatory company-specific data shall be collected for selected processes. Company-specific data may be collected for other processes

(continued)

Table 2 (continued)

	PEF (EC 2021/9332)	PACT V.3 (2025)	Catena-X v.3 (2024)	EN15804:A2 (2019)	Draft DA—CFB-EV[a]
Hierarchy for secondary data selection	EF-compliant datasets as a priority option. If missing, EF-compliant datasets representing a good proxy or ILCD entry-level compliant datasets	Secondary emission factors used shall be compliant with PACT methodology safeguards	As of now, harmonised data does not yet exist and CX will thus require the hierarchical use of secondary data sources as follows 1. Industry association data 2. General LCA data 3. Other documented references	Secondary data shall be valid for the reference year of the study and not older than 10 years from the reference year of the study	(i) The most representative EF-compliant dataset available in LCDN (ii) A representative EF-compliant dataset from any other source (iii) A representative ILCD entry-level compliant dataset either from LCDN or from any other source
Electricity modelling	Market-based	Market-based	Market-based	Market-based or location-based (as in EN 15492)—specific requirements are defined by PO (different approaches are coexisting at European level)	Location-based

(continued)

Table 2 (continued)

	PEF (EC 2021/9332)	PACT V.3 (2025)	Catena-X v.3 (2024)	EN15804:A2 (2019)	Draft DA—CFB-EV[a]
Cut-off rule	3% based on material and energy flows and the level of environmental significance (single overall score)	In aggregate, exclusions shall represent less than 3% of the total cradle-to-gate PCF emissions	Cumulative cut-off rule: may exclude processes contributing to less than 3% of total emissions	1% cut-off per unit process (energy or mass) and a maximum of 5% per module (energy or mass)	1% in mass may be applied to material inputs per system component. The mass gap shall be closed on the system component level by adding the missing mass to the material input flow with the highest specific carbon footprint on the system component level concerned
System boundaries	Cradle-to-grave for final products. Cradle-to-gate for intermediate	Cradle-to-gate	Cradle-to-gate	Cradle-to-grave, mandatory reporting of module A1–A3, module C and module D. Voluntary reporting of module B	Cradle-to-grave

[a] Draft carbon footprint methodology for EV batteries delegated act

They are registered by Programme Operators—POs (e.g. environdec, EPD Norge etc.), each with their own General Programme Instructions (GPIs) and specific PCRs. Whilst EPDs in the construction are standardised due to EN15804, those for the same product under different POs may not be comparable. This standard has also become the reference for declaring the performance of building products under the recently revised CPR (EC 2024/3110).

Draft Delegated Act for the Carbon Footprint of EV Batteries (DA CFB-EV)—A draft methodology under the Batteries Regulation (Article 7) for assessing the CF of EV batteries. Based on PEF (EC 2021/9332) but containing several differences, making this draft delegated act as a separate methodology.

Tables 1 and 2 summarise the main differences and potential similarities between these LCA-based standards and the sectorial guidelines. Table 1 compares LCA-based standards and sectorial guidelines in terms of purpose, sector, alignment with other standards and whether they require additional documents. Table 2 outlines methodological aspects, including the modelling approaches for recycling (both in the production phase and at the End-of-life of the product), requirements for secondary datasets, electricity mix in production stage modelling, cut-off rules, allocation hierarchy, system boundaries and impact assessment methods.

4   Conclusion

The analysis of methodological requirements across different initiatives relevant to the aluminium sector reveals significant variation in origin, purpose and key methodology aspects, sometimes leading to inconsistencies hindering uniform implementation.

As shown in Table 2, whilst some methodological aspects, such as using primary data from the manufacturer, are broadly accepted, practical applications of different initiatives vary considerably, particularly in areas such as recycling modelling and allocation methods, ranging from initiatives that prescribe specific approaches (e.g. cut-off or co-product) to those with insufficient guidance, leaving room for interpretation.

These findings emphasise two main issues. First, the importance of high-quality input (both primary and secondary) data to get sound LCA results, especially when these results inform market comparisons or policy development. Second, the need for consistent data tracking along the value chain, from intermediate to finished products, to ensure methodological approaches are consistently applied and do not provide distorted results. This is increasingly important as industries face growing demands to report EF and CF data at both plant and corporate levels. However, the lack of methodological alignment across initiatives risks overburdening companies and compromises data consistency, especially for small and medium-sized businesses (SMEs).

Although LCA might serve different objectives, such as internal knowledge building, performance measurement or supporting the development of new products, the harmonisation of LCA standards and guidelines is crucial to improving the reliability and policy relevance of environmental assessments. This is particularly critical in the context of market competition and policy development. In the absence of coherent and sector-wide methodological requirements, there is a significant risk that policy processes will be informed by inconsistent or incomparable data, ultimately undermining their effectiveness and credibility.

Whilst the EF framework offers a solid foundation, further efforts are needed to ensure alignment with evolving requirements such as the CPR, ESPR and ongoing updates to the EF method. In this context, three priorities emerge:

- Ensure coherence in environmental requirements across product, plant and corporate levels to prevent overburdening organisations with excessive compliance and reporting activities.
- Provide robust and transparent secondary data for their use in various compliance systems.
- Establish a value-chain-wide governance for environmental data management.

References

Aluminum Association (AA) (2022) https://www.aluminum.org/SustainabilityReports. Accessed 02 June 2025

Catena-X (2024) https://catenax-ev.github.io/assets/files/CX-NFR-PCF-Rulebook_v.3.0-04874a 80a6d27511df06e07ae3049278.pdf. Accessed 06 June 06 2025

European Aluminium (EA) (2023) https://european-aluminium.eu/our-work/standards-life-cycle-assessment/. Accessed 02 June 2025

European Aluminium (EA) (2024) https://european-aluminium.eu/blog/environmental-profile-rep orts/. Accessed 06 June 2025

European Commission, Commission Recommendation (EU) 2021/2279 of 15 December 2021 on the use of the environmental footprint methods to measure and communicate the life cycle environmental performance of products and organisations. http://data.europa.eu/eli/reco/2021/ 2279/oj. Accessed 09 June 2025

European Commission, Commission Recommendation 2021/9332 on the use of the environmental footprint methods to measure and communicate the life cycle environmental performance of products and organisations

European Parliament and the Council, Regulation (EU) 2023/1542 of 12 July 2023 concerning batteries and waste batteries, amending Directive 2008/98/EC and Regulation (EU) 2019/1020 and repealing Directive 2006/66/EC, OJ L 191, 28.7.2023, pp 1–117

European Parliament and the Council, Regulation (EU) No 305/2011 of 9 March 2011 laying down harmonised conditions for the marketing of construction products and repealing Council Directive 89/106/EEC, OJ L 88, 4.4.2011, pp 5–43

European Parliament and the Council, Regulation (EU) 2024/1252 of 11 April 2024 establishing a framework for ensuring a secure and sustainable supply of critical raw materials and amending Regulations (EU) No 168/2013, (EU) 2018/858, (EU) 2018/1724 and (EU) 2019/1020, OJ L 2024/1252, 3.5.2024

European Parliament and the Council, Regulation (EU) 2024/1781 of 13 June 2024 establishing a framework for the setting of ecodesign requirements for sustainable products, amending Directive (EU) 2020/1828 and Regulation (EU) 2023/1542 and repealing Directive 2009/125/EC, OJ L 2024/1781

European Parliament and the Council, Regulation (EU) 2024/3110 of 27 November 2024 laying down harmonised rules for the marketing of construction products and repealing Regulation (EU) No 305/2011, OJ L 2024/3110

Finkbeiner M, Roche L, Holzapfel P (2025) From analysis-LCA to message-LCA: a lost cause? Int J Life Cycle Assess 30(5):803–810

International Aluminium Institute (IAI) (2021) https://international-aluminium.org/resources/aluminium-carbon-footprint-methodology/. Accessed 05 June 2025

International Aluminium Institute (IAI) (2022) http://international-aluminium.org/resources/2019-life-cycle-inventory-lci-data-and-environmental-metrics/. Accessed 05 June 2025

PACT (2025) https://www.carbon-transparency.org/pact-methodology/. Accessed 04 June 2025

Sala S, Crenna E, Secchi M, Sanyé-Mengual E (2020) Environmental sustainability of European production and consumption assessed against planetary boundaries. J Environ Manag 269:110686

Sala S, Amadei A, Beylot A, Ardente F (2021) The evolution of life cycle assessment in European policies over three decades. Int J Life Cycle Assess 26(12):2295–2314

Santero N, Hendry J (2016) Harmonization of LCA methodologies for the metal and mining industry. Int J Life Cycle Assess 21(11):1543–2155

Benchmarking Carbon Emissions and Incorporating Sustainability Metrics in Road Construction Procurement

Nicolás Héctor Carreño Gómez and Pamela Haverkamp

Abstract The construction sector is a significant contributor to environmental impact. Considering this, it has become crucial to develop practical approaches to incentivise the reduction of carbon emissions and other environmental impacts. Scandinavian countries as well as the Netherlands have been successfully using Environmental Product Declarations (EPDs) of road construction materials, to compare materials from different producers and foster competition for more sustainable alternatives in tenders. This has shown a development on low(er) carbon-intensive materials. In Germany, since 2022, various pilot initiatives have been launched to explore similar approaches. A working group within the German Road and Transportation Research Association (FGSV) has developed a guideline with strategies to incorporate sustainability metrics into procurement processes. As part of this effort, benchmarks for asphalt and concrete have been calculated using a cradle-to-delivery (modules A1 to A4) approach, with the intention to expand the scope of this work to include additional materials, life cycle modules and impact categories, such as Global Warming Potential (GWP) in the future. This ongoing development aims to further integrate sustainability into road construction practices and promote the use of environmentally responsible materials.

1 Introduction

Given the extension of the road network and the substantial investments made by European governments, which exceed 300 billion EUR annually over the last decade, the European Union (EU) has deemed energy- and resource-efficient road construction and maintenance, an important policy goal (Garbarino et al. 2016; Brons et al. 2022). Moreover, the road sector is estimated to generate over 15% of Greenhouse

N. H. Carreño Gómez (✉)
VINCI Construction Shared Services GmbH, Bottrop, Germany
e-mail: nicolas.carreno@vinci-construction.com

P. Haverkamp
Institute of Sustainability in Civil Engineering, RWTH Aachen University, Aachen, Germany

M. Traverso et al. (eds.), *Life Cycle Management from Global to Local*,
https://doi.org/10.1007/978-3-032-17987-6_36

Gas (GHG) emissions worldwide, making it a critical sector for achieving global and regional environmental goals (Moins et al. 2025).

To evaluate and improve the environmental performance of roads and their materials, the method of Life Cycle Assessment (LCA) can be used. LCA according to ISO 14040 and ISO 14044 enables the quantification of potential environmental impacts throughout the life cycle (Deutsches Institut für Normung 2021a, b). Whilst LCA has been consistently used in the road construction sector in the last decades and is considered a robust method, some challenges persist. Despite its standardisation, LCA allows considerable flexibility in methodological choices and assumptions, which may lead to inconsistent outcomes and lack of comparability. This issue is particularly relevant when LCA is used as a tool for product comparison or within tendering processes. Additionally, LCA results may be difficult to interpret by non-expert decision-makers.

To address these challenges, some institutions or practitioners resort to the development of benchmarks. According to ISO 21678:2020, benchmarking involves the collection, analysis and comparison of performance data for similar types of construction works (BS ISO 21678:2020 2020). In the construction sector, the generation and use of these values is gaining traction (Mattinzioli et al. 2022). The literature reports several initiatives aimed at developing benchmarks specifically for the road sector. Mattizioli et al. (2022) proposed a benchmarking methodology based on ISO 21678:2020, applicable to construction materials, and introduced four types of benchmark values: limit, reference, short- and long-term.

Moins et al. (2025) defined cradle-to-gate benchmarks for Global Warming Potential (GWP) of asphalt mixtures produced in the Belgian region of Flanders considering more than 600 mixes across nine mixture categories. Moreover, Ashtiani and Munch (2022) used data from 33 projects to define benchmarks for 12 sustainability-related metrics, including project GHG emissions and energy consumption.

A working group within the German Road and Transportation Research Association (FGSV) has developed a working paper to provide guidance for the environmental assessment of asphalt and concrete mixtures in road construction. This document also provides benchmarks for GWP using a cradle-to-delivery (modules A1 to A4) approach. In this regard, the goal of this contribution is to present the methodological background and approach used to define said reference values for asphalt mixtures. This exercise fosters the integration of sustainability considerations into road construction practices and promotes the use of more environmentally responsible materials.

2   Current Practices in Sustainable Road Construction

In the road construction sector, several initiatives have already been undertaken for the harmonisation of LCA practices and the development of benchmarks. At EU level, voluntary Green Public Procurement (GPP) criteria have been defined to support public authorities in the purchase of products, services and works with reduced environmental impacts (Garbarino et al. 2016). A revised version of these GPP criteria for road design, construction and maintenance was published in 2016. Whilst this document does not specify benchmarks for environmental impacts, it introduces criteria that foster an improved environmental performance of roads (Garbarino et al. 2016). The criteria include performance requirements for pavement durability, declaration of the environmental performance of maintenance and rehabilitation activities, the awarding of points based on the improvement in the Carbon Footprint (CF) or LCA performance, as well as the incorporation of recycled content into road elements (Garbarino et al. 2016).

Other initiatives have focused more explicitly on the development of benchmarks. In the United States (US), Environmental Product Declarations (EPDs) have been recognised as suitable tools for benchmarking by the National Pavement Association (NAPA) and the Sustainable Highway Construction Handbook (Mattinzioli et al. 2022). Moreover, in alignment with the goals of the Inflation Reduction Act (IRA), NAPA developed a benchmarking methodology and associated values for asphalt mixtures that account for the sensitivity of GWP to factors beyond the control of asphalt producers (Miller et al. 2024). This methodology aims to ensure a fair transition towards lower-carbon construction materials (Miller et al. 2024). Currently, the approach encompasses the information modules A1–A3 of the material's life cycle, using deterministic values for A1 impacts and distribution-based thresholds for modules A2 and A3.

Additionally, the Colorado Department of Transportation (CODT) developed a benchmarking framework to comply with the Buy Clean Colorado Act (BCCO). This initiative targets eligible materials, such as asphalt, concrete and steel, and defines benchmark values based on the ISO 21678:2020 approach (BS ISO 21678:2020 2020). Initially, CDOT established limit values at the 90th percentile of A1–A3 GWP data collected from EPDs. It is expected that, as industry familiarity increases, this threshold will be lowered to further incentivise progress towards BCCO goals (Senseney et al. 2025).

In the Netherlands, the Environmental Costs indicator (ECI) was developed by the Dutch Ministry of Infrastructure and Water Management. This indicator expresses 11 different environmental impacts in a single monetary value, thus applying a monetarisation approach (Milieudatabase 2022; INDUSA 2025). The ECI reflects the estimated cost of mitigating or reversing the quantified environmental impacts (INDUSA 2025), meaning that the ECI are shadow costs, also considered as societal costs. This indicator is commonly used as an award criterion or contract requirement in civil engineering projects (Flapper 2024). Dutch contracting authorities may establish a maximum ECI value for tender bids, effectively using it as a benchmark for environmental performance. In a tender, bidders with a lower ECI can opt to a fictional bid price reduction, creating and incentive to offer a more sustainable bid. This system has led to more environmentally friendly practices, such as increasing Reclaimed Asphalt Pavement (RAP) content in asphalt mixtures, the search for alternatives to Polymer-modified Bitumen (PmB), amongst others.

A related but distinct approach was developed in Norway, also involving the monetisation of a single environmental impact—GWP. The key difference, however, lies in the valuation method: unlike the Dutch approach, which uses a shadow price, the Norwegian system applies an industry-specific value (Püstow et al. 2023). This price reflects sectoral considerations and market dynamics instead of theoretical externalities. Nevertheless, the overarching objective remains the same: to establish a framework that promotes competition based on costs and sustainability metrics, thereby encouraging the development and adoption of more environmentally friendly road construction practices.

In the Norwegian and Dutch systems, the core of the monetisation process lies on EPDs. Both countries have developed specific Product Category Rules (PCRs) tailored to the road construction sector, which define the calculation methods and assumptions for generating EPDs. The PCRs ensure, to a significant extent, the comparability of environmental data across different producers. Such comparability is essential for the effective functioning of tendering systems that incorporate environmental criteria. Furthermore, the monetisation of environmental impacts is only feasible when these impacts are calculated in a transparent and harmonised manner, leaving no room for subjectivity.

Awarding road construction contracts based not solely on the lowest price, but on a combination of economic and environmental sustainability criteria can lead to higher initial costs for contracting authorities. Experience from early implementations suggests that initial project costs may be up to 10% higher when environmental sustainability is factored into procurement decisions, although this figure varies depending on project scope and local market conditions. At the same time, this reflects a growing commitment amongst public authorities to prioritise climate-conscious infrastructure development. For example, the German construction industry has highlighted how integrating a CO_2 shadow price into procurement decisions encourages bidders to optimise their environmental performance, even if it entails higher upfront investments. In practice, these additional costs are often reinvested by construction firms into more sustainable technologies and measures, enhancing their competitiveness in future tenders. This approach aligns with the intended purpose of such procurement models of driving systemic change in the construction sector and improving its environmental footprint.

3 German Initiatives

In Germany, the development and implementation of environmentally oriented tendering practices in road construction has progressed slowly when compared to countries like the Netherlands and Norway. A significant step was taken in 2022, when the Bavarian State Ministry for Housing, Construction and Transport introduced a guideline that incorporates sustainability-related criteria into public procurement through a structured, point-based evaluation system (Bayerisches Staatsministerium für Wohnen 2022). Unlike the Dutch and Norwegian models, which rely on CO_2

pricing or benchmarking via EPDs, the Bavarian system allows up to 40% of the tender evaluation to be based on technical criteria, including sustainability-related aspects.

Depending on the project, various parameters can be included, such as Concrete Sustainability Council (CSC) certification for concrete structures, quality management systems and digitalisation. Whilst many of these factors may have limited effects on environmental performance, two criteria stand out for their direct impact: transport distance and vehicle emissions and the use of RAP. For transport, the system rewards shorter distances between production and construction sites and the use of cleaner vehicles (e.g. electric or EURO 6 trucks). However, this approach may not always provide the best environmental incentive, as it overlooks the emissions from material production (Module A3), which often outweigh those from transport (Module A4). In contrast, the criterion promoting higher RAP content in asphalt mixtures has a well-documented and significant effect on reducing environmental impact. The guideline allocates points based on the percentage of recycled content. Additionally, the Bavarian model includes a mechanism to incentivise faster project execution through a 'roadway rental' system. This approach aims to minimise traffic disruptions and the associated emissions from congestion, which are known to have substantial environmental consequences.

The Bavarian scoring model has served as a blueprint for other German states, including Hesse and Baden-Württemberg, which have adopted similar systems. The Autobahn GmbH, the German road agency, has also begun piloting alternative tendering models to enhance sustainability. However, none of these systems currently use or are based on benchmarking tools or EPDs to compare bids directly. A notable development in this direction came with a study commissioned by the Federation of the German Construction Industry and conducted by KPMG, which proposed adapting the Dutch and Norwegian models to the German context and recommended the introduction of a CO_2 shadow price in public procurement (Püstow et al. 2023).

After the publication of this study, systems to quantify the carbon emissions of road construction materials were developed by local road authorities and tested by them and the Autobahn GmbH. These systems typically rely on Excel-based tools that only partially follow the principles of LCA and the specifications of EN 15804, limiting their comparability and methodological robustness. In 2024, a significant step towards standardised benchmarking and the use of EPDs was taken by the German Asphalt Association (DAV) with the release of a complementary document to the PCR Part B for asphalt mixtures (c-PCR). This document was designed to harmonise the methodology of underlying LCA studies for EPDs and ensure the transparency and comparability of these declarations across the sector.

However, these initiatives have not yet led to a unified national system. Instead, a variety of pilot tendering models coexist based on different requirements and methodologies. As a result, construction companies must adapt their bids to a fragmented and evolving landscape, leading to significant administrative burdens and may lead to resistance from market participants, potentially slowing the broader adoption of sustainability-based procurement practices.

To support the development and adoption of practical and transparent sustainability-based procurement practices, a dedicated working group was created within the FGSV. The FGSV is an independent network of experts for road and transport research responsible for developing the technical standards and specifications for the road and transport system. Specifically, the working group 4.6.2 was tasked with drafting a working paper to guide both local and national contracting authorities in implementing environmentally conscious tendering procedures. The working group brings together a diverse range of stakeholders, including representatives from academia, local and federal road authorities, the construction industry and interest groups from the asphalt and concrete sectors.

In its initial phase, the FGSV working group focused on developing a methodological framework in the form of a working paper for assessing the environmental impact of road construction materials, specifically targeting raw material extraction, transport to production, manufacturing and delivery to the construction site (Modules A1–A4). The working paper provides detailed guidance for both asphalt and concrete mixtures, without comparing the two materials. This decision reflects the current limitations in LCA expertise across the sector and the complexity of linking environmental performance with technical and functional characteristics of these construction materials. Whilst material comparison may become feasible in the future, the current approach prioritises building foundational knowledge and ensuring methodological consistency.

The document, which will be made available in the autumn of 2025, includes benchmark values for GWP and outlines the calculation methodology used for the most commonly produced asphalt and concrete mixtures in Germany. It explains the assumptions, system boundaries and data sources used in the calculations and clarifies the scope. Additionally, it provides recommendations on how the reference values can be integrated into public procurement processes to promote more sustainable road construction practices.

Importantly, the document follows PCRs developed for asphalt and concrete and aligns with EN 15804 standards. The provided benchmarks serve as a reference for evaluating environmental performance within a given material category. The next chapter of this manuscript focuses on the benchmarking process for asphalt mixtures, including the assumptions and data sources used. Only asphalt is presented here, as the authors were directly involved in the development of the asphalt-related calculations.

4 Benchmarking and Sustainability Metrics

The following chapter outlines the methodology used to benchmark GWP of the most used asphalt mixtures, as part of the working paper mentioned previously. Usually, benchmarking involves collecting large quantities of data to determine the average, best or worst-case scenario of a process or product. Ideally, this data should be diverse and representative of the market. However, in this case, such comprehensive data on the environmental impact of German asphalt mixtures is not available, making this approach unfeasible. Although Germany manages an environmental impact database known as ÖKOBAUDAT, it contains limited information on asphalt mixtures and related materials, as its primary focus is on building construction. In the future, as more EPDs and datasets for asphalt mixtures become available, the database is expected to expand accordingly. This will enhance its relevance for infrastructure projects and eventually allow it to serve as a reliable basis for benchmarking environmental impacts in road construction.

The working group of the FSGV agreed upon including the most produced asphalt mixtures in Germany in the working paper, 27 in total. The selection includes Asphalt Concrete (AC) for surface, binder and base layers, as well as Stone Mastic Asphalt (SMA) for surface and binder layers, Mastic Asphalt (MA) and Porous Asphalt (PA). Furthermore, common rates of RAP content were considered, along with neat bitumen and PmB depending on the asphalt mix.

For module A1 (raw material extraction), the asphalt mixture composition was estimated based on the requirements for asphalt mixtures in Germany stated in the specification TL Asphalt-StB and contrasting this with real type tests from mixing plants of different German regions. From these documents, the bitumen, PmB, filler, sand, aggregates and RAP quantities could be estimated to represent the average composition of the analysed mixtures. RAP enters the system burden free, as stated in the c-PCR developed by the DAV. The environmental impact of the milling, transportation and processing shall be considered in the EoL modules.

Module A2 (raw material transport) was modelled using average transportation distances based on primary transportation data previously gathered by the authors, which were made available to the working group. For certain mixtures, such as SMA, MA and PA, the distances were incremented by different factors since these mixtures usually require higher quality aggregates that are transported longer distances. All assumptions were reviewed and agreed upon by a subgroup of experts from different companies, universities, the German Federal Ministry for Digital and Transport and the Federal Highway Research Institute.

For module A3 (asphalt mix manufacture), data gathered from German EUROVIA mixing plants was used as basis for the energy and water consumption, the standardly used consumables, the waste generated, as well as mixing plant-specific emissions. Additional to the collected data, a thermal model, similar to the ones proposed in Almeida-Costa and Benta (2016), Santos et al. (2018) and Siverio Lima et al. (2020), was developed and used. The thermal model used mainly to differentiate the energy consumption of the different mixtures. In total, the inputs and outputs of 15 mixing plants located in different German regions were used to benchmark the environmental impact of module A3, and assumptions to number of plant starts and moisture level of aggregates and RAP were made and documented. The estimated GWP values for modules A1 to A3 are shown in Table 1.

As mentioned previously, the LCA was done following the c-PCR published by the DAV to ensure future comparability with product-specific EPDs. The database Ecoinvent (3.9) was used as the background database and the data from Eurobitume for neat bitumen and PmB were used in a first approach. Since Eurobitume recently published a new LCA with new and corrected values for neat bitumen, the assessment was conducted taking the new values published by Eurobitume. For PmB, a different approach was implemented, since no values for PmB have been made available by Eurobitume. In this working paper, the EPDs published by Repsol in 2020 were used to consider the delta GWP impact polymer production, transportation and modification process. Specifically, the delta GWP between the EPDs for conventional bitumen and PmB was added to the new Eurobitume GWP value to consider the impact of PmBs for the assessment in the working paper (Table 1).

These values were cross-checked with available EPDs and scientific literature. Given that brown coal dust is still used in Germany for heating and drying aggregates, the results are considered plausible. The obtained results clearly reflect the influence of key parameters such as bitumen type, bitumen content and RAP content on the environmental impact of asphalt mixtures. Mixtures with PmB consistently show

Table 1 GWP values in kg CO_2-eq./t for different asphalt mixtures

Asphalt mixture	Bitumen	RAP content (%)	GWP A1–A3 (kg CO_2-eq./t)
AC surface layer	Neat Bitumen	0	96
		20	87
		40	79
	PmB	0	108
		20	97
		40	87
SMA surface layer	PmB	0	119
		20	108
	Neat Bitumen	0	105
		20	97
MA	Neat Bitumen	0	115
PA	PmB	0	121
AC binder layer	Neat Bitumen	30	75
		50	68
	PmB	30	83
		50	73
AC—SG[a] binder layer	PmB	20	92
		40	83
SMA binder layer	PmB	20	94
		40	85
AC base layer	Neat Bitumen	30	72
		50	67
		80	61

[a] SG: stetig gestuft—continuously graded

higher GWP values than those with neat bitumen, due to the additional processing involved. Conversely, increasing RAP content considerably reduces GWP, particularly in AC surface and binder layers, where emissions drop by up to 20% with higher RAP levels. Mixture type also plays a role: SMA, MA and PA generally exhibit higher GWP values, reflecting their higher binder content and more energy-intensive production. These trends highlight the potential for reducing emissions through material choices and reuse strategies.

For module A4, the working paper includes a calculation method to estimate the environmental impact of this module for each specific project. It provides values to consider the impact of either diesel or EV trucks and is valid for both asphalt and concrete mixtures.

5 Next Steps and Conclusions

The working paper of the FGSV aims to support German road authorities and planners in integrating environmental considerations into project planning and execution. It is not intended for comparing construction methods (i.e. asphalt versus concrete), but rather for promoting transparency and consistency in environmental data, following the principles of EPDs. This document represents a first step towards broader environmental assessments in road construction.

The working group is currently evaluating the scope of a potential subsequent parts of this document. Possible extensions include additional life cycle modules (such as A5, C1–C4 and D), the inclusion of further materials (e.g. unbound granular and foundation layers) or the assessment of additional environmental impact categories (such as abiotic depletion potential or water use, amongst others). The goal is to work dynamically and to rapidly develop additional, similarly structured documents that are concise and easy to understand. As demonstrated by the recently updated LCA published by Eurobitume, these documents must remain adaptable, given the ongoing evolution of LCA practices in road construction.

This working paper introduces a practical and transparent method for estimating the GWP of asphalt mixtures in Germany, focusing on life cycle modules A1 to A4. The results, based on standardised specifications and real-world data, highlight the influence of key parameters such as bitumen type and RAP content on environmental performance.

The present manuscript presents a different approach to standard benchmarking of the environmental impact of asphalt mixtures when facing data constraints. The approach differs in certain ways from established methodologies used in neighbouring countries yet remains aligned in intent and could be seen as a transitional step in the same direction. The end goal is to develop a system based on EPDs, enabling more accurate impact quantification and fostering competition towards more environmentally friendly asphalt mixtures. This would allow sustainability metrics to be meaningfully integrated into public procurement, driving innovation and environmental responsibility in road construction.

References

Almeida-Costa A, Benta A (2016) Economic and environmental impact study of warm mix asphalt compared to hot mix asphalt. J Clean Prod 112

Bayerisches Staatsministerium für Wohnen (2022) Bau und Verkehr, Richtlinien 2270.StB - Zuschlagskriterien Straßenbau- und Wasserwirtschaft

British Standards Institution (2020) ISO 21678:2020

Brons M, Dijkstra L, Ibáñez JN, Poelman H (2022) Road infrastructure in Europe: Road length and its impact on road performance. European Commission, WP 03

Deutsches Institut für Normung e. V (2021) DIN EN ISO 14040

Deutsches Institut für Normung e. V (2021) DIN EN ISO 14044

Flapper J (2024) Protocol berekenen en aantonen MKI-waarde, Rijkswaterstaat Ministerie van Infrastructuur en Waterstaat, Versie 4.1

Garbarino E, Rodriguez Quintero R, Donatello S, Gama Caldas M, Wolf O (2016) Revision of green public procurement criteria for road design, construction and maintenance. European Commission

Indusa Infra (2025) Instrumenten. https://indusa-infra.nl/kennis/instrumenten/1964555.aspx. Accessed 06 Jun 2025

Mattinzioli T, Lo Presti D, Del Jiménez Barco Carrión A (2022) A critical review of life cycle assessment benchmarking methodologies for construction materials. Sustain Mater Technol 33

Milieudatabase SN (2022) Bepalingsmethode Milieuprestatie Bouwwerken, Berekeningswijze voor het bepalen van de milieuprestatie van bouwwerken gedurende hun gehele levensduur, gebaseerd op de EN 15804, Versie 1.1

Miller L, Ciavola B, Mukherjee A (2024) EPD benchmark for national asphalt pavement association, version 2.0

Moins B, Hernando D, Seghers D, van den Bergh W, Audenaert A (2025) Establishing greenhouse gas emission benchmarks for the asphalt industry—are short-term reduction targets feasible using current practice or are more extensive measures needed? Resourc Conserv Recycl 212

Püstow M, Göhlert T, Gielen J, Tenner J (2023) Klimaverträglich bauen mit einem Schattenpreis für CO_2-Emissionen. KPMG Law Rechtsanwaltsgesellschaft mbH

Santos J, Bressi S, Cerezo V, Lo Presti D, Dauvergne M (2018) Life cycle assessment of low temperature asphalt mixtures for road pavement surfaces: a comparative analysis. Resourc Conserv Recycl 138

Senseney CT, Donado-Quintero D, Wieden CA, Goodale H, Meijer J (2025) White paper on the development of CDOT embodied carbon benchmark limit values for the buy clean Colorado Act

Siverio Lima M, Hajibabaei M, Hesarkazzazi S, Sitzenfrei R, Buttgereit A, Queiroz C, Tautschnig A, Gschösser F (2020) Environmental potentials of asphalt materials applied to urban roads: case study of the City of Münster. Sustainability 12

Zokaei Ashtiani M, Muench ST (2022) Using construction data and whole life cycle assessment to establish sustainable roadway performance benchmarks. J Clean Prod 380

LCA-RSIN: The UK LCA Regulatory Science and Innovation Network

Stuart Walker, Katy Armstrong, Lorraine Ferris, Rosa Cuellar Franca, and Rachael Rothman

Abstract Life Cycle Assessment (LCA) can provide evidence to support policy making and regulation, whilst identifying trade-offs between impact categories and potential unintended consequences. However, barriers such as inconsistent methodology, data challenges, lack of transparency, and a skills and knowledge gap hinder its broader application in regulation. Here, we explore the stakeholder engagement undertaken to develop the UK Life Cycle Assessment Regulatory Science and Innovation Network (LCA-RSIN). Initial findings have been synthesised into five research areas (data, bio-based and upcoming materials, methodology, practice and accreditation, and skills) to deliver practical insights, including briefing papers and methodological guidelines, to the UK government on the use of LCA in regulatory science.

1 Introduction

Life Cycle Assessment (LCA) is a powerful tool for assessing the environmental impacts of processes, materials, products and portfolios. To ensure LCAs are informative and contribute to evidence-based decision-making, consistency and transparency of both methodology and data sources are critical. Overarching international standards for LCA (ISO14040 and 14044) outline the framework, process, methodological hierarchies and data requirements for LCA. These are defined relative to the

S. Walker · R. Rothman (✉)
Grantham Centre for Sustainable Futures, School of Chemical, Materials and Biological Engineering, Sheffield, UK
e-mail: r.rothman@sheffield.ac.uk

K. Armstrong
Safety, Environmental and Regulatory Science (SERS), Unilever, Sharnbrook, UK

L. Ferris
Henry Royce Institute, University of Manchester, Manchester, UK

R. C. Franca
Department of Chemical Engineering, University of Manchester, Manchester, UK

© The Author(s) 2026

M. Traverso et al. (eds.), *Life Cycle Management from Global to Local*,
https://doi.org/10.1007/978-3-032-17987-6_37

goal and scope of the study; thereby rendering comparison of LCA results across different studies and contexts challenging (Walker and Rothman 2020).

The flexibility of the approach has historically been seen as a strength, with application of LCA in the industrial context demonstrated over decades, generally informing intra-company decision-making, and environmental product declarations (EPDs) (Hauschild 2018). However, the need for shared understanding of impacts and opportunities across value chains and sectors is increasingly recognised, so that end-to-end environmentally sustainable innovation can be realised. Policy makers are keen to create an enabling environment for sustainable products and consumer participation in the so-called green transition. In this context, LCA and related approaches are acknowledged as key tools to support regulatory decision and policy making.

Whilst there are some good examples of LCA being used in policy and regulation around the world (Sala 2021; Sonnemann 2018), these tend to be immature, adopted unevenly across sectors and often focused on the concept of life cycle thinking, rather than use of LCA in regulation and policy development. Here, we investigate what is needed for the full potential of LCA to be realised in regulatory science, including developments such as harmonising assessment approaches, overcoming implementation challenges, addressing varying levels of robustness in assessments, and upskilling policy makers and those involved in the innovation pipeline are needed.

2 Methods

The UK LCA Regulatory Science and Innovation Network (LCA-RSIN) is funded by Innovate UK, led by the University of Sheffield, and includes UK universities, industrial partners, regulators and policy makers, non-governmental organisations and trade bodies. It was established to provide guidance on adopting robust, harmonised approaches to quantifying environmental impacts of existing and emerging technologies, which can be utilised to develop policy to deliver government net zero and circular economy priorities in the UK. The network focuses on the UK foundation industries (cement, ceramics, chemicals, glass, metals, paper) and their value chains.

During 2024, a six-month multi-stakeholder Discovery phase was undertaken to establish the needs and priority research areas of the network and identify key areas of impact to support a full grant application. Methods used to gather information from stakeholders to support this aim included an in-person workshop, online roundtables (for those unable to attend the workshop), an online survey, and targeted interviews.

2.1 Discovery Phase: In-Person Workshop

The workshop took place in May 2024 and was attended by over 90 participants from across industry, academia, government and other bodies (Fig. 1). Registration to the workshop was free and open to anyone. The LCA skills and experience of the attendees varied, from experienced LCA practitioners to those with no experience of conducting or interpreting LCA studies (Fig. 2).

The workshop was structured with small group breakout sessions to identify the value of an LCA regulatory science network, potential barriers for implementing LCA in regulation and standards, training needs, and long-term network vision. After a short introduction, groups discussed a series of questions relevant to each topic, with key points of the discussions recorded by a notetaker in each group.

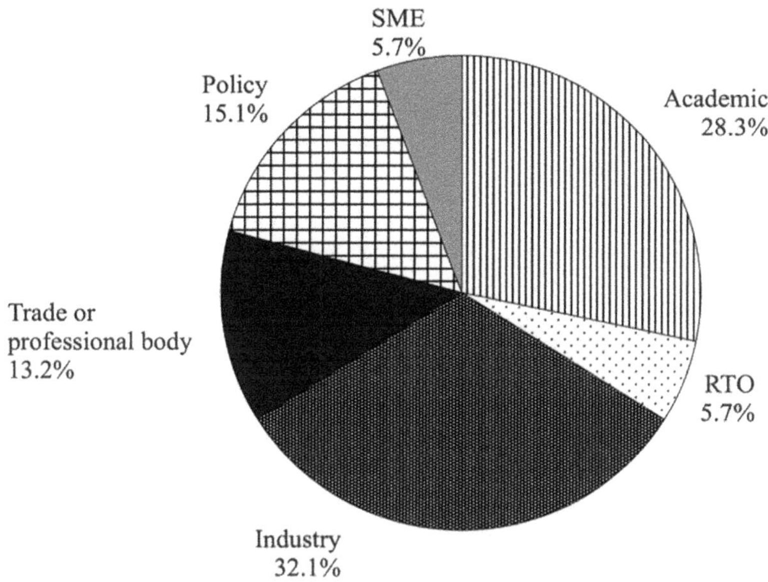

Fig. 1 Workshop attendees by sector. SME = Small and medium-sized enterprise; RTO = Research and technology organisation

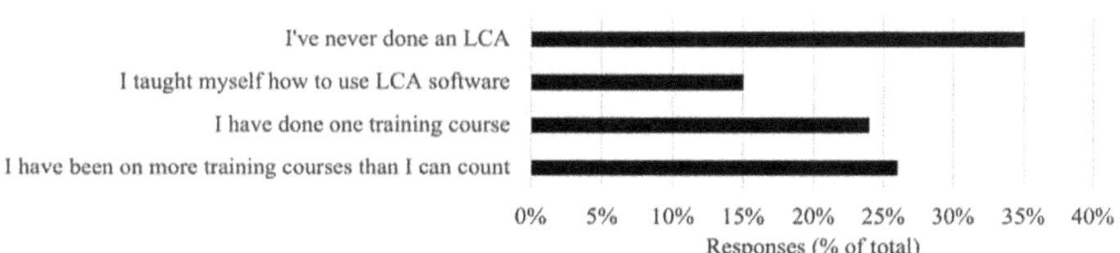

Fig. 2 LCA experience of workshop attendees

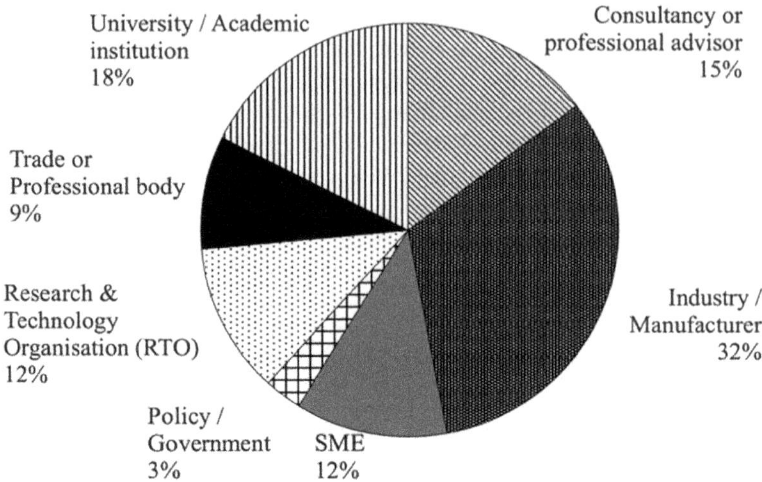

Fig. 3 Survey participant breakdown. SME = Small and medium-sized enterprise

2.2 Discovery Phase: Online Survey

An online survey was distributed to stakeholders via the network mailing list and LinkedIn in June and July 2024. The survey aimed to gather further quantitative and qualitative data on stakeholder views on applying LCA in policy making and regulation and on how industry uses, and could use, LCA to comply with regulation. 29 responses to the survey were received from a range of organisation types (Fig. 3).

2.3 Discovery Phase: Online Roundtables, Interviews, and Discussions

Three online roundtable events were held in June and July 2024 with participants who were unable to join the in-person workshop. Each roundtable had 8–12 participants, with the same topics as the in-person workshop being discussed. Early feedback highlighted the importance of offering both online and in-person events, as representatives from smaller organisations had difficulty committing the time and resources to attend in-person events and thus appreciated the availability of online roundtables.

Semi-structured interviews were subsequently held in July with representatives from four organisations to explore challenges related to the use of LCA in regulatory science in greater depth. Interviewees included a member of a devolved government and three senior figures from across the foundation industries, selected for their diverse backgrounds and experiences.

In-depth discussions were also held with policy advisors from DESNZ (the UK Department for Energy Security and Net Zero) and DEFRA (the UK Department for

Environment, Food and Rural Affairs) to ensure that policy makers' views and needs were fully captured.

2.4 Consolidation and Filtering of Results

The information collected during the Discovery phase engagement was consolidated into topic headings, and then filtered by the LCA-RSIN core team to focus on specific challenges in the implementation of LCA in regulatory science (as opposed to general challenges of conducting LCA).

3 Results

Table 1 gives the consolidated list of challenges and questions for the use of LCA in regulatory science and policy making as identified by stakeholders during the Discovery phase engagement. In many discussions, LCA topics that arose could be considered challenges or strengths dependent on the application context. As LCA-RSIN is focused on enabling successful implementation of LCA in regulation, responses were narrowed to focus on topics essential to ensure consistency and comparability for decision and policy making.

At the start of the workshop, participants' initial thoughts on key issues and challenges in embedding LCA in regulatory science were collected (Fig. 4). Almost 30% of responses indicated challenges related to data, including data quality, reliability, availability and collection. Challenges in choosing suitable proxy data and identifying areas where existing background data was sufficient or requires updating were identified as key issues to address for effective regulatory use of LCA.

Consistent application of methodology is essential when applying LCA in a regulatory context to enable effective decision making. Issues highlighted included inconsistent methods in the comparison of bio-based and fossil-based products, challenges with inconsistent allocation when considering recycling, along with a lack of background data for recycling processes.

Harmonisation across sectors was highlighted as something that would be beneficial, with one participant commenting "plastic has their standards data, and then construction comes and try to use them, but it is not possible".

Workshop discussions emphasised the need for industry, regulators, and policy makers to ensure transparency in LCA. Participants cited examples where incomplete inventory reporting and lack of methodological detail made LCA results irreproducible and incomparable. Concerns about greenwashing were raised, with the issue seen as incentivised by market pressures related to emission reduction targets and competition.

Workshop participants identified a skills gap in LCA knowledge and methodology. This skills gap was felt to be across all sectors. They emphasised the need for life

Table 1 Consolidated challenges for the implementation of LCA in regulation identified through stakeholder engagement

Challenge	Example challenges and questions identified by stakeholders for using LCA in regulation and policy
Data	• Data transparency • Data quality and outdated data • Lack of data/data availability • Can new approaches help address the data gaps?
Credibility/confidence	• Lack of transparency • Lack of comparability/need for harmonised methodologies • Vested interests • Misuse of conclusions • How can policy makers have confidence in LCA?
Transparency	• Lack of clear guidelines • Hard to access information • Clear system boundaries not applied • Incomplete outputs • Balance of commercial confidentiality with transparency
Costs and time	• Cost of software and data and LCA as a whole • Time consuming to perform studies • Time-specific bespoke studies are difficult to update • Can AI/new approaches reduce costs and time requirements and improve dynamic nature of LCAs?
Methodology	• Lack of harmonisation/standardisation in methodology • LCA practitioner-dependent choices • Reproducibility • Varying uncertainty in methods and impact assessment • Differing approaches for end of life/recycling • Lack of consensus on emerging topics e.g. biogenic carbon, new models—mass balance chain of custody
Skills	• Technical complexity • Lack of in-house skills • Hard to access training • Recruitment issues
Regulatory preparedness	• Lack of understanding of future regulation such as CBAM • Lack of expertise to engage in policy development process • Data availability

Fig. 4 Workshop attendee responses to question on sources of key challenges in LCA for regulatory science (larger words indicate a greater number of identical responses)

cycle thinking before any LCA training and highlighted the importance of LCA interpretation skills. It was highlighted that 'commissioning an LCA' is a skill in itself, as it is critical to carefully define the question being answered so as to set the goal and scope appropriately. These will be key considerations for designing future regulatory frameworks utilising LCA.

Skills and training gaps were further evidenced in survey responses. Training in both conducting and interpreting LCA was recognised as crucial, with cost and time to do training highlighted as barriers. Among those participants who had completed LCA training, the type of training most frequently undertaken was online training, followed by courses offered by software providers or universities. Nevertheless, 61% of those trained reported that the training did not meet their needs or was unsatisfactory. Consequently, 82% of respondents expressed the belief that current training is inadequate, and 97% indicated that additional training is certainly or probably required. Training in understanding and assessing data quality and in application of LCA to regulation were highlighted as important, as was technical training for practitioners and introductory training for non-LCA practitioners who need to use LCA information in either business or policy development. The survey also provided insights into the challenges encountered in finding suitable training in LCA interpretation, and the particular gaps identified by different sizes and types of organisations. Discussion during the roundtables captured specific challenges for SMEs (small and medium-sized enterprises) concerning lack of employees with the required expertise, time and resources needed to commission LCAs. These issues need careful consideration regarding the practicality of increasing the use of LCA in regulation.

Survey participants unanimously supported the network's creation and identified several key benefits (Fig. 5). They ranked all potential benefits as important, scoring above 3.5 out of 5, showing strong support for the network. Benefits most strongly cited as important were:

- The opportunity to hear from government about upcoming regulatory developments.
- The opportunity to develop evidence-based approaches to resolve issues of consistency in application of LCA.

The findings from the Discovery phase stakeholder engagement revealed clear challenges and gaps for the use of LCA methodology/studies in regulatory science to inform innovation and policy. This provides opportunities for a network to act as an enabler to bring together academia, industry, regulators, policy makers and other organisations to co-develop strategies and solutions.

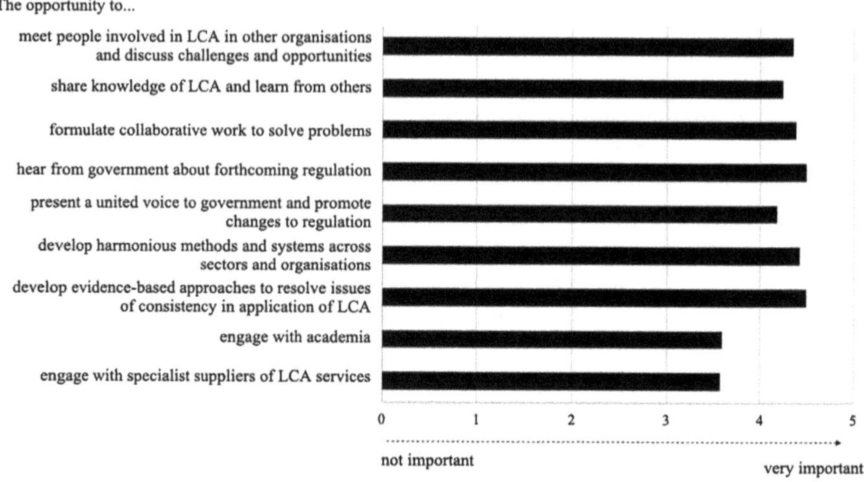

Fig. 5 Importance of a range of key benefits anticipated from the network, scored by survey respondents

4 Discussion

The Discovery phase of LCA-RSIN identified challenges (Table 1) requiring research and development to support the use of LCA in policy making and regulation in the UK. The LCA-RSIN core team subsequently synthesised these findings and co-developed a strategic plan with DESNZ and DEFRA for the Implementation phase (2025) of the network. A two-pronged approach was developed, including research to tackle the challenges for embedding LCA in future policy making and regulation, and knowledge exchange to upskill industry to comply with that regulation. Five key research topic areas emerged: data, bio-based and upcoming materials, methodology, practice and accreditation, and skills. These form the basis of five Working Groups (WGs) for the Implementation phase of the network (Table 2). The core team along with DESNZ and DEFRA developed key questions for each WG to investigate, with the aim of presenting research findings at the end of the Implementation phase. Although each WG has a distinct focus, there is clear crossover between them. For example, if AI is to be used to address data gaps (WG1), LCA practitioners will require skills in AI and interpretation of results (WG5), considering potential uncertainty created by the use of AI approaches, and policy makers will need confidence in the practice and reliability of using AI in LCA (WG3, WG4).

The need to develop a community and provide a forum for knowledge exchange came through strongly during the Discovery phase. In response to this, the Implementation phase also proposes Special Interest Groups (SIGs) on the following topics:

- Policy and regulation.
- Development and use of LCA tools.

Table 2 LCA-RSIN Working Groups for 2025, with the remit to research and deliver findings to answer specific questions related to the successful use of LCA in policy making and regulation

Working Group (WG)	Key questions being investigated by the WG
1. Data	• How does the use of default data impact confidence in LCA results? Where is good quality default data lacking? • How should new data be incorporated into regulatory science LCA, and should specific guidance be made available for this process? • How could AI combined with life cycle thinking improve decision making, innovation and regulation to decarbonise industry?
2. Bio-based and upcoming materials	• What guidelines, standards or regulations are needed to assess and compare LCA of bio-based or recycled materials with fossil-based materials? • Does current LCA practice allow assessment of circular economy goals; if not what guidelines and methodology developments are needed?
3. Methodology	• How can life cycle thinking and assessment be used to develop policies, benchmarks or targets, and how can those support innovation, e.g. of carbon intensive goods? • How might adopting a standard methodology for measuring environmental performance affect UK industry? Would the EU PEF be appropriate?
4. Practice and accreditation	• What would give policy makers greater confidence in using LCA results and/or approaches to inform and evaluate policy? • Should there be accreditation of LCA and LCA practitioners, how would this be developed?
5. Skills	• What is the skills gap in LCA for regulatory science? • What training and education is needed in the UK, for policy makers, industry, students, innovators?

- LCA Practitioner technical forum.
- Environmental sustainability reporting (including Environmental Product Declarations, EPDs).
- Forthcoming Carbon Border Adjustment Mechanism (CBAM).
- Social LCA and the integration of social/environmental LCA.

5 Conclusions

During its Discovery phase, LCA-RSIN undertook stakeholder engagement using workshops, online round tables, surveys and interviews. This engagement revealed an array of challenges to embedding LCA in regulation and policy making to promote sustainable innovation, including data availability and quality, harmonisation of methodologies, consistent accounting of biogenic carbon, allocation methodology for recycled materials and multifunctional systems, transparency, reliability and trustworthiness of outputs and the need for skills development.

To address these challenges and remove barriers to the advancement of LCA in regulation and policy making, the LCA-RSIN network was formally established in 2025 (Implementation phase). The network encompasses mechanisms to address each identified need, along with knowledge exchange mechanisms to increase expertise and share expertise. Five Working Groups convene diverse teams of expert stakeholders to conduct research on the specific challenges identified and engage directly with policy makers. Special Interest Groups bring together policy advisors, academics, and industry experts, to exchange information and knowledge. External engagement activities, such as webinars on CBAM and annual network meetings, ensure knowledge is shared and reach amplified throughout LCA and policy communities in the UK.

Acknowledgements The authors would like to acknowledge Innovate UK for funding of the work under grants 10114584 and 10139850.

References

Hauschild M, Rosenbaum R, Olsen S (2018) Life cycle assessment theory and practice. Springer

Sala S, Amadei A, Beylot A, Ardente F (2021) The evolution of life cycle assessment in European policies over three decades. Int J Life Cycle Assess 26:2295

Sonnemann G, Gemechu E, Sala S, Schau E (2018) Life cycle thinking and the use of LCA in policies around the world. In: Hauschild M, Rosenbaum R, Olsen S (eds) Life cycle assessment. Springer, Cham

Walker S, Rothman R (2020) Life cycle assessment of bio-based and fossil-based plastic: a review. J Clean Prod 261:121158

From Voluntary Initiatives and Collaboration to Regulation: Embodied Carbon Progress in Australia and New Zealand

Barbara Nebel

Abstract Since 2014, collaboration across industry, government, green building councils, and LCA communities in Australia and New Zealand has driven significant progress in addressing embodied carbon. Key milestones include the launch of the EPD programme by ALCAS and LCANZ, the 2018 report revealing buildings and infrastructure contribute up to 20% of NZ's emissions, and MBIE's 2020 Building for Climate Change programme. In Australia, thinkstep-anz and GBCA warned in 2021 that embodied emissions could reach 85% by 2050 without action. Recent developments include NABERS' embodied carbon rating consultation and new regulatory moves in 2024. Despite challenges like data gaps and inconsistent methods, joint efforts are advancing carbon reduction in the built environment.

1 Introduction

Until 2014, the Green Building rating tools by the Australian and New Zealand Green Building Councils did not consider Life Cycle Assessment (LCA) or Environmental Product Declarations (EPD). The lack of a scheme for the development and registration of Environmental Product Declarations was one of the barriers for implementing this.

There also was the perception that manufacturers were not undertaking LCA studies, since no such studies were published. However, companies doing LCAs for internal improvement processes and seeking to differentiate the environmental performance of their products through science-based, third party verified reporting had no opportunity to communicate their work in a credible manner (LCANZ 2013).

A major consequence of this information gap was that embodied carbon—the emissions associated with the materials and construction phase of buildings—was largely ignored. Due to the lack of publicly available information on embodied carbon of building products, estimates published by the NZ Government suggested that

B. Nebel (✉)
thinkstep-anz, Porirua, New Zealand
e-mail: barbara.nebel@thinkstep-anz.com

M. Traverso et al. (eds.), *Life Cycle Management from Global to Local*,
https://doi.org/10.1007/978-3-032-17987-6_38

the built environment accounted for only 2% of national emissions (New Zealand Productivity Commission 2017). These figures primarily considered direct emissions and did not encompass the full life cycle of buildings and infrastructure.

Similarly, Australia's built environment has traditionally been addressed in climate policy through the lens of operational emissions, associated with electricity and fuel consumption during a building's use phase.

The lack of published, credible data on life cycle impacts and embodied carbon hindered the integration of LCA and EPDs into green building tools and policies. As a result, policy makers and rating systems overlooked comprehensive environmental impacts, focusing primarily on operational emissions and underestimating the true carbon impact of the built environment.

2 Approach

To overcome the lack of data-driven credits in green building tools and the broader absence of credible, publicly available information on embodied environmental impacts, the Life Cycle Association of New Zealand (LCANZ) and the Australian Life Cycle Assessment Society (ALCAS) took the initiative to establish EPD Australasia in 2014 (LCANZ 2013). Recognising that their role extended beyond promoting LCA best practices to enabling industry-wide transparency, these two organisations responded to the clear gap in credible communication mechanisms. Under licence from the International EPD® System, EPD Australasia was launched as the region's first programme operator for Environmental Product Declarations (EPDs). Its primary aim was to provide a harmonised, standardised, and independently verified framework for communicating the environmental impacts of products across their life cycles. By doing so, it enabled manufacturers already undertaking LCAs to credibly disclose their performance aligned with international standards. EPDs for most building products in New Zealand provide a solid basis to provide information that is relevant to policy makers.

2.1 Embodied Carbon in the Built Environment

2.1.1 The Contribution of the Built Environment to New Zealand's Greenhouse Gas Emissions

The report 'The Carbon Footprint of New Zealand's Built Environment: Hotspot or Not?' (Vickers et al. 2018) reassessed the contribution of New Zealand's built environment to the nation's total greenhouse gas emissions by applying a life cycle approach, addressing the significant underestimation of emissions in previous assessments that focused only on operational emissions.

The report employed a consumption-based approach, allocating emissions to sectors at the point of consumption rather than production. The consumption-based approach accounts for the emissions driven by a New Zealand's consumption, regardless of where the emissions occur. It includes imported goods and accounts for the export of goods as well. The production-based approach analyses greenhouse gas (GHG) emissions based on where they physically occur. All agricultural emissions for example would be attributed to the country's emissions, even if most of the agricultural goods are exported.

Additionally, the study incorporated a life cycle perspective, considering emissions from the extraction of raw materials, manufacturing, construction, operation, and end-of-life treatment of buildings and infrastructure.

The report found that New Zealand's built environment may be responsible for approximately 20% of the country's gross carbon footprint (Fig. 1c). This significant increase was attributed to the inclusion of indirect and embodied emissions in the analysis. International trade was taken into account, because New Zealand's economy is dominated by agricultural production for export. If international trade was excluded, the contribution of the built environment would contribute 13% (Fig. 1b).

There are two main reasons why the New Zealand figure for the carbon footprint for energy use in buildings appeared to be four times smaller than the global average of 19% (Lucon et al. 2014): (1) low-carbon electricity and the use of wood for most of building heating, and (2) the percentages are distorted by New Zealand's large share of direct agricultural emissions, which account for roughly half of the nation's total gross carbon footprint (Fig. 1a).

The findings underscore the necessity of integrating embodied carbon considerations into national climate strategies. To achieve meaningful reductions in GHG emissions, policies must extend beyond operational efficiencies to encompass the entire life cycle of buildings and infrastructure. This includes promoting the use of low-carbon materials, encouraging sustainable construction practices, and implementing regulations that account for both direct and indirect emissions.

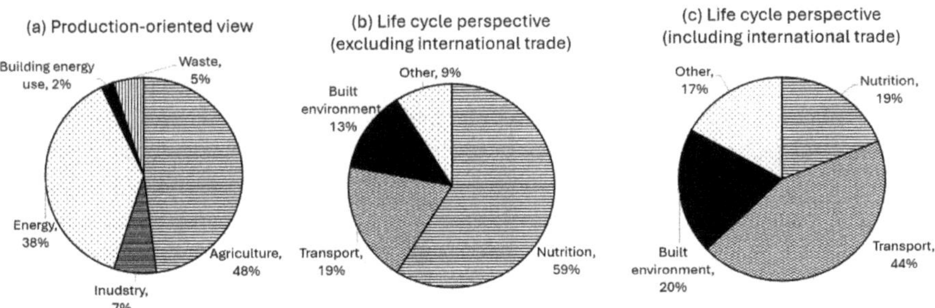

Fig. 1 A breakdown of New Zealand's carbon footprint in 2025 from **a** a production perspective, **b** a life cycle perspective, and **c** a life cycle consumption perspective (Vickers et al. 2018)

2.2   Embodied Carbon in the Built Environment in Australia

The impact of embodied carbon in Australia's built environment was quantified in two key reports. The first report established the impact of embodied carbon vs. operational carbon (GBCA and thinkstep-anz 2021), while the second report assesses the impact of emissions related to the built environment in relation to the overall emissions in Australia (Infrastructure Australia 2024).

2.2.1   Embodied Versus Operational Emissions of Australia's Built Environment

The 2021 Embodied Carbon & Embodied Energy in Australia's Buildings report establishes a 2019 baseline for embodied carbon and energy in commercial and residential buildings and projects future emissions under a business-as-usual (BAU) scenario (Fig. 2) (GBCA and thinkstep-anz 2021).

Over the entire Australian building stock, this report finds that the share of whole-life building carbon emissions from embodied carbon was 16% in 2019 (Fig. 2). Under business-as-usual (BAU), this share is expected to climb to 85% by 2050. These calculations include operational emissions from the entire building stock in 2019 and 2050, but only the embodied emissions from new builds and substantial renovations in those same years.

As the electricity grid decarbonises and operational efficiencies improve embodied carbon, the emissions arising from the manufacture, transport, and assembly of building materials are set to become the dominant contributor to emissions related to the built environment.

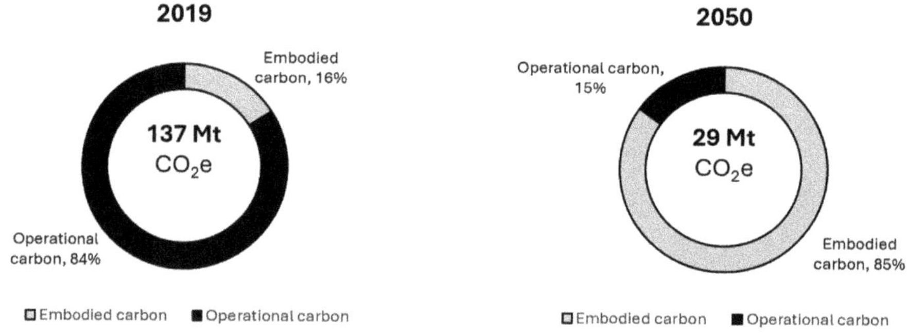

Fig. 2   Carbon emissions from Australia's building stock in 2019 and 2050

2.2.2 Contribution of the Construction Industry to Australia's Total Greenhouse Gas Emissions

The 2024 Embodied Carbon Projections for Australian Infrastructure and Buildings report uses Infrastructure Australia's Market Capacity Program data and a hybrid life cycle analysis methodology to forecast emissions from the construction pipeline between 2022–2023 and 2026–2027 (Infrastructure Australia 2024). Commissioned by the Australian Department of Climate Change, Energy, the Environment and Water, the report aims to inform decarbonisation policies aligned with Australia's Climate Change Act 2022 and Net Zero 2050 goals.

Both reports adopt a process-based Life Cycle Assessment (LCA) approach, with the latter integrating environmentally extended input–output data to account for project uncertainties and missing data in smaller-scale developments.

Australia's built environment directly contributes nearly one-third of national emissions and indirectly over half. Unlike operational emissions, embodied carbon is locked in at the construction phase and cannot be reversed. Therefore, tackling embodied emissions is crucial for achieving near- and long-term climate targets. The report identifies significant and immediate opportunities for carbon abatement through material substitutions and construction process optimisations.

3 Results

The establishment of EPD Australasia as a regional platform helped shift the focus of policy and rating tools towards life cycle thinking and embodied impacts, laying the groundwork for more comprehensive and science-based decision-making in the built environment.

3.1 *Implementation of LCA and EPD Credits in Rating Tools*

Shortly after the launch of EPD Australasia the Green Building Council of Australia (GBCA) incorporated Life Cycle Assessment (LCA) and Environmental Product Declarations (EPDs) into its Green Star rating tool with the launch of Green Star—Design & As Built v1 in 2014 (GBCA 2014). This marked a strategic shift towards performance-based assessment of materials and promoted the use of EPDs for transparency and comparability. From 2017, NZGBC explicitly recognised EPDs for use in their Green Star NZ and Homestar tools to encourage sustainable procurement and material choices. EPDs contribute towards material credits and LCA-based performance improvements. Both councils, as well as the Infrastructure Sustainability Council (ISC) aligned their frameworks with international best practices to incentivise data-driven, environmentally responsible design and procurement.

In May 2025 the 2000st EPD has been registered under EPD Australasia (EPD Australasia 2025).

3.2 Government Response NZ

In 2020 the New Zealand Ministry of Business, Innovation and Employment (MBIE) launched the Building for Climate Change (BfCC) programme, aiming to transform how buildings are designed, constructed, and operated in New Zealand to significantly reduce greenhouse gas emissions and improve resilience to climate impacts (MBIE 2020).

The programme was launched to address the built environment's substantial contribution to New Zealand's carbon footprint of 20% of the country's gross emissions when measured using a consumption-based Life Cycle Assessment (Vickers et al. 2018). Historically, policy and regulation had focused predominantly on operational energy efficiency, overlooking the emissions embedded in construction materials and processes—commonly known as embodied carbon.

BfCC was introduced in response to this oversight, with the following primary goals: (1) lower the whole-of-life carbon footprint of new buildings, including both operational and embodied emissions, (2) improve building climate resilience, enabling new constructions to better withstand extreme weather events and long-term environmental changes, (3) support the transition to a low-emissions economy, aligning with New Zealand's obligations under the Climate Change Response (Zero Carbon) Amendment Act 2019, which commits the country to net-zero carbon emissions by 2050 (New Zealand Parliament 2019).

BfCC is structured around two core workstreams: (1) Whole-of-Life Embodied Carbon Emissions Reduction. This was designed to require new buildings to report and then reduce embodied carbon across their life cycle. It includes development of emissions caps and regulatory pathways for low-carbon materials and construction methods, (2) Operational Efficiency Improvements, with a focus on reducing emissions from energy used to heat, cool, and operate buildings. This aims to encourage passive design, renewable energy integration, and energy efficiency.

The Building for Climate Change programme marked a significant shift in New Zealand's regulatory approach to buildings, embedding climate considerations across the full building life cycle. It was initiated to fill regulatory gaps, meet statutory climate obligations, and modernise the building sector in alignment with both domestic targets and global best practice. Its success will be instrumental in achieving a low-carbon, climate-resilient built environment.

3.3 Government's Response in Australia

The Australian National Construction Code (NCC) is progressively integrating embodied carbon considerations into its regulatory framework, marking a pivotal shift in sustainable building policy. From 2025, the NCC will introduce a voluntary pathway for commercial buildings to measure and report embodied carbon emissions, using the National Australian Built Environment Rating System (NABERS) methodology as a foundation. This initiative aims to build industry capability and data transparency (Australian Government Department of Industry, Science and Resources 2024). In a further step, the Australian Building Codes Board (ABCB) has been directed to develop mandatory minimum standards for embodied carbon to be included in the 2028 NCC update (Australian Government, Department of Climate Change, Energy, the Environment and Water 2025). This policy evolution is aligned with the national Trajectory for Low Energy Buildings, which highlights the importance of addressing both operational and embodied emissions to meet Australia's net zero commitments (Australian Government, Department of Climate Change, Energy, the Environment and Water 2024). These planned changes reflect broader industry and government recognition of the significant climate impact of embodied carbon in the built environment.

4 Conclusion

The transition from voluntary industry-led initiatives to formal policy-making in addressing embodied carbon in Australia and New Zealand underscores the critical role of collaborative frameworks in driving systemic change. Initially, the absence of regionally coordinated mechanisms for recognising LCA and EPDs hindered the ability of manufacturers and designers to quantify and communicate environmental performance. The establishment of EPD Australasia in 2014 by two organisations driven by volunteers filled this void, enabling transparent, standardised, and credible communication of environmental impacts. This, in turn, catalysed the integration of embodied carbon metrics into major sustainability rating systems such as Green Star (GBCA n.d.) and the Infrastructure Sustainability (IS) rating scheme (Infrastructure Sustainability Council n.d.).

In New Zealand, the Building for Climate Change (BfCC) programme represents a paradigmatic shift in regulatory thinking, embedding life cycle emissions within the policy architecture to meet statutory net-zero targets. Similarly, in Australia, robust research has provided a credible evidence base that now informs impending changes to the National Construction Code (NCC). The planned voluntary embodied carbon reporting in 2025, followed by potential regulatory standards in 2028, highlights a policy trajectory founded on data, stakeholder engagement, and international best practice alignment.

Crucially, the evolution of embodied carbon policy across both countries illustrates the relationship between voluntary leadership and formal governance. Industry, NGOs and government agencies have collaborated around a shared recognition of the built environment's climate impact. This collaborative momentum has laid the groundwork for enforceable policy instruments.

This is a compelling example of how voluntary action, supported by technical rigour and institutional partnerships, can evolve into comprehensive, forward-looking policy that supports national and global climate objectives.

References

Australian Government, Department of Climate Change, Energy, the Environment and Water (2025) Commercial buildings. https://www.dcceew.gov.au/energy/energy-efficiency/buildings/commercial-buildings. Accessed 25 May 2025

Australian Government (2024) Department of climate change, energy, the environment and water. Trajectory for low energy buildings national construction code 2025 & 2028. https://www.dcceew.gov.au/sites/default/files/documents/trajectory-for-low-energy-buildings-national-construction-code-2025-2028-presentation.pdf. Accessed 25 May 2025

Australian Government, Department of Industry, Science and Resources (2024) Building ministers' meeting: Communiqué in June 2014. https://www.industry.gov.au/news/building-ministers-meeting-communique-june-2024. Accessed 25 May 2025

EPD Australasia (2025) 2000 EPDs and counting. https://epd-australasia.com/2025/05/2000-epds-and-counting-a-milestone-for-transparency/. Accessed 25 May 2025

GBCA, thinkstep-anz (2021) Embodied carbon and embodied energy in Australia's buildings. Sydney. https://www.thinkstep-anz.com/assets/Whitepapers-Reports/Embodied-Carbon-Embodied-Energy-in-Australias-Buildings-2021-07-22-FINAL-PUBLIC.pdf. Accessed 25 May 2025

GBCA (2014) Green star—design & as built v1 technical manual. Green Building Council of Australia

GBCA (n.d.) Green star rating system. https://new.gbca.org.au/green-star/rating-system/. Accessed 20 July 2025

Infrastructure Australia (2024) Embodied carbon projections for australian infrastructure and buildings. https://www.infrastructureaustralia.gov.au/reports/embodied-carbon-projections-australian-infrastructure-and-buildings. Accessed 25 May 2025

Infrastructure Sustainability Council (n.d.) IS rating scheme. https://www.iscouncil.org/is-ratings/. Accessed 20 July 2025

LCANZ (2013) The Australasian EPD scheme. https://lcanz.org.nz/2013/10/14/the-australasian-epd-scheme/. Accessed 25 May 2025

Lucon O, Ürge-Vorsatz D, Zain Ahmed H, Akbari H, Bertoldi P, Cabeza LF, Vilariño MV. Buildings. In: Edenhofer O, Pichs-Madruga R, Sokona Y, Farahani E, Kadner S, Seyboth K, Adler A, Baum I, Brunner S, Eickemeier P, Kriemann B, Savolainen J, Schlömer S, von Stechow C, Zwickel T, Minx JC (eds) Climate change 2014: mitigation of climate change. Contribution of working group iii to the fifth assessment report of the intergovernmental panel on climate change. Cambridge, United Kingdom and New York, NY, USA: Cambridge University Press

Ministry of Business, Innovation & Employment (MBIE) (2022) Emissions reduction plan: building and construction sector initiatives. https://www.mbie.govt.nz/dmsdocument/23616-emissions-reduction-plan-building-and-construction-initiatives-proactiverelease-pdf. Accessed 25 May 2025

Ministry of Business (2020) Innovation & employment (MBIE). Building for climate change: transforming the building and construction sector. https://www.mbie.govt.nz/building-and-ene rgy/building/building-for-climate-change/. Accessed 25 May 2025

New Zealand Parliament (2019) Climate change response (zero carbon) amendment act 2019. https://www.legislation.govt.nz/act/public/2019/0061/latest/LMS183736.html. Accessed 25 May 2025

New Zealand Productivity Commission (2017) Low-emissions economy: issues paper. https://www.treasury.govt.nz/sites/default/files/2024-05/pc-inq-lee-low-emissions-economy-issues-paper.pdf. Accessed 25 May 2025

Vickers J, Fisher B, Nebel B (2018) The carbon footprint of New Zealand's built environment: hotspot or not? https://www.thinkstep-anz.com/assets/Whitepapers-Reports/Built-environment-carbon-footprint-v15-updated_BN.pdf. Accessed 25 May 2025

The Role of LCA in Policy for Enabling Sustainable Consumption and Production: A Case Study from South Africa

Valentina Russo, Kolobe Chaba, Taahira Goga, William Stafford, and Anton Nahman

Abstract This paper explores the evolving role of Life Cycle Assessment (LCA) in shaping public policy to enable sustainable consumption and production (SCP), with a focus on the South African context. It highlights the dual function of LCA: informing policy decisions through science-based evidence and enhancing life cycle sustainability; in turn, in some cases, the need for LCA studies is driven by policy requirements. Drawing from global examples and focusing on South Africa as a case study, the paper illustrates how LCA has influenced regulations, procurement strategies, and sustainability standards. In South Africa, the integration of LCA into Extended Producer Responsibility (EPR) Regulations marks a significant milestone, despite challenges such as limited local data and LCA expertise in the country. The paper reviews key LCA studies that have informed national policies, such as those on plastic packaging, green hydrogen, and waste management. It also discusses how the development of national LCA guidelines and LCA database could help supporting LCA implementation. The findings highlight the potential of LCA to drive a Just Transition (JT) and support long-term environmental and socio-economic goals.

1 Introduction

The relationship between Life Cycle Assessment (LCA) and policies aimed at promoting sustainable consumption and production (SCP) is a two-way street. On one hand, LCA can inform policies targeting high-impact products and activities, by identifying key policy issues, comparing solutions (ex-ante), and evaluating implementation (ex-post). On the other hand, policies can promote the use of LCA and improvements across the product life cycle, through voluntary or mandatory assessment, incentives, limits, or thresholds for compliance (Sonnemann et al. 2018; Sanyé-Mengual and Sala 2022; Jegen et al. 2024).

V. Russo (✉) · K. Chaba · T. Goga · W. Stafford · A. Nahman
CSIR, Stellenbosch, South Africa
e-mail: vrusso@csir.co.za

© The Author(s) 2026

M. Traverso et al. (eds.), *Life Cycle Management from Global to Local*,
https://doi.org/10.1007/978-3-032-17987-6_39

Despite its growing relevance and being a well-established analytical framework, LCA remains underutilised in public policy and is not yet universally embedded in policy frameworks. Nevertheless, its adoption is increasing in contexts where quantitative environmental impact data is essential; with examples ranging from environmental regulation—where it informs emissions standards and product bans based on full life cycle impacts—to sustainable procurement, waste management, circular economy strategies, and climate policy, particularly in assessing sectoral carbon footprints (CFs) to guide mitigation efforts. Even when full LCA implementation is not feasible, the integration of Life Cycle Thinking (LCT) can enhance the quality and longevity of policy decisions (Seidel 2016; Sonnemann et al. 2018; Jegen et al. 2024). The CIRAIG's *LCA in Public Policy Database* (2020) documents the use of life cycle approaches across 17 jurisdictions within the Organisation for Economic Co-operation and Development (OECD), including Australia, Canada, the European Union, and the United States. However, due to its exploratory nature, it is not exhaustive, lacks recent developments, and excludes data from non-OECD countries.

Europe leads globally in integrating LCA into policy, reflecting a shift towards accounting for environmental externalities. LCA aids policymakers assessing full life cycle impacts, supporting eco-design, taxation, and labelling aligned with the polluter-pays principle (Sala et al. 2019; European Commission 2021). A key initiative is the Product Environmental Footprint (PEF), a standardised LCA-based method covering 16 impact categories, central to European regulations such as the Ecodesign for Sustainable Products Regulation (Sala et al. 2019; European Commission 2022). LCA and PEF also inform Green Public Procurement (GPP), guiding choices in sectors like construction and electronics through sector-specific benchmarks (SAPIENS Network 2023).

The European Union has been increasingly integrating LCA into its regulatory and financial frameworks to support sustainability transitions. The EU Taxonomy Regulation (European Parliament and Council 2020), a key part of the European Green Deal, uses LCA to shape technical screening criteria and ensuring activities support climate goals without harming other environmental objectives. In the Circular Economy Action Plan (CEAP), LCA informs regulations on product durability, reparability, and recyclability, guiding both policy and consumer decisions. LCA metrics are also being considered for integration into macroeconomic systems: efforts like the System of Environmental-Economic Accounting (SEEA) aim to align with the System of National Accounts (SNA) to reflect life cycle-based resource flows. Though not yet embedded in the SNA, initiatives such as SEEA-EEA, INEXT, and EXIOBASE have advanced environmentally extended input-output (EE-IO) models that link consumption to global environmental impacts (Tukker et al. 2013; Eurostat 2020).

Outside Europe, LCA has informed for example climate and energy policy in North America. In the United States, the Renewable Fuel Standard (EPA 2005) and California's Low Carbon Fuel Standard (CARB 2009) uses LCA to assess fuel emissions and to guide incentives; whilst Canada has piloted LCA-based procurement to cut embodied carbon in infrastructure. In the Asia-Pacific region, Japan integrates

LCA into green procurement, eco-labelling, and carbon neutrality strategies—e.g. the EcoLeaf program (JEMAI, n.d.); Australia and New Zealand promote LCA through building codes, infrastructure planning, and product stewardship schemes; while Singapore uses LCA in its Green Mark certification for sustainable buildings (Building and Construction Authority [BCA] 2021).

In South Africa, LCA has a relatively well-established base; with academics, researchers, and consultants generating LCA studies since the mid-1990s, mainly in the private sector (Harding et al. 2021; Notten and von Blottnitz 2018). LCA has, to date, been less actively applied in the South African public sector, where its role is centred around evaluating environmental impacts across a product's life cycle. Nevertheless, with an increasing emphasis on evidence-based policymaking and on green/circular economy, the need for credible evidence to support South Africa's transition to a sustainable society is increasingly being realized. Therefore, the aim of this paper is to pay key attention to the body of LCA research in South Africa which is having an impact by informing policy, action plans, and strategies in the last 5 years, in order to showcase how LCA is being mainstreamed within South African policy.

2 Method

Review studies aimed at assessing the current state of LCA research in South Africa have previously been conducted by Harding et al. (2021), Karkour et al. (2021) and Isah et al. (2024) (see Sect. 2.1). These foundational reviews provided the basis for a more refined analysis focused on identifying emerging LCA research that contributes providing the evidence base for policymaking, particularly in the context of the green and circular economy with a specific focus on sustainable consumption and production (SDG12). Also, being the LCA community in South Africa well-established but rather small, LCA experts from research organizations, academic institutions, and environmental consulting companies were approached to provide information regarding existing, completed, ongoing, or recently initiated LCA-related studies with a clear policy-oriented outcome in order to have the most up-to-date information. Additional studies were identified through internet searches (Google/Google Scholar and Scopus) using combinations of the terms '*LCA and policies in South Africa*' and '*Mainstreaming LCA in policies in South Africa*'.

The results, which include both peer-reviewed scientific publications and projects' reports, are limited to the past five years (2020–2025). These are summarized in Table 1, which outlines each study, the methodology employed, and the specific policy outcomes—whether the study has informed, is expected to inform, or has contributed input to policy development.

2.1 Life Cycle Assessment in South Africa

South Africa leads LCA-related research in Africa over the past 30 years, followed by Egypt, Tunisia, and Algeria (Karkour et al. 2021; Isah et al. 2024). Since the mid-1990s, South African academics have applied LCA across sectors such as energy, chemicals, retail, agriculture, and mining (Harding et al. 2021). Industry-focused LCAs have largely been conducted by local and international consultants, often in collaboration with academia, the National Cleaner Production Centre (NCPC-SA) especially for SMMEs, WWF-SA, and more recently, the Council for Scientific and Industrial Research (CSIR) (Notten and von Blottnitz 2018). Industry associations like the Clay Brick Association (CBA 2016), the International Platinum Association (IPA 2025), and the South African Sugar Association (Mashoko et al. 2010) have also been active in the LCA space.

With the increasing need of evidence-based policymaking and the green/circular economy, LCAs are increasingly shaping policy decisions. South Africa recently became one of the first countries to mandate LCAs under its Extended Producer Responsibility (EPR) Regulations (DFFE 2021). Producers in six sectors must now conduct LCAs to reduce material use, redesign products for reuse, recycling or recovery, and lower the environmental toxicity of post-consumer waste (DFFE 2021). This regulatory requirement is expected to significantly boost LCA uptake in the country. A first example of how LCAs have already influenced policy is provided by the Life Cycle Sustainability Assessment (LCSA) study on grocery bags (Russo et al., 2020) which led to amendments in plastic carrier bag regulations, including targets for increased recycled content (DFFE 2021; South Africa Government 2020).

3   Results: Mainstreaming LCA in Policy in South Africa

LCA was firstly introduced by the then Department of Environmental Affairs and Tourism (DEAT) in 2004 as part of South Africa's Integrated Environmental Management framework (DEAT 2004). The National Development Plan 2030 (NDP 2030) emphasises environmental sustainability and climate change, advocating for indicators and tools like LCA to guide policy (The Presidency 2011). Chapter 5 of the NDP 2030 further supports the use of such tools to enhance environmental resilience. Similarly, the Waste Research Development and Innovation (RDI) Roadmap includes LCA under its *'Modelling and Analytics'* cluster, highlighting its role in national sustainability efforts (DST 2014). Figure 1 presents a timeline of LCA developments in South Africa.

Table 1 summarizes LCA studies conducted in South Africa over the past five years that have influenced policies, action plans, and broader strategies. These include assessments of single-use plastics and food packaging (Russo et al. 2020; Stafford et al. 2022; Russo et al. 2023), disposable diapers (Chitaka and Nel 2025), and emerging e-fuels (Stafford et al. 2024; Russo and Goga 2025). While some related

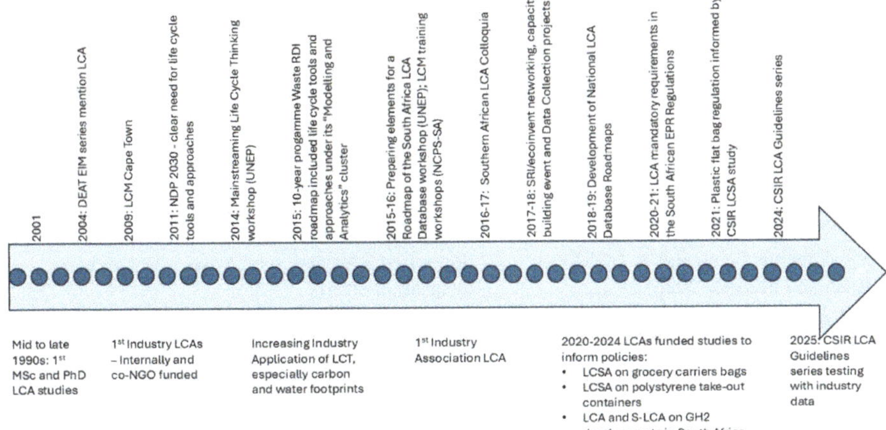

Fig. 1 Timeline of LCA developments and significant LC-related events held in South Africa (Notten and von Blottnitz (2018) updated by the authors)

LCA research has yet to directly impact policy, it underscores the urgent need for a national LCI database (Notten and von Blottnitz 2018), particularly in the context of South Africa's EPR regulations (Botha and Harding 2024).

Since 2019, the CSIR has played a leading role in LCA research by informing policy development. In 2020, the CSIR conducted a Life Cycle Sustainability Assessment (LCSA) study on grocery bags sold in South African retail stores (Russo et al. 2020; Stafford et al. 2022). The research highlighted that reusability and high recycled content were more effective mitigation strategies than the choice of material itself. These findings gained significant attention from industry, consumers, and government, directly influencing amendments to the Plastic Carrier Bags and Plastic Flat Bags Regulations under the National Environmental Management Act (Act No. 107 of 1998) (DFFE 2021). Specifically, the CSIR study informed targets for increasing recycled content in plastic carrier bags (South Africa Government 2020).

Between 2019 and 2022, the CSIR collaborated with UNIDO, the University of the Witwatersrand, and the Japanese government to explore sustainable alternatives to conventional plastics in South Africa. The project aimed to (1) develop an evidence-based action plan for transitioning to more sustainable materials and (2) enhance plastic waste collection by supporting waste picker integration. A key input to the project action plan was the LCSA study on 11 types of take-out containers (Russo and Stafford 2023), which guided material selection for plastic alternatives. This study expanded conventional LCA by incorporating ad-hoc indicators for material persistence and environmental pollution potential (Stafford et al. 2022).

In response to the LCA requirements outlined in the EPR Regulations (DFFE 2021)—specifically Regulation 5, sub-regulations (1)(k) and (1)(l), which mandate producers, brand owners, and importers of identified products to conduct LCA studies—the CSIR developed the *CSIR LCA Guidelines* (Russo et al. 2024) (Fig. 2). These guidelines offer clear guidance to South African producers on how to conduct

Table 1 Summary of LCA studies and the policy, action plan, and strategy informed by them in the past 5 years

Case study	Methodology	Policy outcome
Grocery bags in South Africa (Russo et al. 2020)	LCSA/LCA	Informed recycled content targets in plastic bag regulations (DFFE 2021) and supported policy on plastic pollution
Plastic bag alternatives (Stafford et al. 2022)		Provided evidence for SA's position in UN treaty negotiations on plastic pollution
Food containers and cups (Russo and Stafford 2023)	LCSA	Supported SA's action plan and waste picker integration efforts
LCA guidelines (Russo et al. 2024)	LCA	Enabled harmonised LCA practices for EPR compliance (ex-ante and ex-post)
Green ammonia for transport (Stafford et al. 2024)	LCA	Informed project design and contributed to green hydrogen certification development (DFFE)
Saldanha Hydrogen Hub (Russo and Goga 2025)	S-LCA	Supported Just Energy Transition (JET) alignment (ex-ante)
National LCA database roadmap (Notten and von Blottnitz 2018)	LCT/LCM	Provided a pathway for SA's national LCA database and evidence-based policy
International LCA data in EPR (Botha and Harding 2024)	LCT/LCM	Assessed use of secondary data to streamline LCA for EPR compliance
Diaper LCA in rural context (Chitaka and Nel 2025)	LCA	Highlighted geographic relevance in LCA and supported SARChI Chair's municipal waste research
Power-to-X and GHG reduction (Stafford et al. 2025a)	LCA (GHG focus)	Recommended PtX inclusion in future NDCs to meet Net Zero by 2050 (RSA 2021)
Circularity in NDCs (Stafford et al. 2025a, b)		Piloted UN Toolbox; identified circular economy potential in construction sectors (UNEP et al. 2023)

an LCA in an effective, practical, and consistent way. Given the wide range of products listed in the EPR Notices (DFFE 2021), the numerous producers involved, the limited pool of LCA practitioners, and the paucity of local LCA data, the *CSIR LCA Guidelines* also propose a pragmatic, cost-effective approach to operationalise the LCA requirements, while ensuring that LCA studies are robust and scientifically credible, and that the results are meaningful. Similar to the PEF approach, the guidelines aim to standardise key methodological choices, promoting relevance, consistency, completeness, comparability, and transparency in South African LCA practice.

More recently, the CSIR team contributed to the '*Promoting a South African Green Hydrogen Economy (H2.SA)*', funded by the Deutsche Gesellschaft für Internationale Zusammenarbeit (GIZ) GmbH, which supports South Africa's public and

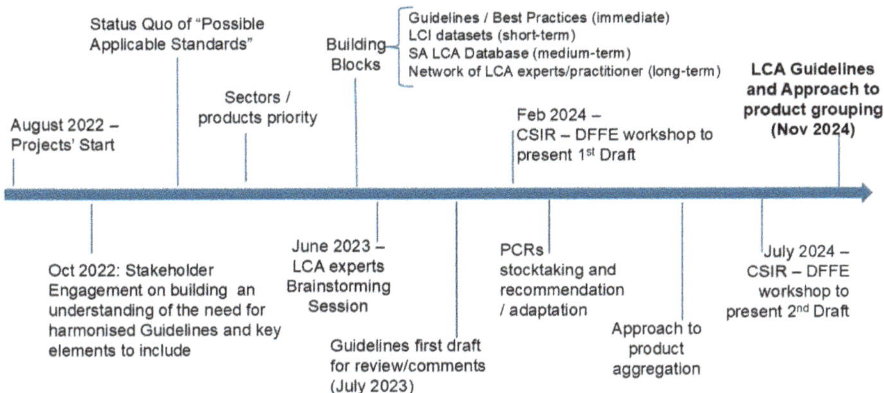

Fig. 2 CSIR LCA Guideline timeline. *Source* Authors

private sectors in developing a sustainable green hydrogen (GH_2) economy. Two key studies emerged: a LCA of green ammonia produced at a coastal South African facility for heavy-duty transport in Germany (Stafford et al. 2024), and a social LCA (S-LCA) of Power-to-X (PtX) products at the Saldanha Bay Hydrogen Hub (Russo and Goga 2025). The green ammonia LCA aims to inform project design and the creation of harmonised standards for certifying GH_2 and its derivatives as low-carbon fuels. The findings are also informing the development of a national GH_2 certification standard led by the Department of Forestry, Fisheries and the Environment (DFFE) (Stafford et al. 2024). The S-LCA study emphasises the importance of contextualising clean energy and low-carbon development within local socio-economic realities. It highlights both opportunities and challenges for the GH_2 sector in Saldanha Bay and across South Africa more in general, underscoring the need for policy recommendations that maximise positive impacts and mitigate risks (Russo and Goga 2025).

Beyond the CSIR, other research institutions have actively contributed to advancing the use of local data in LCA within South Africa. Notably, the roadmap report and implementation plan by Notten and von Blottnitz (2018), developed through collaboration between the University of Cape Town (UCT) and the National Database Working Group, outlines a vision for a national life cycle database as a '[…] *repository of credible datasets useful in evidence-based policy- and decision-making advancing sustainable development*' (ibid, 7). This document moves the conversation forward by offering a concrete framework for database development, including actionable steps, potential partnerships, funding strategies, and a risk management overview to clarify abstract concepts. Furthering this discourse, Botha and Harding (2024) examined the relevance of local data in LCA studies under South Africa's EPR Regulations (DFFE 2021). Using the paper sector as a case study, they tested various life cycle inventory (LCI) modelling approaches—local, international, and hybrid datasets—while adjusting uncertainty levels to reflect their alignment with South African industrial conditions. Their findings offer practical guidance on when fully local data is essential and how international datasets can be adapted for local relevance.

Ongoing research continues to build the evidence base for future national strategies, even if not yet reflected in policy. For example:

- Chitaka and Nel (2025) conducted a LCA study on diapers' disposal in rural South African communities. This study forms part of broader research by the SARChI Chair in Waste and Society, which aims to enhance South Africa's global competitiveness in waste-related research and innovation (SARChI, n.d.).
- The CSIR investigated greenhouse gas (GHG) mitigation potential in hard-to-abate sectors using LCA-based approaches. Their work includes prospective studies on PtX technologies (Stafford et al. 2025a) and material circularity opportunities in building and civil engineering construction (Stafford et al. 2025a, b). Both studies offer sector-specific recommendations for inclusion in the next iteration of South Africa's Nationally Determined Contributions (NDC).

4　Discussion

Mainstreaming LCA in public policy across Africa is still in its early stages, but there are promising developments and growing momentum, particularly in countries like South Africa. Despite being fragmented, LCA started being embedded in policy frameworks and evidence-based LCA findings have started informing policies and strategies going forward (Table 1).

South Africa's only example of integrating LCA into public policy is found in the EPR Regulations. Specifically, Regulation 5, sub-regulations (1)(k) and (1)(l), mandate producers in six sectors to conduct LCAs aimed at reducing material use, redesigning products for reuse or recycling, and minimising environmental toxicity in post-consumer waste (DFFE 2021). Given the country's limited LCA capacity and few existing studies, this requirement is expected to significantly drive LCA adoption. To address challenges such as scarce local data and the lack of methodological guidance in the EPR Regulations, research institutions have developed the *CSIR LCA Guideline* series (Russo et al. 2024) and explored local adaptations of international LCI data (Botha and Harding 2024). These entities are now collaborating with industry to pilot testing the CSIR LCA Guideline and refine both the methodology and product grouping approach. Beyond this policy, LCA continues to serve as a science-based tool supporting evidence-informed decision-making in South Africa.

Several government departments in South Africa actively support LCA initiatives. The Department of Forestry, Fisheries and the Environment (DFFE) has played a key role since publishing the LCA information series in 2004 as part of its Integrated Environmental Management tools. Drawing on evidence from the CSIR's LCSA study on grocery carrier bags, DFFE set targets for increased recycled content in plastic bags through amendments to the Plastic Carrier Bags and Plastic Flat Bags Regulations under the National Environmental Management Act (Act No. 107 of 1998) (DFFE 2021). In 2020, DFFE also gazetted the EPR Regulations, mandating LCA to promote long-term product sustainability and material circularity. The Department of Science, Technology and Innovation (DSTI) has further advanced LCA adoption through its 10-year '*Waste RDI Roadmap*' programme, which incorporates life cycle tools under the '*Modelling and Analytics*' cluster (DST 2014). This initiative has supported numerous LCA research studies (Waste RDI Roadmap, n.d.), many of which have informed policy and contributed to implementing LCA requirements in EPR regulations.

The waste sector plays a key role in supporting SCP and the circular economy. Recognized by local government for its socio-economic potential, it contributes through waste prevention and diversion from landfills towards value recovery. Research in this area has largely focused beyond specific LCAs. More recently, the energy sector, particularly the emerging green hydrogen industry, has begun integrating LCA into ex-ante pre-feasibility studies. These efforts aim to guide PtX developments in South Africa, support renewable energy strategies, and align economic, social, and climate goals to ensure a Just Energy Transition (JET). They also inform the development of the South African Green and low-carbon hydrogen standard, which defines the requirements for green hydrogen in order to increase interoperability with global green hydrogen markets and incentivise local industry to reduce carbon emissions and switch to cleaner production. Additionally, LCA-based approaches are used in prospective studies on the greenhouse gas (GHG) mitigation potential, in order to identify sectors and sub-sectors for inclusion in future updates of South Africa's Nationally Determined Contributions (NDC), in line with the 2050 Net Zero target (RSA 2021).

5   Conclusion

South Africa became a founding member of the African Circular Economy Alliance (ACEA) in 2017, alongside Nigeria and Rwanda. ACEA promotes life cycle thinking and circular economy strategies across Africa, aligned with the African Union's Agenda 2063 for inclusive and sustainable development (African Union 2015; ACEA 2017). Nationally, the White Paper on Science, Technology and Innovation (DST, 2019) highlights the role of STI in advancing environmental sustainability. It advocates for systemic innovation to tackle socio-economic and ecological issues, referencing life cycle approaches and aligning with both the National Development Plan 2030 and the Sustainable Development Goals.

Life Cycle Assessment (LCA) is increasingly recognized as a valuable tool for evidence-based policymaking in South Africa, particularly in the context of sustainable consumption and production. Its integration into national policy—most notably through the EPR Regulations—marks a significant milestone in advancing environmental sustainability. However, broader adoption of LCA in South Africa faces several hurdles. A key issue is the limited local expertise, which restricts broader application of LCA studies. Additionally, integrating LCA into policymaking is complicated by its technical nature—policymakers often lack the skills to interpret results—and by the variability of outcomes, which depend heavily on methodological assumptions. Although research institutions have made progress by developing guidelines and harmonised modelling approaches, the lack of local life cycle data continues to limit relevance for national policy. Moreover, environmental indicators alone are insufficient; socio-economic metrics reflecting South Africa's unique challenges, such as inequality, are also needed for a more comprehensive sustainability assessment. In this regard, Life Cycle Sustainability Assessment is a developing framework to integrate socio-economic aspects with environmental LCA and inform decisions for sustainable development.

References

ACEA (2017) About the African Circular Economy Alliance. https://www.aceaafrica.org/about-acea. Accessed 16 Jun 2025

African Union (2015) Agenda 2063: the Africa we want. https://au.int/agenda2063. Accessed 16 Jun 2025

Botha EE, Harding KG (2024) Significance of international life cycle data in South African extended producer responsibility. South African J Sci 120(11/12). https://doi.org/10.17159/sajs.2024/16384

Building and Construction Authority (2021) Green Mark 2021 certification standard. https://www1.bca.gov.sg/buildsg/sustainability/green-mark-certification-scheme/green-mark-2021. Accessed 16 Jun 2025

California Air Resources Board (2009) Low Carbon Fuel Standard. https://ww2.arb.ca.gov/our-work/programs/low-carbon-fuel-standard. Accessed 16 Jun 2025

Chitaka TY, Nell C, Schenck C (2025) A critical view of applying life cycle assessment on disposable diapers in a rural context. South African J Sci 121(3/4). https://doi.org/10.17159/sajs.2025/18211

CIRAIG (2020) Life Cycle Assessment in Public Policy Database. https://ciraig.org/index.php/project/lca-in-public-policy/. Accessed 11 Jun 2025

Clay Brick Association of South Africa (2016) Life cycle assessment of clay brick walling in South Africa. ISBN 978-1-77592-113-4.https://www.claybrick.org/lca-life-cycle-assessment-clay-brick-walling-south-africa. Accessed 16 Jun 2025

Department of Environment, Forestry and Fisheries (DFFE) (2021) National Environmental Management Act: amendment of plastic carrier bags and plastic flat bags regulations

Department of Environment, Forestry and Fisheries (DFFE) (2021) National Environmental Management: Waste Act—Regulations and notices regarding extended producer responsibility. https://www.dffe.gov.za/registration-terms-regulations-regarding-extended-producer-responsibility-2020. Accessed 16 Jun 2025

Department of Environmental Affairs and Tourism (DEAT) (2004) Integrated Environmental Management Information Series: Life Cycle Assessment. https://www.dffe.gov.za/sites/default/files/docs/series9_lifecycle_assessment.pdf. Accessed 16 Jun 2025

Department of Science and Technology (DST) (2014) A Waste Research, Development and Innovation Roadmap for South Africa (2015–2025): Summary Report. Department of Science and Technology, Pretoria

Environmental Protection Agency (2005) Renewable fuel standard program. https://www.epa.gov/renewable-fuel-standard-program. Accessed 22 Jun 2025

European Commission (2022) Proposal for a regulation establishing a framework for setting ecodesign requirements for sustainable products and repealing Directive 2009/125/EC (ESPR), COM(2022) 142 final, Brussels. https://eur-lex.europa.eu/legal-content/EN/TXT/?uri=CELEX:52022PC0142. Accessed 20 Jun 2025

European Commission (2021) Transition pathway for the proximity and social economy ecosystem. Directorate-General for Internal Market, Industry, Entrepreneurship and SMEs, Brussels

European Commission (2020) Regulation (EU) 2020/852 of the European Parliament and of the Council on the establishment of a framework to facilitate sustainable investment. J Europ Union L 198:13–43. https://eur-lex.europa.eu/legal-content/EN/TXT/?uri=CELEX:32020R0852. Accessed 20 Jun 2025

Eurostat E (2020) Environmental Accounts—establishing the links with national accounts. Publications Office of the European Union, Luxembourg

Harding KG, Friedrich E, Jordaan H et al (2021) Status and prospects of life cycle assessments and carbon and water footprinting studies in South Africa. Int J Life Cycle Assess 26:26–49. https://doi.org/10.1007/s11367-020-01839-0

International Platinum Association (IPA) (2025) The life cycle assessment of platinum group metals—reference year 2022. https://www.ipa-news.com/assets/contentimg/sustainability/ipa-lca-3-fact-sheet-final-april-2025.pdf. Accessed 16 Jun 2025

Isah ME, Zhang Z, Matsubae K et al (2024) Bibliometric analysis and visualisation of research on life cycle assessment in Africa (1992–2022). Int J Life Cycle Assess 29:1339–1351. https://doi.org/10.1007/s11367-024-02313-x

Japan Environmental Management Association for Industry (n.d.) EcoLeaf environmental label. http://www.ecoleaf-jemai.jp/eng/. Accessed 16 Jun 2025

Jegen M (2024) Life cycle assessment: from industry to policy to politics. Int J Life Cycle Assess 29:597–606. https://doi.org/10.1007/s11367-023-02273-8

Karkour S, Rachid S, Maaoui M, Lin CC, Itsubo N (2021) Status of life cycle assessment (LCA) in Africa. Environments 8:10. https://doi.org/10.3390/environments8020010

Mashoko L, Mbohwa C, Thomas V (2010) LCA of the South African sugar industry. J Environ Plan Manage 53:793–807. https://doi.org/10.1080/09640568.2010.488120

Notten P, von Blottnitz H (2018) Development of national LCA database roadmaps, including further development of the technical helpdesk for national LCA databases: Deliverable D 4.3. Final roadmap report for South Africa. https://helpdesk.lifecycleinitiative.org/wp-content/uploads/2020/11/d_4.3_final_roadmap_report_za.pdf. Accessed 16 Jun 2025

Oelofse S, John M, Samson M (2022) Supporting the transition from conventional plastics to more environmentally sustainable alternatives

Regulation (EU) (2020) 2020/852 of the European Parliament and of the Council of 18 June 2020 on the establishment of a framework to facilitate sustainable investment, and amending Regulation (EU) 2019/2088. http://data.europa.eu/eli/reg/2020/852/oj. Accessed 05 Aug 2025

Republic of South Africa (RSA) (2021) South Africa's First Nationally Determined Contribution—Updated September 2021. https://unfccc.int/sites/default/files/NDC/2022-06/South%20Africa%20updated%20first%20NDC%20September%202021.pdf. Accessed 16 Jun 2025

Russo V, Goga T (2025) Incorporating a Just Transition Mindset in Addressing the Climate Emergency: A Focus on Green Hydrogen Production in South Africa, TIPS Forum 2025, available on-line at: Paper_Incorporating_a_Just_Transition_mindset_in_addressing_the_climate_emergency_A_focus_on_green_hydrogen_production_in_South_Africa.pdf

Russo V, Stafford W (2023) Life cycle sustainability assessment (LCSA) of material alternatives for food take-out containers and cups, Japan–UNIDO–CSIR–Wits Project Report. https://www.csir.co.za/sites/default/files/Documents/CSIR%20PS%20LCA%20report_Final_May%202023%20update.pdf. Accessed 16 Jun 2025

Russo V, Stafford W, Nahman A (2020) Comparing grocery carrier bags in South Africa from an environmental and socio-economic perspective: evidence from a life cycle sustainability assessment, waste research development and innovation roadmap research project. DST & CSIR. https://wasteroadmap.co.za/wp-content/uploads/2020/05/22-CSIR-Final-LCSA_Bags_Final-Report-vs2.pdf. Accessed 16 Jun 2025

Russo V, Goga T, Stafford W, Nahman A (2024) Best practice guideline for conducting life cycle assessment (LCA) studies in South Africa. CSIR's LCA Guideline Series. CSIR, Stellenbosch. https://wasteroadmap.co.za/wp-content/uploads/2025/04/57-CSIR_Final-Report_LCA_Guideline-1.pdf. Accessed 16 Jun 2025

Sala S, Pant R, Hauschild M, Pennington D (2019) Life cycle assessment in the EU: supporting decision-making for sustainable production and consumption. J Clean Prod 208:733–745. https://doi.org/10.1016/j.jclepro.2018.09.179

Sanyé-Mengual E, Sala S (2022) Life cycle assessment support to environmental ambitions of EU policies and the sustainable development goals. Integr Environ Assess Manag 18(5):1221–1232. https://doi.org/10.1002/ieam.4586

SAPIENS Network (2023) Integrating life cycle assessment and circularity assessment into EU public procurement: Opportunities, challenges and future directions. https://sapiensnetwork. eu/integrating-life-cycle-assessment-and-circularity-assessment-into-eu-public-procurement-opportunities-challenges-and-future-directions/. Accessed 06 Aug 2025

SARChi Chair in Waste and Society (n.d.) Waste & Society. Accessed 16 Jun 2025

Seidel C (2016) The application of life cycle assessment to public policy development. Int J Life Cycle Assess 21:337–348. https://doi.org/10.1007/s11367-015-1024-2

Sonnemann G et al (2018) Life cycle thinking and the use of LCA in policies around the world. In: Hauschild M, Rosenbaum R, Olsen S (eds) Life cycle assessment. Springer, Cham. https://doi. org/10.1007/978-3-319-56475-3_18

South Africa Government (2020) Minister Barbara Creecy: plastic colloquium feedback session. https://www.dffe.gov.za/minister-barbara-creecy-hosts-2020-feedback-session-plastics-colloquium-12-november-2020, (Accessed 16.06.2025).

Stafford W, Russo V, Nahman A (2022) A comparative cradle-to-grave life cycle assessment of single-use plastic shopping bags and various alternatives available in South Africa. Int J Life Cycle Assess 27:1213–1227. https://doi.org/10.1007/s11367-022-02085-2

Stafford W, Chaba K, Russo V, Goga T, Nahman A (2024) Life cycle assessment of green ammonia produced at a coastal facility in South Africa and utilised for heavy-duty transport in Germany. Deutsche Gesellschaft für Internationale Zusammenarbeit (GIZ). Available on-line at: GH_-LCA-Green-Ammonia_4Feb_draft1.pdf

Stafford W, Chaba K, Russo V, Goga T, Nahman A, Villatico F (2025a) Potential of Power-to-X products to reduce greenhouse gas emissions and inform South Africa's Nationally Determined Contributions to mitigate climate change, Final draft

Stafford W, Padayachi Y, Chaba K, Russo V, Oelofse S, Kissoon S, Thambiran T, Naidoo S, van Reenen C, Nahman A (2025b) Building Circularity into Nationally Determined Contributions: Piloting the UN Toolbox in South Africa, Available online at: https://www.circulareconomy.co. za/research/grant-061/

The Presidency (South Africa) (2011) The National Development Plan. https://www.gov.za/doc uments/national-development-plan-2030-our-future-make-it-work. Accessed 16 Jun 2025

Tukker A, Poliakov E, Heijungs R, Hawkins T, Neuwahl F, Rueda-Cantuche JM, Giljum S, Moll S, Oosterhaven J, Bouwmeester M (2013) EXIOPOL—development and illustrative analyses of a detailed global MR EE SUT/IOT. Econ Syst Res 25(1):50–70. https://doi.org/10.1080/095 35314.2012.712075

UNEP, UNDP, UNFCCC Secretariat, Building Circularity into Nationally Determined Contributions (NDCs) (2023) A Practical Toolbox, Learning for Nature. https://www.learningfornature. org/en/building-circularity-into-nationally-determined-contributions/. Accessed 16 Jun 2025

Waste Research, Development and Innovation Roadmap (n.d.) Waste research, development and innovation roadmap—a waste R&D and innovation programme for South Africa. Accessed 16 Jun 2025

GPSR Compliance
The European Union's (EU) General Product Safety Regulation (GPSR) is a set
of rules that requires consumer products to be safe and our obligations to
ensure this.

If you have any concerns about our products, you can contact us on

ProductSafety@springernature.com

In case Publisher is established outside the EU, the EU authorized
representative is:

Springer Nature Customer Service Center GmbH
Europaplatz 3
69115 Heidelberg, Germany